KAUFMAN FIELD GUIDE TO INSECTS OF NORTH AMERICA

ERIC R. EATON

AND

KENN KAUFMAN

with the collaboration of
RICK and NORA BOWERS

Illustrated with
more than 2,350 images
based on photos
by more than
120 top photographers

HOUGHTON MIFFLIN COMPANY

PICTORIAL TABLE OF CONTENTS

WHEN YOU SEE AN UNFAMILIAR INSECT:

1. Is it really an insect? If so, it should have six legs and three distinct body regions. Some are legless, though, and some larval forms appear to have more than six legs, with additional stumpy "prolegs."

2. Is it an adult? If it has wings, it is. If not, it might be a younger stage, which might make identification more difficult. Most insects hatch from eggs. After that, their development follows one of three typical patterns:

Complete metamorphosis. Larval stage looks very different from adult and is followed by an inactive pupal stage from which adult emerges.

Partial metamorphosis. Larval stage is very different from adult, but adult emerges directly from larva without passing through pupal stage.

Simple metamorphosis. Youngest stages look like small wingless versions of the adults.

3. Can you match your mystery insect to a group in this Pictorial Table of Contents? Overall shape, details of structure, and habits are far more important than color or pattern. Don't expect an exact match here.

4. If your insect seems to fit one of these groups, follow the color tabs or page numbers to that section of the book, and look for the best match. Be sure to read the text for more information; don't just match pictures.

5. Don't be discouraged if you have to leave some insects unidentified. No expert can name them all. Just try again with the next interesting insect!

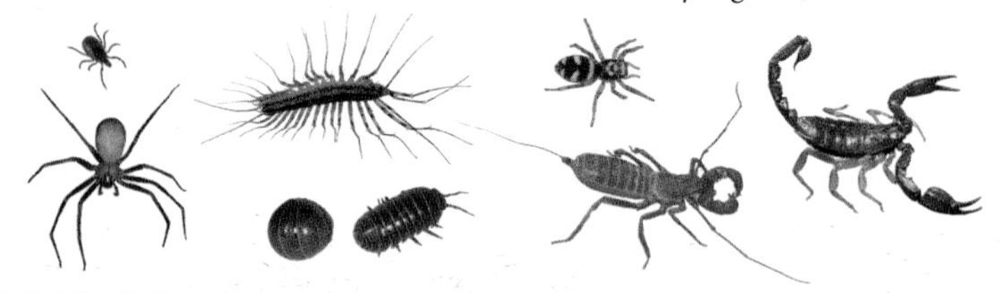

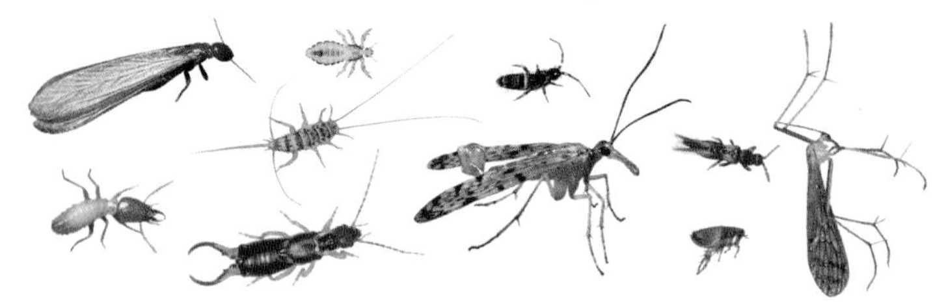

DRAGONFLIES AND DAMSELFLIES, PAGES 42–53

Two pairs of wings, both sets membranous and of equal length. Narrow body, large eyes, tiny antennae. Aquatic larvae. Partial metamorphosis.

MAYFLIES, CADDISFLIES, STONEFLIES, PAGES 54–61

Two pairs of membranous wings (some mayflies have one pair), those of caddisflies often fuzzy. Metamorphosis partial (stoneflies, mayflies) or complete (caddisflies). Larvae aquatic, adults usually near water.

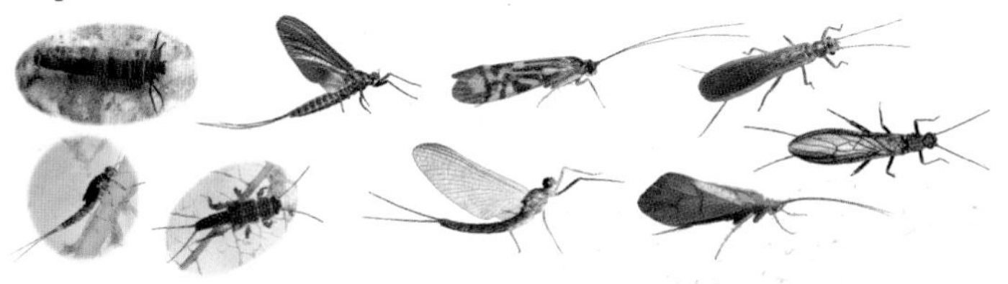

ROACHES, MANTIDS, WALKINGSTICKS, PAGES 62–67

Winged or wingless, front pair of wings leathery. Front legs modified (mantids), body flattened (roaches), or body sticklike. Short appendages (cerci) at tip of abdomen. Chewing mouthparts. Simple metamorphosis.

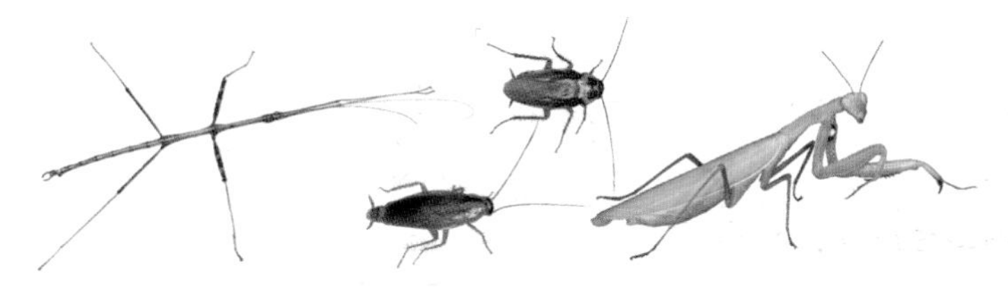

GRASSHOPPERS, KATYDIDS, CRICKETS, PAGES 68–85

Winged or wingless. If winged, two pairs, front pair of wings leathery, hind pair folded or rolled at rest. Hind legs usually modified for jumping. Chewing mouthparts. Simple metamorphosis.

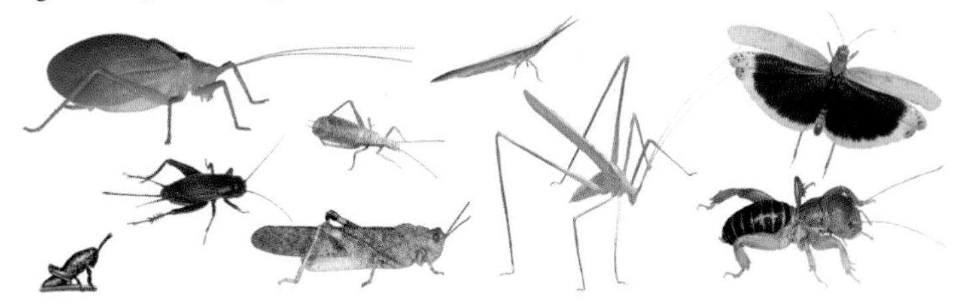

CICADAS, APHIDS, AND THEIR RELATIVES, PAGES 86–105

Cicadas are *large*, most others *very small*. Two pairs of wings when present, both membranous. Piercing-sucking mouthparts. Simple or partial metamorphosis. Antennae often short. Included are scales and aphids that look unlike insects, plus galls (plant growths stimulated by an insect or mite).

TYPICAL TRUE BUGS, PAGES 106–127

Winged or wingless, front pair of wings half leathery (at base), half membranous (at tips). Wings folded in X pattern flat on back. Piercing-sucking mouthparts. Simple metamorphosis. Antennae usually obvious, long.

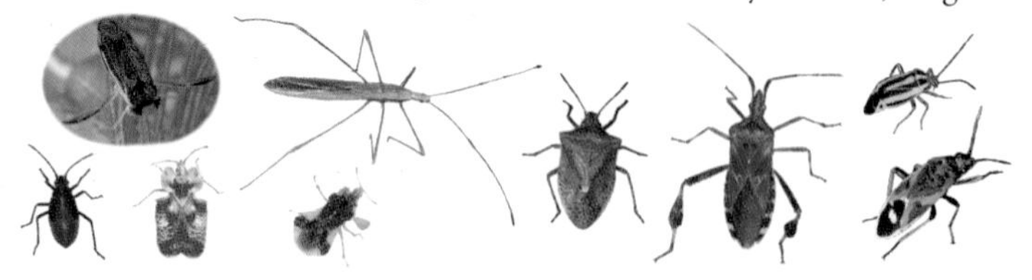

BEETLES, PAGES 128–219

Winged or flightless. If winged, front pair usually modified into hard plate-like covers, hind wings membranous. Body shapes highly diverse. Chewing mouthparts. Complete metamorphosis. Larvae usually hidden.

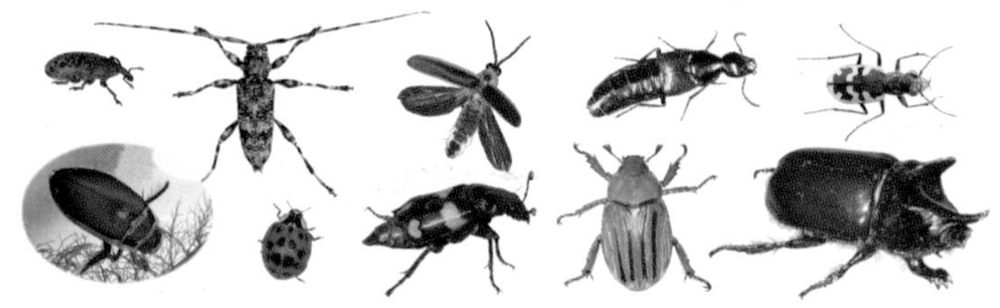

LACEWINGS, ANTLIONS, DOBSONFLIES, PAGES 220–227

Winged (two pairs). Both sets of wings membranous, equal or nearly equal in length. Long, narrow body. Antennae usually obvious, long. Chewing mouthparts. Complete metamorphosis.

BUTTERFLIES AND MOTHS, PAGES 228–271

Two pairs of wings, usually covered with tiny scales. (A few moths are wingless.) Complete metamorphosis. Larvae (caterpillars) have chewing mouthparts and often feed in the open; adults have sucking mouthparts.

TRUE FLIES, PAGES 272–317

One pair of wings (rarely wingless). Eyes often very large, antennae usually short. Mouthparts lapping, sucking, or piercing-sucking. Many are very similar to wasps or bees. Complete metamorphosis, larvae usually hidden.

WASPS, BEES, ANTS, PAGES 318–365

Two pairs of wings, or wingless. Antennae are usually long. Chewing mouthparts. Solitary or social. Females of many possess a stinger or a long, whiplike ovipositor. Complete metamorphosis. Larvae of sawflies often feed in the open, most others are usually hidden.

Dedication

To my mentors: past, present, and future
— E.R.E.

To Kim, and the abundance of life
—K.K.

Visit our Web site: www.houghtonmifflinbooks.com.

LIBRARY OF CONGRESS CATALOGING-IN-PUBLICATION DATA

Eaton, Eric R.
Kaufman field guide to insects of North America
/ Eric R. Eaton and Kenn Kaufman;
with the collaboration of Rick and Nora Bowers; illustrated with more than 2,350 images based on photos by more than 120 top photographers.
p. cm.—(Kaufman field guides)
ISBN-13/EAN: 978-0-618-15310-7
ISBN-10: 0-618-15310-1
1. Insects—North America—Identification. 2. Insects—North America—Pictorial works. I. Kaufman, Kenn. II. Title.

QL473. E28 2006
595.7097—dc 22
2006029511

Book design by Anne Chalmers
Typefaces: Minion, Univers Condensed
Materials for this guide were produced in Rocky Ridge, Ohio, and Tucson, Arizona, by Hillstar Editions L.C. and Bowers Photo.

PRINTED IN SINGAPORE

TWP 10 9 8 7 6 5 4 3 2 1

APPRECIATING INSECTS

A Note from Kenn Kaufman

If variety is indeed the spice of life, then insects are the spiciest creatures on earth. Their seemingly endless variety is almost impossible to comprehend. It is also utterly unknown to many people. For example, many would think of "the moth" as a plain, drab thing, but in North America alone there are almost 11,000 kinds of moths, most of them crisply patterned and many bearing a rainbow of colors. Many people regard "the fly" as a pesky presence, but we have more than 16,000 kinds of flies on this continent, many with bizarre shapes and bright colors. Beetles are unbelievably diverse, with over 24,000 kinds known in North America, many looking like living jewels. No one person could hope to see all of these amazing creatures in one lifetime, or in a dozen lifetimes. But those of us with an interest in insects are guaranteed to have days and nights filled with dizzying, dazzling diversity.

My own fascination with insects was sparked at the age of 13, when I had already been a rabid bird enthusiast for seven years. We were visiting distant cousins, and the boys were proving their macho status by catching bumble bees in jars in my aunt's garden, when I noticed a "bumble bee" that wasn't. It was, as I eventually figured out, a fuzzy yellow and black moth with clear wings. This brilliant deception captured my imagination, and I set out to learn to identify the insects I saw, as I had earlier with birds. After innumerable trips to the library, I came to this conclusion: I will never be an expert on these creatures, but I will always be amazed by them.

That dual prediction has held up well, and my amazement at insects increased all the more when I had the privilege of working on this field guide with Eric R. Eaton. A scholar of the first order, Eric has worked as an entomologist for the Cincinnati Zoo and on contract for the Smithsonian and other organizations, but his foremost skill is not as a researcher but as a communicator. In writing about insects, he combines a professional's level of knowledge with an amateur's sense of wonder and delivers it in beautifully clear prose. The only painful thing for me was having to rein in this elegant prose to fit the tight limits of the field guide format. This guide could have been 10,000 pages long without exhausting Eric's supply of knowledge or the fascination of the subject. If any professional entomologists find that their favorite insect group has been given short shrift, or that the treatment is oversimplified to the point of error, they should blame me, not Eric Eaton.

Professional entomologists have their own technical references, of course, and this book is for everyone else, for those who are just curious about the natural world. This is most emphatically not a pest-control book—plenty of those are available already. Insects are not pests to me. They are symbols of diversity, gems of novelty, a magical kingdom open to anyone who will take the time to look. I hope that this field guide will help you to make your own discoveries about the endless variety of the real world.

WHAT THIS GUIDE INCLUDES

This book treats the insects of Canada and the United States, with the exception of Hawaii and offshore territories. It should also prove useful in nearby areas such as Bermuda and northern Mexico. Many definitions of "North America" do include Mexico, of course, but in this book, when we refer to North America, we mean that part of the continent lying north of the Mexican border.

Within this vast area, nearly 90,000 species of insects are known to live, and of course a portable field guide cannot attempt to discuss each species. We have aimed for a "naked-eye entomology," treating those insects that you are most likely to notice through simple observation.

FINDING INSECTS

Most insects are easy to find, and some will find you. Day or night, at any season of the year, there will be activity. Look everywhere.

Concentrations of flowers may yield a bonanza of bees, butterflies, flies, and other pollinators; weedy patches are likely to be more productive than formal gardens, because flower varieties bred for their looks may be poor in nectar. In-ground swimming pools, especially if suffering from neglect, will quickly be colonized by aquatic insects. Crawling insects will fall into the water in the course of their ramblings. Turning on your porch light will bring moths, beetles, true bugs, flies, and other insects to your doorstep, particularly on moonless, humid nights. Aphid and scale insect colonies, especially those on trees, attract an amazing number of insect predators and parasites. Ants, bees, wasps, and flies come for the sweet "honeydew" secreted in copious amounts by aphids and scales as a waste product. Look under eaves of buildings for paper wasps, mud daubers, or yellowjackets setting up housekeeping. Unless disturbed frequently by banging doors, they will be peaceful neighbors. Gardeners will find insects in the soil, on flowers, on foliage, in the compost heap, in the birdbath, flying overhead, and crawling underfoot. Examine light fixtures and you may turn up smaller species seldom seen. Clean out window wells to find beetles, crickets, and other insects trapped inside. Check firewood for wood-boring beetles clinging to the logs. Those still inside the wood may emerge indoors later. Look for insects hitchhiking on produce and houseplants you bring home from the market.

Visit a variety of habitats in vacant lots, parks, and wilderness areas to see an even greater diversity of insects. Even inhospitable, desolate terrain will yield insects adapted to the harsh conditions. Turn over rocks to find beetles and insect larvae, but be sure to replace them so no creatures are left homeless. Watch tree trunks for the parade of ants and ant-mimicking wasps, beetles, and true bugs. A great many insects feed only on certain plants (like milkweed) or are predators or parasites of specific animals. Some are associated only with decaying fungi, fermenting tree sap, or other decomposing matter. In short, the more places you look, the more interesting insects you will find.

INSECT REPRODUCTION AND DEVELOPMENT

There is no end to the absurd rituals undertaken by insects to ensure the survival of future generations and, most importantly, to pass along a given individual's genetic material. Some female insects do not even need the benefit of males to generate fertile eggs. This form of reproduction is called parthenogenesis, and several walkingsticks, some cockroaches, and other insects have this ability. All insects begin life as an egg, though the female sometimes hatches them internally, giving birth to "live" offspring. This is the case with several flies and aphids. Females of some tiny parasitic wasps are able to lay one egg that gives rise to many embryos in a phenomenon called polyembryony.

Metamorphosis. Some insects, such as grasshoppers and true bugs, hatch from the egg and simply grow larger, developing wings and sex organs as they mature. This is called "simple," or "incomplete," metamorphosis. Many other insects undergo "complete" metamorphosis, making dramatic transformations from youth to adult. The advantages are many: Immature insects can exploit food resources different from their adult counterparts. Insects can tolerate heat, cold, drought, and food scarcity better in the egg, larva, or pupa stage. Larval insects can even dwell in mediums inhospitable to most adult insects, such as water, or even pools of crude oil in the case of one fly species. Just as an individual insect's body parts serve different functions, the different stages of metamorphosis perform different jobs. The egg is the incubation stage, insulating the embryo from environmental extremes. The larva that hatches from the egg is quite literally an eating machine, dedicated to accumulating enough fat and potential energy to get through the next stage. Some caterpillars can grow over a thousandfold in only a few weeks. Wood-boring species, and some others, may spend many months, even years, as larvae. Despite their flexibility, larvae still have an exoskeleton and must molt a number of times to achieve larger sizes. Eventually, the larva molts a final time into a pupa. The pupa appears to be a lifeless blob; inside, however, the cellular structure of the creature is being rearranged and reprogrammed, as latent genes are turned on and obsolete genes turned off. This helps explain why monarch butterfly caterpillars don't try to walk to their Mexican wintering grounds. Pupae are often enclosed in nests, underground cells, the last larval skin casing, or silken cocoons. These protective coverings, like the eggshell, help insulate and disguise the living creature inside. The adult stage is the dispersal and reproductive agent. For many, if not most, insects, the adult stage is fleeting, enduring for only a few weeks or even days. Adults, if they eat at all, frequently feed on nectar, fermenting sap, and other sweet carbohydrates to fuel their frenetic activities.

A fair number of insects undergo a type of metamorphosis halfway between simple and complete. Dragonflies, stoneflies, mayflies, and other aquatic insects, as well as cicadas, all forgo the pupal stage—the active larval stage giving birth to an active adult. In aquatic species this larval stage is often referred to as a naiad. The terrestrial version is known as a nymph.

INSECT ANATOMY

Insects are well built for survival. Their skeleton is on the outside of their body, hence the term "exoskeleton." This hardened cuticle is essentially waterproof, but confining, so growth is achieved by periodically molting this shell during the insect's immature stages. The animal becomes larger in the interval before the fresh exoskeleton hardens. "Albino" insects are specimens that have just molted, with pigments not yet apparent.

Insects have three main body sections, each with a particular function. The head is the major sensory center, where antennae, eyes, and mouthparts are located. Antennae act mostly as a nose, often sensitive to chemicals in the most dilute concentrations per volume of air or water. Antennae also respond to tactile stimulation. The large compound eyes of insects are composed of many (often hundreds) of individual lenses, or facets, which allow insects to register motion far better than the human eye can. A fly would see a movie as a series of still photos. The wraparound eyes of many species also give them a greater field of vision than most organisms. Just the same, insects probably cannot resolve images of stationary objects with any degree of clarity. The bottom line is that scientists have a fuzzy picture of insect vision. We do know that many insects can see in the ultraviolet portion of the spectrum and can navigate using polarized light. In addition to compound eyes, most insects have small "simple eyes," called ocelli, atop the head. These organs distinguish between darkness and light. Eyes of larval insects are usually reduced or even absent. Insects have several types of mouthparts. The dominant form is chewing mouthparts, like those of beetles, grasshoppers, and ants. The mandibles (jaws) chew food prior to ingestion. Pairs of palps (which resemble short antennae) help analyze and taste a potential meal. In other insects, the mouthparts are fused into tubes. Insects that feed on blood, plant sap, or other liquids under pressure, have piercing-sucking mouthparts. Butterflies and moths have a proboscis that actively pumps liquid nutrients from open-air sources. Many flies have similar mouthparts, modified into a spongelike pad.

The insect's middle body region is the thorax. It is the locomotion center to which all six legs and one or both pairs of wings are attached. In many

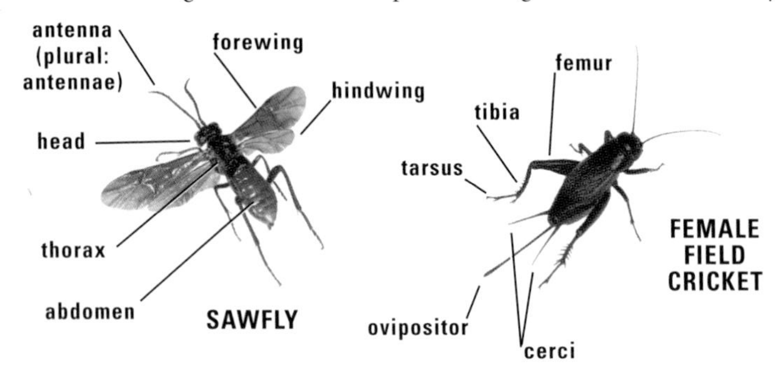

insects, all these appendages are highly modified to fit the creature's lifestyle. Burrowing insects often have rakelike spines or spade-shaped flanges on the legs to help them dig. Aquatic insects may have flattened legs with fringes of stiff hairs to help them stroke through water. Wings also come in a variety of styles. Most beetles have the first pair hardened into sheathlike covers for the second, membranous pair. We are only beginning to comprehend the mechanics of insect flight. They are the most maneuverable of any aerial organisms. Dragonflies generate turbulence, then use it for lift, contradicting conventional theories of flight for our fixed-wing aircraft.

Bringing up the rear of the insect body plan is the abdomen. This is the metabolic, excretory, and reproductive headquarters. The digestive system is housed mostly in the abdomen, along with the sex organs. Various tail-like filaments, forceps, or frightening spear- and bladelike structures may protrude from the abdomen. They can do you no harm. Stingers of wasps, bees, and ants are retractable, unseen until firmly planted and causing pain. Stingers once served the same purpose as most of those other devices on the bug's rear end: inserting eggs into the ground, wood, plants, or other organisms. These egg-laying organs are called ovipositors. In mayflies, stoneflies, cockroaches, crickets, termites, and silverfish, the tail-like appendages on the abdomen are sensory in function and are called cerci.

Technical identification of insects often involves fine structural details—the arrangement of certain wing veins, for example, or the shape of specific segments of the legs or antennae. Since this book is a field guide, we mostly omit such in-hand details. For describing many insects, however, it is often useful to recognize the three most obvious parts of the leg: the femur, tibia, and tarsus. These are illustrated on the facing page.

Internally, insects are just as strange as they are externally. They have what passes for a brain, and sometimes more than one heart. Their circulatory systems are open, meaning all the organs are bathed in blood (hemolymph). Their blood does not carry oxygen, however. Branching tracheae furnish air directly to each cell from the outside. Insects breathe through tiny holes in their abdomen (and thorax) called spiracles. This arrangement is one reason insects cannot reach the gigantic proportions of B-movie bugs. They would collapse under the weight of their tracheae, if the tracheae could even move air over that large an internal volume. Some insects have more than one nerve center, allowing them to bypass the brain and react instantly to threats to life and limb. A roach literally runs out of reflex, its tail sensors (cerci) connected to a ganglion in the thorax that sends the legs into action at the slightest disturbance. Do insects feel pain? Scientists have not found pain receptors in insects, suggesting that they do not. Most insects are, however, sensitive to excessive heat or cold.

Insects interpret their environment in a largely chemotactile fashion. They may find the opposite sex by following the faintest odor trail. Others communicate through vibrational signals. Butterflies have "taste" receptors in their feet, the better for females to find appropriate food plants for their offspring. Social insect colonies are mostly governed by an array of pheromones communicating everything from alarm to food resources to suppressing the reproductive potential of subordinate female workers.

IDENTIFYING INSECTS

People who have pursued other aspects of nature study may have to adjust their expectations when it comes to identifying insects. An experienced birder might name every bird species seen on a day's outing, and a skilled field botanist might recognize every species of local plant, either on sight or by keying it out in a manual. But that level of identification is not possible with insects. There are so many of them, the differences among species are often so subtle, and the published information is so fragmentary that it's simply unrealistic to expect to identify most insects to species.

There are exceptions: species so uniquely marked that they can be named with ease. The Luna Moth (p. 238), the Horse Lubber grasshopper (p. 68), and the beetle *Chrysina gloriosa* (p. 144) are examples. We have tried to show these distinctive creatures whenever possible. More often, though, we can only illustrate representatives. For example, North America has about 173 species of blister beetles of the genus *Epicauta*. We illustrate four of them, but as we mention in the text (p. 202), there are others that look almost identical to the pictured ones. Our treatment might help you to make an educated guess that you've found an *Epicauta*, but you won't be able to rule out the other 169 species not illustrated or discussed here!

Books are available that allow identification of a few groups of insects, even in the field: most butterflies, large moths, dragonflies, grasshoppers, and members of some groups of beetles. But beyond that it becomes far more difficult. Often the only information consists of highly technical keys published in scattered scientific journals, and for some groups of insects there has never been anything published on their identification. For certain groups of insects, the only way to get a definite identification may be to send a carefully prepared specimen to the one scientist somewhere in the world who is a specialist on that group, and hope that he or she will have time to compare your specimen to a research collection and give you a determination.

For most groups of insects, serious identification is not a matter of comparing pictures. It involves having a prepared dead specimen of the insect and being able to examine it closely, often under a microscope, for the technical details that define each order, family, genus, and species. Correctly diagnosing the family of a certain fly, for example, might involve looking at the pattern of three different wing veins, checking to see if the tibia of the hind leg lacks bristles near the tip, noting that the third segment of the antenna is elongated, and confirming the shape of the bristles on the top of the head. And that's just to get to family—determining the genus or species would require a study of smaller details!

Still, for the person who simply wants to be an informed naturalist rather than a professional entomologist, there are many shortcuts. The simplified treatment in this field guide should give you a good idea of what you are seeing most of the time. Opposite, we list some things to consider to avoid going astray in identifications.

Color may be the first thing we notice about an insect, but it is usually the last thing to consider in making an identification.

Structure is vastly more important. Study the Pictorial Table of Contents on pages 2–5 of this guide for a quick overview of the structure of different groups of insects. Soon you'll be noticing these things automatically when you see an unfamiliar insect: does it have one pair of wings, or two, or none? If it has two pairs of wings, are the forewings similar to the hindwings, or different? Are the antennae short or long? What about the shape of the abdomen, thorax, or head, or the size or position of the eyes? All of these structural points will help to determine the group to which an insect belongs.

Size. The type of development, or metamorphosis, of each group of insects will affect the variation that we see in their sizes. Grasshoppers, for example, have simple metamorphosis, so little wingless grasshoppers may grow up to be big winged ones. But little dragonflies, cicadas, beetles, and moths do not grow up to be bigger. For a dragonfly or a cicada (with partial metamorphosis) or for a beetle or a moth (with complete metamorphosis), all the growing is done by the larval form, and once it emerges as an adult and its wings are fully dried, it has reached its final size. We do see individual variation in the sizes of adult insects of a given species. In some cases, the amount of food available to the larva will affect the size of the individual when it becomes adult. There is often a major size difference between males and females, or among individuals born at different seasons.

Still, size is an excellent clue. Most sphinx moths are large, most leaf beetles are small, most fleas are tiny. To indicate general sizes, we have included a silhouette shown at actual size for one insect on each color plate in this guide. The other species on the plate will be shown at the correct scale relative to the image that corresponds to the silhouette.

Habitat is sometimes a major clue. Some insects are found mainly near (or in) water. Some inhabit forests, others open fields. Many insects are likely to be found in association with particular plants, because they are often highly specialized in what they will eat.

Behavior. With experience, you may be able to identify some insects at a distance merely by what they are doing. Even when you are seeing something for the first time, its behavior may help to reveal its identity. Tiger beetles, for example, are fast-running predators of flat open ground, while tortoise beetles clamber about sluggishly on plants. Feeding behavior is critically important: if an insect has punctured a plant with its mouthparts and is sucking out the juices, for example, it's not a beetle. This is all the more reason to read the text in this guide, where we try to describe habitat and behavior whenever appropriate. You can't get this information just by looking at the pictures.

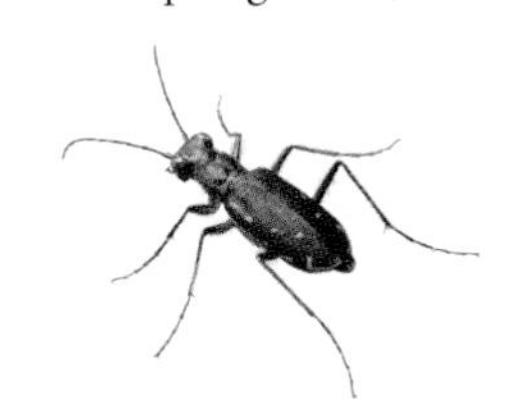

A tiger beetle: alert and fast on open ground

HOW INSECTS ARE CLASSIFIED AND NAMED

The sheer variety of nature is wonderful, but it can also be very confusing. To make sense of this diversity, scientists classify living things into categories. The animal kingdom is divided into a number of broad groups, each called a phylum (plural: phyla), and insects are placed within the phylum Arthropoda (arthropods) along with centipedes, millipedes, crustaceans, arachnids, and various other groups, a few of which are discussed on pages 22–27 of this guide. Within this phylum, insects make up a distinct class, called either class Hexapoda or class Insecta; there is debate about some primitive six-legged creatures that might be considered either insects or "non-insect hexapods." Regardless of how these are delineated, the insects are divided into a number of orders, and some of these correspond to groupings that most of us would recognize; for example, the beetles make up the order Coleoptera, while the dragonflies and damselflies make up the order Odonata. An order may be divided among a number of families. North America has about 130 families of beetles, for example, with the lady beetles making up the family Coccinellidae, the scarab beetles making up the family Scarabaeidae, and so on. A family may contain any number of genera (singular: genus). The lady beetle family includes more than 480 species in North America, of which 18 are classified in the genus *Hippodamia*. One of the most abundant is *Hippodamia convergens*, the Convergent Lady Beetle.

These are the basic categories, but some subcategories are used at times. An order may be divided into suborders. Within the order Odonata, the dragonflies make up the suborder Anisoptera, the damselflies the suborder Zygoptera, with each of these suborders containing several families. A family may be divided into subfamilies, and this can be quite useful in making sense of large groups: the May beetles, dung beetles, flower scarabs, shining leaf chafers, and rhinoceros beetles all make up separate subfamilies within the huge family of scarab beetles. A subfamily may be divided into tribes, a genus may be divided into subgenera, and various other arrangements are possible. All of these categories are to some extent artificial, but they all represent a serious attempt to understand real relationships. If the classification is done correctly, all the species within a group will share a common ancestor, and they will be more closely related to each other than to members of other groups.

Species of insects. Whole books have been written to define exactly what a species is, and several alternate definitions have support in the scientific community. No definition will fit perfectly, because there will always be borderline cases, populations that appear to be in the process of becoming species but are not yet quite distinct enough. In general, though, members of a species are considered to be isolated from other species in terms of reproduction. Different species often *can* interbreed (and may even produce fertile offspring), but they generally *don't.* The clues that insects use to recognize members of their own species may not be apparent to us. Field crickets (p. 80) that appear identical to human eyes turn out to represent several species, kept distinct by differences in the males' "songs" and probably by chemical cues. On the other hand, differences that seem obvious to us may not be important to the insects. The Asian Multicolored Lady Beetle (p. 156) shows extreme variation in color and pattern, but all these variants interbreed freely, so they are all clearly the same species.

Scientific names are applied to every species that has been formally recognized. Mainly Latin or Latinized Greek, these names are accepted by scientists working in any language. The names are written in italics: *Lygaeus kalmii* is the Small Milkweed Bug. The first word in the name, always capitalized, is the genus: *Lygaeus.* North America has five other species in this genus, but the Large Milkweed Bug *(Oncopeltus fasciatus)* is not one of them, despite the similarity of its English name. Scientific names may look confusing and unpronounceable at first, but a glance at the name of the genus can quickly tell us if some species are considered to be close relatives. And for many insects, the scientific name is the only name, so we have no choice.

English names. Many naturalists begin by watching birds and get used to the idea that all birds have official English names established by committees of scientists. It can come as a shock to realize that hardly any insects have English names that are standardized or "official" in any sense of the word, and that most have no English names at all.

Only one order of insects in North America has a complete set of official names. A committee of scientists from the Dragonfly Society of the Americas has established a list of all our dragonflies and damselflies, and these English and scientific names are now being almost universally followed. Enthusiasts from the North American Butterfly Association have proposed a list of English names for butterflies, and these are now being widely followed. A committee of the Entomological Society of America has endorsed official names for a handful of other insects, mostly those with economic importance. Still, in English, most insects are nameless.

We have used English names wherever such were available, because we wanted to make this guide as user-friendly as possible for newcomers to entomology. We resisted the temptation to coin new names, however, so all of the names used here have appeared in print before. Occasionally we had to choose among a number of possibilities. *Hesperotettix viridis* is called Snakeweed Grasshopper in this book, but we found at least four other names in use for it.

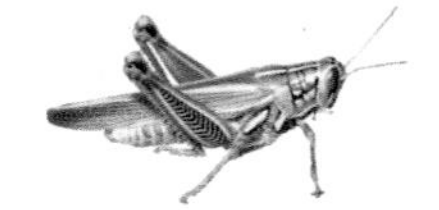

Snakeweed Grasshopper: also known as Green-striped, Green-streak, Purple-striped, or Meadow Purple-striped Grasshopper

In this guide, as in others in the Kaufman Field Guide series, we follow the convention of capitalizing the names of full species. These are proper names: the name "Southern Yellowjacket" refers to just one species, *Vespula squamosa.* The name "green lacewing" applies to any of the species in the family Chrysopidae, so it is not capitalized.

ACTIVITIES INVOLVING INSECTS

Plant a pollinator garden. Many native bees and other pollinating insects are on the decline as we replace native vegetation with lawns and exotic ornamentals. Try planting native flora to bring back those beneficial insects. The solitary wasps you attract will also prey on caterpillars and flies. Many beetles that visit flowers eat aphids, too. Solitary bees also need help with housing. Simply drill deep holes of various diameters into a block of wood and hang this "trap nest" in a protected spot. It will provide leafcutter, mason, and other bees with cozy homes.

Join an organization. Try joining, or volunteering for, organizations such as 4-H, which have an entomology division. Consider taking up apiculture and joining a local association of beekeepers. Investigate national or international nonprofits such as the Xerces Society, which is devoted to the conservation of threatened and endangered invertebrates. The Young Entomologist's Society has been serving aspiring entomologists, and many adult amateurs, since 1965. Its headquarters in Lansing, Michigan, now features a "Minibeast Zooseum."

Collect insect memorabilia. Instead of collecting specimens, collect insect stamps, models, replicas, toys, puzzles, art, T-shirts, jewelry, and other items featuring insects. Make your own insect crafts, like origami dragonflies or Japanese insect kites. Collect stories about insects, such as Native American myths and legends and regional folklore. Look for these items at insect fairs, including the annual exhibition at the Los Angeles County Muesum of Natural History.

Take up insect photography. Pursuing insects with a camera is rewarding and sometimes challenging. At the simplest extreme, large moths attracted to lights may be easy to photograph, as they often sit still and present a relatively flat plane that will all be in sharp focus. For insects that are smaller and less flat, a macro lens is a must, a small aperture (higher f-stop) is important for increasing the depth of field, and a ring or twin flash may be necessary. For fast-moving insects, remember that most of them will slow down if they're cold. A number of the individuals pictured in this guide spent a few minutes in KK's refrigerator before being photographed on the kitchen table and then released outside again. Before you try this bugs-in-the-fridge routine, make sure it's acceptable to other family members. (You may be blessed, as KK was, in having a spouse who will actively pursue insects and bring in some of the jazziest subjects to be photographed.) Today, some relatively simple point-and-shoot digital cameras can take remarkably good closeup photographs, and they promise to make insect photography easier and more popular.

Visit insect zoos. Increasingly, zoological parks and museums are recognizing that live insects make attractive exhibits and help complete the spectrum of animal life usually displayed at such facilities. Please patronize them. They offer a great way to familiarize yourself with local species and learn about the many beautiful, exotic species found in other parts of the world. Visit frequently, as the short lifespan of arthropods means the exhibits change often.

Pest insects. "Pest" is a label we ascribe to any organism that competes for "our" resources. It is an artificial concept. Nature recognizes no ownership. Even mosquitoes understand you do not own your blood. The rise of insects to pest status is also an artifact of man's intervention. Nearly all important pests in North America were introduced, intentionally or accidentally, from foreign countries. Our drive toward a global economy ensures this trend will accelerate. In cultivating gardens and crops, we set the table for insect species that dine on a particular type of plant. Monoculture provides acres of corn available to the European corn borer, for example. Pesticide use is often of questionable efficacy and can even result in new pests by suppressing competing ones. Our attempts to spray the boll weevil into oblivion once allowed the tobacco budworm to replace the beetle as the major pest on many cotton plantations. Pesticides also kill beneficial insects that, left alone, will hold populations of pest species to tolerable levels. Many pest species also develop resistance to chemical applications over time. There are some insects for which there are no longer any effective pesticides. There are, however, practical alternatives.

Integrated Pest Management (IPM) is both a philosophy and a strategic action plan for solving the dilemma of troublesome species. Simple cultural controls can often eliminate a pest problem. For instance, keeping a clean kitchen discourages cockroaches. Crop rotation can disrupt the life cycle of an agricultural pest. Failing such measures, baits can be used for roaches, and traps can assess population levels of a crop pest. Pesticides are used as a last resort. Major differences between IPM and conventional treatments are in immediacy and duration. Pesticides have an immediate impact, but IPM measures last longer. Consult your local cooperative extension agent, usually affiliated with your state's land grant university. He or she will identify your problem critter and recommend appropriate controls that will not harm you, your family, or pets. Personnel in the entomology departments of universities and museums can provide similar services. Many commercial pest control technicians have little, if any, background in entomology and insect identification.

Beneficial insects. The vast majority of insects are inconsequential to our personal lives, but of critical importance to humanity as a whole. Pomace flies and flour beetles are valuable research subjects in the field of genetics. Blow flies help forensic scientists solve homicides. Some insects are employed to control invasive weeds. Chemical compounds produced by insects are increasingly useful in medicine and other fields. Some fireflies, for example, manufacture a chemical that is effective against herpes. Other insects produce products for consumption or industrial use: beeswax, silk, honey, dyes, and shellac, for example. Insects also inspired the invention of such items as paper and the chain saw. Excepting western cultures, insects are heavily exploited as a source of human food. Aquatic insects are excellent indicators of water quality and make great fishing bait; and what would a summer night be without the symphony of crickets and katydids or the flashing of fireflies?

HEALTH, SAFETY, AND INSECT-WATCHING

Insect-borne diseases. It is true that insects act as vectors, or carriers, for a variety of diseases. Some, such as malaria and plague, have had serious consequences for humanity. When a new disease arrives in the North American consciousness, as in the recent case of West Nile virus, it may create a certain hysteria. It is important to remember that none of these diseases is carried by "insects" in a broad sense but rather by only one or a very few species. An overreaction that calls for wholesale chemical destruction of insect life in general will not make us any safer. If common sense prevails, we will follow a strategic approach that targets only the vector and cuts down on the risks of contact.

What bit me? Sometimes the answer to that question will be "nothing." The vast majority of "bug bites" and "spider bites" are misdiagnosed, both by the patient and, to a lesser degree perhaps, by the medical profession. A variety of dermatological conditions can result in minor to severe skin lesions. Symptoms incurred overnight, with no interruption of sleep and no arthropod suspect in sight, are likely the manifestation of some other condition. Believe it or not, one of the most common causes of "bites" is ingrown hairs. Small red itchy spots are often caused by this phenomenon. More than 50 other conditions can also mimic bites. Please consult your physician to treat any symptoms that concern you, but remember that insects may not have been the causative agents.

Dangerous insects. In encouraging people to observe insects and other arthropods, we need to offer a word of warning. Though few of these creatures can cause serious injury to an average, healthy adult human, pain and illness are no fun, either, and it pays to be able to recognize potentially hazardous species. Most female wasps and bees can sting, but only social species do so with any regularity, in defense of their nests. Many species of ants, not just fire ants, can also sting. Predatory true bugs, both terrestrial and aquatic, can inflict excruciating bites to defend themselves if carelessly handled. Not all dangerous insects are easily recognized, however. Blister beetles may look innocuous, but pinch one or squash it, and the beetle exudes droplets of a chemical irritant that can raise painful blisters on sensitive areas of your skin. Many caterpillars appear deceptively cute and fuzzy, but don't cuddle up to one. The long hairs on some species can cause alarming allergic reactions on the skin, mucous membranes, or both. In fact, caterpillars of the introduced Brown-tailed Moth are so allergenic that two entomologists have died from working to control this pest. Fortunately, only small populations of that species remain in the northeastern U.S. Avoid spiny caterpillars, too, as some are venomous. Saddleback caterpillars and their relatives have retractable spines.

In all cases, knowledge is power: knowing which insects to avoid and which are harmless. The educated naturalist realizes that most bugs are not going to bite and that many of the "bees and wasps" we see are actually harmless mimics, unable to sting. This level of understanding can make the outdoors seem much less threatening and more inviting.

CONSERVATION

Most people, when they think about wildlife conservation, think first about large animals such as whales, eagles, and pandas. But the vast majority of the world's wildlife, in terms of both varieties and individuals, consists of the smaller creatures — especially insects. A more realistic perspective would accord great value to them. Insects, in the words of eminent biologist E. O. Wilson, are "the little things that run the world." Our planet's essential functions would surely collapse without them. The importance of insects can even be described in economic terms. A recent study by Cornell University and the Xerces Society found that ecological services provided by wild insects and related arthropods have an enormous economic price tag. Dung removal, pollination, pest control, and wildlife nutrition alone amount to a conservative estimate of $57 billion (yes, with a "b") every year in the United States alone. That kind of statistic suggests that conservation of insect populations should be a concern for everyone, not just a preoccupation for nature lovers.

Insects vary so much in their requirements that some will survive no matter what we do. Some insect species undoubtedly would prosper even if we made this planet uninhabitable for ourselves. So mere survival of some insects is not a measure of anything. A better goal would be to maintain healthy, functioning natural habitats and a high diversity of species.

A focus on individual species may work well with larger animals but is extremely difficult with insects because there are so many of them. A few insects are getting some attention under the U.S. Endangered Species Act. For example, the American Burying Beetle (p. 198) was once widespread over most of the eastern half of the continent but has disappeared from 90 percent of its former range. Scientists are working to restore populations and, just as importantly, to understand the causes of the decline, because these could have implications for many other species.

In fact, the role of individual insect species in conservation could be viewed in this way: they are not so much the target of conservation efforts as they are indicators, measuring how well we're doing. For example, some species of mayflies, stoneflies, and caddisflies are very sensitive to water quality. If they start to vanish from our streams, it's a clear sign that the water is polluted; if we clean up the streams, that will benefit all species, not just these particular insects.

At the moment we cannot begin to monitor populations of all our insect species. We don't even know all of them yet — species new to science are still discovered every year, and the work of insect collectors will continue to be necessary in documenting these unknowns. To maintain a high diversity of species, including the ones we haven't discovered yet, we need to protect examples of as many natural habitats as possible. Two fine organizations deserving support are the Xerces Society for Invertebrate Conservation, 4828 SE Hawthorne Blvd., Portland, OR 97215 (www.xerces.org) and The Nature Conservancy, 4245 N. Fairfax Dr., Suite 100, Arlington, VA 22203 (www.nature.org).

SOURCES OF FURTHER INFORMATION

This guide is necessarily compact because it was designed for easy use in the field. Here are some sources to check for more information.

For classification and names of families and orders of insects in this guide, we followed *Borror and Delong's Introduction to the Study of Insects, 7th edition*, by C. A. Triplehorn and N. F. Johnson (Thomson Brooks/Cole, 2005). This standard textbook is the best current overview of North America's total insect fauna. In the Peterson series, *A Field Guide to the Insects* by D. J. Borror and R. E. White (Houghton Mifflin, 1970) is somewhat out of date but valuable for its concise treatments of characteristics for all North American insect families, including those that are distinguishable only under a microscope.

Books that we recommend for deeper insight about insects include *Bugs in the System* by May R. Berenbaum (Addison-Wesley, 1995), *For Love of Insects* by Thomas Eisner (Belknap, 2003), *Broadsides from the Other Orders: A Book of Bugs* by Sue Hubbell (Random House, 1993), and *What Good Are Bugs?* by Gilbert Waldbauer (Harvard University Press, 2003).

Books on specific orders and families of insects are far too numerous to list, although, ironically, there are not nearly enough of them. One potential source for finding out about such books is the BugGuide Web site listed below, which includes helpful lists of references and is kept up to date.

The Internet is an unsteady medium from which Web sites come and go daily. As a general rule, for reliable information about insects, use ".edu" or ".gov" sites. Commercial (".com") sites may give erroneous facts. University Web sites, such as www.ent.iastate.edu/list and www.colostate.edu/Depts/Entomology/ent.html, are an excellent place to start and find links to other resources. A new kind of Web site with many contributors is BugGuide (www.bugguide.net), a rapidly growing site with outstanding images of many North American insects.

ACKNOWLEDGMENTS

Our first debt is to tens of thousands of researchers who have studied insects and amassed the information distilled in this guide. Over the years, the authors both have learned from innumerable authorities, far too many to list. However, the following individuals helped directly with information included in this guide on specific groups of insects: Bob Androw, George Ball, Troy Bartlett, Robert A. Behrstock, May Berenbaum, Craig Brabant, Brian Brown, Horace R. Burke, Bob Carlson, Shawn M. Clark, Theodore J. Cohn, Gary Coovert, Whitney Cranshaw, Liz Day, Lewis Deitz, Hume Douglas, Gary Dunn, Arthur V. Evans, Neal Evenhuis, Richard Fagerlund, Ed Fuller, David H. Funk, Stephen Gaimari, Eric Grissell, Jeffrey Gruber, Steve Halford, Andy Hamilton, Wilford J. Hanson, Martin Hauser, Thomas J. Henry, Steve Hopkin, David J. Horn, Michael E. Irwin, Tony Irwin, John A. Jackman, Peter H. Kerr, Takumasa Kondo, Faith B. Kuehn, Peter W. Kovarik, Paul K. Lago, Allen Norrbom, Charles O'Brien, Cheryle O'Donnell, James O'Hara, Carl Olson, Kristian Omland, John D. Oswald, Foster Purrington, Mike Quinn, Bret C. Ratcliffe, David Rider,

Jacques Rifkind, Mark J. Rothschild, Catherine Seibert, Scott R. Shaw, Matthew Shepherd, Bradley Sinclair, David R. Smith, Bill P. Stark, Margaret K. Thayer, F. Christian Thompson, Charles Triplehorn, Natalia Vandenberg, Eileen VanTassell, Rick Villegas, Matthew Wallace, Bill Warner, Marius Wasbauer, Donald W. Webb, Richard Westcott, Terry Whitworth, Michael Wigle, Alex Wild, Kipling Will, Dave Williams, Doug Yanega, Chen Young, Daniel K. Young, and James R. Zimmerman. ERE would also like to thank Anita Buck for editorial help and the following for inspiration: Robert M. Pyle, Jim Anderson, the late Howard E. Evans, and George C. Ferguson. In addition, ERE thanks the wonderful international community of entomologists on the Entomo-l listserv, hosted by Dr. Peter Kevan, University of Guelph, and the bugnet listserv hosted by Rosmarie Kelly. He also thanks his supervisors and coworkers at JupiterImages.com for accommodating this project by allowing extensive time off for writing as needed.

As with previous books in the Kaufman Field Guide series, the visual approach and compact format of this guide required a lot of work. The staggering task of tracking down photos was ably handled by Nora Bowers and Rick Bowers. They contributed many images from their own outstanding collection of BowersPhoto.com, and their highly organized approach to photo research made it possible to juggle thousands of images and settle on the ones used here. Much of the early work on images was done by Amanda Chatterton, and Brendan O'Rourke designed a number of the color plates. Stacy M. Fobar did much of the organizational and editorial work on large sections of the guide. Others who provided important help with this book project in a variety of ways include Jim P. Brock, Hannah Frey, Tyson Frey, Tad Jewell, Sam Macomber, and Mark Shieldcastle.

Of course, this book would not have been possible without the cooperation of many fine photographers who not only allowed us to use their images but also gave permission for these pictures to be digitally edited, often very extensively. The photographers are acknowledged individually in a section beginning on p. 366. In addition to the individuals mentioned there, particular thanks to Troy Bartlett, founder of bugguide.net, and Chuck Bargeron of the Bugwood Network and ForestryImages.com, for helping us to track down many important images.

Once again it was a pleasure to work with the professionals at Houghton Mifflin. Our editor, Lisa White, patient and powerful, improved every aspect of our approach. Super-designer Anne Chalmers not only came up with the attractive and functional layout for the series, but helped immensely with finalizing all of this book's details. Thanks also to Shelley Berg, Nancy Grant, Taryn Roeder, Mimi Assad, Katrina Kruse, Becky Saikia-Wilson, and Bill McCormick. We also thank Wendy Strothman, Dan O'Connell, and Madelyn Medeiros at the Strothman Agency for support during this project.

Finally, Kenn offers deepest thanks to Kim Kaufman for the inspiration to complete this book. Final stages of any book project can be excruciating, but Kim made it possible to see past the pain of details and deadlines to the potential joy of sharing our sense of wonder in the natural world.

SPIDERS

(order Araneae) are not insects; they belong to a separate class, the arachnids. Spiders have two body regions (insects have three), eight legs (insects have six), and spinnerets (insects have none). Spiders lack antennae and wings. Most spiders have eight eyes. All spiders are predatory, but the vast majority are harmless to humans. These fascinating and varied creatures (close to 4,000 species in North America) are outside the scope of this insect guide, but well worth appreciating in their own right.

CELLAR SPIDERS (family Pholcidae) are common in human dwellings and build extensive, haphazard webs. Disturb a cellar spider and it may shake violently in place. The **Long-bodied Cellar Spider** *(Pholcus phalangioides)* is our most abundant species. **TARANTULAS (Theraphosidae),** not illustrated, are giant, hairy, primitive spiders native to the southwest. Our species live in silk-lined burrows, the males wandering in the fall in search of mates. **SIX-EYED SPIDERS (Sicariidae)** include the **Brown Recluse** *(Loxosceles reclusa)*. Many specimens have a violin-shaped mark on the cephalothorax. These are spiders of the southern midwest and southwest. Moving stored objects can result in encounters with the recluse. Bites range in severity from minor inflammation to extreme tissue necrosis. Many other medical conditions are misdiagnosed as spider bites.

SHEETWEB WEAVERS (Linyphiidae) spin distinctive webs, usually outdoors. The **Bowl-and-doily Spider** *(Frontinella communis)* is named for the shape of its web. Most spiders in this family are tiny and live in leaf litter. **COBWEB SPIDERS (Theridiidae)** are common household invaders. The **House Spider** *(Achaeranea tepidariorum)* is a familiar sight in basements and garages, resting upside down in a tangled silken snare. The **Southern Black Widow** *(Latrodectus mactans)* is recognized by the red hourglass marking on the underside of the abdomen. Widow venom is neurotoxic, resulting in violent and prolonged muscle cramping. Four other widow species range over most of the U.S. and southern Canada. **ORB WEAVERS (Araneidae)** make circular webs. The **Cross Spider** *(Araneus diadematus)* is abundant, especially in fall. The **Golden Garden Spider** *(Argiope aurantia)*, not illustrated, and the similar **Banded Argiope** *(Argiope trifasciata)* are large enough to be startling but are completely harmless.

CRAB SPIDERS (Thomisidae) include the **Flower Crab Spider** *(Misumena vatia)*. Females can change color from white to yellow to match the flowers where they wait in ambush for flies, bees, and other pollinators. Many crab spiders hunt beneath the bark of trees or on the ground. **WOLF SPIDERS (Lycosidae)** are ground-dwelling hunters, mostly nocturnal. Many larger species reside in burrows, the entrance ringed with a turret of silk and debris. The female carries the egg sac, and the newly hatched spiderlings ride on her back. **JUMPING SPIDERS (Salticidae)** are keen-eyed daytime hunters that often stray indoors. Most are small, compact, and as cute as spiders get. They do not build webs but trail draglines for safety during their leaps. Males perform dances to pacify the aggressive females or to repel rivals. The **Bold Jumping Spider** *(Phidippus audax)* and the **Zebra Jumping Spider** *(Salticus scenicus)* are two common species.

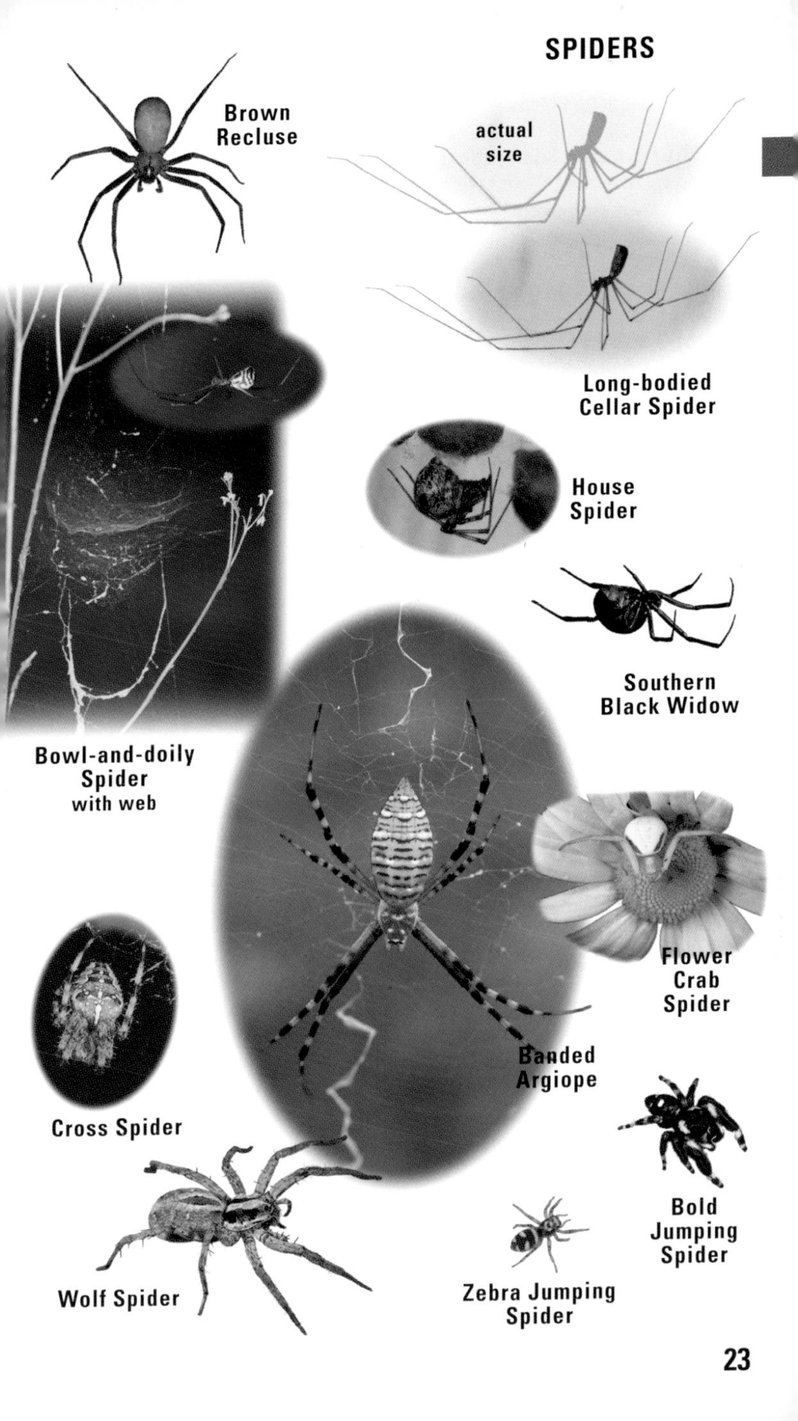
Brown Recluse
actual size
Long-bodied Cellar Spider
House Spider
Southern Black Widow
Bowl-and-doily Spider with web
Flower Crab Spider
Banded Argiope
Cross Spider
Bold Jumping Spider
Wolf Spider
Zebra Jumping Spider

OTHER ARACHNIDS

Spiders (previous page) are the most familiar arachnids, but many other orders belong to this class. Unlike insects, most have eight legs, not six.

SCORPIONS (order Scorpiones) are nocturnal predators. About 90 species occur here, mostly in the south and west. All possess a stinger at the end of the "telson," an elongation of the abdomen. Only one Arizona species is potentially lethal. Scorpions use their pincers to grab prey and to grasp each other during courtship. During the day, scorpions hide under stones or bark or in burrows they dig themselves. Females give live birth, and the babies ride on mom's back until their first molt. Look for scorpions at night with a black light: their bodies glow blue or green under the ultraviolet light. **SOLPUGIDS (order Solifugae),** also known as camel spiders or windscorpions, are fast-running nocturnal hunters of deserts. While not venomous, they have the largest jaws for their size of any invertebrate, the better to shred prey of all kinds. Like shrews, they eat constantly to maintain their frenetic metabolism. By day they hide in burrows or beneath stones, boards, and debris. **WHIPSCORPIONS (order Uropygi)** are mostly tropical, represented here by the vinegaroons of Arizona and Florida. They look menacing, but are harmless. The palps are modified into claws used to grab prey and dig burrows. The first pair of legs resemble antennae. **HARVESTMEN (order Opiliones)** are also known as daddy-longlegs, though this name is also applied to cellar spiders (previous page). They are not venomous, despite urban myths to the contrary. They prey on insects and other arthropods but also scavenge dead insects and drink plant juices. **PSEUDOSCORPIONS (order Pseudoscorpiones)** are tiny and rarely seen. Harmless to humans, these predators live under bark and in leaf-litter, soil, and caves. Some hitch rides on insects.

MITES AND TICKS (order Acari) are incredibly abundant and diverse. Virtually nothing on the planet is mite-free, including our own hair follicles. **HARD-BACKED TICKS (family Ixodidae)** are giant mites, parasitic on vertebrate animals. Females become bloated when engorged with blood, eventually dropping from their host. Ticks are capable of waiting years, perhaps even decades, for an appropriate host to come within reach as they lurk at the tip of a grass blade or twig. **Deer ticks** (genus *Ixodes*) are best known as the principal vectors of Lyme disease. **Wood ticks** (genus *Dermacentor*) are common, especially in the west. **VELVET MITES (family Trombidiidae)** can be conspicuous, especially in the arid southwest, where they often appear in numbers after rains. **SPIDER MITES (family Tetranychidae)** include the **Clover Mite** *(Bryobia praetiosa)*, not shown, which sometimes invades houses. They leave a bright red stain if crushed but do no harm to people. Normally, they thrive outdoors feeding on plants. **Red spider mites** (genus *Tetranychus*) are similar. **HARVEST MITES (family Trombiculidae)** make their presence known mainly in the larval stage, better known as chiggers. These larvae feed readily on human tissues. They do not burrow into the skin but excrete enzymes from their mouthparts that dissolve cells and cause extreme itching. A dusting of sulphur around boots, socks, and waistband may effectively repel chiggers.

ARACHNIDS GREAT AND SMALL

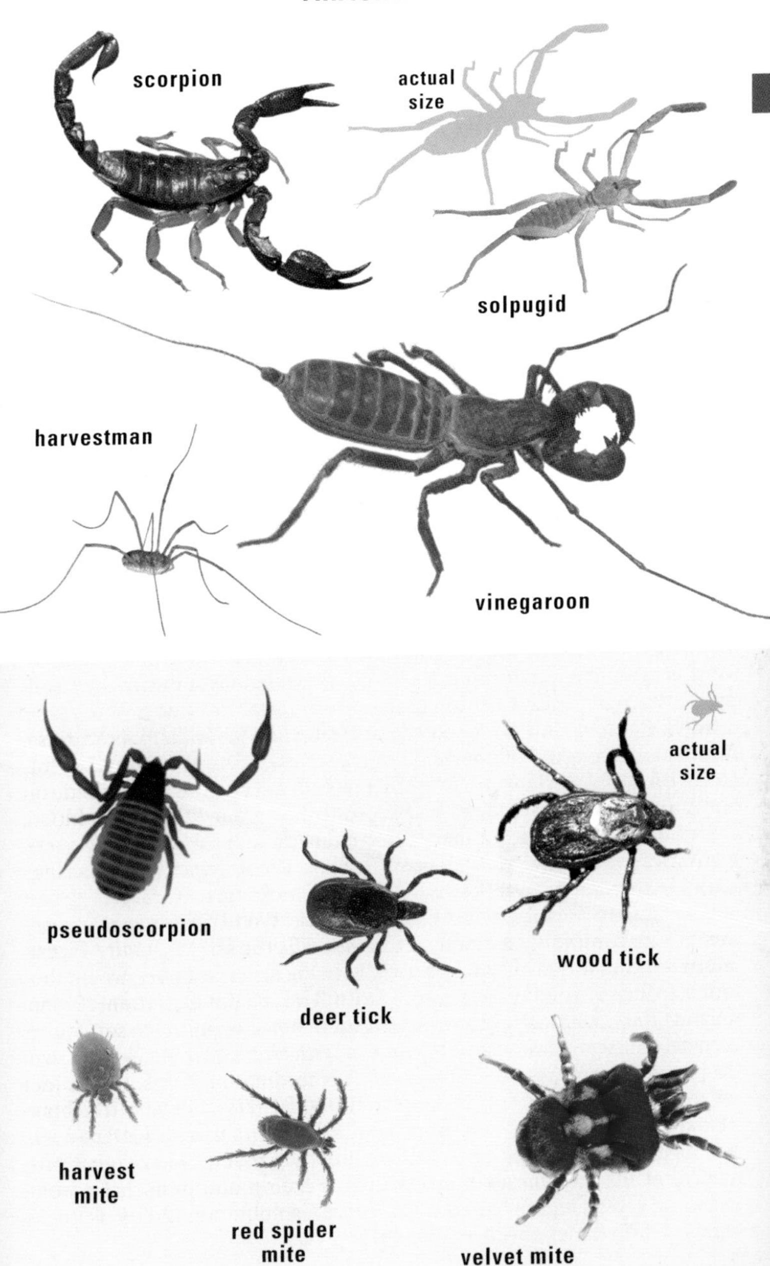

CENTIPEDES, MILLIPEDES, ISOPODS

The creatures here belong to the same phylum (Arthropoda) as insects and spiders, but at the next taxonomic level (class) they are distinct.

CENTIPEDES (class Chilopoda) are flat-bodied, fast-moving predatory creatures found in a variety of habitats. They bear one pair of legs per body segment, with the total leg count ranging from 30 to over 300, depending on species. The first pair of legs is modified into "fangs" with which they inject venom into prey. Some females protect their eggs and newly hatched offspring, curling around them in a ball. So little is known of North American fauna that there is no accurate estimate of the number of species, but they are currently divided among four orders. **HOUSE CENTIPEDES (order Scutigeromorpha)** includes only one North American species, the **House Centipede** *(Scutigera coleoptrata)*. It is common in human dwellings, generally nocturnal, fast, and able to scale walls and run across ceilings. Despite its rather startling appearance, it is harmless to humans. **SOIL CENTIPEDES (order Geophilomorpha)** are subterranean centipedes, looking rather snakelike or wormlike, and often whitish or yellowish. They have 29 or more pairs of legs. **STONE CENTIPEDES (order Lithobiomorpha)** are common outdoor centipedes with 15 pairs of legs. When threatened, they fling droplets of a sticky liquid from their last pair of legs. Some of the largest native centipedes of the **order Scolopendromorpha** (not illustrated), found mainly in the southern states, can inflict painful bites.

MILLIPEDES (class Diplopoda) are slow-moving, usually cylindrical, and feed on decaying vegetation. There are two pairs of legs per body segment. Most millipedes burrow in the soil and curl into a tight spiral when threatened. A few secrete repulsive liquids from the sides of their body segments. There are about 1,400 species in North America north of Mexico, divided among as many as 10 orders. **FLAT-BACKED MILLIPEDES (order Polydesmida)** are highly diverse. They are transcontinental in the U.S. and much of Canada. Many are brightly colored, advertising a toxic nature. Some even secrete cyanide in self-defense. The **order Spirostreptida** includes the world's largest millipedes, cylindrical and robust. Members of this order are mostly tropical, but three species of desert millipedes are found in the American southwest.

The familiar and abundant **PILLBUGS and SOWBUGS (order Isopoda)** are actually terrestrial crustaceans. There are also aquatic versions of these animals found in springs and streams. Many of these creatures are important decomposers of rotting wood and other organic material. **Pillbugs** *(Armadillium vulgare)* and **sowbugs** (genera *Porcellio* and *Oniscus*) found commonly around houses and yards in North America are introduced species from the Old World. Pillbugs, known to children as roly-poly bugs, will roll up in a ball when threatened. Sowbugs are very similar but lack the roly-poly habit and have two short "tails" on the posterior end. **Beach hoppers,** also known as sand fleas, are amphipods that crawl or hop along the surface of the sand in the intertidal zone. Feeding as scavengers, they are active mainly at night. They have gills that function almost as lungs, and they enter the water only occasionally.

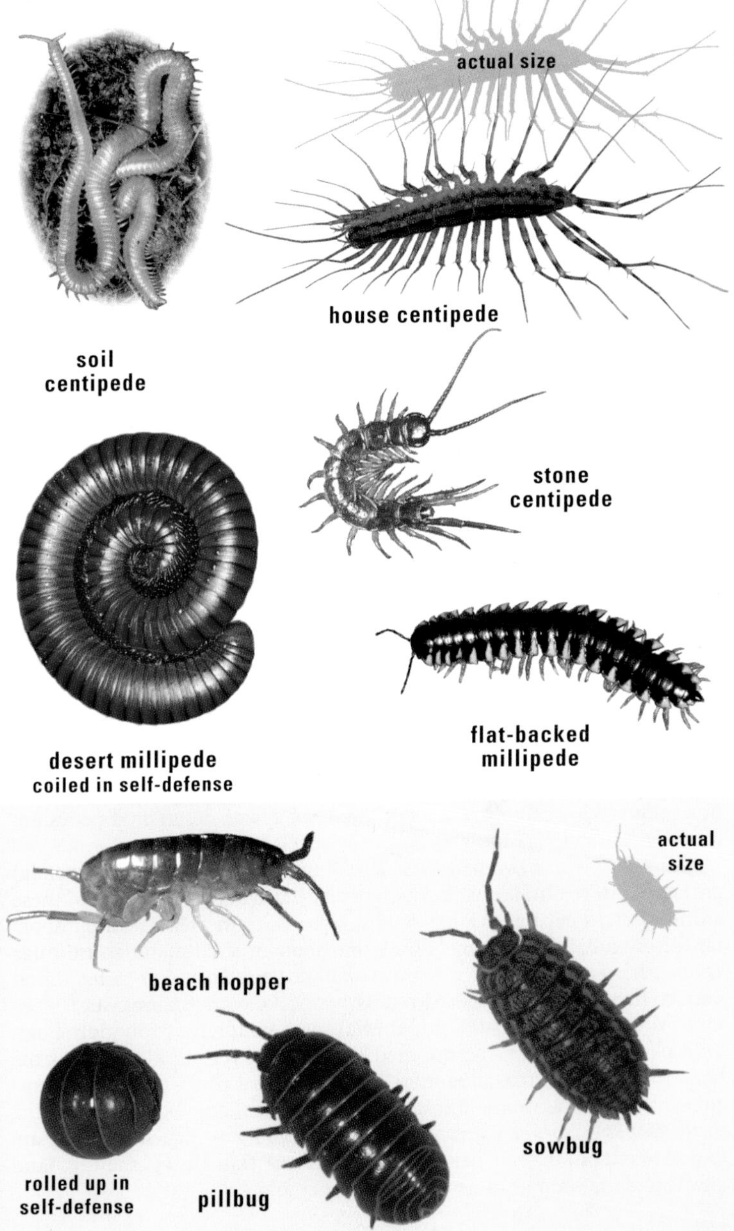

house centipede

soil
centipede

stone
centipede

desert millipede
coiled in self-defense

flat-backed
millipede

beach hopper

sowbug

rolled up in
self-defense

pillbug

EARWIGS, BRISTLETAILS, DIPLURANS

EARWIGS (order Dermaptera) are plagued by superstition. Contrary to myth, they do not habitually enter the ears of slumbering humans. The name may be derived from the shape of the wing in some species: hence, "ear wing" was the intended name, later corrupted. A few species exhibit parental care, and most are scavengers or predators that do no harm.

COMMON EARWIGS (family Forficulidae) include the **European Earwig** *(Forficula auricularia)*, introduced here and now locally abundant coast to coast. It first became established in Rhode Island around 1912. Portland, Oregon, declared a state of emergency and established the Bureau of Earwig Control in 1924. Normally predatory on small insects, it becomes a pest of flowers when prey is scarce. Females guard their clutch of eggs and the newly hatched nymphs. **RED-LEGGED EARWIGS (Carcinophoridae)** include the **Maritime Earwig** *(Anisolabis maritima)*. This wingless species ranges along all our coasts, north to southern Canada and inland to Ontario. Look for them under debris on dry beaches, where they capture sand fleas. Females dig brood chambers in sand and feed the newly hatched nymphs the maggots of shore flies and other small prey. The **Ring-legged Earwig** *(Euborellia annulipes)* is cosmopolitan in range. This mostly predatory species occasionally damages stored organic matter in warehouses and the roots of live plants in greenhouses. Females guard their eggs. The **Riparian Earwig** *(Labidura riparia)* is the only North American representative of the **Labiduridae, LONG-HORNED EARWIGS.** Individuals are most common along beaches and riverbanks, from Texas to North Carolina. They defend themselves with a stinky odor. **LITTLE EARWIGS (Labiidae)** include nine species in North America. The **Least Earwig** *(Labia minor)*, not illustrated, ranges throughout the U.S. and southern Canada. It flies well, is partial to manure, and comes to lights at night.

SILVERFISH (order Thysanura) are occasional pests in libraries, households, and other human hangouts, where they feed on starchy materials such as the dried paste used in bookbindings. They are delicate, scale-coated, with three prominent "tails." Silverfish and their kin have an elaborate courtship whereby the male deposits sperm packets, then guides his potential mate to the area. In some species the male issues silk from mandibular (mouth) glands, gently binding her until she takes up a sperm packet. Found in buildings throughout North America (and much of the world) are the **Silverfish** *(Lepisma saccharina)*, which favors cool, damp situations, and the **Firebrat** *(Thermobia domestica)*, often found in warm spots, such as near furnaces. Several other species also occur in buildings, such as the very large **Urban Silverfish** *(Ctenolepisma urbana)*.

JUMPING BRISTLETAILS (order Microcoryphia) were formerly considered part of the order Thysanura. They have a shrimplike "broken-back" appearance, with three streaming "tails." Look for them under rocks and logs in damp areas on the forest floor.

DIPLURANS (order Diplura) are now usually regarded as "non-insect hexapods." Looking like pale silverfish with only two "tails," they are usually found in damp situations such as under logs or in caves.

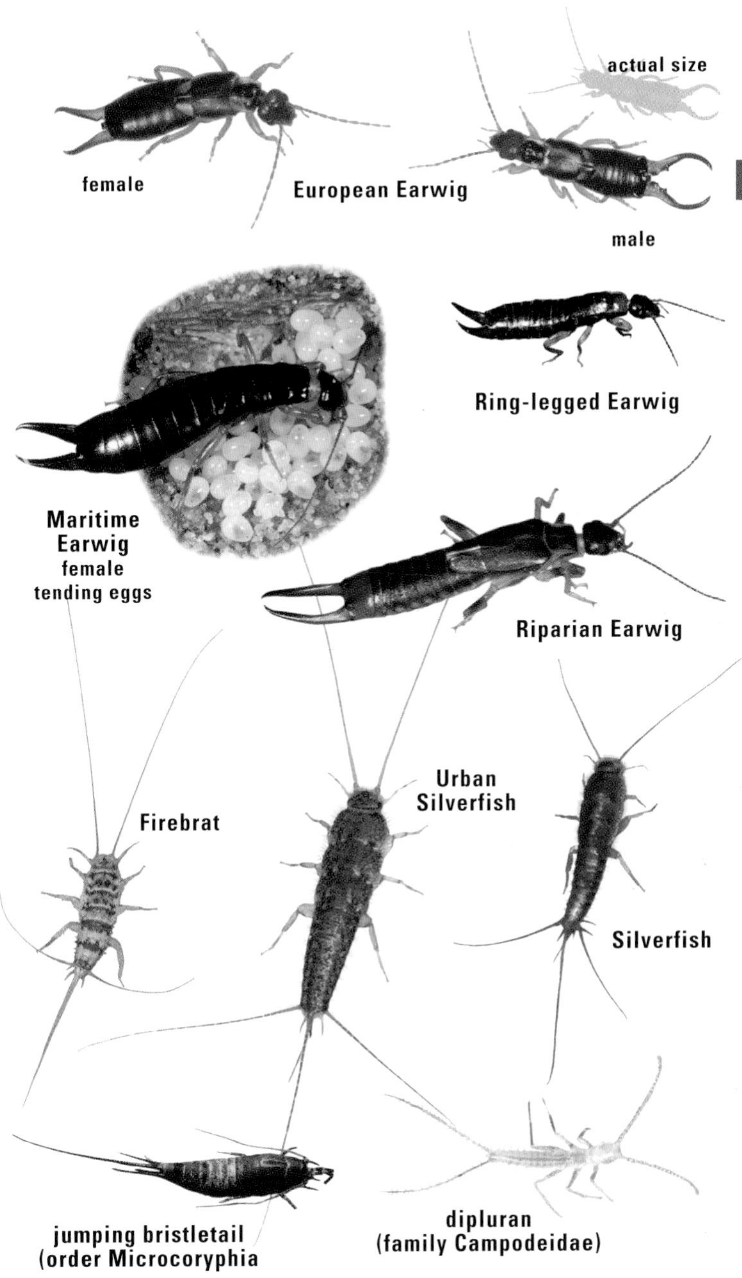
actual size
female
European Earwig
male
Ring-legged Earwig
Maritime
Earwig
female
tending eggs
Riparian Earwig
Urban
Silverfish
Firebrat
Silverfish
jumping bristletail
(order Microcoryphia
dipluran
(family Campodeidae)

SPRINGTAILS

(order Collembola) are now considered "non-insect hexapods," or even placed in their own class, but traditionally were treated as an order in the class Insecta. They are among the most abundant invertebrates, especially in soil, but can be encountered anywhere from snowfields to the surface of ponds, to your bathtub. The common name reflects the ability of many species to leap. The forked, tail-like furcula normally rests in a latch called the tenaculum, on the underside of the abdomen. The sudden release of the tenaculum propels the furcula downward and backward, catapulting the springtail into the air, often many times the body length of the animal. There are about 675 species in America north of Mexico.

SLENDER SPRINGTAILS (family Entomobryidae) include about 138 species. The genus ***Entomobrya*** has about 36 species in North America. ***Pogonognathellus flavescens*** is huge for a springtail, averaging a quarter inch long. Look for it under stones, logs, and other debris. ***Seira domestica*** belongs to one of several genera known to disperse passively on the wind, sometimes up to 9,000 feet or higher. The genus ***Orchesella*** includes 14 species in North America. ***Lepidocyrtus cyaneus*** is one of 15 species in its genus found here. In ***Heteromurus nitidus,*** as in many springtails, individuals are covered in detachable scales, enabling a slippery escape from small predators that are left with only a mouthful of scales.

ELONGATE-BODIED SPRINGTAILS (Hypogastruridae) include the species most likely to be seen in swarms. These insects are tiny (1–2 mm), and many do not possess the jumping organ. Many lack compound eyes as well. Some are aquatic, such as ***Anurida maritima,*** the "seashore springtail," which lives in intertidal zones along the northeast Atlantic coast, and also British Columbia, often in great numbers. They survive high tide clustered in air pockets beneath rocks. ***Hypogastrura nivicola*** is the best-known "snow flea." Other species can be encountered in the soil, under bark, in commercial mushroom-growing facilities, or even in the percolating filters of some sewage treatment plants. This is the largest family of springtails, with over 200 species in North America alone.

WATER SPRINGTAILS (Poduridae) include only ***Podura aquatica,*** common in aggregations on the surface of ponds, ditches, even on puddles after rain, over most of the northern hemisphere. **GLOBULAR SPRINGTAILS (Sminthuridae)** are rotund in shape and 1–3 mm in length. They can be confused with some barklice (p. 34). They are among the best jumpers among all springtails. Species of ***Sminthurides*** are usually found on the surface of freshwater ponds. **SMOOTH SPRINGTAILS (Isotomidae)** lack scales on the upper surface. Some occur in extreme habitats such as deserts and the Antarctic. ***Folsomia candida,*** eyeless and without pigment, often colonizes potting soil and might be found indoors under potted plants. This species is often used in tests for toxicity of pesticides on nontarget soil organisms. The genus ***Proisotoma*** includes 33 species in North America. ***Isotomurus palustris*** can be abundant on the surface of water or on waterside plants. ***Isotoma viridis*** is a veritable giant, averaging 6 mm in length, and common across the entire continent, especially east of the Rockies.

SPRINGTAILS

actual size

Pogonognathellus flavescens

Entomobrya albocincta

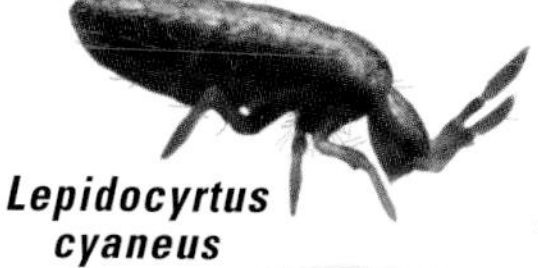

Lepidocyrtus cyaneus

Orchesella sp.

Seira domestica

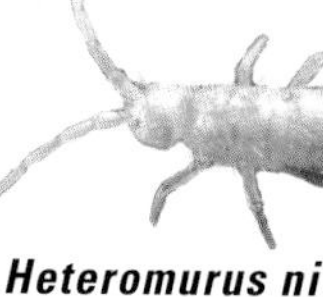

Heteromurus nitidus

snow fleas (Hypogastruridae)

Anurida maritima

family Sminthuridae

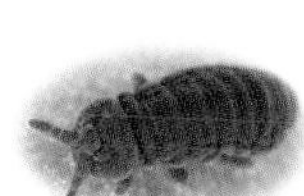

Podura aquatica

Proisotoma minuta in swarm

Sminthurides aquaticus

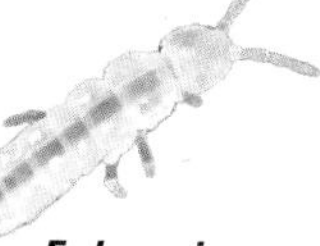

Folsomia candida

Isotomurus palustris

Isotoma viridis

TERMITES

(order Isoptera), famous for eating wood, can do extensive damage to homes and buildings, though in their natural role as decomposers they are highly beneficial. They host protozoans in their gut that enable the digestion of cellulose. Termites are social, living in colonies usually composed of several castes. Workers do the foraging, care for siblings and the queen, and repair or expand the nest. Soldiers defend the colony with huge jaws and/or "squirt gun" heads. The queen is an egg-laying machine, the king her lifelong mate. Once each year a large number of potential kings and queens is launched. These winged individuals, called alates, are spawned like living confetti from the nest. They are often confused with winged ants (see p. 360), but in termites both pairs of wings are equal in size, and the body is not distinctly segmented. Birds, dragonflies, and other predators feast heavily on such swarms. The presence of alates can be one sign of an infestation, mud tubes extending up walls being another. North America has about 44 species of termites, in four families.

SUBTERRANEAN TERMITES (family Rhinotermitidae) nest in the soil, and foraging for wood often results in extensive damage of human structures. One indicator of an infestation is the presence of mud tubes breaching the foundation, suspended from ceilings, or stretching across walls. The genus ***Reticulitermes*** includes six species in North America, collectively widespread, ranging north to Ontario and British Columbia. The introduced **Formosan Subterranean Termite** *(Captotermes formosanus)* has done major damage in New Orleans in the past and has earned the nickname of "condo-eater" in southern California.

DRYWOOD TERMITES (Kalotermitidae) are also known as powder-post termites. Colonies are contained within wood rather than headquartered in the soil. There are 18 species found in the southern U.S., some of the most widespread being in the genus ***Incisitermes.*** Their dry fecal pellets are a telltale sign of an infestation. Winged adults (alates) swarm mostly in fall, on hot, sunny days. Virtually all wooden structures and objects are vulnerable to attack, including live trees. Two species in the genus ***Cryptotermes*** (not illustrated) occur in Florida. Soldiers have a block-headed appearance. These insects are tiny (soldiers only 3 mm) but can be highly destructive, sometimes honeycombing wood and leaving only a paper-thin exterior shell. Alates swarm in May and June.

DAMPWOOD TERMITES (Termopsidae) include one genus, ***Zootermopsis,*** with three species in western North America. They are primitive, with only reproductives and soldiers; what resemble workers are the nymphal stages of the other castes. Colonies can number up to 3,500 individuals and occupy logs and stumps in contact with moisture. Fence posts and utility poles may sometimes be damaged. **HIGHER TERMITES (Termitidae)** include the tropical varieties that build the enormous mounds in Africa and Australia. Our 15 North American species are not so architecturally inclined. Soldiers of the two southwestern species of ***Tenuirostritermes*** (not illustrated) have strange nozzlelike heads from which they squirt irritating chemicals, highly effective against ants, their chief predators.

TERMITES

(not all figures are to scale)

swarm of winged adults

subterranean termites
(Reticulitermes)

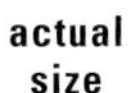

actual size

soldier close-up

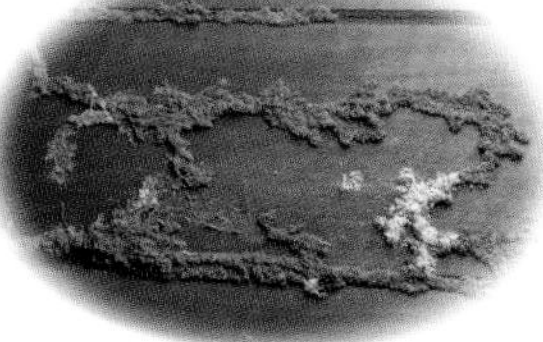

mud tubes along wall

workers and soldiers

soldier

Formosan Subterranean Termite

damage

drywood termites
(Incisitermes)

workers and soldiers

dampwood termites
(Zootermopsis)

winged adult

soldiers and nymphs

FLEAS, LICE, BARKLICE

FLEAS (order Siphonaptera) are extraordinary parasites of mammals and birds. Superbly adapted to their lifestyle, they are flattened side to side for easy penetration of fur or feathers. Their leaping ability is legendary, but thanks to special combs and spines all over their bodies, they stay put once they land on a host. Some have visited catastrophe on humanity by ferrying disease organisms to human hosts. Flea larvae feed on organic debris in the nests of the host, while adults feed on blood.

Only 16 species of **COMMON FLEAS (family Pulicidae)** live in North America, but they include some of our best known. The **Dog Flea** *(Ctenocephalides canis)* is not restricted to dogs. Cats, rats, rabbits, foxes, and people are also recorded as hosts. The **Human Flea** *(Pulex irritans)* is relatively rare, its populations on the decline in the U.S. The **Oriental Rat Flea** *(Xenopsylla cheopis)* is the plague flea that transmits the bacillus that causes both plague and murine typhus. The **Sticktight Flea** *(Echidnophaga gallinacea)* is a serious pest of poultry in tropical climates, less so in temperate regions. **BIRD and RODENT FLEAS (Ceratophyllidae)** make up our largest family of fleas, with 125 species here. The **Western Chicken Flea** *(Ceratophyllus niger)* can be a poultry pest. **RODENT FLEAS (Hystrichopsyllidae)** include seven species north of Mexico, all restricted to rodents or shrews.

LICE (order Phthiraptera) include chewing lice that occur on birds or mammals and feed on flakes of skin, fur, or feathers, while sucking lice live on mammals and feed on blood. Virtually no warm-blooded vertebrate escapes being a host. Even walruses are plagued by lice. The 77 species of **MAMMAL CHEWING LICE (family Trichodectidae)** include the **Dog Chewing Louse** *(Trichodectes canis)*. It commonly afflicts very young, old, or sickly hosts. **MARSUPIAL CHEWING LICE (Boopidae)** include only the **Australian Dog Louse** *(Heterodoxus spiniger)* in North America. It probably evolved as a parasite of kangaroos before making the jump to dingos and then domestic dogs. **HUMAN LICE (Pediculidae)** include only the **Human Louse** *(Pediculus humanus)*, of which some populations infest only the scalp, others only body hair. A case could be made that this is one of the world's deadliest animals, given its record of spreading disease, including the microbe that causes typhoid fever. **POULTRY LICE (Menoponidae)** infest many kinds of birds. The **Large Poultry Louse** *(Menacanthus stramineus)* was probably restricted to North America's wild turkey until domesticated fowl were introduced. The **family Haematopinidae** includes species that afflict cattle, deer, and their relatives. The **Hog Sucking Louse** *(Haematopinus suis)* is enormous by louse standards, females exceeding 5 mm.

BARKLICE and BOOKLICE (order Psocoptera) are easily mistaken for psyllids (p. 100) or aphids (p. 98), but note their chewing mouthparts and long antennae. Many are colonial. Adults may be winged, wingless, or have reduced wings. The majority graze on lichens, molds, or fungi, or scavenge on detritus. There are 28 families represented north of Mexico, but most often seen are **COMMON BARKLICE (Psocidae),** with 62 species, often on trunks of maples. Nymphs of some species gather in "herds" on tree trunks in spring, hence a name for the entire family: "tree cattle."

"herd" of barklice on tree

common barklice (2 examples)

THRIPS (order Thysanoptera) is both singular and plural, so one is a "thrips," and several are likewise "thrips." These minute insects are often abundant in flowers and fungi. They have asymmetrical mouthparts, with only a left mandible, used in a "punch and suck" feeding method. The mandible lances a plant cell or other morsel, and the other mouthparts vacuum out the contents. The life cycle defies classification. "Larvae" hatch from eggs, but they are not unlike the adult in appearance, sprouting wingpads in the case of species with wings as adults. They eventually go through a "propupa" stage, followed by one or two pupal stages, and finally adults emerge. The damage that thrips do to plants is often the only obvious sign of their presence: air entering drained plant cells gives leaves a silvered appearance. At least half of all thrips species feed on fungi. Others are predatory on mites and tiny insects; others are opportunistic predators but feed principally on plants or fungi. Even wingless thrips disperse readily on the wind. Many species have been spread nearly globally through trade in crops and ornamental plants. Like most insects, thrips cannot be deemed good or bad, as their role in pollination, and predation on other pests, offsets their damage to plants.

PREDATORY THRIPS (family Aeolothripidae) include some species that are not strictly predatory but feed on pollen as well. Those that are predatory feed mostly on mites or other thrips. ***Aeolothrips*** includes 31 species, several of which have contrasting bands on the wings. **COMMON THRIPS (Thripidae)** include most of our true pest species. The **Red Grass Thrips** *(Aptinothrips rufus)*, originally from Europe, is now found globally. It is common in pastures but does little if any damage to cereal crops. The genus ***Frankliniella*** includes 40 species here, some considered pests. The **Western Flower Thrips** *(Frankliniella occidentalis)* is a native of California that has spread globally. The **Tobacco Thrips** *(Frankliniella fusca)* is the vector of tomato spotted wilt virus and also damages cotton and tobacco. The **Greenhouse Thrips** *(Heliothrips haemorrhoidalis)* has spread worldwide from its native South America, infesting a wide range of plants. ***Echinothrips americanus*** has been spread to Europe by the trade in poinsettia plants. The **Privet Thrips** *(Dendrothrips ornatus)*, most noticeable for the leaf damage it causes, jumps actively when disturbed. The genus ***Thrips*** includes over 30 species north of Mexico. The **Onion Thrips** *(Thrips tabaci)* feeds not only on leaves but also on pollen and the eggs of mites.

TUBE-TAILED THRIPS (Phlaeothripidae) have diverse feeding habits. The **Cuban Laurel Thrips** *(Gynaikothrips ficorum)* is actually a pest of fig trees. Its feeding activities cause leaf galls, which are also home to other insects, throughout the thrips's native tropical range.

WEBSPINNERS (order Embiidina) spin silk from glands on their front feet, and they spend most of their lives in silken galleries hidden under leaf litter or under bark. Females are wingless, while males usually have wings. They are easily mistaken for other insects such as earwigs or winged termites, but are usually less common. North America has 11 species. ***Oligotoma nigra,*** an introduced species, is common in the southwest.

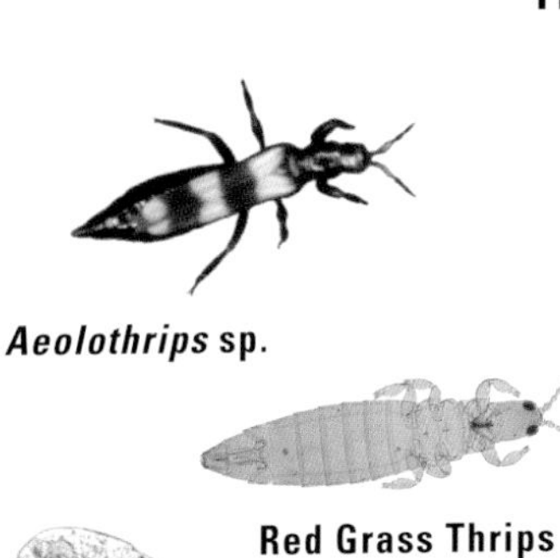

Aeolothrips sp.

actual size

Western Flower Thrips

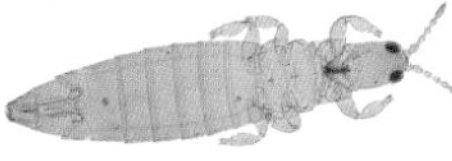

Red Grass Thrips

Tobacco Thrips

Greenhouse Thrips

Echinothrips americanus

damage caused by Privet Thrips

Onion Thrips
adult (top) and immature

Cuban Laurel Thrips

webspinner (*Oligotoma nigra*)

HANGINGFLIES AND SNOW SCORPIONFLIES

belong to the **order Mecoptera,** along with the scorpionflies on the next page. These bizarre insects are now considered to be most closely related to the fleas (p. 34). Despite their varied appearance, most families in this order can be readily identified with a close view by the long beaklike "face." They have chewing mouthparts, and they undergo a complete metamorphosis. Look for winged forms in wooded areas and waterside habitats. Adults of some families in this order scavenge for dead insects, sometimes even robbing them from the webs of spiders.

HANGINGFLIES (family Bittacidae) are easily mistaken for the crane flies (p. 276), but they have two pairs of wings, not one pair like crane flies. Hangingflies are often found in open woodlands and forest edges, flying slowly and hanging from twigs, grasses, or leaf edges by their front feet, or by the front and middle pairs of feet. Their prehensile feet (tarsi) are unsuited to standing on flat surfaces, but they have other skills: the hangingflies use their hind feet to capture insect prey, either while hanging or while flying upward along plant stems. Captured insects are grasped firmly in the hind tarsi, held up to the face, and eaten. In addition to feeding on their small prey, males also use captured or scavenged dead insects to attract females (along with the broadcast of a seductive pheromone). Mating takes place when she accepts the meal. Larvae live mostly in leaf litter or soil and scavenge dead insects.

The **Wingless Hangingfly** *(Apterobittacus apterus)* occurs in central California. Crane flies are among its preferred prey items. The **Black-tipped Hangingfly** *(Hylobitticus apicalis)*, not illustrated, is an eastern species with dark wingtips. It rests with the wings outstretched to the sides, while all the remaining species hold their wings folded down rooflike over the abdomen at rest. The genus ***Bittacus*** includes seven species collectively ranging in the eastern U.S. and California. ***Bittacus strigosus*** and ***Bittacus pilicornis*** are two common examples. They are sometimes attracted to lights at night. There is one other genus, ***Orobittacus,*** with one species in California.

SNOW SCORPIONFLIES (Boreidae) are now considered to be the members of the Mecoptera that are the closest living relatives of fleas. They are small (2–7 mm) and flightless and feed on mosses and liverworts in boreal habitats. Adults and larvae share the same diet. Adult males have reduced, hardened wings, each with a sharp terminal spine used to grasp the female during mating. The wingless females ride atop males when coupled. There are three genera in North America, but only ***Boreus*** is common and widespread, with two eastern species and eight representatives in the west. ***Boreus notoperates*** and ***Boreus californicus*** are two of the western species. Look for them in alpine habitats and northern latitudes, where they are typically active on snow in late winter and early spring.

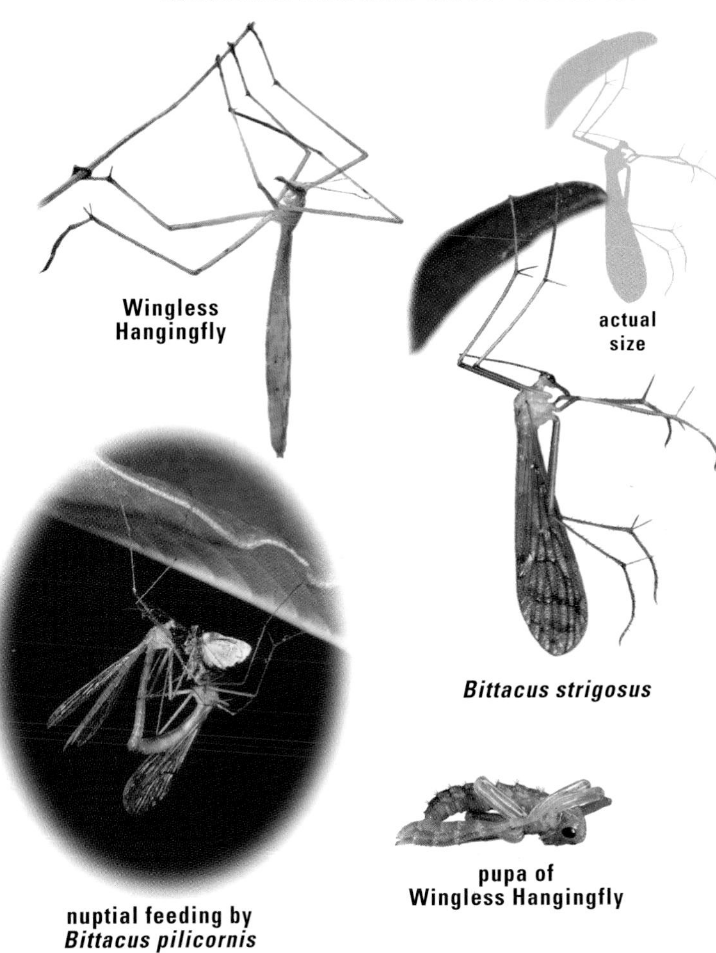

Wingless Hangingfly

Bittacus strigosus

pupa of Wingless Hangingfly

nuptial feeding by *Bittacus pilicornis*

snow scorpionflies
(at different scale from hangingflies)

Boreus notoperates **female**

Boreus californicus

SCORPIONFLIES

(family Panorpidae) look utterly different at first glance from the hangingflies and the snow scorpionflies on the preceding page, but they are classified in the same order (Mecoptera). These insects are named for the appearance of the male, his abdominal claspers enlarged and modified into a wicked-looking "tail" like that of a scorpion. Despite their appearance, scorpionflies neither sting nor bite and are completely harmless.

Scorpionflies usually are uncommon insects, but at times they can be found in fair numbers, sitting on leaves in the undergrowth of open woods or in overgrown old fields. Larvae resemble the caterpillars of moths or butterflies, but are unique in having compound eyes. They pupate in a small chamber that they make just below the surface of the soil. Both larvae and adults scavenge for dead insects, adults sometimes even feeding on prey trapped in spider webs. Males attract females with a morsel of food or a salivary secretion and an airborne sexual scent called a pheromone. The pair form a V-shaped mating posture, the male coupling with his mate's terminal segments and grasping the edge of her front wing with a special device on the top of his abdomen.

Currently 54 species of scorpionflies are recognized in North America, all restricted to the east, from southeastern Canada south to Florida and Texas, and all classified in the genus ***Panorpa*** (which occurs in the Old World as well). ***Panorpa nuptialis*** is one of the largest species and one of the most striking, reddish with strong black and amber banding on the wings. It ranges widely but is most common in the south-central U.S. ***Panorpa helenae*** is very widespread in the east, but many scorpionflies have much more limited ranges.

SHORT-FACED SCORPIONFLIES (Panorpodidae) occur in the Appalachian Mountains (three species) and in the Pacific northwest (two species). ***Brachypanorpa*** (not illustrated) is the only genus. Males resemble typical scorpionflies but have shorter "faces," lacking the long-beaked appearance, and have less patterning, usually looking a more uniform yellowish brown. Their abdomens are thickened at the tip but are not curled up and forward over the back in scorpion style. Females have shortened wings and are flightless. The larvae, looking somewhat like beetle larvae, live in the soil and probably feed on plant material.

EARWIGFLIES (Meropeidae) are perhaps the most bizarre of the scorpionflies because they are not easily recognized as such. Our sole species, ***Merope tuber*** (not illustrated), occurs over much of the east. Adults are yellowish brown and about half an inch long. The common name of the family derives from the male's enormous claspers that suggest the cerci on an earwig. The female's abdomen tapers and lacks any obvious extremities. The larvae of this insect are completely unknown. Adults were once thought to be rare, but are now collected with some frequency with special flight-intercept devices called Malaise traps. Look for this insect in the same habitats as most of the more common scorpionflies.

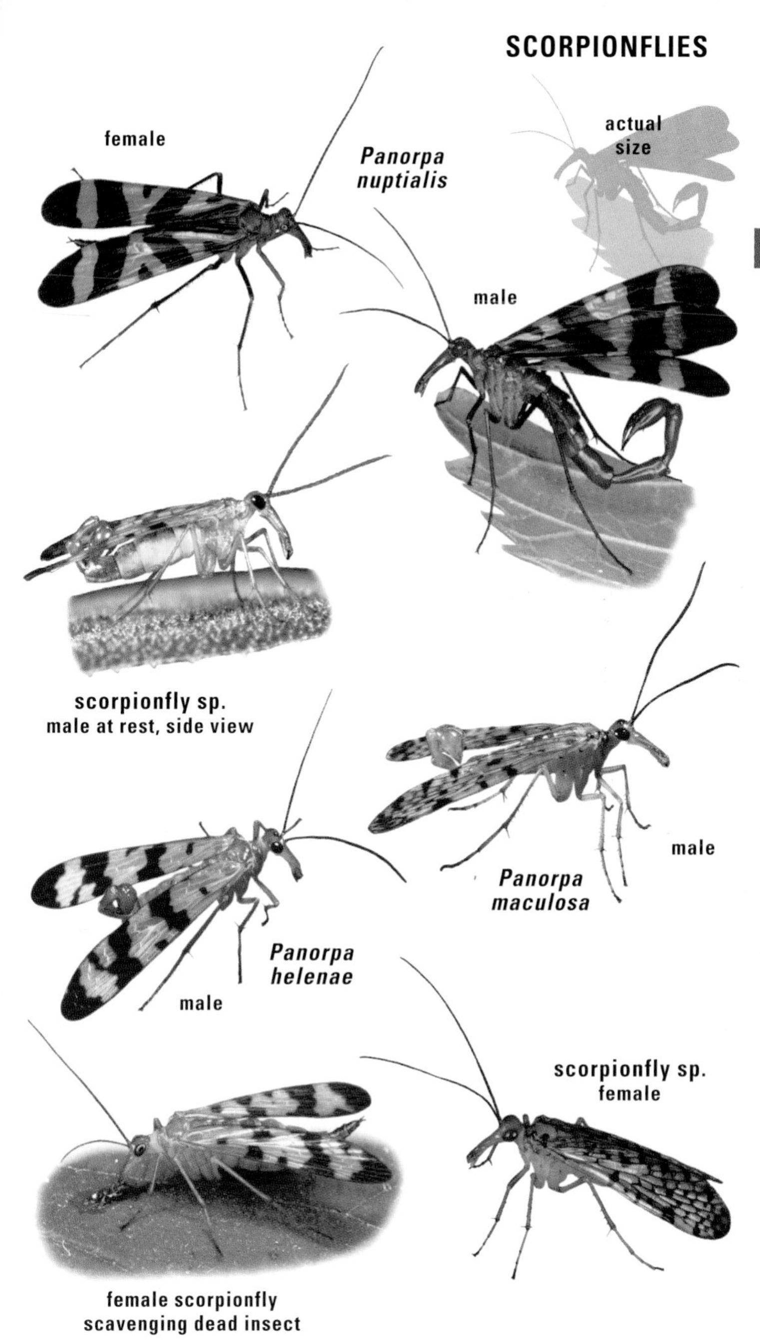
female
Panorpa nuptialis
actual size
male
scorpionfly sp.
male at rest, side view
male
Panorpa maculosa
Panorpa helenae
male
scorpionfly sp.
female
female scorpionfly
scavenging dead insect

DRAGONFLIES

and damselflies **(order Odonata)** have long captured popular imagination, as reflected in nicknames such as "snake doctors" and "devil's darning needles." Despite their appearance, they are harmless to humans. They do not sting. They may attempt to bite if handled, but only the larger species can deliver a noticeable pinch. Adults are highly aerial, larvae are aquatic, and all ages are strictly predatory. They undergo partial metamorphosis, the mature larva clambering out of the water (usually at night) and the winged adult emerging from the split exoskeleton. North of Mexico there are about 300 species of dragonflies **(suborder Anisoptera)** and about 130 species of damselflies **(suborder Zygoptera).**

DARNERS (family Aeschnidae) include our largest dragonflies. They have huge eyes that meet on the tops of their heads, their wings are usually unmarked, and their abdomens are long and narrow, perhaps suggesting the shape of a darning needle. They may spend much of their time flying, and at rest they hang vertically rather than perching horizontally like many dragonflies. Their larvae develop in a variety of waters, from bogs and lakes to flowing streams, depending on the species. In some northern species, larvae may take several years to mature. About 40 species occur in North America north of Mexico.

larva of Common Green Darner

The **Common Green Darner** *(Anax junius)* is one of our most widespread and numerous dragonflies, found throughout the U.S. and southern Canada and south to Costa Rica. Adult males show a contrast between the blue abdomen and green thorax. Females and younger males are less distinctive in color but are known by the black "bull's-eye" pattern on the forehead. Common Green Darners are strongly migratory, and big flights may be seen in spring and fall. The **Comet Darner** *(Anax longipes),* an uncommon but widespread and strikingly colored eastern species, usually breeds in small ponds but ranges widely over the countryside.

Two of our largest dragonflies are the **Regal Darner** *(Coryphaeschna ingens),* of marshes and swamps of the southeast, and the **Swamp Darner** *(Epiaeschna heros),* widespread in the east around slow forest streams and wooded swamps. The latter species has been noted as entering buildings surprisingly often, perhaps because the deep shadows indoors suggest its usual shady forest haunts.

More than half of our darners belong to the genus ***Aeschna,*** or **mosaic darners.** The **Blue-eyed Darner** *(Aeschna multicolor)* is widespread in the western and central states and southwestern Canada, ranging from the plains well up into the foothills. The **Variable Darner** *(Aeschna interrupta)* is one of the most widespread species, found throughout most of Canada and southern Alaska and well south into the western and northeastern U.S., usually around marshy ponds.

Common
Green Darner
actual
size
female or
younger male
adult
male
Comet
Darner
Regal
Darner
Blue-eyed
Darner
Swamp Darner
Variable Darner

CLUBTAILS AND PETALTAILS

The large eyes of most dragonflies meet on the top of the head, but in these two families, the eyes are widely separated. **CLUBTAILS (family Gomphidae)** are named for the thickened tip of the abdomen, more pronounced in males. Almost 100 species live in North America, most in eastern or central regions. Generally they live around rivers and streams, usually perching on the ground or other flat surfaces, where their dull colors and camouflaged patterns make them hard to spot. Their larvae mostly burrow in sand or mud at the bottom of flowing streams.

Over one-third of our clubtails belong to the genus ***Gomphus,*** including the **Lancet Clubtail** *(Gomphus exilis),* a typically dull-colored species very common east of the Mississippi. It is numerous partly because, unlike many species in the clubtail family, it regularly breeds in still ponds and marshes as well as in flowing rivers. The **Black-shouldered Spinyleg** *(Dromogomphus spinosus)* is widespread in the east, especially in forested areas along streams and rivers, where it may perch and forage in deep shade. Like the other two species of spinylegs, it has a relatively small head and long legs, but (despite the name) the spines on the legs are hard to see. The **Grappletail** *(Octogomphus specularis)* is an odd clubtail restricted to the far west, living mainly along clear, rocky streams. Its abdomen is mostly black and its thorax is yellowish to gray-green with a wide black stripe. **Snaketails** (genus ***Ophiogomphus***) include almost 20 species, most with repeating "snakeskin" patterns on the abdomen. They live mostly along clear, fast-flowing rivers and streams, and when away from the water they may perch on the ground, on low twigs, or even in treetops. The **Pale Snaketail** *(Ophiogomphus severus),* paler than most of its relatives, is widespread in the western U.S. and western Canada. The genus ***Stylurus*** includes about 11 species, most with pale markings near the tip of the dark abdomen (which is strongly thickened or "clubbed" in males but not in females). The **Russet-tipped Clubtail** *(Stylurus plagiatus)* is widespread in the east and much of the southwest. **Sanddragons** (genus ***Progomphus***) are mostly tropical. We have only four species, including the **Common Sanddragon** *(Progomphus obscurus),* which is widespread in the eastern and central U.S., usually around rivers and streams with sandy shores.

Our largest clubtail is the **Dragonhunter** *(Hagenius brevistylus),* a fearsome beast that specializes in eating other dragonflies as well as other large insects. Widespread along forested rivers in the east, it is as large as some darners (previous page), but it perches horizontally, not vertically, and its eyes are widely separated, not meeting at the top of the head.

PETALTAILS (Petaluridae) are primitive dragonflies, with only 10 surviving species in the world, including two in North America. Unlike most dragonflies, the **Gray Petaltail** *(Tachopteryx thoreyi)* often perches on tree trunks. It is locally distributed over much of the east, mainly around springs and seeps in forest. The only other petaltail in our area, the **Black Petaltail** *(Tanypteryx hageni),* lives in northwestern bogs.

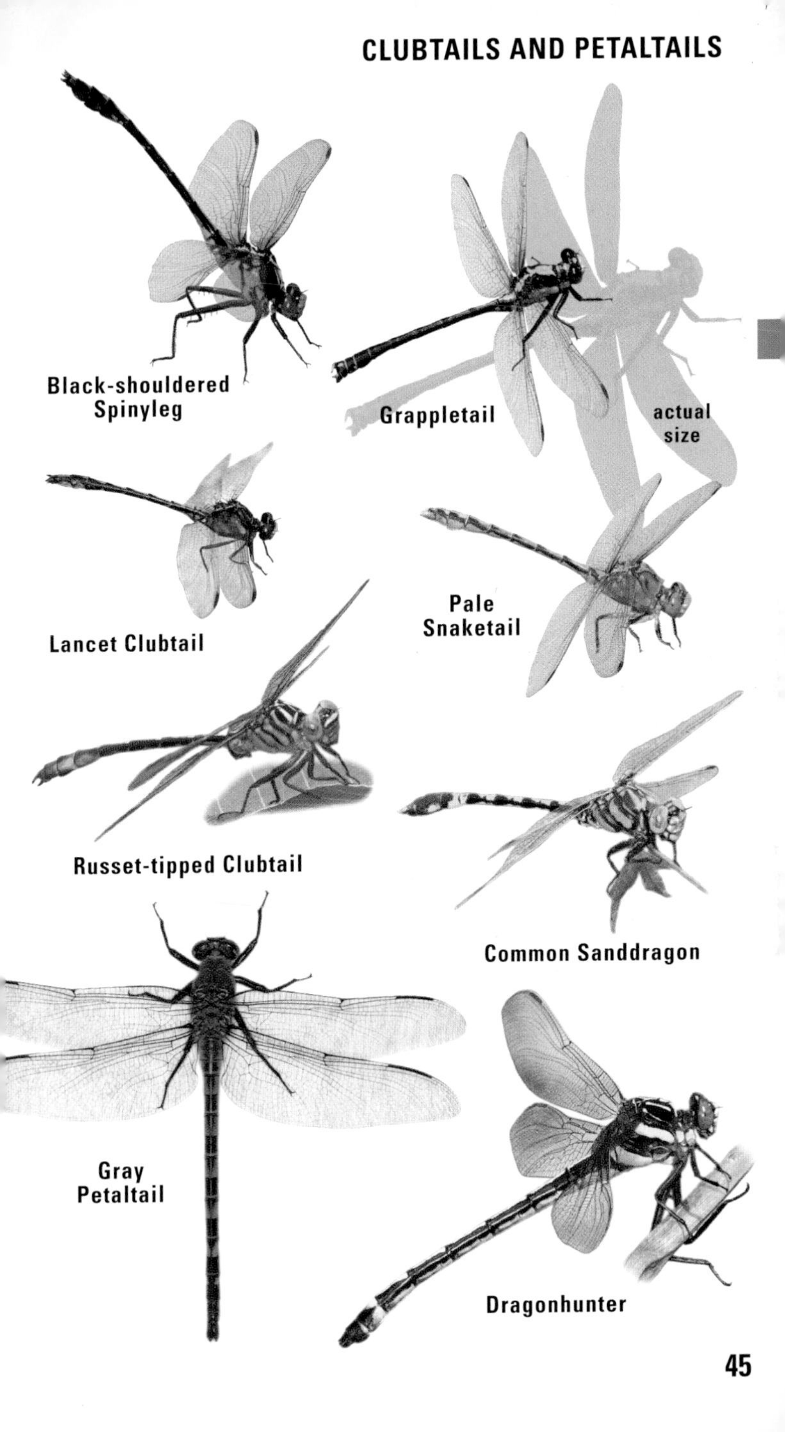
Black-shouldered Spinyleg
Grappletail
actual size
Lancet Clubtail
Pale Snaketail
Russet-tipped Clubtail
Common Sanddragon
Gray Petaltail
Dragonhunter

EMERALDS, CRUISERS, SPIKETAILS

Dragonflies belonging to these three families are usually not very common or conspicuous. We have about 50 species of **EMERALDS (family Corduliidae)**, about nine of **CRUISERS (family Macromiidae)**, and about nine of **SPIKETAILS (family Cordulegastridae)**.

Emeralds tend to wear subtle colors, except for their eyes, which are bright emerald green in many species. Many also have glossy reflections of bronze or greenish on their bodies. They tend to spend a lot of time in the air, and when perched they usually hang vertically from twigs or stems. About half of our species of emeralds belong to the genus ***Somatochlora,*** including the **Ringed Emerald** *(Somatochlora albicincta)*, which is widespread in the far north, south to parts of New England and the northwestern states. Like many of its relatives, it favors boggy or marshy ponds, sometimes slow-moving streams with marshy edges. The **American Emerald** *(Cordulia shurtleffii)* might better be called the "Canadian Emerald," as it ranges over most of Canada, as well as parts of the northern U.S. and Alaska.

The **baskettails** (genus ***Epitheca***) form a distinctive group of 10 emeralds, mostly with eyes brown instead of green. Several have distinct markings on the wings, suggesting dragonflies of the skimmer family (next page). Most strikingly marked is the **Prince Baskettail** *(Epitheca princeps)*, found almost throughout the eastern U.S. and southeastern Canada. It lives mainly around slow streams, lakes, and ponds and usually flies non-stop when hunting. Also widespread and numerous in the east is the **Common Baskettail** *(Epitheca cynosura)*. This species is most numerous early in the season, from late spring into early summer, and often feeds in swarms in the late afternoon.

Shadowdragons (genus ***Neurocordulia***) are elusive emeralds that forage mostly in the evening, just before dusk. All seven species are eastern, especially southeastern, including the poorly known **Alabama Shadowdragon** *(Neurocordulia alabamensis)*. Our two smallest emeralds are the **boghaunters,** including the elegant little **Ebony Boghaunter** *(Williamsonia fletcheri)*, found locally in bogs and fens in southeastern Canada and the northeastern U.S.

Of our nine species in the cruiser family, one of the most numerous is the **Stream Cruiser** *(Didymops transversa)*. It is widespread in the east, mainly around slow-moving rivers in wooded areas, sometimes around lakes. The **Illinois River Cruiser** *(Macromia illinoiensis)* cruises in the air above streams and rivers, the males patrolling regular territories throughout the day. This species is widespread in the east and extends far out onto the Great Plains.

Members of the spiketail family are usually uncommon and usually found along small streams. Most of our nine species are eastern, but the **Pacific Spiketail** *(Cordulegaster dorsalis)* is widespread in the mountains of the Great Basin region and the Pacific states.

larva of Twin-spotted Spiketail *(Cordulegaster maculata)*

actual
size
Prince
Baskettail
American
Emerald
Common
Baskettail
Ringed
Emerald
Alabama
Shadowdragon
Stream
Cruiser
Ebony
Boghaunter
Illinois
River
Cruiser
Pacific
Spiketail

SKIMMERS

(family Libellulidae) are the quintessential dragonflies, flashy denizens of ponds and meadows on summer afternoons. They make up our largest family of Odonata, with over 100 species in North America. This group includes many of our best-known species as well, since many are easily recognized by their color patterns. Most of our dragonflies with strong patterning on the wings belong to this family. Larvae of most of the skimmers develop in still waters of all kinds, not flowing streams.

Found at ponds, marshes, and meadows almost throughout the U.S. and in extreme southern Canada, the **Common Whitetail** *(Plathemis lydia)* is one of our most recognizable dragonflies. Older dragonflies of many species develop a pale powdery look on the body, called pruinosity, but this is most pronounced in some of the skimmers and famously so in this species. Mature male Common Whitetails may hold their white abdomens high as a threat display against intruding males. Females resemble those of the Twelve-spotted Skimmer but look shorter-bodied and have different markings along the sides of the abdomen.

Members of the genus ***Libellula*** are often called king skimmers. North America has about 18 species, most readily identified to species by their strong color patterns. The beautiful **Twelve-spotted Skimmer** *(Libellula pulchella)* was formerly called the Ten-spotted, based on a count of the white spots in the wings of the male instead of the dark spots! This species is common over most of southern Canada and the U.S. except for parts of the southwest and far south. Usually seen patrolling in rapid flight over ponds, it also strays far from the water. Common across most of the U.S., except for the northern Rockies and Great Basin region, is the **Widow Skimmer** *(Libellula luctuosa)*. Its larvae live in still waters of ponds, lakes, and marshes, but adults range widely in open country.

Pondhawks (genus ***Erythemis***) are aptly named for their active predatory behavior, regularly capturing other insects as large as themselves, including other pondhawks. Males defend territories along the water's edge, patrolling back and forth, resting on low plants or on the ground. **Common Pondhawks** *(Erythemis simplicicollis)* are often split into two species, Eastern and Western Pondhawks, but these forms apparently intergrade in some places where their ranges meet in the center of the continent. Mature males are dull powdery blue, while females and younger males are more strikingly patterned, bright green with black rings on the abdomen.

Whitefaces (genus ***Leucorrhinia***) make up a group of seven species of relatively small skimmers, all mostly dark but with white faces. They are mainly northern, with four of the species found as far north as central Alaska. The most widespread species south of the Canadian border is the **Dot-tailed Whiteface** *(Leucorrhinia intacta)*, common around marshes, slow streams, and farm ponds, and ranging as far south as Tennessee and Colorado. The pale dot on the dark abdomen is a good field mark.

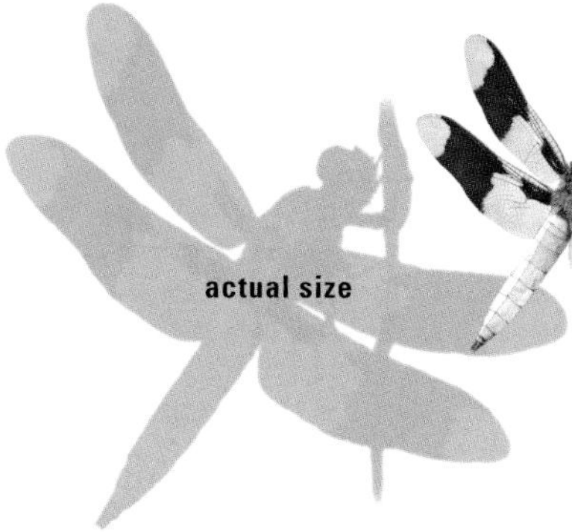

actual size

male

Common Whitetail

female

Twelve-spotted Skimmer

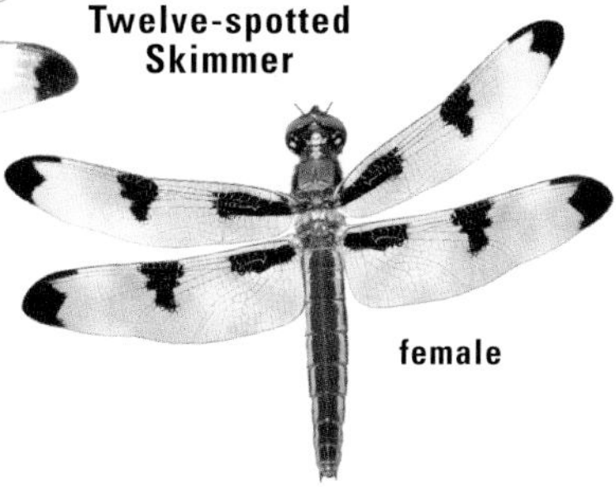

male

female

male

Widow Skimmer

female

mature male

Common Pondhawk

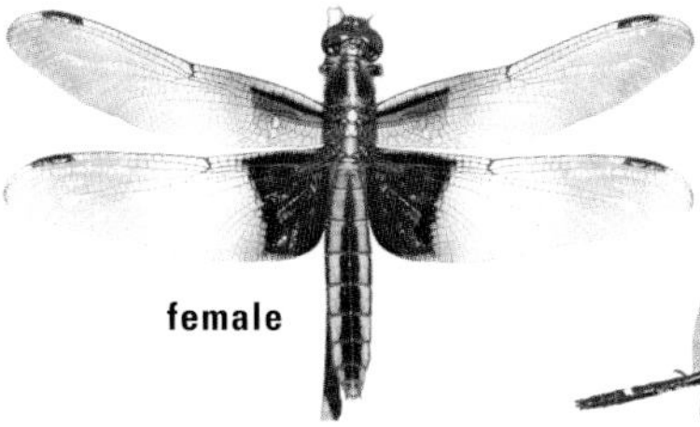

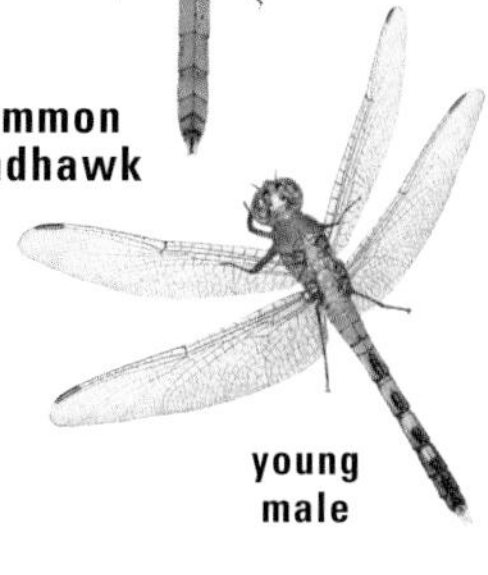

young male

male

Dot-tailed Whiteface

Saddlebag gliders (genus ***Tramea***) get their fanciful name from the darkened area at the base of the hindwings, which also makes them recognizable in flight at a long distance. These are moderately large skimmers that spend much of their time in the air, only occasionally perching. Seven species occur in North America, four restricted to subtropical areas and three ranging more widely. The most widespread, the **Black Saddlebags** *(Tramea lacerata)*, is highly migratory, and large concentrations may be seen moving south in early fall. The **pennants** (genus ***Celithemis***) are eight small species of skimmers, all found in the east. Several have markings on the wings. The most strikingly marked species is also the most widespread, the **Halloween Pennant** *(Celithemis eponina)*, common at ponds and marshes east of the Rockies and as far north as extreme southeastern Canada. Like other members of this group, it is often seen perched at the tip of a weed stalk, waving in the breeze like a pennant.

Meadowhawks (genus ***Sympetrum***) are more than a dozen species of small to medium-sized dragonflies often found in open meadows far from water. Males of most (and females of some) are marked with red; small, bright red dragonflies seen in late summer or early fall are probably members of this group. The **Variegated Meadowhawk** *(Sympetrum corruptum)* is the species most often seen in the west and the Great Plains, and it sometimes wanders long distances, appearing anywhere in the east.

For wandering, nothing beats the **Wandering Glider** *(Pantala flavescens)*. Superbly adapted for a nomadic lifestyle, this broad-winged skimmer can drift on the wind for days, even across oceans or deserts, until it finds a suitable spot for breeding. Females may lay their eggs in temporary rain pools, where the larvae may develop to adulthood in as little as five weeks. If the water dries up faster than that, larvae can survive for months in the dry mud, waiting for rains to fill up the puddle again.

Common in southern areas and often appearing quickly at temporary habitat is the **Roseate Skimmer** *(Orthemis ferruginea)*. As with many dragonflies, males change color as they mature; brown at first, they gradually develop the distinctive lavender thorax and bright pink abdomen.

Although not as conspicuous as some larger skimmers, **Blue Dashers** *(Pachydiplax longipennis)* are often the most numerous dragonflies at marshes and ponds over much of the U.S. Males patrol territories along the water's edge, chasing away other males and waiting to intercept females, while females are often away from the water except when they come there for mating and egg-laying.

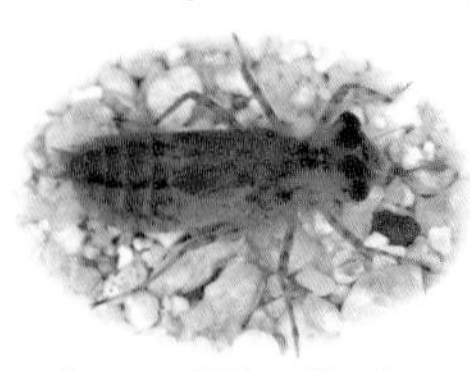

larva of Blue Dasher

Amberwings (genus ***Perithemis***) are small but fast-flying skimmers of ponds and streams. Of our three species, only the **Eastern Amberwing** *(Perithemis tenera)* is widespread. Males have amber wings, while those of females are clear with brown spots. Both sexes can look surprisingly similar to wasps as they perch on twigs low over the water.

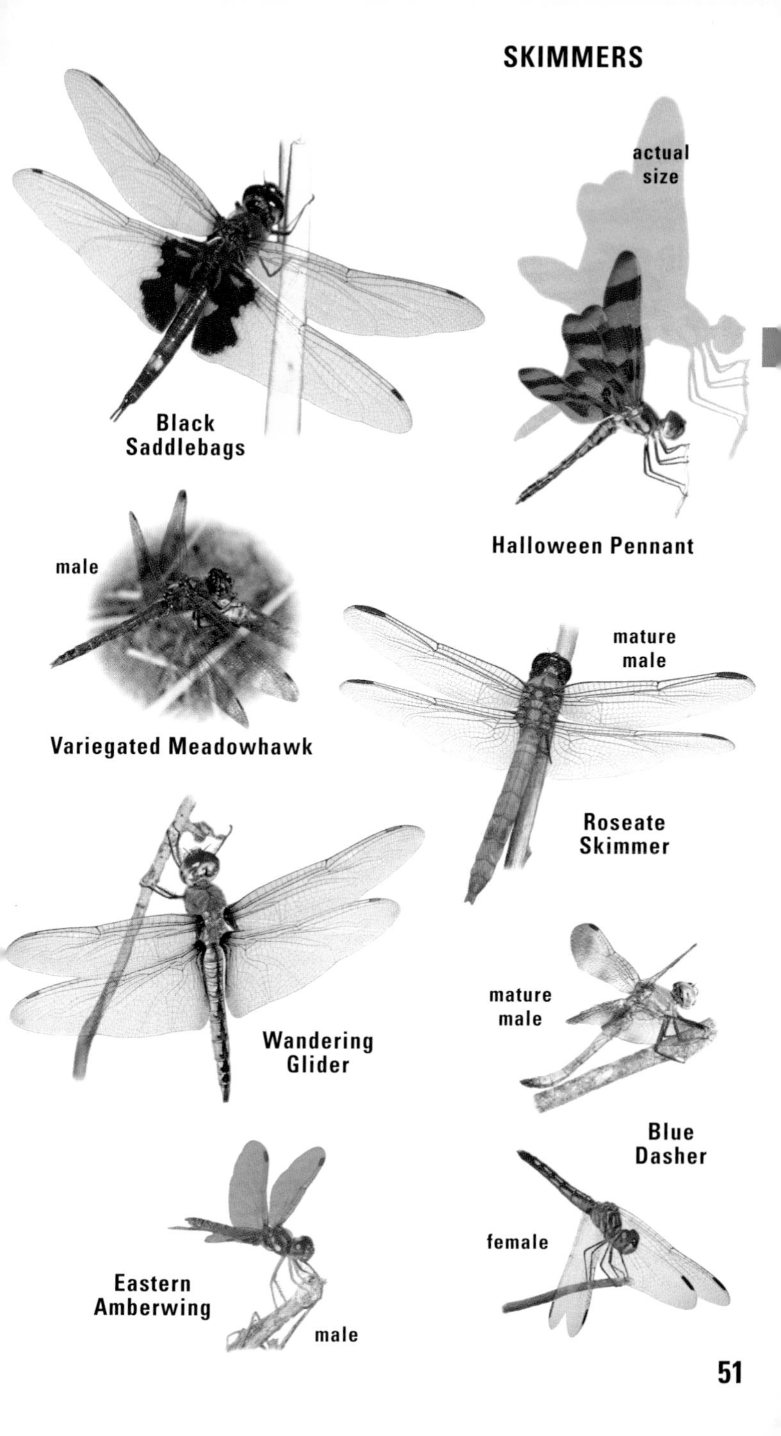

Black Saddlebags

Halloween Pennant

Variegated Meadowhawk

Roseate Skimmer

Wandering Glider

Blue Dasher

Eastern Amberwing

could be considered the "kid sisters" of the dragonflies. They make up a separate suborder **(Zygoptera)** within the order Odonata, with almost 130 species in North America. Damselflies average smaller and more slender-bodied than dragonflies, with forewings and hindwings about the same shape (dragonflies have the hindwings broader at the base).

The tiny, slender insects that hover almost unnoticed along pond edges and in tall grass are **POND DAMSELS (family Coenagrionidae).** They fold their wings together along or just above the abdomen at rest. Males are often marked with turquoise or blue. Females of many species may have two or even three different color forms, blue or brown or green, and they can be very difficult to identify. Our 31 species of **dancers** (genus ***Argia***), unlike most in the family, often live along flowing streams. When perched, they usually hold their wings slightly above the abdomen. The **Blue-fronted Dancer** *(Argia apicalis)* is common along muddy rivers and other waters east of the Rockies. The **Vivid Dancer** *(Argia vivida)* lives mainly around small streams and spring-fed ponds in the west. **Bluets** (genus ***Enallagma***) include more than 35 North American species. One of our most numerous damselflies is the **Familiar Bluet** *(Enallagma civile)*, found almost everywhere from southern Canada southward. It is quick to show up at new or temporary ponds. Not all bluets are blue, as illustrated by the **Orange Bluet** *(Enallagma signatum)*, common over most of the east.

Forktails (genus ***Ischnura***), so called for a small structure at the tip of the abdomen in males, are tiny and often brightly colored. The **Eastern Forktail** *(Ischnura verticalis)* is abundant in eastern Canada and the northeastern U.S., where it has a long flight season, spring to fall. **Firetails** (genus ***Telebasis***) make up a large tropical group, with two species extending north to the U.S. The **Desert Firetail** *(Telebasis salva)* is fairly common around marshy ponds and streams in the southwest, north to Kansas.

BROAD-WINGED DAMSELS (Calopterygidae) include only eight large distinctive species north of Mexico, usually found along streams. The **American Rubyspot** *(Hetaerina americana)*, one of three members of its genus, is common over most of the U.S. and parts of southeastern Canada. **Jewelwings** (genus ***Calopteryx***) have metallic bodies and often dark markings on the wings, taken to the extreme in the beautiful black-winged male **Ebony Jewelwing** *(Calopteryx maculata)*.

nymph of Great Spreadwing

SPREADWING DAMSELS (Lestidae) are recognized by their resting posture, with wings splayed out to the sides. Our 17 members of the genus ***Lestes***, including the **Common Spreadwing** *(Lestes disjunctus)*, are generally found along the edges of ponds, and several of the species are quite similar to each other. The **Great Spreadwing** *(Archilestes grandis)*, which lives along slow-moving streams as well as ponds over most of North America, often can be identified by its large size.

actual size

Vivid Dancer

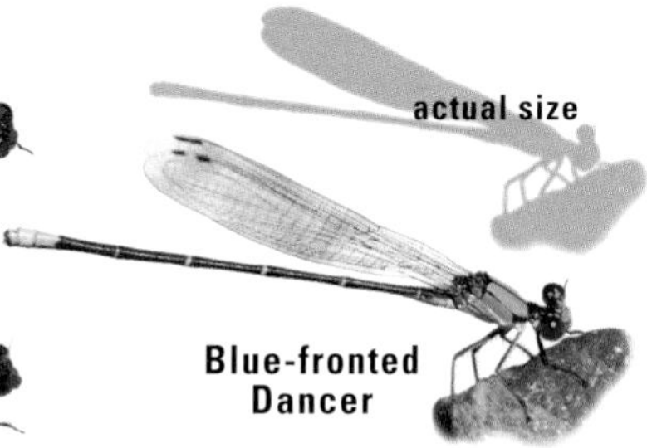

Blue-fronted Dancer

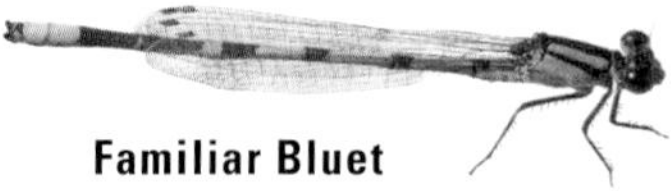

Familiar Bluet

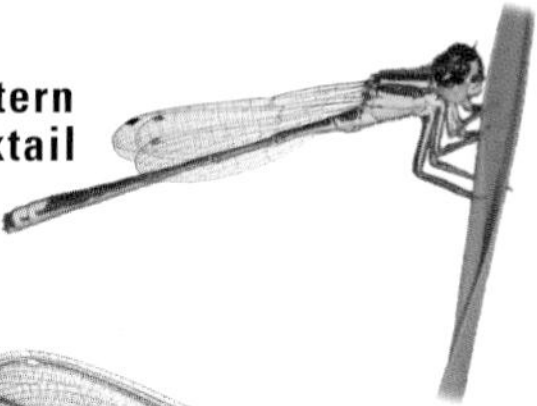

Orange Bluet

Eastern Forktail

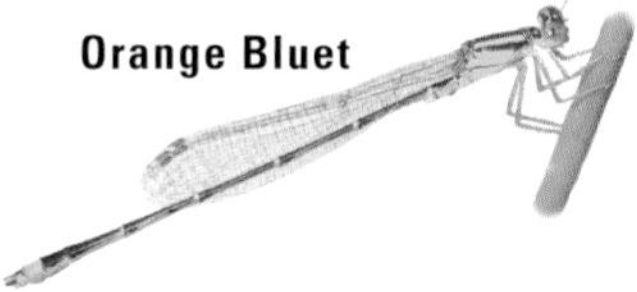

Desert Firetail

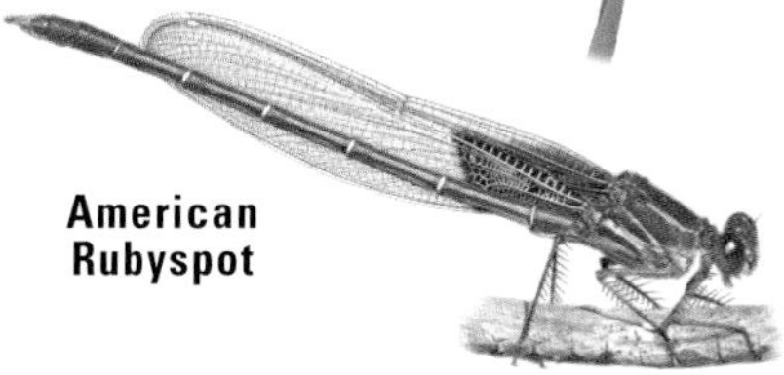

American Rubyspot

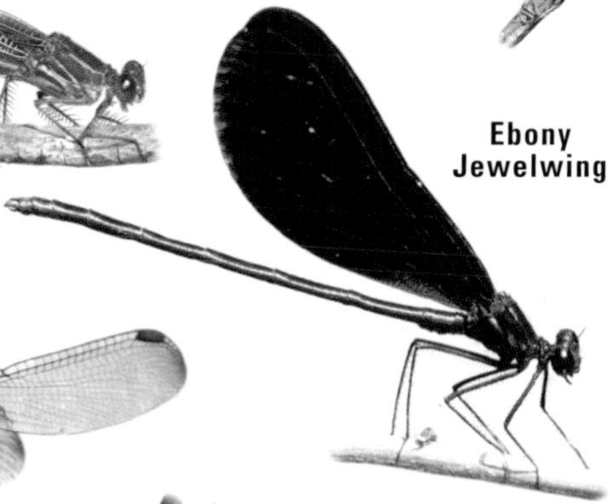

Ebony Jewelwing

Great Spreadwing

Common Spreadwing

MAYFLIES

(order Ephemeroptera) are famed, and scientifically named, for the ephemeral lives of adults. Mayfly adults normally live only a day or two. With only vestigial mouthparts, they do not eat. Larvae, often called nymphs or naiads, live underwater in various situations, from swift streams to large lakes; most feed on plants or organic debris, but some are predatory. They have gills along the sides of the abdomen and three or sometimes two "tails" (caudal filaments). Stonefly larvae (p. 60) have gills on the thorax and only two "tails." Larval mayflies may live a year or two, or even up to four years, molting (shedding their exoskeletons) several times before they leave the water. The first winged stage, called the subimago, has a dull appearance. Unlike any other insects, mayflies molt again after reaching this winged stage, to a fully adult stage called the imago. (Anglers call the subimagos "duns" and the imagos "spinners.") The imagos live just long enough to mate and for the females to lay eggs. North America has about 600 mayfly species, currently divided into 21 families. Because larvae of many have specific habitat requirements and are sensitive to pollution, some species are endangered or have recently become extinct.

English names of mayfly families refer mostly to the structure or habits of the larvae. **SMALL MINNOW MAYFLIES (family Baetidae)** have slender, minnowlike larvae that live in a wide variety of waters. Adults are quite small and have very small (or absent) hindwings, and males have divided eyes, the upper half swollen. This is our largest family of mayflies, with more than 130 species in North America. **PRONG-GILL MAYFLIES (Leptophlebiidae)** include more than 60 species, the larvae living mostly in streams, although some live in still waters. **FLAT-HEADED MAYFLIES (Heptageniidae)** include more than 120 species, mostly medium-sized. The larvae are flattened creatures that often look very dark (although freshly molted larvae of these and other mayflies generally look pale); their shape is an adaptation to living under rocks or logs in flowing streams.

PRIMITIVE MINNOW MAYFLIES (Ameletidae, Isonychiidae, and Siphlonuridae) have slim, minnowlike larvae that often live in small swift-flowing streams, although a few are found in ponds or swamps. Adults have well-developed hindwings. **SPINY CRAWLER MAYFLIES (Ephemerellidae)** are common and widespread, with more than 90 species. Larvae live in a variety of waters but are especially prevalent in trout streams.

COMMON BURROWING MAYFLIES (Ephemeridae) include only about a dozen species here, and the larvae are inconspicuous because they burrow in sediment at the bottoms of lakes and streams, but adults are among the largest mayflies and are sometimes among the most noticeable of all insects. Species of ***Hexagenia*** are abundant in the southern Great Lakes, and emergences of millions of adults in early summer formerly left huge drifts of dead insects on sidewalks in places such as Toledo, Chicago, and Green Bay. These huge hatches disappeared by the 1960s, for a sinister reason: the lakes were too polluted. Recently, massive flights have reappeared around Lake Erie, and the seeming "nuisance" created is really a welcome sign that the lake has been cleaned up to healthy levels again.

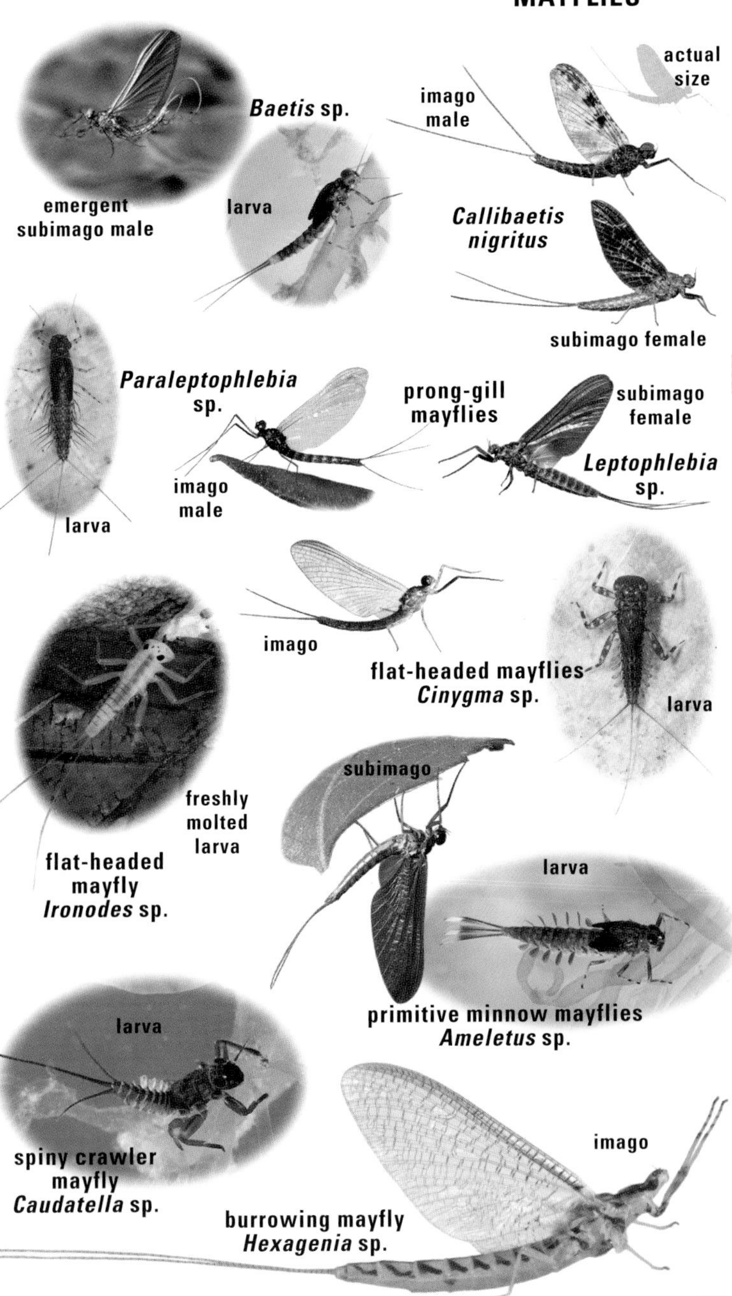
emergent
subimago male
Baetis sp.
larva
imago
male
actual
size
Callibaetis
nigritus
subimago female
larva
Paraleptophlebia
sp.
imago
male
prong-gill
mayflies
subimago
female
Leptophlebia
sp.
imago
flat-headed mayflies
Cinygma sp.
larva
freshly
molted
larva
flat-headed
mayfly
Ironodes sp.
subimago
larva
primitive minnow mayflies
Ameletus sp.
larva
spiny crawler
mayfly
Caudatella sp.
imago
burrowing mayfly
Hexagenia sp.

CADDISFLIES

(order Trichoptera) resemble sleek moths as adults, but the aquatic larvae build cases or retreats of nearly infinite design. All stages are models for fly fishermen who closely follow the hatches of various species. Many caddisflies are sensitive to low levels of pollution and are among the species used to monitor the health of rivers, streams, and lakes. Larvae may be "shredders" that tear pieces from plants, "collectors" that filter particulates from water currents, "scrapers" that graze the film of algae or detritus coating underwater objects, or predators on other aquatic animals. Metamorphosis is complete. Eggs are usually laid in a gelatinous mass, sometimes ring-shaped. Larvae usually molt six or seven times, and they may change the architecture of their cases as they grow. The pupae row themselves to the surface with oarlike legs just prior to emergence of the adult. Adults are mostly active at dusk. There are 18 families with over 1,260 species in North America, most diverse in the mountain west.

NET-SPINNING CADDISFLIES (family Hydropsychidae) live in flowing waters where the larvae weave sturdy meshlike nets. Some species spin fine mesh to snag tiny particles drifting on slow currents. Others spin coarse nets to catch larger prey in swift currents. Larvae of many produce sound by rasping the underside of the head against a ridge on each front leg. The sound is to ward off competitors, as they vigorously defend their foraging territories. There are roughly 150 species in North America. ***Hydropsyche*** larvae, often with ornate contrasting patterns on the head, are major predators of black fly larvae. **Zebra Caddis** *(Macrostemum zebratum)* adults are often common in the northeast and upper midwest in late June.

TRUMPET-NET CADDISFLIES (Polycentropodidae) construct silken retreats as larvae and actively prey on other organisms. Some are significant predators of black fly larvae. We have about 70 species. Larvae of ***Neureclipsis*** build sinuous trumpet-shaped silken retreats, anchoring their filtering nets on submerged roots and tree branches. **PRIMITIVE CADDISFLIES (Rhyacophilidae)** are unique in that the free-ranging larvae do not build cases. Almost all of our species (about 125) are in the genus ***Rhyacophila.*** Most are predatory on other aquatic creatures, including larvae of other caddisflies. A crude pupal case is composed of small stones. **SADDLE CASE-MAKERS (Glossosomatidae)** make saddle- or turtle-shaped cases. They occur mostly in cool, flowing water, though some are found along wave-tossed shallows of large lakes. ***Glossosoma*** includes 23 species here, most in western montane areas. The genus ***Anagapetus*** includes six western species, usually in cooler headwaters of mountain streams. The larval cases are composed of rock fragments of roughly uniform size.

LARGE CADDISFLIES (Phryganeidae) build tubular cases. Larvae generally begin as vegetarians but later become voracious predators consuming other insects and invertebrates, even salamander eggs. The genus ***Agrypnia*** includes eight North American species. Larvae construct tidy cases of neatly trimmed pieces of leaves and bark in a spiral pattern. They live in marshes, lakes, and slow-moving rivers. The four species of ***Ptilostomis*** range across the continent, adults flying from May to August.

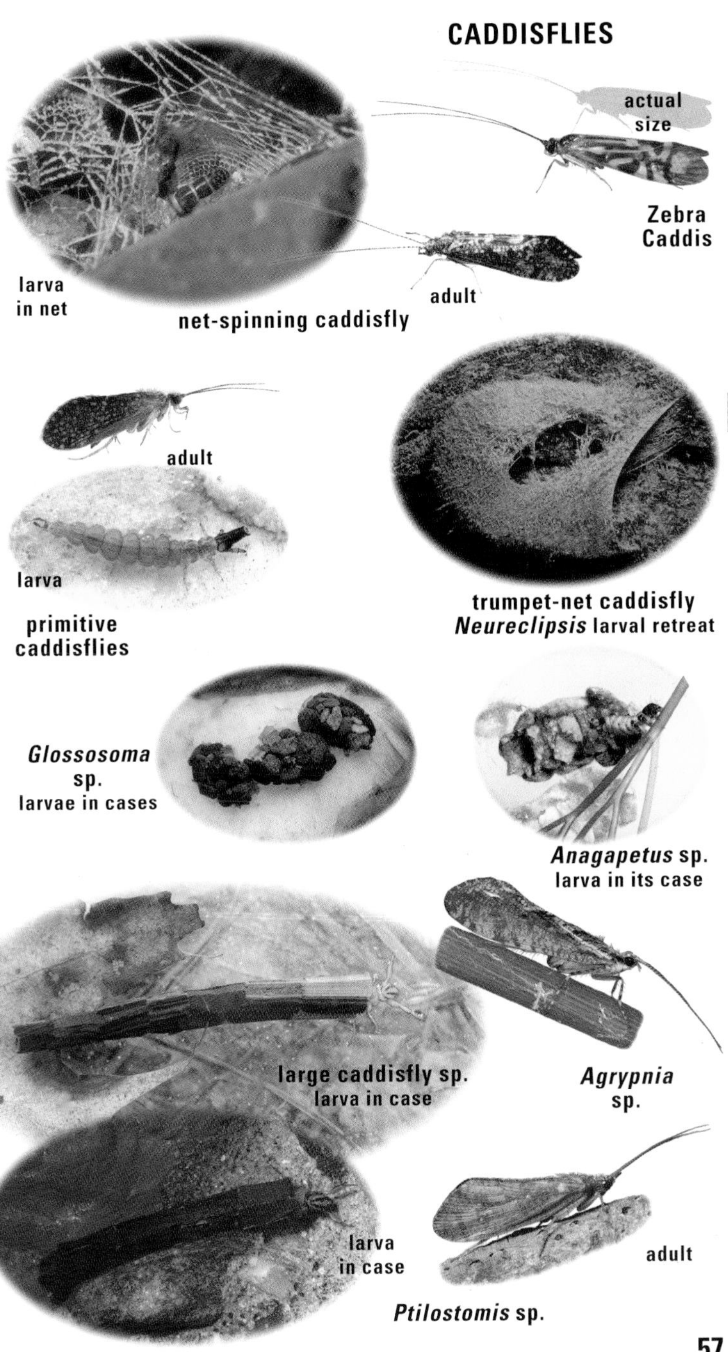
actual
size
Zebra
Caddis
larva
in net
adult
net-spinning caddisfly
adult
larva
primitive
caddisflies
trumpet-net caddisfly
Neureclipsis larval retreat
Glossosoma
sp.
larvae in cases
Anagapetus sp.
larva in its case
large caddisfly sp.
larva in case
Agrypnia
sp.
larva
in case
adult
Ptilostomis sp.

Larvae of **NORTHERN CADDISFLIES (family Limnephilidae)** build cases of varying styles. This family is most abundant in higher latitudes and elevations, with over 200 species in North America. While most larvae are "shredders" of plant matter, others are "collectors" of fine particles. Adults of at least a few genera produce a foul odor when handled. Larvae of the five species of ***Dicosmoecus*** build cases of plant material at first, but change to fine gravel as they mature. Adults emerge in late summer and early fall. The genus ***Hesperophylax*** includes six North American species, only one found east of the Rockies. Larvae build cases of uniformly sized gravel particles in small streams. The **Giant Golden Caddis** *(Hesperophylax designatus)* has a peak flight period in early July, the adults emerging at night. ***Platycentropus radiatus*** larval cases are crisscrossed collections of twigs and fragments of grass. This species inhabits a wide range of habitats, from cool streams to warm ponds. Our 15 species of ***Pycnopsyche*** live chiefly in woodland streams in the east. Larval cases begin as a collection of various organic particles, then change to fragments of bark or rock as the larva matures. ***Hydatophylax argus*** is a very large caddis, with a wingspan of up to 2 inches, ranging in the northeastern U.S.

HUMPLESS CASEMAKERS (Brachycentridae) usually build boxlike cases of neatly trimmed twig segments. There are about 30 species north of Mexico. Feeding behavior varies considerably according to genus, some actively filter-feeding while others graze on organic particulates. A few live in hot springs. The genus ***Brachycentrus*** occurs throughout much of the northern hemisphere, with 12 species on our continent, the larvae making squarish cases. Larvae of ***Micrasema*** (18 species) make more conical cases. Larvae of **SCALEMOUTH CADDISFLIES (Lepidostomatidae)** are consumers of fallen conifer needles and tough evergreen leaves such as rhododendron. They are also "gougers" that eat submerged, decaying wood. Most of our 80 species are in the genus ***Lepidostoma.*** These mostly make boxlike cases of uniform fragments of bark when they are nearing maturity but make different cylindrical cases of sand when they are young.

COMB-LIPPED CASEMAKERS (Calamoceratidae) are mostly tropical, with only five species north of Mexico. Larvae make many different types of cases. ***Heteroplectron californicum*** simply uses a hollowed-out stick, open at both ends to allow water to flow over the larva's gills. The southeastern ***Anisocentropus pyraloides*** is the only species in its genus. Larvae fashion cases of two pieces of leaves cut out and sewn together with silk.

LONG-HORNED CADDISFLIES (Leptoceridae) include about 100 species in North America. Adults of some species gather in enormous mating swarms. Feeding and casemaking behaviors are all over the map. ***Nectopsyche*** includes 14 species here. Larval cases are mostly long and skinny, often with twigs extending beyond the retreat. ***Triaenodes*** includes 25 species in North America, most diverse in the east but with several transcontinental in distribution. The larvae build amazing cases of uniform snippets of green plants, woven together in a long tapering spiral. The **Black Dancer** *(Mystacides sepulchralis)* is one of three species in the genus.

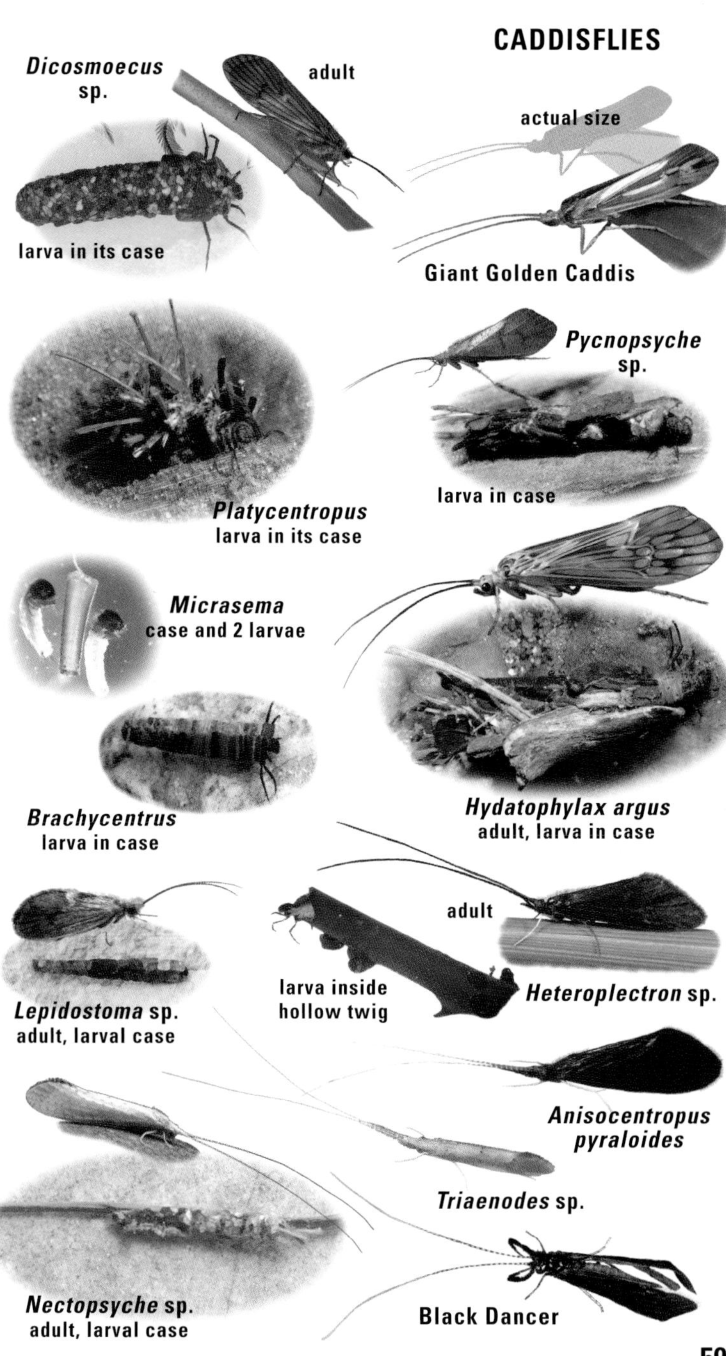
Dicosmoecus sp.
adult
larva in its case
actual size
Giant Golden Caddis
Pycnopsyche sp.
larva in case
Platycentropus larva in its case
Micrasema case and 2 larvae
Brachycentrus larva in case
Hydatophylax argus adult, larva in case
adult
larva inside hollow twig
Heteroplectron sp.
Lepidostoma sp. adult, larval case
Anisocentropus pyraloides
Triaenodes sp.
Nectopsyche sp. adult, larval case
Black Dancer

(order Plecoptera) are aquatic as larvae, often referred to as naiads (NYE-ads). They can be sensitive to pollution, and together with mayflies (**E**phemeroptera) and caddisflies (**T**richoptera), are the "EPT" indicator species used to assess the health of rivers, streams, and other watercourses. Metamorphosis is partial, mature naiads simply climbing from the water to facilitate emergence of the adult. Look for the ghostly shed "skins" clinging to rocks above the water. Fly fishermen often closely monitor large "hatches" of the adults. There are nine families in North America, with more than 600 species. Males of many stonefly species drum their abdomens against branches and foliage to call nearby females.

COMMON STONEFLIES (family Perlidae) emerge in summer. They include our most common species. Larvae are predatory; adults do not feed. The two western species of ***Hesperoperla*** are common, the larvae occupying various flowing waters. The genus ***Perlesta*** ranges over most of the U.S. and southern Canada. Larvae feed chiefly on midge larvae. **PREDATORY STONEFLIES (Perlodidae)** now include 50 species of ***Isoperla,*** formerly in their own family. Larvae of some are omnivores, others predators. The genus ***Cultus*** includes four widespread species, while the genus ***Skwala*** includes two species in the west. **GIANT STONEFLIES (Pteronarcyidae)** are sometimes called salmon flies for the pinkish color of freshly emerged adults. Larvae are mostly "shredders" of decaying leaves. ***Pteronarcys dorsata*** is widespread in the north and east. Adults typically emerge in large hatches between March and June. **GREEN STONEFLIES (Chloroperlidae)** are mostly pale yellow or green, resembling recently emerged individuals of other stonefly families. ***Alloperla*** is our largest genus, with 17 species. The genus ***Suwallia*** includes five species, mostly northern. Larvae live in creeks and small rivers. **SPRING STONEFLIES (Nemouridae)** are small and brownish, with an X pattern of venation toward the wingtips. Most emerge as adults in spring, a few in fall. ***Amphinemura*** is our most common genus, with 13 species. ***Zapada*** includes eight species living mainly west of the Rockies. Larvae are "shredders" of fallen leaves. **WINTER STONEFLIES (Taeniopterygidae)** are medium-sized and dark and emerge in winter or early spring. The genus ***Taeniopteryx*** includes nine eastern and northern species. ***Taenionema*** is a common western genus, but one species occurs in the northeast. Adults emerge between February and June.

SMALL WINTER STONEFLIES (Capniidae) are usually very small and blackish. Adults are often found walking over the snow or climbing on streamside plants or on bridges. Some adults have short, nonfunctional wings. About 130 species live north of Mexico. ***Allocapnia*** (38 species) is the genus commonly seen in the east. Adults can be active at temperatures as low as 20 degrees F. Adults of the genus ***Isocapnia*** may be common near rivers in late winter and early spring. The genus ***Capnia*** includes 50 species, all found from the Rockies westward. Adult **ROLLED-WINGED STONEFLIES (Leuctridae)** wrap their wings around the sides of the body. There are 45 species known for the U.S. and Canada. The five species of ***Megaleuctra,*** unusually large for this family, live around montane seeps.

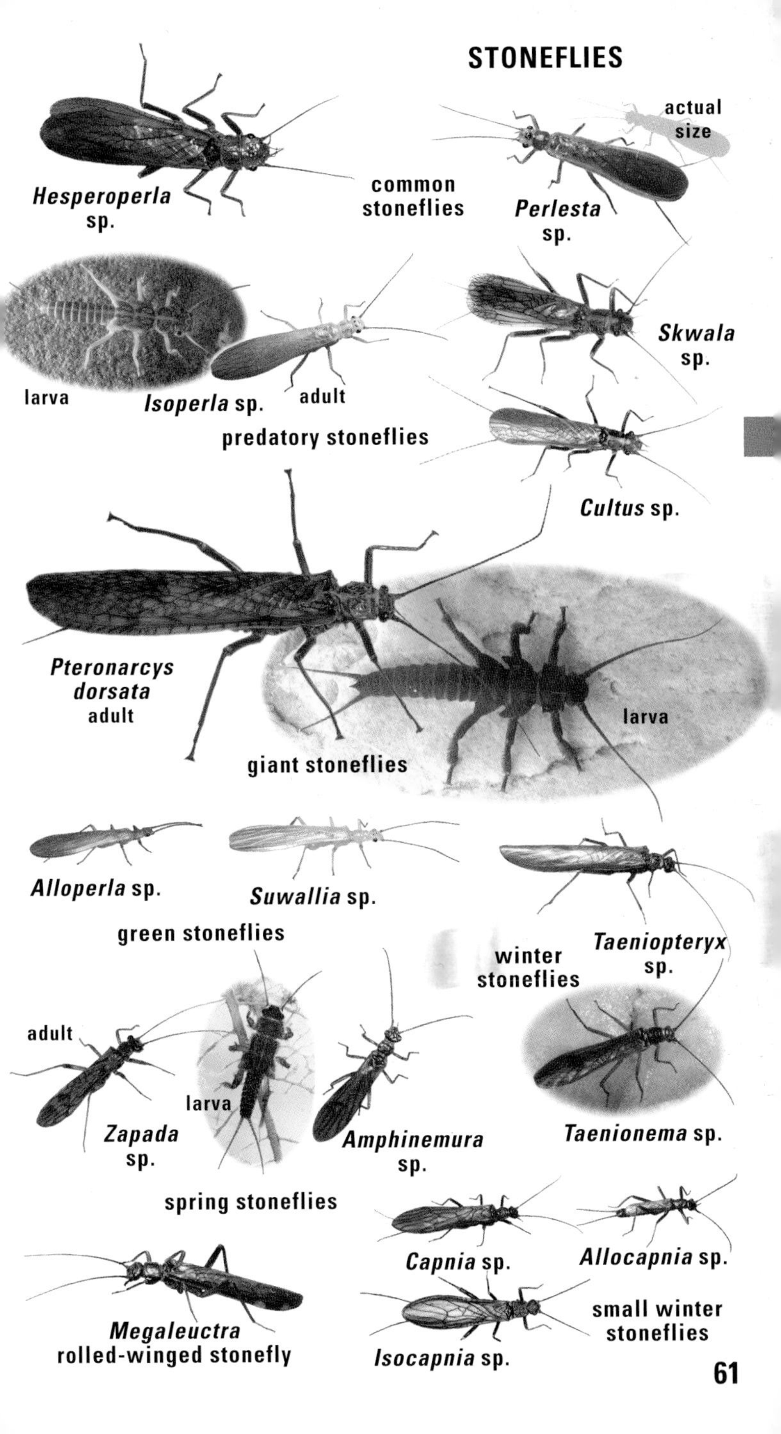
actual
size
Hesperoperla
sp.
common
stoneflies
Perlesta
sp.
Skwala
sp.
larva
Isoperla sp.
adult
predatory stoneflies
Cultus sp.
Pteronarcys
dorsata
adult
larva
giant stoneflies
Alloperla sp.
Suwallia sp.
green stoneflies
Taeniopteryx
sp.
winter
stoneflies
adult
larva
Zapada
sp.
Amphinemura
sp.
Taenionema sp.
spring stoneflies
Capnia sp.
Allocapnia sp.
small winter
stoneflies
Megaleuctra
rolled-winged stonefly
Isocapnia sp.

COCKROACHES (order Blattodea) are reviled in part because their presence suggests sloppy housekeeping. Pest species are suspected, not proven, disease carriers. Females of most species produce egg capsules (oothecae), depositing them in cracks or crevices. Metamorphosis is simple, nymphs resembling adults. Their diet includes almost anything organic. Such generalist habits have helped them thrive for over 280 million years according to the fossil record. In natural habitats they are valuable decomposers. There are about 68 species in North America. Roaches have a flattened profile, long, filamentous antennae, and two tail-like appendages called cerci.

Family Blattidae includes the **Oriental Cockroach** or "water bug" *(Blatta orientalis).* Native to Asia, it is now widespread throughout North America. Sluggish for a roach, it uses plumbing to access homes. The **American Cockroach** or "palmetto bug" *(Periplaneta americana)* probably originated in northern Africa. This swift, strong-flying cosmopolitan species is often seen outdoors, especially in the south.

Family Blattellidae includes native species and a few invaders. Females carry egg cases before depositing them. The **German Cockroach** *(Blatella germanica),* actually from northern Africa, is our worst household pest. **Brown-banded Cockroach** *(Supella longipalpa)* probably originated in Africa. This "TV roach" favors warm places like the inside of electronic appliances. ***Ischnoptera deropeltiformis*** is a forest-dwelling native of the eastern U.S. The 12 species of ***Parcoblatta*** are native wood cockroaches. Males are winged and often come to lights; females are flightless.

Family Blaberidae includes the **Death's-head Cockroach** *(Blaberus craniifer),* native to Cuba but now established in Key West, Florida. The **Suriname Cockroach** *(Pycnoscelus surinamensis)* ranges worldwide in the tropics and turns up in greenhouses, transplanted in potting soil. In North America, females reproduce without mating, giving birth to live young.

Family Polyphagidae includes small southern roaches, such as the genus ***Arenivaga*** of the desert southwest, in which males and females look very different. The **Brown-hooded Cockroach** *(Cryptocercus punctulatus),* often placed in its own family **(Cryptocercidae),** is a unique social insect of northwestern mountains and the Appalachians. It lives in family groups in rotting wood, females giving live birth to three or four offspring. These nymphs feed on adult feces, consuming protozoans that help digest cellulose. They take six years to mature.

ROCK CRAWLERS (order Grylloblattodea) are wingless, nocturnal creatures found at high altitudes in western North America and northeast Asia. Here there are 11 species, all included in the **family Grylloblattidae,** genus ***Grylloblatta.*** They prefer talus slopes, the edges of snowfields, lava tubes, and similar habitats, and they are so sensitive to heat that the warmth of a human hand can kill them. Rock crawlers probably scavenge, or prey on windblown insects rendered inactive by the cold. Metamorphosis is simple, and the life cycle may take years.

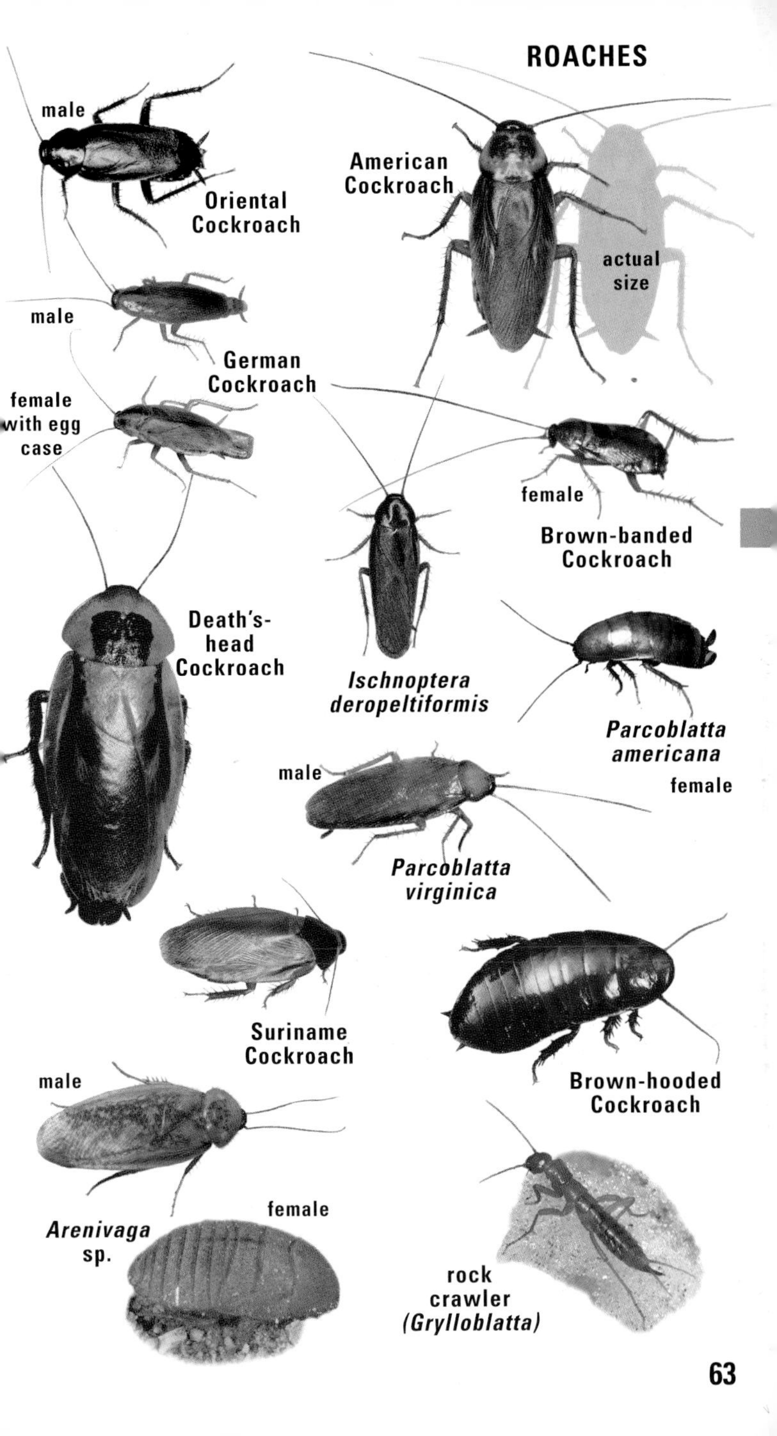
male
Oriental Cockroach
American Cockroach
actual size
male
German Cockroach
female with egg case
female
Brown-banded Cockroach
Death's-head Cockroach
Ischnoptera deropeltiformis
Parcoblatta americana
female
male
Parcoblatta virginica
Suriname Cockroach
Brown-hooded Cockroach
male
female
Arenivaga sp.
rock crawler (Grylloblatta)

(order Mantodea) are generalist predators that cannot be categorized as helpful or harmful. The front legs are modified into viselike appendages for grasping and holding prey, which is eaten alive. At rest, the folded weapons give the impression of a posture of prayer, hence "praying mantis." Cannibalism of males by females is famous but perhaps not the rule; it may often be an artifact of captive matings that leave the male no escape. Females lay eggs in a mass (ootheca), often on a vertical twig, with a foamlike material that hardens to prevent water loss. Metamorphosis is simple, with even the newly hatched nymphs resembling tiny versions of adults. Larger species can prey on small vertebrates, including hummingbirds. There are almost 20 North American species; all but one (and all those discussed here) belong to the family **Mantidae.**

The **European Mantis** *(Mantis religiosa)*, native to the Mediterranean region and temperate Asia, has been introduced to our continent and may be common in parts of the northeast and midwest. Look for the black-ringed white spot on the inside of the foreleg as a field mark. Another introduced species, the **Chinese Mantis** *(Tenodera aridifolia sinensis)*, entered the U.S. as early as 1896. It is still deliberately introduced to many areas as a biological control agent. Its efficiency is questionable, as it is an indiscriminate hunter, taking beneficial insects as well as pests. Both sexes fly well and come to lights at night in search of prey. Specimens may be brown, green, or both.

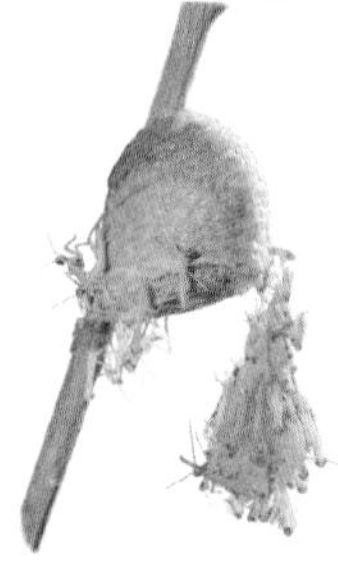

mass of young mantids hatching from egg case

Stagmomantis is a genus of six native species. The **Carolina Mantis** *(Stagmomantis carolina)* ranges throughout the east and west to Utah. Look for the robust, flightless females on flowers and shrubs. They may be green or mottled gray. The slender males fly well and often come to lights at night. The remaining species in the genus are chiefly southwestern.

Native to southern Europe and northern Africa, the **Mediterranean Mantis** *(Iris oratoria)* has been introduced into southern California and is spreading eastward. Its scientific name translates to "talking eye," in reference to its defensive display, which reveals an ominous dark spot on each hindwing. The abdomen may be scraped against the wing veins to add sound effects to the bluff.

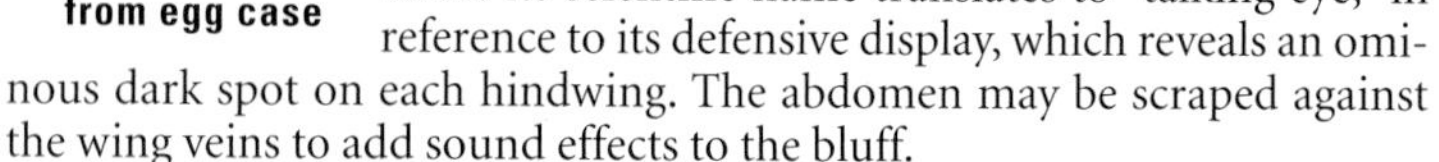

There are several genera of smaller mantids. The **Grizzled Mantis** *(Gonatista grisea)* is perfectly camouflaged on tree trunks, where it ambushes prey. It is found from South Carolina to Florida and Cuba. The abdomen of the wingless female is lobed along the edges, further breaking up her outline and adding to the camouflage. **Scudder's Mantis** *(Oligonicella scudderi)* is a slender species, not uncommon on grasses and other weedy vegetation from the Great Plains to Florida. Females are wingless. The **Minor Ground Mantis** *(Litaneutra minor)*, barely more than an inch long, is widespread in the west, from Canada to Mexico and east to the Great Plains. Males fly, but females are wingless.

MANTIDS

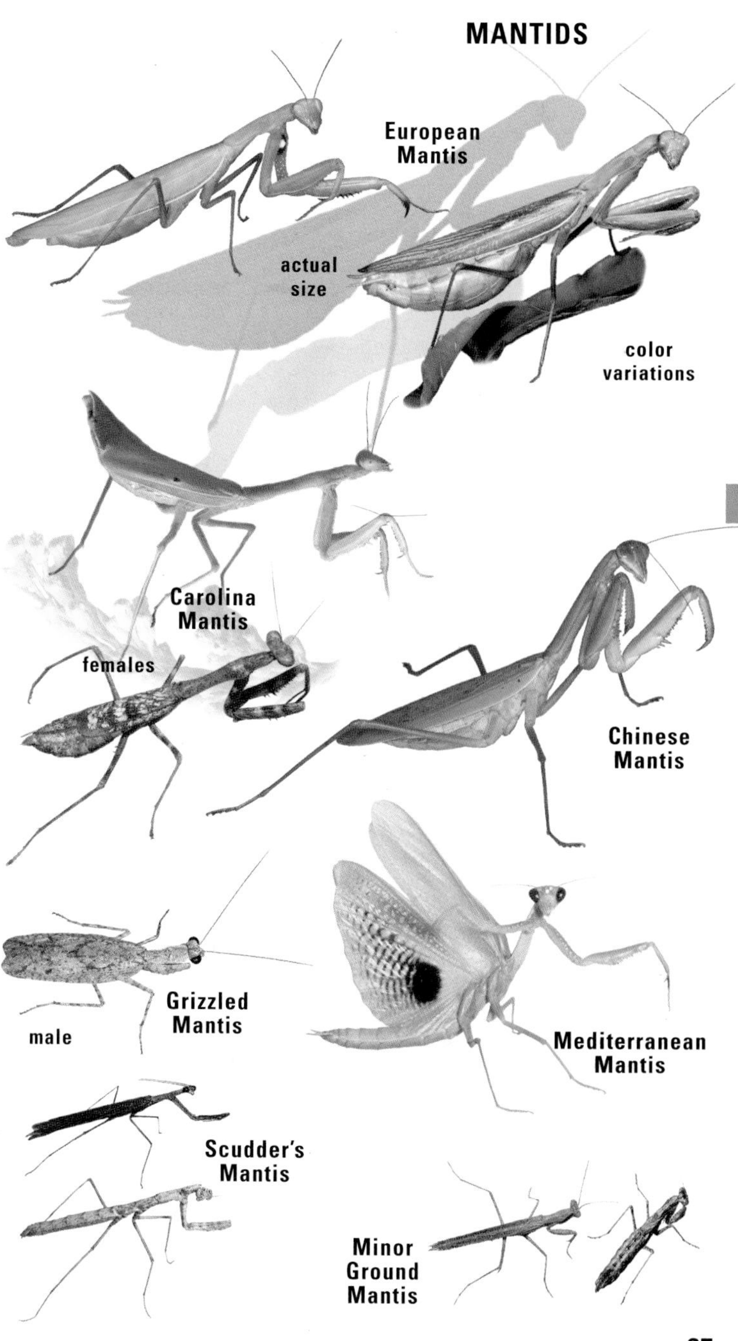
European
Mantis
actual
size
color
variations
Carolina
Mantis
females
Chinese
Mantis
Grizzled
Mantis
male
Mediterranean
Mantis
Scudder's
Mantis
Minor
Ground
Mantis

STICK INSECTS

(order Phasmatodea) are vegetarians, well disguised as twigs or stems, some tropical species bearing an uncanny resemblance to leaves. They usually move very slowly, and at rest they align the front and rear legs with their body, heightening their camouflage. Metamorphosis is simple, nymphs resembling adults. Males are often smaller than females, and mating pairs may remain together for days, or longer. Some species are parthenogenic, meaning females can reproduce without males. A walkingstick that loses a leg may be able to grow it back, a feat impossible for most insects. All walkingsticks in North America are wingless except for *Aplopus mayeri* of southern Florida (not illustrated), our only member of the family Phasmatidae. Most of the 2,000-plus species in this order live in the tropics, with only about 33 (in four families) in North America.

The **family Heteronemiidae** includes our most typical (or most "sticklike") walkingsticks. All of the varieties in our area are wingless. Their structure would seem to make them unique, but some other insects are occasionally mistaken for small walkingsticks, such as the thread-legged assassin bugs (p. 118). ***Diapheromera*** is a genus of eight species collectively distributed over much of the eastern and southern United States. The **Northern Walkingstick** *(Diapheromera femorata),* found from southeastern Canada to Florida and Arizona, is perhaps our most common and widespread walkingstick. Occasionally it undergoes population explosions and becomes extraordinarily abundant in hardwood or mixed forests. The **Giant Walkingstick** *(Megaphasma denticrus)* is our longest insect in North America, sometimes getting up to 7 inches long (some tropical members of this order can be up to 12 inches long). It is sometimes common locally on grape vines, oak, or grasses in the midwest and the southeast. The genus ***Parabacillus*** includes two species of short-horned walkingsticks in the western and midwestern United States. ***Parabacillus hesperus,*** which is sometimes more than 3 inches long, is found on range grasses and other vegetation in dry areas from Arizona to Oregon.

We have only two species in the **family Pseudophasmatidae,** both included in the genus ***Anisomorpha.*** They are restricted to the southeastern states and are especially prevalent in Florida. Also known as devil-riders or musk-mares, males and females pair for extended periods (sometimes for days), the much smaller male riding on the female's back. In defense, they squirt a milky cocktail of irritating chemicals called terpenes. This can cause temporary blindness should the liquid reach the eyes of an attacker. Look (cautiously!) for this species on grasses or brush, or under bark or boards, in the southeast and Florida.

The **family Timematidae** includes only 10 species in North America, all included in the genus ***Timema.*** Shorter and thicker than most walkingsticks, they might be mistaken for some altogether unrelated insects such as earwigs (p. 28), but they are usually green, sometimes brown or even pink. Look for them on foliage of trees and shrubs (such as ceanothus) in California, Nevada, and Arizona.

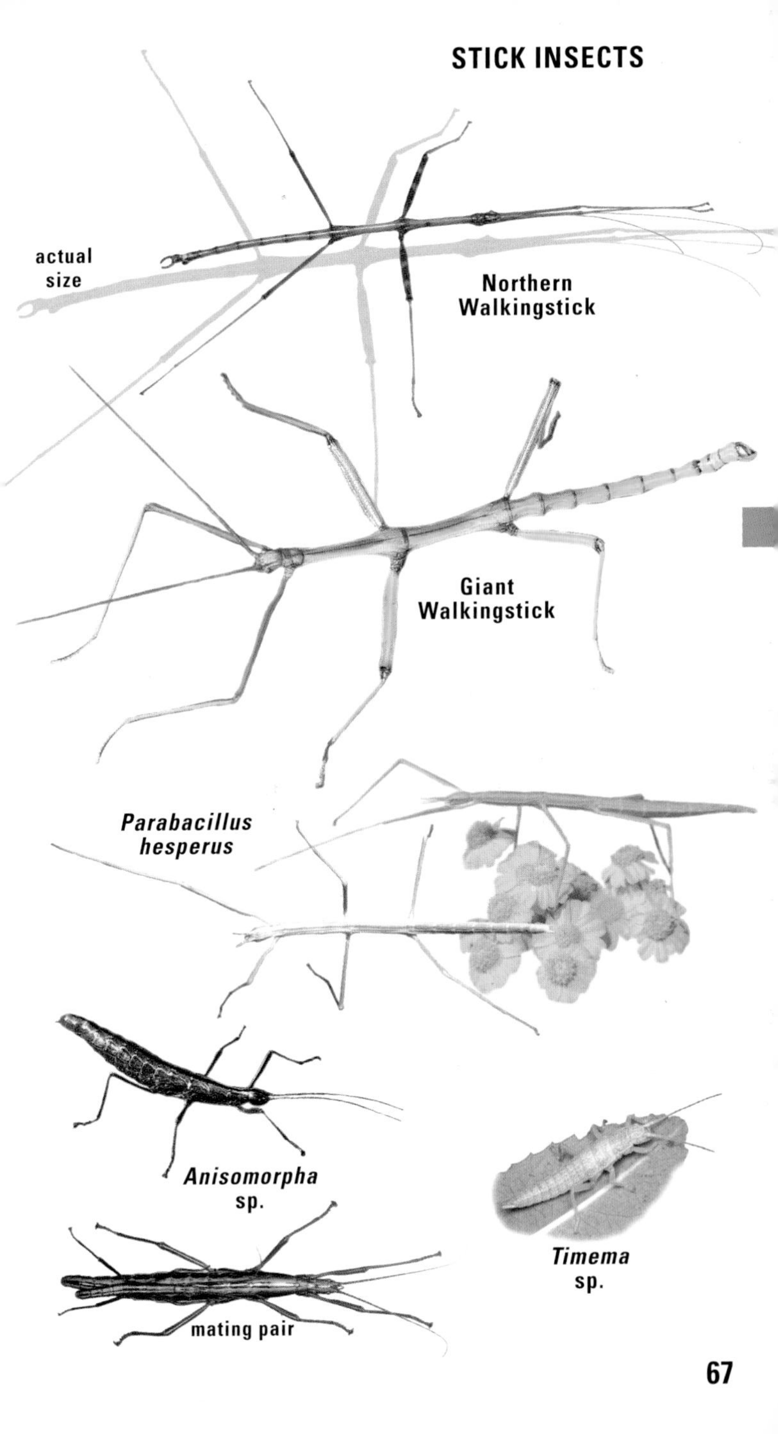
actual
size
Northern
Walkingstick
Giant
Walkingstick
*Parabacillus
hesperus*
Anisomorpha
sp.
mating pair
Timema
sp.

GRASSHOPPERS, KATYDIDS, CRICKETS

(order Orthoptera) are among our best-known insects. While short-horned grasshoppers are often considered crop and rangeland pests, katydids and crickets are beloved for their summer night serenades. In actuality, most members of this order are omnivorous, grazing on plants but scavenging dead insects (and even taking live prey or engaging in cannibalism) when the opportunity presents itself. Members of this order go through a simple metamorphosis, with the younger stages usually resembling small (and wingless) versions of the adults.

Many members of this order produce sounds. Male grasshoppers rub the inside of the hind tibia ("thigh") against raised veins on the forewings. Male crickets and katydids have highly modified front wings designed not only to produce sound but to amplify and broadcast it.

The two families below include our largest and smallest grasshoppers.

LUBBER GRASSHOPPERS (family Romaleidae) are 10 species of enormous lumbering insects. The **Southeastern Lubber** *(Romalea guttatus* or *R. microptera)* is widespread in the southeastern states, adults appearing from June to November, or at any season in southern Florida. When threatened, they display red hindwings and ooze a foul-smelling, frothy liquid from pores in the thorax. They are fond of amaryllis foliage but otherwise cause little damage. Shrikes are among their few predators.

Horse Lubbers *(Taeniopoda eques)* occur in the southwestern U.S. They can be common locally, often marching across roads and getting smeared in the process. They raise their hindwings in defense, as well as secreting a noxious froth from their thoracic spiracles. Young nymphs are gregarious, the better to reinforce their "hands-off" warning colors. Look for adults of the **Plains Lubber** *(Brachystola magna)* from August to October throughout the center of the continent, from Minnesota and Montana to Arizona and Texas. They favor short-grass prairie and open range habitats but feed on various broad-leaved plants.

GROUSE LOCUSTS (Tetrigidae), also known as pygmy grasshoppers, are recognized by the way the top of the thorax projects over the length of the abdomen, hiding the hind pair of wings in winged adults. The front wings are abbreviated and flaplike. These insects can be common; look for them in muddy places, where they graze on microscopic algae or small pieces of living plants. There are about 29 North American species.

Tettigidea lateralis, one of four species in its genus, is found east of the Rockies, north to Manitoba and Nova Scotia. It occupies various habitats including dry, open woodlands. The white face and thoracic band are distinctive. The genus ***Tetrix*** includes eight species. ***Tetrix subulata*** ranges across Alaska and Canada, south to Arizona and Pennsylvania. Adults are usually winged and are found in spring and fall. They favor moist habitats but are sometimes found in drier situations such as forest edges. ***Paratettix*** is represented in North America by seven species. ***Paratettix cucullatus*** is widespread in the north and west. Fully winged adults are seen in spring and fall along the edges of streams and lakes, especially on wet, sandy soil.

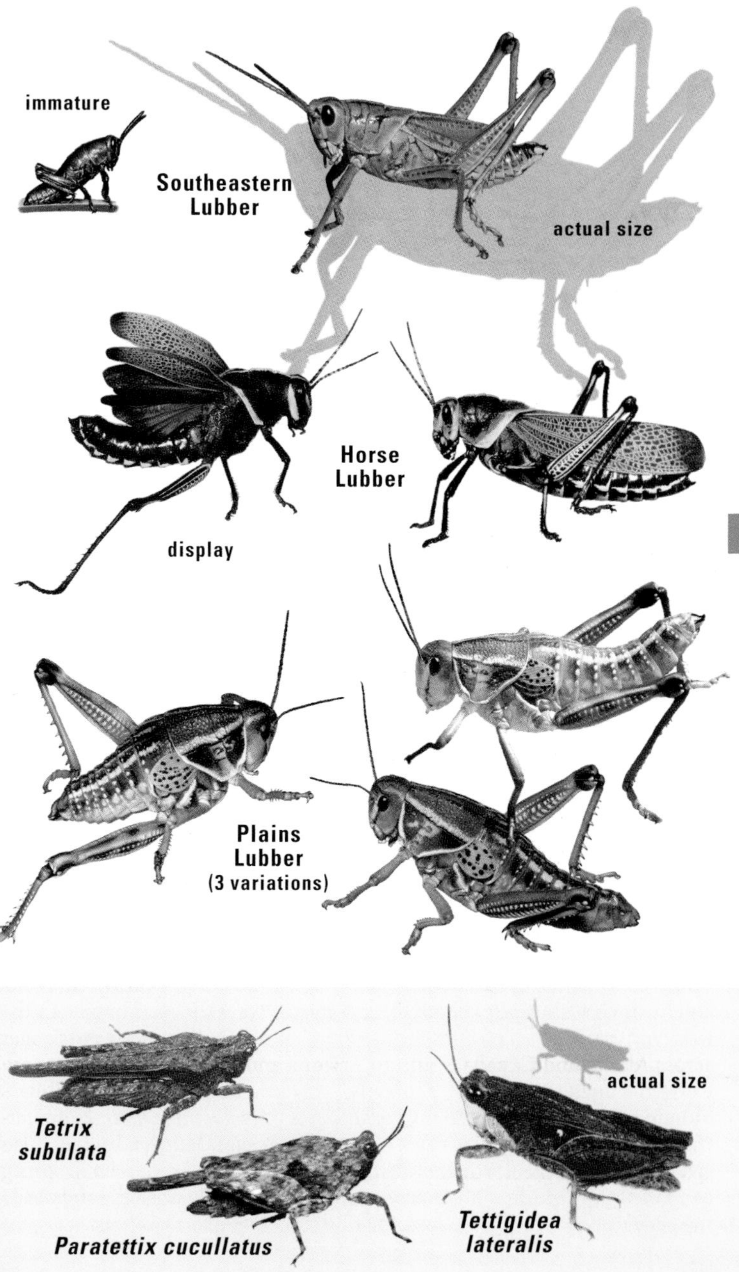
immature
Southeastern Lubber
actual size
display
Horse Lubber
Plains Lubber
(3 variations)
Tetrix subulata
Paratettix cucullatus
actual size
Tettigidea lateralis

SHORT-HORNED GRASSHOPPERS

(family Acrididae) are named for their short antennae. Acridids include some of our most colorful (band-winged grasshoppers), best camouflaged (slant-faced grasshoppers), and most pestiferous (spur-throated grasshoppers) insects. Many species fly well, while others are wingless. Males of most species make sound by rubbing the inside "thigh" of the hind leg against veins on the front wing. Females are frequently larger than males. There are 630 species in North America.

Slant-faced grasshoppers are disguised by their shape and are often striped to further their resemblance to grasses. Like many grasshoppers, they dodge to the opposite side of a grass blade at your approach. **Short-winged Green Grasshopper** *(Dichromorpha viridis)*, common throughout the east, may be green or brown. **Marsh Meadow Grasshopper** *(Chorthippus curtipennis)* ranges from Alaska to Mexico. Look for it in tall grasses in areas such as lakeshores and mountain meadows. The life cycle may take two years in northern latitudes. The genus ***Chloealtis*** includes five species with red or orange "shins" on the hind legs. **Sprinkled Grasshopper** *(Chloealtis conspersa)* ranges widely in the northern U.S. and southern Canada. Dry forest edges and brushy pastures are preferred habitats, females laying their eggs in decaying wood. The male's song is a series of *zeeek* sounds.

The **Spotted-winged Grasshopper** *(Orphulella pelidna)* ranges throughout the U.S. and southern Canada. Look for it in a variety of habitats from midsummer to fall. Color varies among individuals, from green to brown. ***Eritettix*** includes three North American species, mostly southern in distribution, such as **Obscure Slant-faced Grasshopper** *(Eritettix obscurus)*. The **Desert Clicker** *(Ligurotettix coquilletti)* of the Sonoran Desert is more often heard than seen, males calling day and night from creosote bushes.

Boopedon includes four North American species. **Ebony Grasshopper** *(Boopedon nubilum)* ranges from the Canada border to Mexico in shortgrass prairies. Females are much larger than the all-black males. Adults occur in late summer and fall, sometimes invading grainfields. ***Syrbula*** is a genus with two species north of Mexico. **Handsome Locust** *(Syrbula admirabilis)* is common over much of the eastern and southwestern U.S. Look for it in old fields and vacant lots. Courtship is elaborate, involving calls and posturing. ***Mermiria*** includes four U.S. species, mostly southern. **Two-striped Mermiria** *(Mermiria bivittata)* is most common on the southern Great Plains. **Sumichrast's Toothpick Grasshopper** *(Achurum sumichrasti)* is one of three oddly elongated species in its genus. The **Creosotebush Grasshopper** *(Bootettix argentatus)* is widespread in the southwest and perfectly camouflaged on the host plant. **Bunch Grass Locust** *(Pseudopomala brachyptera)* occurs across the northern states and southwestern Canada. Adults are generally short-winged, though fully winged individuals occur. **Green Fool Grasshopper** *(Acrolophitus hirtipes)* ranges from Canada to Mexico, just east of the Continental Divide. **Clipped-wing Grasshopper** *(Metaleptea brevicornis)* ranges from the Great Lakes south to Florida and Mexico. It can be locally common in sedges and grasses at the edge of ponds, in salt marshes, and in other wetland habitats.

SLANT-FACED GRASSHOPPERS

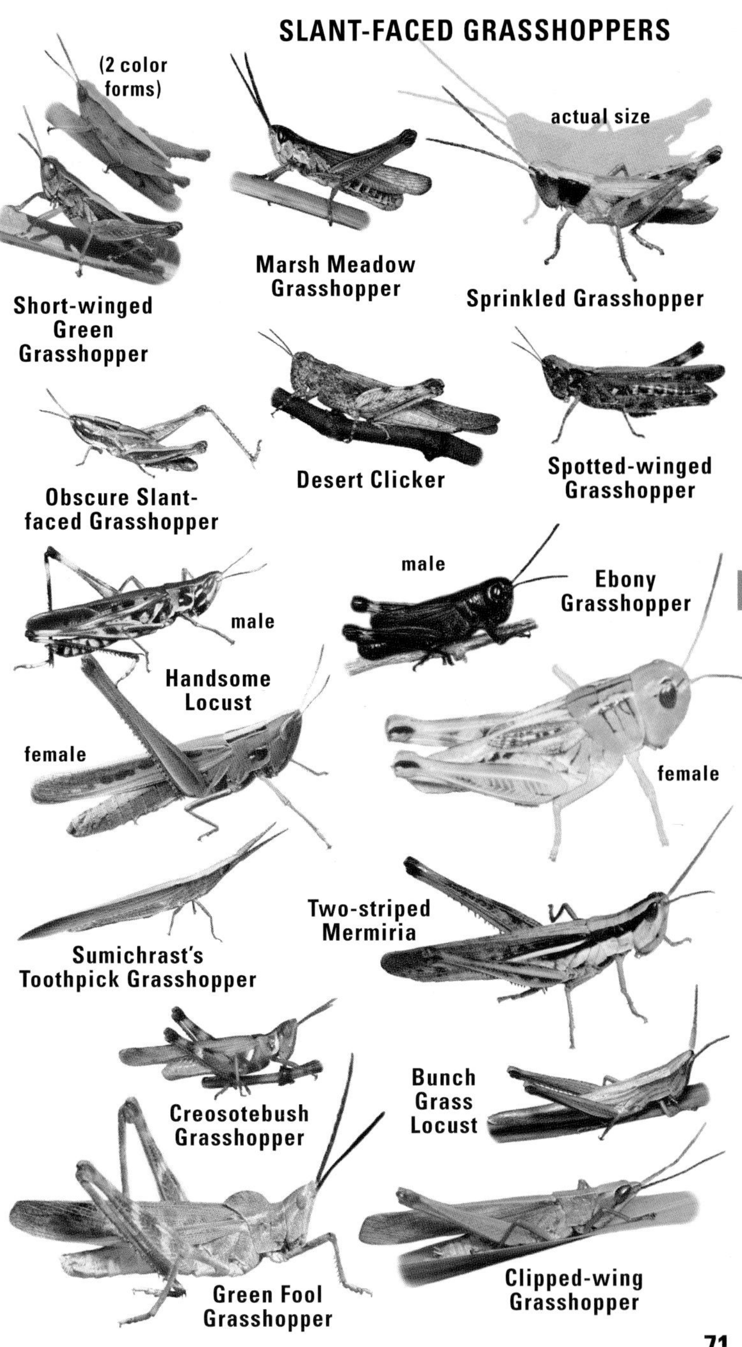

BAND-WINGED GRASSHOPPERS

(subfamily Oedipodinae) are perfectly camouflaged at rest but expose often brilliant hindwings in flight. This startling display is frequently accompanied by loud crackling and snapping noises called crepitations. On hot surfaces, they lift alternate feet to avoid scorching. The bulk of species are not destructive in their numbers or appetites for crops.

The **Carolina Locust** *(Dissosteira carolina)* is abundant across the U.S. and southern Canada. In flight, these large insects may be mistaken for Mourning Cloak butterflies (p. 234). Males perform sustained, hovering courtship flights that attract rival males as well as females. Look for this species in all habitats, including roadsides, vacant lots, and gravel quarries. The genus ***Pardalophora*** has four species in North America. In the **Orange-winged Grasshopper** *(Pardalophora phoenicoptera),* the inside of the hind "thigh" is blue. This species is common in the southeast, with adults present from early spring to July. The **Great Crested Grasshopper** *(Tropidolophus formosus)* ranges in the western Great Plains and the southwest. Nicknamed dinosaur grasshopper, it feeds on low shrubs in dry grasslands.

The **Wrinkled Grasshopper** *(Hippiscus ocelote)* ranges east of the Rockies and southwest to Arizona. This is among our most common "spotted" grasshoppers in waste places and vacant lots as well as native prairies. The hindwings are usually yellow, sometimes orange or red. Two species of ***Cratypedes*** occur in the west. ***Cratypedes neglectus*** is the most wide-ranging. In some localities it has a two-year life cycle. Habitat varies from sagebrush plains to forest openings. ***Conozoa*** is represented by five species north of Mexico, including the southwestern ***Conozoa carinata.***

The **Clear-winged Grasshopper** *(Camnula pellucida)* is widespread across Canada and the northern and western U.S. Also called the warrior grasshopper, it is known to migrate in the nymphal stage, swarms marching like armies across fields and pastures. At peak populations, this insect can be destructive to crops and rangeland. Interestingly, habitats overgrazed by livestock are favored by females for egg-laying. Two species of ***Hadrotettix*** occur north of Mexico. ***Hadrotettix trifasciatus*** is the most widespread, in the middle third of the continent. Look for adults in summer and early fall in areas with rocky soils and sparse vegetation.

The genus ***Arphia*** includes 11 species north of Mexico. In profile, the thorax is without notches. The **Sulphur-winged Grasshopper** *(Arphia sulphurea)* is widespread in the east. Look for adults from spring to midsummer in a variety of dry habitats. In summer and fall this species is replaced by the **Autumn Yellow-winged Grasshopper** *(Arphia xanthoptera).*

Leprus includes two species in the southwest. In **Cockerell's Blue-winged Grasshopper** *(Leprus intermedius),* the hindwings vary from turquoise to blue-gray. Look at higher elevations in rocky areas, as well as in rolling grasslands.The genus ***Xanthippus*** is represented by four species in North America. The life cycle takes two years, at least in northern latitudes. ***Xanthippus corallipes*** is the most variable and widely distributed species, ranging in the west from southern Canada to Mexico. Adults are most common in May and June in a variety of habitats including rocky areas.

BAND-WINGED GRASSHOPPERS

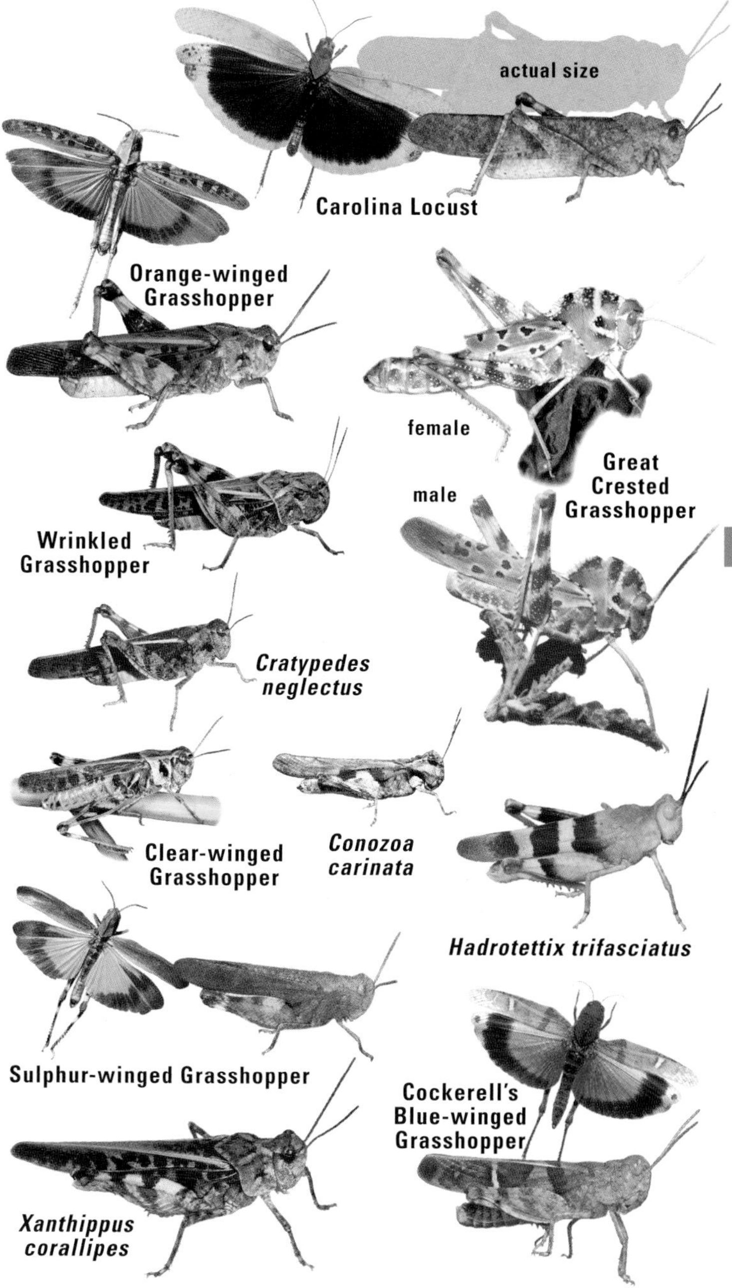

actual size

Carolina Locust

Orange-winged Grasshopper

female

Great Crested Grasshopper

male

Wrinkled Grasshopper

Cratypedes neglectus

Clear-winged Grasshopper

Conozoa carinata

Hadrotettix trifasciatus

Sulphur-winged Grasshopper

Cockerell's Blue-winged Grasshopper

Xanthippus corallipes

Trimerotropis is a genus with 39 species north of Mexico. Hindwing color is never red or orange. **Pallid-winged Grasshopper** *(Trimerotropis pallidipennis)* is abundant in the desert southwest but ranges as far north as British Columbia. **Blue-winged Grasshopper** *(Trimerotropis cyaneipennis)* is locally common in rocky habitats of the west and northwest. ***Circotettix*** includes eight western species. Males make sustained, hovering, sometimes acrobatic flights, rattling like a novice typist on a manual machine. **Carlinian Snapper** *(Circotettix carlinianus)* is widespread on grasslands. **Longhorn Band-winged Grasshopper** *(Psinidia fenestralis)* is widespread in the east in summer and fall in sandy situations. **Kiowa Grasshopper** *(Trachyrhachys kiowa)* is widespread but concentrated in the midwest. Look for it along roadsides and rocky slopes. There is great regional variability in this species. The hindwings vary from clear to yellow with a heavy black band. Depending on latitude, adults can be found from June to November. ***Mestobregma*** includes three southwestern species. **Platte Range Grasshopper** *(Mestobregma plattei)* is the most widespread, in rocky terrain with sparse grasses from South Dakota to Arizona.

Bird grasshoppers (genus ***Schistocerca***) include 12 species in America north of Mexico. **American Bird Grasshopper** *(Schistocerca americana)* is a migratory species of the southeastern states, though occasionally seen as far north as Ontario. In large numbers it may be very destructive. **Leather-colored Bird Grasshopper** *(Schistocerca alutacea)* is known from Massachusetts to Florida and Arizona. **Green Bird Grasshopper** *(Schistocerca shoshone)* ranges from California to Colorado and Texas.

SPUR-THROATED GRASSHOPPERS (subfamily Melanoplinae) are named for a peglike process protruding from their "neck." Some of our worst pests are in this group, but at the other extreme is a formerly abundant species that is now extinct (Rocky Mountain Grasshopper, *Melanoplus spretus*). Many species are wingless in the adult stage. ***Melanoplus*** is a genus with at least 238 species in North America. Accurate identification is beyond difficult, heavily reliant on differences in the male genitalia. Several species are economically important, including the appropriately named ***Melanoplus devastator*** of the Pacific states. Some species have reduced or absent wings as adults, or both winged and wingless individuals in a given population. **Differential Grasshopper** *(Melanoplus differentialis)* is recognized by its herringbone pattern of black on the hind "thighs." This insect ranges coast to coast in a variety of habitats, including vacant lots. **Two-striped Grasshopper** *(Melanoplus bivittatus)* is widespread across the U.S. and southern Canada in thick low growth during summer and fall.

Rainbow Grasshopper *(Dactylotum bicolor)* ranges through most of the interior of the west, on gravelly soil with sparse grasses. **Post Oak Grasshopper** *(Dendrotettix quercus)* occurs over much of the central U.S. Adults may be long-winged or short-winged and are found in trees, especially oaks. **Snakeweed Grasshopper** *(Hesperotettix viridis)* is widely distributed and highly variable in color, with green or purple stripes.

GRASSHOPPERS

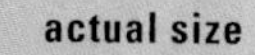

actual size

Pallid-winged Grasshopper

Blue-winged Grasshopper

Carlinian Snapper

Longhorn Band-winged Grasshopper

Kiowa Grasshopper

Platte Range Grasshopper

Leather-colored Bird Grasshopper

American Bird Grasshopper

Green Bird Grasshopper

Differential Grasshopper

Two-striped Grasshopper

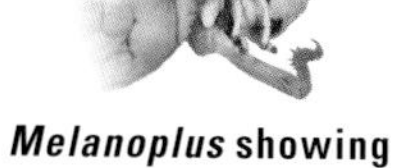

Melanoplus showing spur on throat

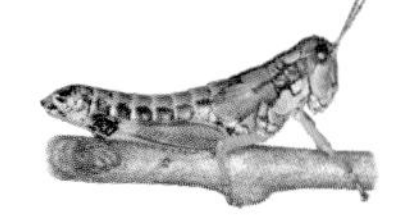

Post Oak Grasshopper

Snakeweed Grasshopper

Rainbow Grasshopper

(family Tettigoniidae) are also called longhorned grasshoppers. Only one genus makes the *katy-did, katy-didn't* call. Species can be reliably identified only by examining the male cerci (short paired appendages) or female ovipositor (a sickle- or sword-shaped organ for laying eggs), or by analysis of the song. Most of these insects are large and bright foliage green, or, predictably enough in late summer, brown. During mating, males transfer bubblelike sperm packets to the female, loaded with protein for her to feast on and use in the development of her eggs. Males may invest up to 40 percent of their body weight in these "gifts."

The **True Katydid** *(Pterophylla camellifolia)* is the famed singer of *katy did, katy didn't.* Males broadcast their loud raspy calls from the crowns of trees on hot, humid summer nights. Females can reply to the males' calls, unlike many katydid species. Both sexes are fully winged but are rather poor fliers, mostly gliding, rather than flying, from tree to tree. Eggs are laid in stems and tree bark. This insect ranges south of the Great Lakes, west to Kansas and Texas.

Members of the genus ***Microcentrum*** are the angle-wing katydids. Of the six species, the **Greater Angle-wing** *(Microcentrum rhombifolium)* is found coast to coast in the south, north to New York, Minnesota, and Utah, and has also been introduced in British Columbia. Songs of the males include both a slow, widely spaced series of single lisps and a very rapid ticking series. The **Lesser Angle-wing** *(Microcentrum retinerve)* is smaller and occurs mainly in the southeast. Eggs of these katydids are laid shinglelike, in rows on twigs.

eggs of angle-wing katydid

The nine species of ***Scudderia*** are the bush katydids. Males have distinctive tail plates that can be used to identify the species, but identification of females requires association with males. Bush katydids are abundant and call day and night. All species lay their eggs between the layers of leaves. The **Fork-tailed Bush Katydid** *(Scudderia furcata)* is common coast to coast and border to border.

The genus ***Amblycorypha*** includes 14 species of round-headed katydids. Typically they are green like most katydids, but they can be pink or occasionally yellow. The **Rattler Round-winged Katydid** *(Amblycorypha rotundifolia)* ranges from the Great Lakes and New York south to Georgia. The **Oblong-winged Katydid** *(Amblycorypha oblongifolia)* is common throughout the eastern half of the U.S., exclusive of the Florida peninsula.

Our 14 species of thread-legged katydids (genus ***Arethaea***), recognized by their absurdly long legs, are mostly restricted to Texas, the southwest, and the southern Great Plains, with only one in the southeast. ***Insara*** is a genus of six species, including **Creosotebush Katydid** *(Insara covilleae).* Look for it on that plant in southern Arizona, California, and Nevada. ***Neobarrettia*** is a genus of two arid-land katydids found from southwestern Kansas to south Texas and Arizona. They frequent open habitats from desert scrub to oak-juniper woods and are at least partly predatory.

KATYDIDS

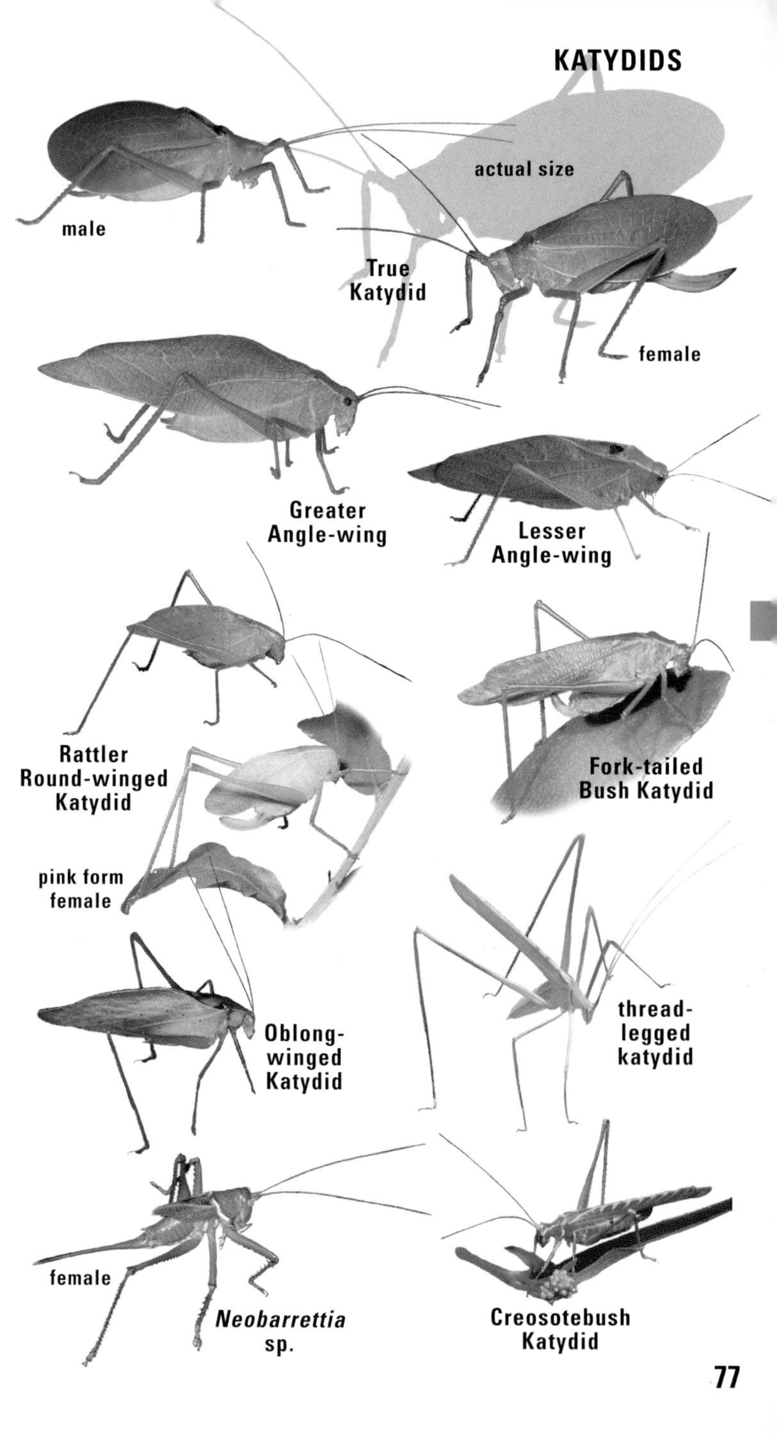
actual size
male
True
Katydid
female
Greater
Angle-wing
Lesser
Angle-wing
Rattler
Round-winged
Katydid
pink form
female
Fork-tailed
Bush Katydid
thread-
legged
katydid
Oblong-
winged
Katydid
female
Neobarrettia
sp.
Creosotebush
Katydid

The 14 katydids of the genus ***Neoconocephalus*** are called "coneheads" for obvious reasons. Males call at night, usually from tall grass or tangled low cover. Both sexes feed chiefly on grass seed and can bite hard. The **Slightly Musical Conehead** *(Neoconocephalus exiliscanorus)* has the longest "nose" of all our species. Males sing in synchronous choruses in thickets, canebrakes, and cornfields of the eastern states from July to September. The **Round-tipped Conehead** *(Neoconocephalus retusus)* sings at night and late afternoon, August to October, from weedy fields and roadsides of the east. The five short-winged species of coneheads in the genus ***Belocephalus*** are mainly southeastern, with some restricted to Florida.

The smaller meadow katydids of the genus ***Conocephalus*** include 20 species found mostly in lush grasses. Males sing in daylight. The **Slender Meadow Katydid** *(Conocephalus fasciatus)* is common across the entire U.S. and southern Canada. The **Straight-lanced Meadow Katydid** *(Conocephalus strictus),* named for the female's long ovipositor, ranges from Montana to southern Ontario and southwest to Arizona.

In the genus ***Orchelimum*** are 19 species of larger meadow katydids. Males sing in daylight, from tall grasses. The **Red-headed Meadow Katydid** *(Orchelimum erythrocephalum)* is common in summer in the southeast. The **Black-legged Meadow Katydid** *(Orchelimum nigripes)* lives mostly west of the Appalachians, from the Great Lakes to the western Gulf Coast, but is also locally common near Washington, D.C.

The genus ***Atlanticus*** includes seven eastern species of shield-backed katydids. These are largely ground-dwelling insects of hardwood forests. They are mostly predators and scavengers of other insects, but they also feed on live vegetation.

Of the four species of flightless shield-backed katydids in the genus ***Anabrus,*** by far the most famous is the so-called **Mormon Cricket** *(Anabrus simplex)* that periodically ravages the Rocky Mountain states in marching hordes. Legend has it that when a plague of them threatened the first crops of the Mormon settlers in 1848, a flock of gulls saved the day by eating the offending insects. Mormon Crickets are large (up to 62 mm long) and may be brown, yellowish, green, or black. The species ranges well into Canada, south to New Mexico, and east to the Dakotas.

The genus ***Capnobotes*** is widespread in the west. Of the eight species, the **Sooty Longwing** *(Capnobotes fulginosus)* has the broadest range, from central Nevada and Utah south to southern California and New Mexico and east to San Antonio. This species is partly carnivorous. Startled individuals display their dark hindwings. ***Eremopedes*** is a genus of 10 lanky species limited to the southwest; some are widespread there, while others have very restricted ranges, such as a single desert dune system. The songs of males are usually a series of soft buzzy notes.

Metrioptera is a genus of two boreal species. Males trill softly in summer from the edges of meadows, trails, and roads. **Roesel's Katydid** *(Metrioptera roeselii),* introduced from Europe, is now established in New England and adjacent Quebec. Males may be long- or short-winged.

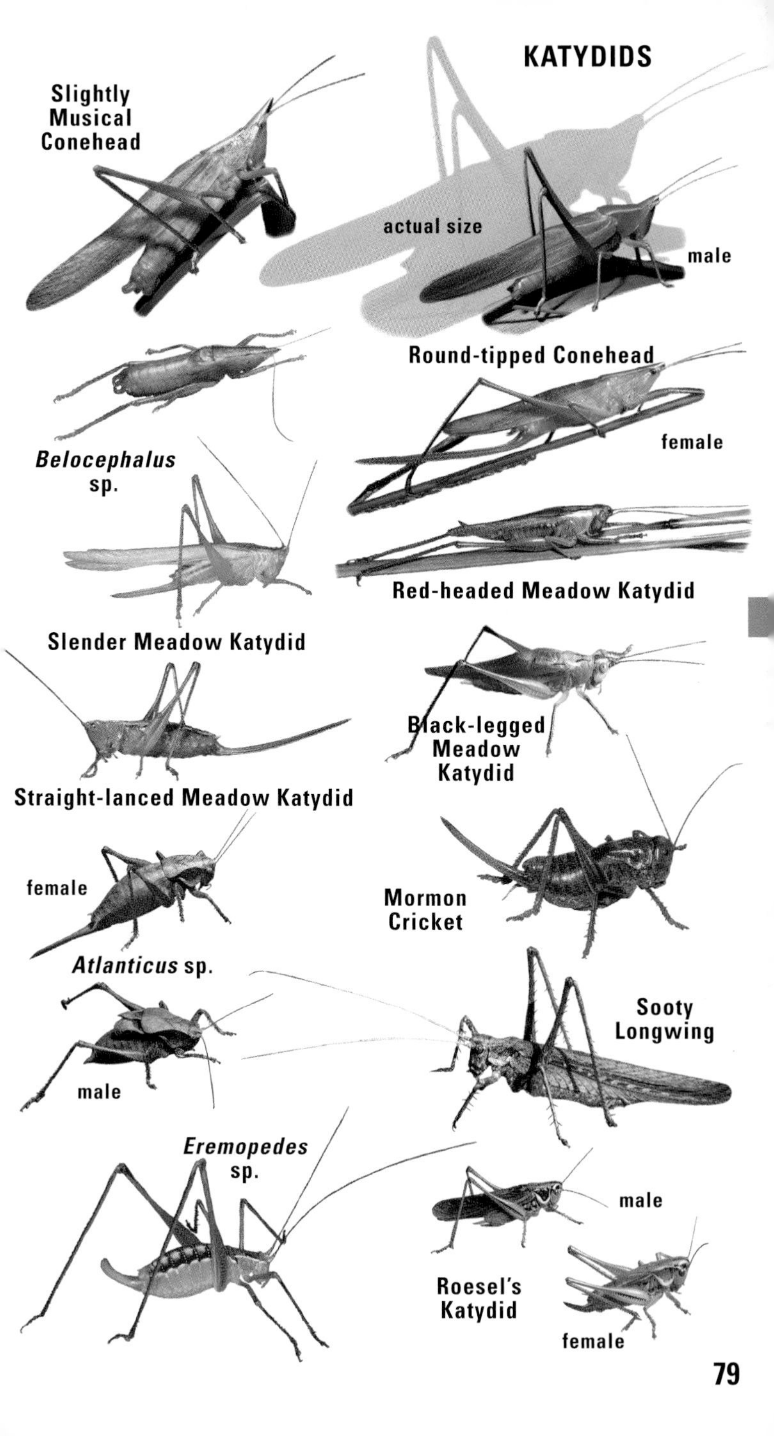
KATYDIDS
Slightly Musical Conehead
actual size
male
Round-tipped Conehead
female
Belocephalus sp.
Red-headed Meadow Katydid
Slender Meadow Katydid
Black-legged Meadow Katydid
Straight-lanced Meadow Katydid
female
Mormon Cricket
Atlanticus sp.
Sooty Longwing
male
Eremopedes sp.
male
Roesel's Katydid
female

TRUE CRICKETS

(family Gryllidae) are celebrated songsters of the insect realm. Only now are scientists fully appreciating the complex acoustic repertoire of male crickets and learning how diverse our fauna truly is. Males have highly modified forewings with a "scraper"and "file" on each wing that produce, amplify, and broadcast sound. Females have a spear- or sticklike ovipositor extending from the tip of the abdomen. There are 25 genera and at least 115 species in North America.

Gryllus is a genus that includes anywhere from 16 to 32 or more species of field crickets. What was once considered a handful of species turns out to be several, separated by seasonality and song patterns. Males have a "calling song" to attract a female, a "love song" to woo her once she is near, and a "rivalry song" to ward off competing suitors. The calling song can also draw a female parasitic fly, *Ormia,* that plugs the cricket with an egg in what amounts to a drive-by shooting. Most species of field crickets are glossy black and 14–30 mm in length. They can be abundant under stones, boards, and other debris, and in rock walls, animal burrows, and other sheltered spots.

The **Common House Cricket** *(Acheta domesticus)* has become even more common because it is sold in bait shops and pet stores as a "feeder" animal. Native to southwest Asia and northern Africa, it now lives freely throughout the eastern U.S. and adjacent southeastern Canada, as well as in the Los Angeles area. Females may be fully winged or "semi-dealated," meaning the wings are rudimentary and nonfunctional.

The **Indian House Cricket** *(Gryllodes sigillatus,* formerly called *Gryllodes supplicans)* is widespread in the tropics and is now becoming more abundant across the southern U.S., mainly in cities.

The genus ***Hapithus*** includes three species of flightless bush crickets. The best-known of these, the **Restless Bush Cricket** *(Hapithus agitator),* occurs in the eastern half of the U.S. south of the Great Lakes. Look for it in the understory of moist or wet woodlands. Many populations of this insect do not have a calling song. After mating, a male will permit the female to feed on his forewings, allowing the sperm packet to empty before she can eat that, too.

Three species of ground crickets make up the genus ***Eunemobius*** (formerly part of *Nemobius).* The best known of these is the **Carolina Ground Cricket** *(Eunemobius carolinus),* which occurs over most of the U.S. (except parts of the northwest) as well as in southeastern Canada. Look for it in damp grassy areas such as around lakes and streams. Members of the genus ***Allonemobius*** (formerly included in *Nemobius)* are 10 species of robust ground crickets.

The **Red-headed Bush Cricket** *(Phyllopalpus pulchellus)* is common over much of the eastern one-third of the U.S. Listen for its surprisingly loud, high-pitched trill from low shrubs, grasses, and short trees. The palps, looking like little boxing gloves, are in perpetual motion when the insect is excited.

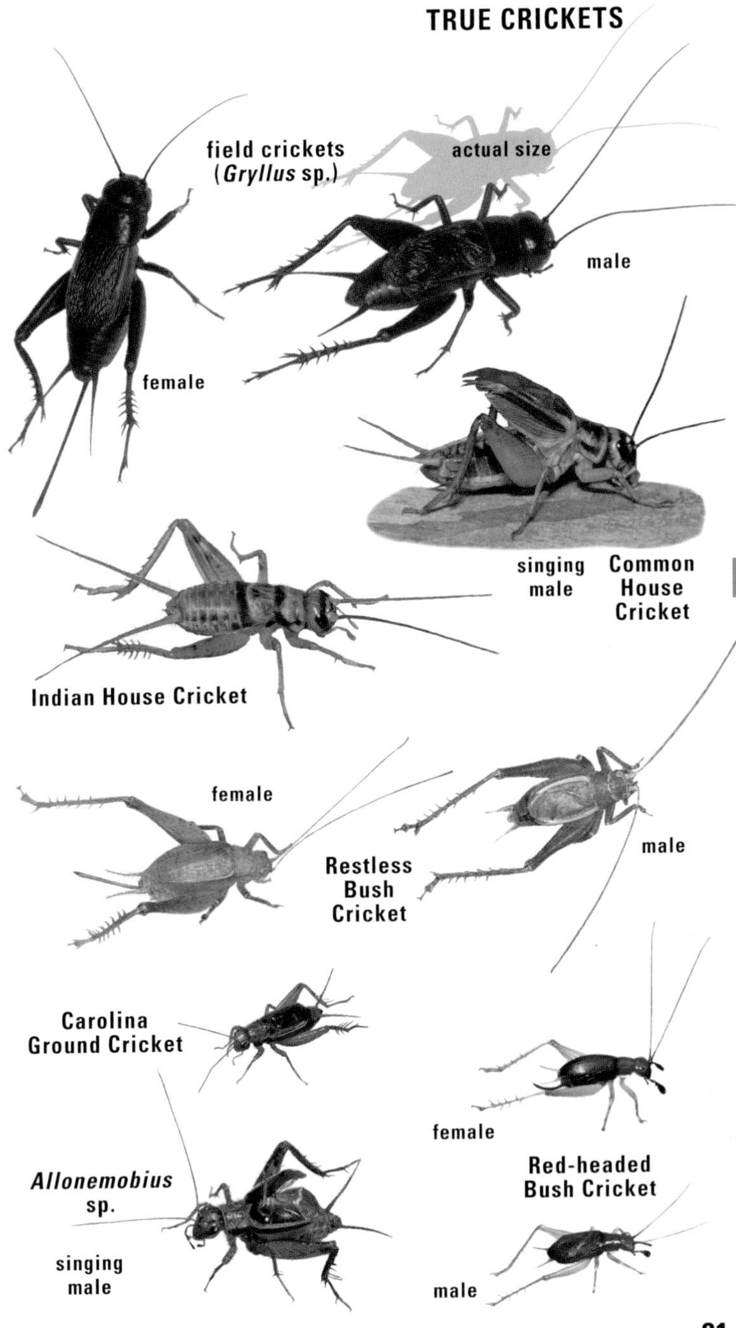
field crickets
(*Gryllus* sp.)
actual size
male
female
singing male
Common House Cricket
Indian House Cricket
female
Restless Bush Cricket
male
Carolina Ground Cricket
female
Red-headed Bush Cricket
Allonemobius sp.
singing male
male

TREE CRICKETS AND HUMP-WINGED GRIGS

The tree and bush crickets below are in the **family Gryllidae,** like those on the previous page. Also below are two small but distinctive families.

The genus ***Oecanthus*** includes at least 16 species of tree crickets. These small creatures might be mistaken for green lacewings (p. 220) by sight, or simply overlooked, but their sounds are hard to miss. "If moonlight could be heard, it would sound like that," wrote Nathaniel Hawthorne of the trilling choruses of males. These tree crickets feed mostly on aphids and other soft-bodied insects, sometimes nibbling on foliage and fruits. The male offers his mate secretions from paired thoracic glands during mating. Females can damage canes by leaving egg scars in the process of ovipositing. The **Snowy Tree Cricket** *(Oecanthus fultoni)* is also known as thermometer cricket because the rate of its chirping (as with many other singing insects) speeds up in higher temperatures. Dolbear's formula asserts that the air temperature in degrees Fahrenheit equals the number of Snowy Tree Cricket chirps in 1 minute, minus 40, divided by 4, and plus 50. This species ranges across much of North America. **Western Tree Cricket** *(Oecanthus californicus)* is widespread in the west, from southern British Columbia to southern California, east to Colorado and New Mexico. **Riley's Tree Cricket** *(Oecanthus rileyi)* is common from California to British Columbia and locally east to Arizona. The **Black-horned Tree Cricket** *(Oecanthus nigricornis)* of the northern U.S. and southern Canada is recognized by its blackish antennae and legs.

The **Two-spotted Tree Cricket** *(Neoxabea bipunctata)* is common in the eastern half of the U.S. Look for it at lights at night. After mating, the male suspends himself and his mate from foliage, allowing the female to dine on secretions from glands in his thorax, beneath his front wings. Members of the genus ***Orocharis*** are loud-singing bush crickets with six North American species. The **Jumping Bush Cricket** *(Orocharis saltator)* is most common, ranging widely in the east. Listen for these insects in broad-leaved trees, from August to October in northern latitudes.

The **family Gryllacrididae** is represented in North America by just one species, **Leaf-rolling Cricket** *(Camptonotus carolinensis)*. Widespread in the east, from Indiana to Florida, it is active at night, preying on aphids. It fashions a daytime retreat for itself by rolling leaves and securing them with silk threads spun from glands in its mouth. This insect is the main prey of a species of solitary wasp, *Sphex nudus.*

HUMP-WINGED GRIGS (Prophalangopsidae) are known today only from northwestern North America and central Asia, but many fossils of these odd insects have been found over a wider area. All three North American species are in the genus ***Cyphoderris.*** Males call at night, with a short high-pitched trill, delivered while sitting head downward on a tree trunk; they spend the day in burrows or under rocks. **Great Grig** *(Cyphoderris monstrosa)* occurs in evergreen forests in the Pacific northwest, while **Sagebrush Grig** *(Cyphoderris strepitans)* may be in forest or sagebrush flats.

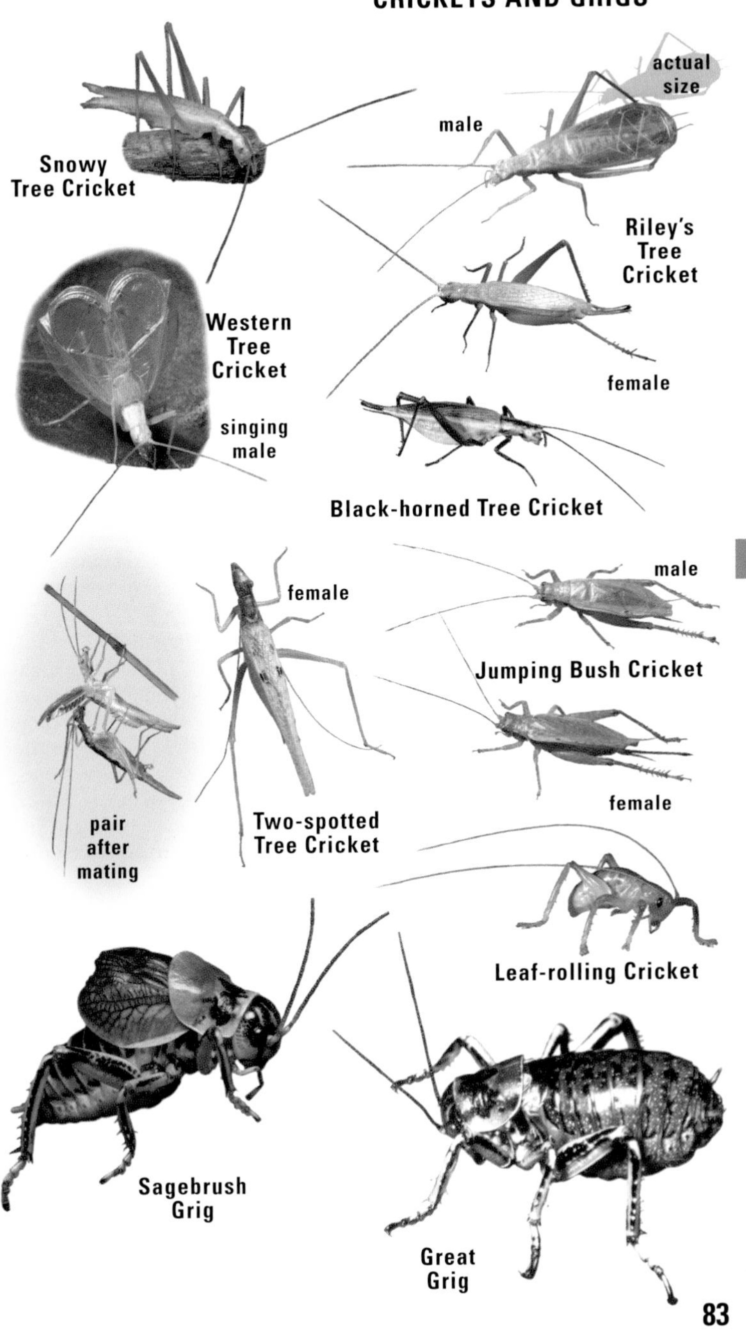
actual
size
male
Snowy
Tree Cricket
Riley's
Tree
Cricket
Western
Tree
Cricket
female
singing
male
Black-horned Tree Cricket
male
female
Jumping Bush Cricket
female
pair
after
mating
Two-spotted
Tree Cricket
Leaf-rolling Cricket
Sagebrush
Grig
Great
Grig

MOLE, CAMEL, AND JERUSALEM CRICKETS

Three families of highly distinctive orthopterans, not closely related to each other, nor to the true crickets on p. 80.

MOLE CRICKETS (family Gryllotalpidae) are nocturnal subterranean insects. The seven North American species include some introduced pests. They are well built for their underground existence, complete with spade-like "hands" and "forearms" on their front legs. Adult males of some species make U-shaped tunnels with dual "speakers" at each end, the better to broadcast their songs to females.

The **Northern Mole Cricket** *(Neocurtilla hexadactyla)* ranges throughout the eastern U.S. Look for this species in saturated soil along the edges of ponds and lakes. This insect has a two-year life cycle. There are two other species in the genus. The call is harsh and froglike.

The genus ***Scapteriscus*** includes three nonnative species that are now widespread in the southeastern U.S. The **Southern Mole Cricket** *(Scapteriscus acletus)* is an occasional pest in vegetable gardens from North Carolina to Florida and Texas. Wasps of the genus *Larropsis* (p. 338) are important predators of these crickets.

CAMEL CRICKETS (Rhaphidophoridae) are wingless, hump-backed creatures commonly found in cellars, old wells, abandoned mine shafts, caves, and other dark, damp habitats. A few are "sand treaders" with broad, spiny feet that give them traction and make them more effective at burrowing in dunes. Those found outside of caves are nocturnal and can be lured by laying a trail of oatmeal. Females have a short, knifelike ovipositor.

Most of our camel crickets belong to ***Ceuthophilus,*** an enormous genus with at least 89 North American species. The species can be separated reliably only by subtle structural differences. These are the camel crickets most often seen in eastern North America, but collectively they are more diverse in the west.

The bizarre ***Tropidischia xanthostoma*** may have a legspan of up to 8 inches and very long antennae, all the better to feel its way around in the dark. It occurs in old wells, under bridges, and in similar niches in coastal forests from British Columbia to California.

JERUSALEM CRICKETS (family Stenopelmatidae) are perhaps our most maligned and misidentified insects. Steeped in superstition, from Native American mythology to urban legend, their aliases include "chaco," "potato bug," and Niña de la Tierra (Child of the Earth). They are not crickets but more closely related to the wetas of Australia and New Zealand. Found mostly west of the Rockies, they are nocturnal, burrowing in loose soil in dry habitats. They are probably predatory and seldom if ever reach the status of being pests. While not venomous, they deliver a wicked bite if carelessly handled. The life cycle may take four or five years to complete. Recent research has indicated that as many as 40 new species of the genus ***Stenopelmatus,*** from various dune systems in southern California, await formal description.

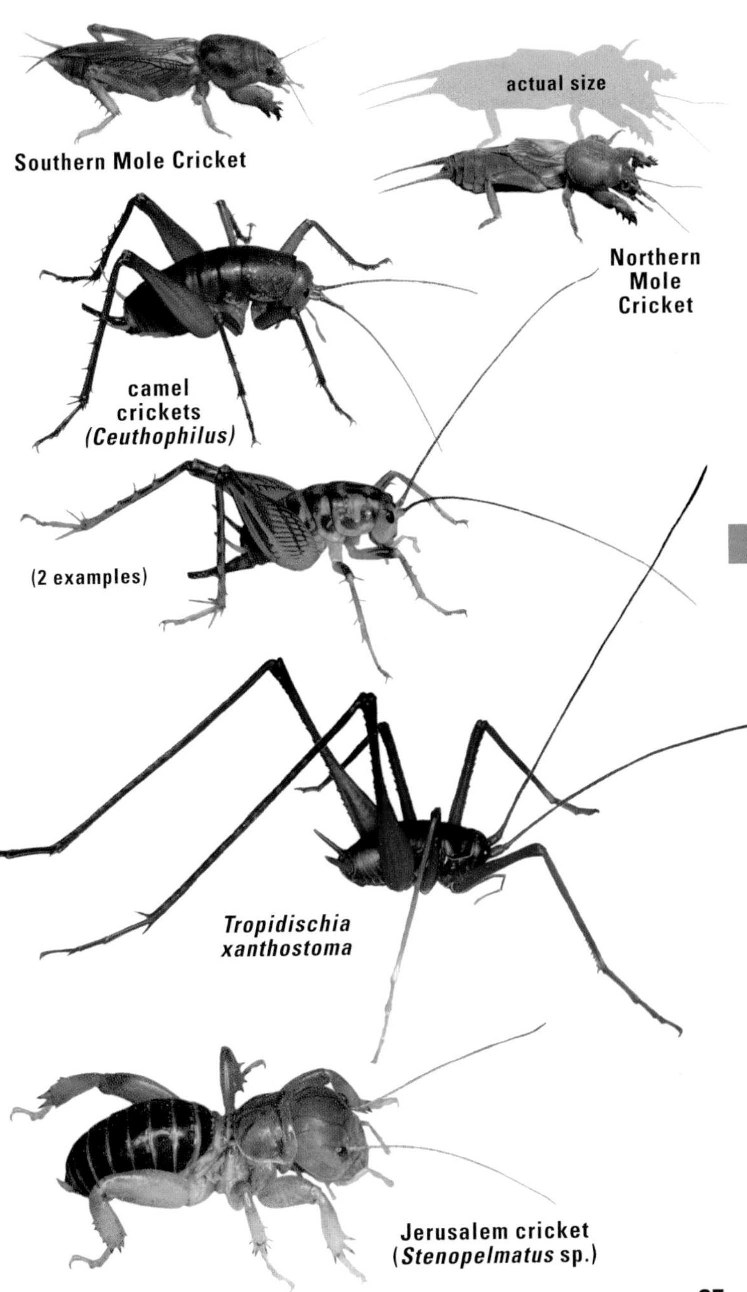

Southern Mole Cricket

Northern Mole Cricket

camel crickets *(Ceuthophilus)*

(2 examples)

Tropidischia xanthostoma

Jerusalem cricket (*Stenopelmatus* sp.)

TRUE BUGS

Members of the **order Hemiptera** are insects that you can call bugs without offending any entomologists! They are distinguished by having sucking, beaklike mouthparts, tucked beneath the insect's "chin" and "chest" when not in use. Metamorphosis is incomplete, with even the youngest nymphs typically resembling miniature versions of the adults, although they are wingless and often have different color patterns. All other characters of the order are highly variable. The cicadas, aphids, spittle bugs, scale insects, and various small "hoppers" were once placed in their own separate order, the Homoptera. Their merger with the remainder of the true bugs is now considered standard practice.

To make sense of this large and highly diverse assemblage, experts commonly refer to three distinct suborders within the order Hemiptera. In this guide we have separated these insects into two sections: the current one, labeled "Cicadas, aphids, and their relatives" (pp. 86–105), and the following section, labeled "Typical true bugs" (pp. 106–127). The three major suborders are divided among these two sections as follows:

Suborder Auchenorrhyncha: Cicadas, leafhoppers, sharpshooters, treehoppers, planthoppers, spittlebugs, and the like (pp. 88–97). Cicadas are mostly large insects, while the others in this suborder are mostly quite small, but all are active as adults (with strong jumping or flying abilities). The adult males of many are capable of making sounds, but only the tones of male cicadas are generally audible to humans.

Suborder Sternorrhyncha: Aphids, mealybugs, whiteflies, psyllids, phylloxerans, adelgids, and scale insects (pp. 98–105). These are mostly small to very small insects, and they are mostly not very active, with the extreme example of adult female scale insects not moving at all. Also included here are galls, made by many kinds of insects, most not related to true bugs.

Suborder Heteroptera: The "original" true bugs, including such familiar forms as stink bugs, squash bugs, water bugs, and assassin bugs (pp. 106-127). Many are mimics of other insects, and the majority of these may be mistaken for beetles (see discussion on p. 128).

Many kinds of true bugs, especially aphids, scale insects, and treehoppers, secrete a sweet liquid waste product called honeydew. This is the "manna from heaven" of the Bible (harvested from mealybugs by the Israelites). Ants often tend these insects, protecting them from parasites and predators in exchange for honeydew. Some true bugs exhibit parental care behavior, one or both genders guarding clusters of eggs and/or the newly hatched nymphs. Certain stink bugs, leaf-footed bugs, giant water bugs, and treehoppers are among them.

While many insects in this order are considered pests of trees, crops, or people, a few provide us with products for our use and consumption. Natural shellac, for example, is produced by the lac insect, a type of scale insect found in India. Next time you chew on a gumball or jawbreaker, remember that they are coated with this substance. Meanwhile, a bright red dye is obtained by crushing cochineal scales. The scales have been farmed for ages, the British redcoats owing their cardinal garb to this insect. Today, cranberry juice, cosmetics, and other products still utilize cochineal for their coloring.

The crop pests may cause damage simply by feeding, but a more serious threat comes from the transmission of viral or bacterial agents that cause plant diseases.

Many hemipterans also produce sound. Cicadas are the loudest of all insects, producing their racket with a complicated percussion organ. Treehoppers, leafhoppers, and their allies may communicate by drumming various body parts on the stem or leaf to which they are clinging. When under distress, assassin bugs make a squeaking sound by rubbing their beak along a ridged groove.

CICADAS

(family Cicadidae) are large and loud (the males, anyway). These are the only insects that have a true percussion mechanism for sound production, one contained on each side of the thorax and/or abdomen. A special ribbed membrane called a tymbal is vibrated by a large muscle to create the "song," which is echoed and amplified in the male's air-filled abdomen. Lidlike plates cover the "speakers." The bulk of cicadas' lives are spent as nymphs, which hatch from eggs laid in twigs of living trees. The nymphs plummet to earth, then dig underground, where they tap into roots and suck sap, usually for several years. They eventually crawl to the surface, usually at night, clamber up any vertical object with their grappling-hook forelegs, and split their exoskeleton, allowing the winged adult to emerge. Look for the ghostly, gnomelike husks hanging on tree trunks and fenceposts. There are 22 genera and more than 160 species of cicadas in North America.

Members of the genus ***Tibicen*** are the "dog day cicadas" or "harvestflies," named for their peak abundance in the hottest days of summer and early fall. Indeed, their grinding, whining buzzes are among the characteristic sounds of summer over much of this continent. They are also known as annual cicadas, since adults are present every year, in contrast to periodical species. However, the life cycle of an individual takes two to five years, so the name "annual" is something of a misnomer. The 30 species of ***Tibicen*** north of Mexico range everywhere but the far west. Adults emerge between May and November, depending on latitude. The **Swamp Cicada** *(Tibicen chloromera)* is a relatively small, dark species common over much of the east. **Linne's Cicada** *(Tibicen linnei)* is another common eastern species. The one illustrated here is freshly emerged and looks pale, but pigments will darken within a few hours as its exoskeleton hardens. The **Silver-bellied Cicada** *(Tibicen pruinosa)* is widespread in woodlands and suburbs over much of the east. ***Tibicen dorsata*** is a large ornate species often found in grassland rather than woodland in the central and eastern U.S., while ***Tibicen superba*** is found in the south-central U.S., west as far as New Mexico.

Cacama valvata is one of five species in its genus found in the southwestern U.S. Often called cactus dodgers, males may "sing" only a few times from one spot before flying to another location, rather than singing continuously from one place for long periods like some cicadas.

Members of the genus ***Okanagana*** (about 50 species) range over much of the northern U.S. and adjacent Canada, but most are western, and these are the most common cicadas in the Pacific northwest. Listen for adults between May and August. The ear-splitting *SSSSSSSSSSSSSSS* . . . will get your attention. The **Hieroglyphic Cicada** *(Neocicada hieroglyphica)* is an eastern species, most common in the southeastern states. Males sing in small localized choruses, especially from oak trees. The song begins as a series of whines, progressively softer, and ending with a long, even whine; others have likened the sound to a model aircraft engine revving up. This is often the first cicada species to be heard in the spring.

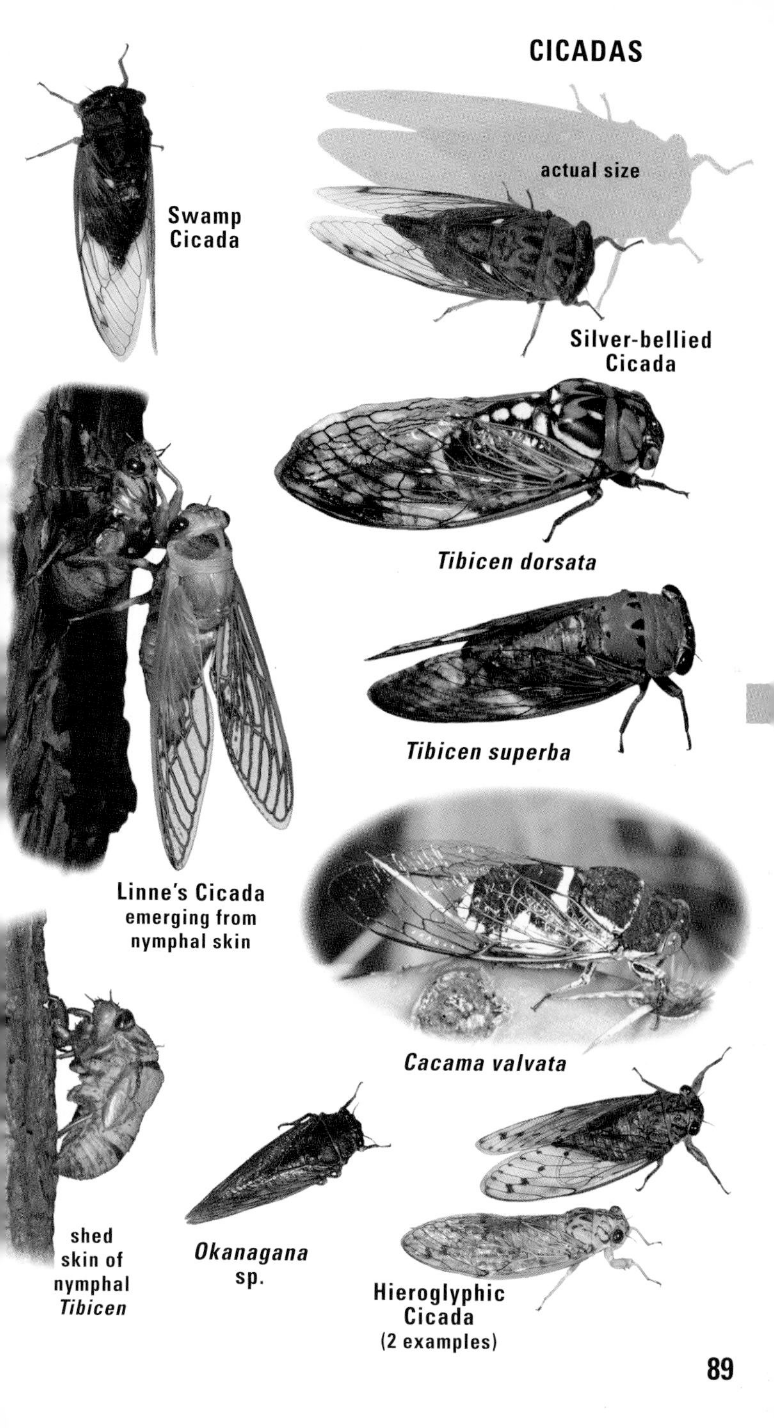

Swamp Cicada

Silver-bellied Cicada

Tibicen dorsata

Tibicen superba

Linne's Cicada emerging from nymphal skin

Cacama valvata

shed skin of nymphal *Tibicen*

Okanagana sp.

Hieroglyphic Cicada (2 examples)

PERIODICAL CICADAS

These are classified in the genus *Magicicada,* and the name seems entirely appropriate for these uniquely North American insects. While most cicada species have staggered generations, so that some adults emerge every year, periodical cicadas have synchronized broods. Once every 13 years or once every 17 years, depending on the species, a local brood stages a massive emergence, briefly becoming the most conspicuous of insects before disappearing again.

The genus ***Magicicada*** includes three species on the 17-year cycle in the eastern and northern U.S. *(M. septendecim, M. cassini,* and *M. septendecula),* and four on the 13-year cycle in the southern and central U.S. *(M. tredecim, M. tredecassini, M. tredecula,* and the recently discovered *M. neotredecim,* described to science in 2000). Different broods, numbered by scientists with Roman numerals, emerge in different areas in different years. Some broods are extremely local (one has even become extinct) while others are widespread. Each brood may contain more than one species, with different proportions of abundance. They "hatch" in overwhelming numbers, generating more decibels than a jackhammer. The performance of the entire ensemble has been likened to the sound effects for a flying saucer in an old sci-fi movie. The size and sheer abundance of the cicadas never fails to instill panic in the media and in the uninformed human populace. Amazingly, these are essentially harmless insects. Females do some damage in the course of laying eggs, inserting their rodlike ovipositor into twigs of live trees, occasionally with enough force and in enough quantity to break the stem. The result is a characteristic "flagging" of tree branches as the terminal leaves beyond the break die prematurely. Smaller trees can be killed outright in this manner, but homeowners can cover vulnerable saplings with protective cloth. Additionally, some yard-care machinery generates noise similar enough to that of male cicadas that it draws huge flocks of females. The din and the stench (dead cicadas rot and smell) end roughly a month after the initial emergence, between mid-May and the end of June. The remainder of the life cycle is spent in the nymphal stage, underground, feeding on roots.

Exactly why they have such a bizarre life cycle remains largely a mystery. Perhaps the cicadas evolved the odd-numbered emergence to outlast some type of persistent parasite. Even so, a fungal disease slaughters thousands of the insects every time they do reach adulthood. Birds, various mammals, and other insects certainly feast on as many cicadas as they can stomach. So do some people, who claim stir-fried cicada nymphs taste much like canned asparagus or shrimp. Historically, the Iroquois harvested the cicadas for food.

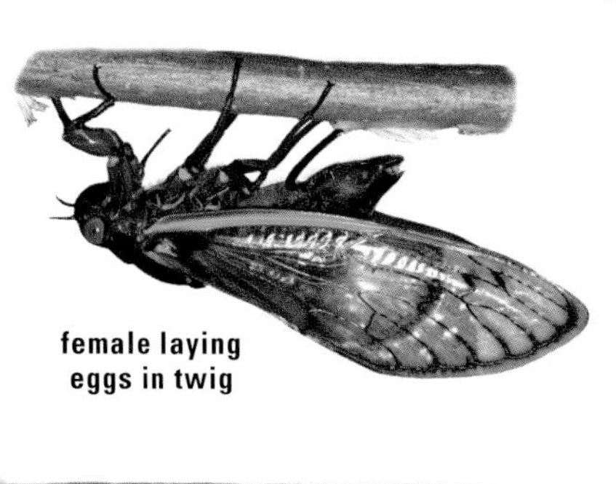

female laying eggs in twig

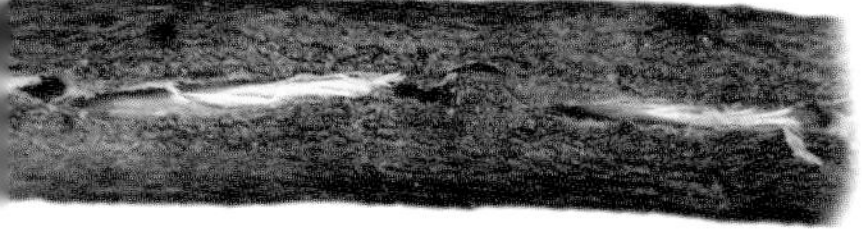

close-up of twig with egg-laying scars

actual size

adult feeding

freshly emerged adult

adult in process of emerging from nymphal skin

emergence "chimney"

nymph

TREEHOPPERS

(family Membracidae) are often barely recognizable as living creatures, let alone as insects. Many species mimic thorns, with spikes, horns, crests, or other bizarre modifications to their thoracic anatomy. Many species occur on oak trees (look in the crooks of twigs and branches), and a few come to lights at night. Females of some species guard their eggs and nymphal offspring. Communication between individuals during courtship and parental care is accomplished by vibration of the abdomen against stems and foliage. There are close to 260 species north of Mexico.

Ceresa diceros is one of about 16 species of "buffalo treehoppers" typically found in fields, savannas, and prairies rather than woodlands. The **Three-cornered Alfalfa Treehopper** *(Spissistilus festinus)* is one of nine North American species in its genus. It is more often a pest of soybeans than alfalfa, adult and nymphal individuals sometimes girdling the stem of young plants in the course of feeding, causing the plant to be vulnerable to collapse. The **Locust Treehopper** *(Thelia bimaculata)* is very common on black locust trees. Their family groups are often tended by the ant *Formica obscuripes,* which "milks" them for their secretions of honeydew. One other species of ***Thelia*** occurs north of Mexico.

The genus ***Glossonotus,*** with six species in North America, includes some impressive thorn mimics. ***Telamona monticola*** is a common species that feeds on oak. There are 25 additional species in this genus known from North America. ***Smilia fasciata,*** one of two species in its genus here, is common on oak trees in eastern North America. ***Archasia belfragei*** occurs on white oak and locust trees. The height of the crest varies considerably among individuals, but most specimens have a high crest. ***Ophiderma flavicephala*** is one of 14 species in its genus in North America.

The **Black Locust Treehopper** *(Vanduzea arquata)* is very abundant on its host tree over much of eastern North America. There are four other species in the genus north of Mexico. ***Entylia carinata*** (formerly *E. bactriana*) is very common on giant ragweed throughout eastern North America. Adults and nymphs are often tended by ants. Four other species of ***Entylia*** are known north of Mexico. ***Publilia concava*** is found on the stems and leaves of sunflowers and thistles in shady conditions. The species ranges throughout the eastern U.S. and northwest to Washington. Look for clusters of adults and nymphs on the host plant. This is one of five North American species in the genus.

The ornately marked ***Platycotis vittata*** is large, up to half an inch including the horn on the thorax, though not all individuals develop this growth. This species feeds on birches. ***Campylenchia latipes,*** a good thorn mimic, is rather common on weedy members of the sunflower family. Also thornlike is the **Two-marked Treehopper** *(Enchenopa binotata),* which has been collected from black cherry and related trees. This may actually be a complex of nine or more species that have yet to be described individually by scientists. The **Thorn Bug** *(Umbonia crassicornis)* is impressively marked, but groups may be passed off as mere thorny twigs.

Ceresa diceros

Three-cornered Alfalfa Treehopper

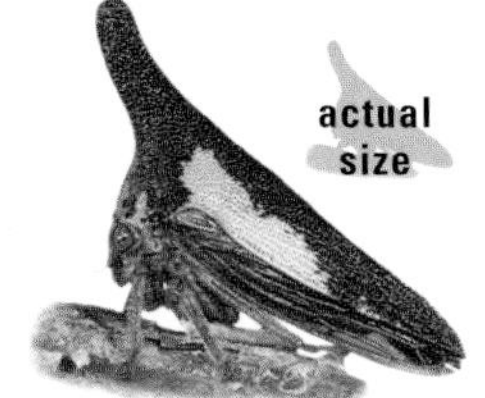

Locust Treehopper

Glossonotus sp.

Telamona monticola

Smilia fasciata

Black Locust Treehopper

Ophiderma flavicephala

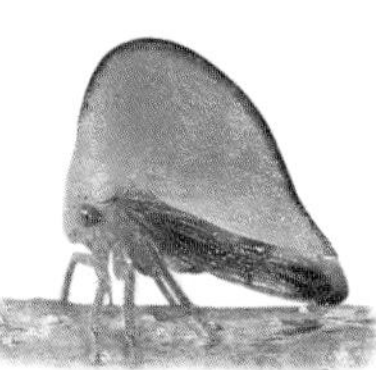

Archasia belfragei

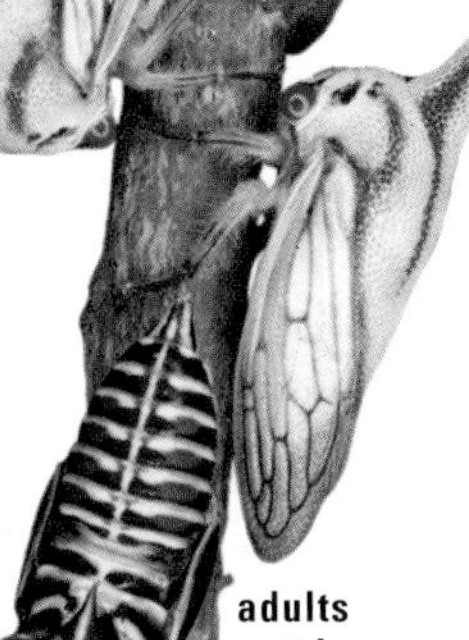

Platycotis vittata

adults and nymph

Publilia concava

Entylia carinata

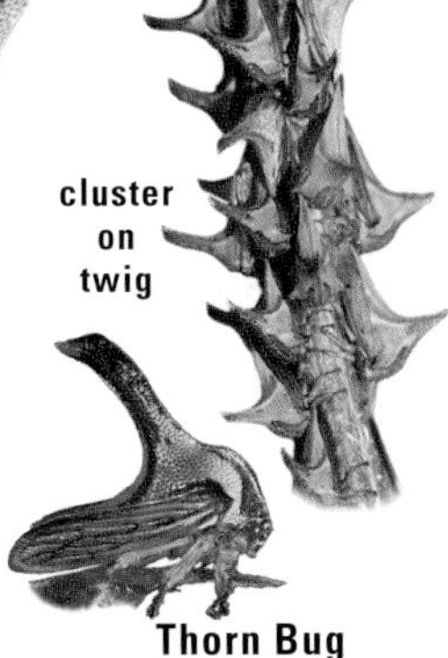

cluster on twig

Thorn Bug

Campylenchia latipes

Two-marked Treehopper

LEAFHOPPERS (family Cicadellidae) are incredibly diverse, with at least 2,500 species on our continent. They can be separated from similar families by the row of spines along the outer edge of the hind tibia ("shin"). They often cock those hind legs to better facilitate escape from predators with a quick, long leap. The longest leaper in the animal kingdom (over 40 times its body length) is in this family. Many species anoint themselves with special protein particles called brochosomes, manufactured by the excretory system. Females of some species pile brochosomes into oval patches on their wings, which they later wipe onto egg scars to help prevent the ova from drying out; these temporary white patches are not permanent field marks. They lay their eggs in plant stems.

The **Glassy-winged Sharpshooter** *(Homalodisca vitripennis)*, native to the southeastern U.S., has wreaked havoc on vineyards in California where it is introduced. It and some others in this family earn the name sharpshooters for their habit of expelling excess watery waste with such force that it spurts a fair distance with an audible popping noise. **Candy-striped Leafhoppers** *(Graphocephala coccinea)* are common on blackberry canes and various ornamental plants throughout most of the continent. This genus includes 17 other species in North America, some of them very similar.

Oncometopia orbona is the only widespread eastern species in its genus. The genus ***Draeculacephala*** includes 20 species north of Mexico, many recognizable by the sharply pointed "nose." Look for members of the genus ***Tylozygus*** in grassy habitats. The **Saddled Leafhopper** *(Colladonus clitellarius)*, one of 51 species in its genus, is found on willows and many other plants. ***Cochlorhinus pluto*** is one of eight species in its genus occurring north of Mexico, while the genus ***Scaphytopius*** includes 72 species in North America. ***Gyponana*** includes 49 species in North America. Most are yellow or green, but some specimens may appear pink. They usually have a dense network of veins at the wingtips. ***Ponana pectoralis*** is sometimes attracted to lights at night. There are 26 other species in the genus in North America. ***Idiocerus alternatus*** is one of 73 species in its genus found north of Mexico. It ranges east of the Rockies and is found mostly on willows. ***Coelidia olitoria*** varies in color and pattern, but females have conspicuous pale bands on the front wings. This is a very common species on trees and shrubs throughout the eastern U.S. and adjacent southern Canada.

SPITTLEBUGS (family Cercopidae) make the familiar masses of "spit" in the nymphal stage. They mix liquid waste products with a mucous secretion, whipping air bubbles into the broth with fingerlike appendages at the tip of the abdomen. The foam helps hide the soft-bodied insect from predators and parasites, and also keeps the bug moist, preventing dessication. Adults are sometimes called froghoppers for their squat appearance and jumping ability. There are about 54 species in the U.S. and Canada. Many are somewhat specialized; for example, **Pine Spittlebugs** (*Aphrophora* sp.) feed mainly on conifers. The **Meadow Spittlebug** *(Philaenus spumarius)* is widely distributed, with a wide variety of adult color forms. The **Two-lined Spittlebug** *(Prosapia bicincta)* is common in the eastern U.S.

Glassy-winged Sharpshooter

actual size

Candy-striped Leafhopper

Oncometopia orbona
with and without brochosomes

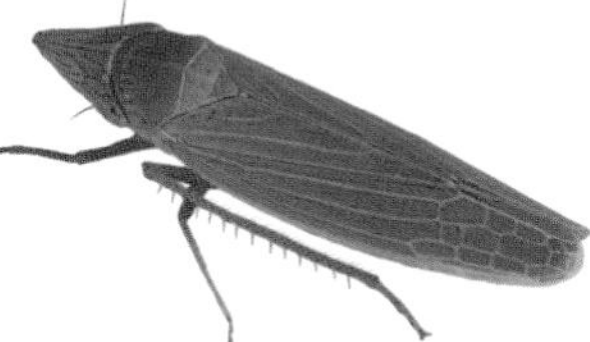

Draeculacephala sp.

Saddled Leafhopper

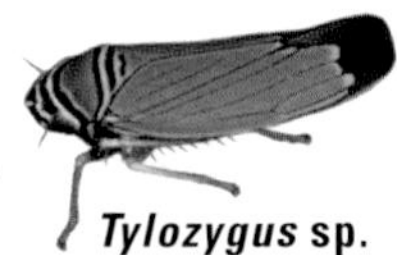

Tylozygus sp.

Cochlorhinus pluto

Gyponana sp.

Ponana pectoralis

nymph

Scaphytopius sp.

Idiocerus alternatus

Coelidia olitoria
nymph and adult

nymph in "nest"

Pine Spittlebug

Meadow Spittlebug
(2 variations)

Two-lined Spittlebug
(2 variations)

VARIOUS PLANTHOPPERS

All of the families below are included in the **superfamily Fulgoroidea.** Mostly less common than leafhoppers (previous page), they have fewer spines on the hind legs and often have more interesting head shapes.

CIXIID PLANTHOPPERS (family Cixiidae) are represented by 172 species in North America. Look for adults on shrubs and trees; nymphs feed on roots and overwinter in soil. ***Cixius*** and ***Bothriocera*** are two common genera. **FULGORIDS (Fulgoridae)** are known as lantern bugs in the tropics, where some are very large, with elongated protrusions from their heads. Our 18 domestic species are considerably less spectacular. Some species fly to lights at night. ***Poblicia fuliginosa*** occurs in the southwest. About 145 species of **DELPHACIDS (Delphacidae)** occur in North America. They are small, averaging 2–9 mm, and even adults may be short-winged, giving them the appearance of nymphs of many other types of planthoppers. Many delphacids have prominent antennae with thickened basal segments. All may be distinguished from similar families of insects by a single broad, flexible spur on the inside tip of the hind tibia. Delphacids are perhaps most diverse in moist grassy habitats such as meadows and streamsides. The genus ***Megamelus*** includes 10 North American species.

ACHILID PLANTHOPPERS (Achilidae) include 46 North American species. Their life histories remain largely a mystery. There is surprising diversity in temperate coniferous forests, where the nymphs occur under bark and in decaying logs, possibly feeding on fungi. The genus ***Epiptera*** includes 12 species. **ISSID PLANTHOPPERS (Issidae)** are sometimes known as weevil-like planthoppers for their slightly elongated "noses" and often shortened wings. They are fairly common in old fields. The genus ***Bruchomorpha*** includes 20 species in North America, while ***Dictyssa*** includes 15. **ACANALONIID PLANTHOPPERS (Acanaloniidae)** includes 18 North American species in the genus ***Acanalonia,*** sometimes considered part of the Issidae. The semicircular shape of the forewings is distinctive. ***Acanalonia bivittata*** is common on grasses in the eastern and central U.S.

FLATIDS (Flatidae) are wedge-shaped insects with 33 species north of Mexico. The **Citrus Planthopper** *(Metcalfa pruinosa)* is covered with a whitish waxy coating. It feeds on many woody plants. Its egg-laying activities sometimes cause damage and stunted growth of twigs. ***Ormenis saucia*** is one of four species in its genus found mostly in the southwest. ***Ormenaria rufifascia*** can be fairly common on fronds of palms and palmettos in Florida. **DICTYOPHARID PLANTHOPPERS (Dictyopharidae)** include common species easily recognized by the long "snout." There are 64 species north of Mexico. The genus ***Scolops*** includes 32 species, widely distributed in prairie habitats. The three eastern species of ***Rhynchomitra*** are green fully winged insects with a pronounced snout. Look for them at lights at night. **DERBID PLANTHOPPERS (Derbidae)** resemble small moths or caddisflies but have a pronounced broad snout. They feed on woody fungi, and adults sometimes fly to lights at night. The genus ***Cedusa*** includes 16 species in North America. The widespread ***Otiocerus degeeri*** is one of 10 species in its genus, mostly eastern.

VARIOUS PLANTHOPPERS

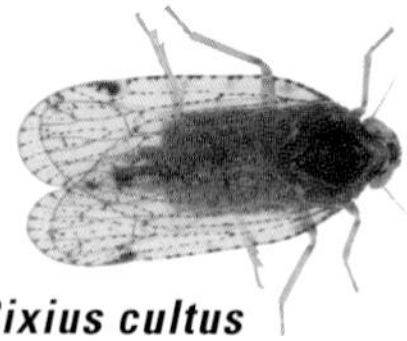
Cixius cultus

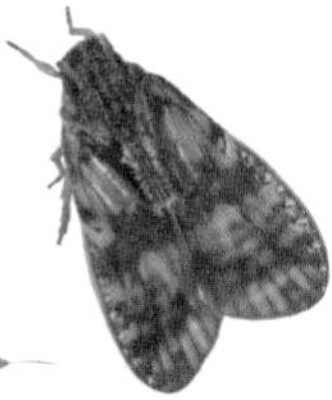
Bothriocera sp.

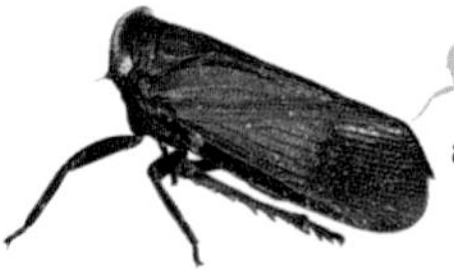
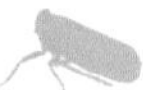
actual size

Poblicia fuliginosa

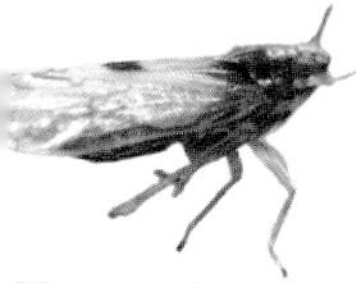
Megamelus sp.

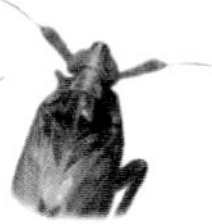
from above

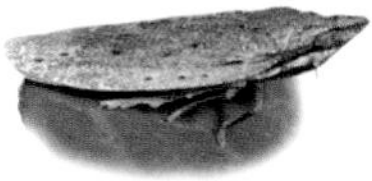
Epiptera floridae

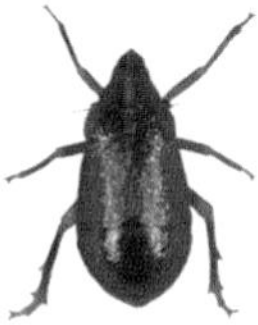
Bruchomorpha sp.

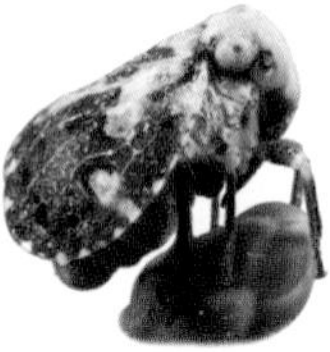
Dictyssa sp.

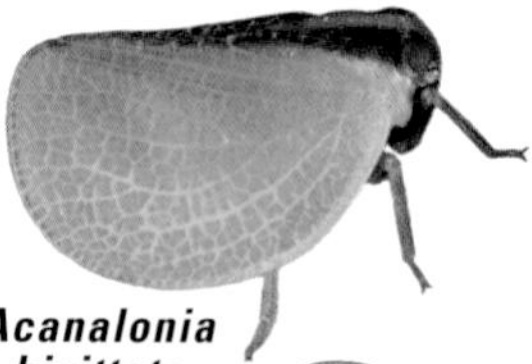
Acanalonia bivittata

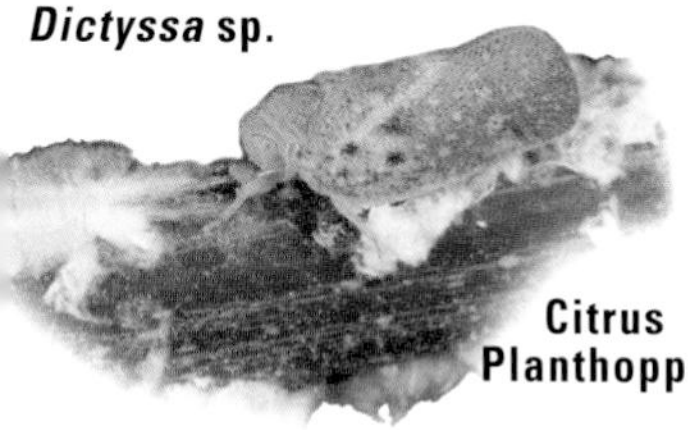
Citrus Planthopper

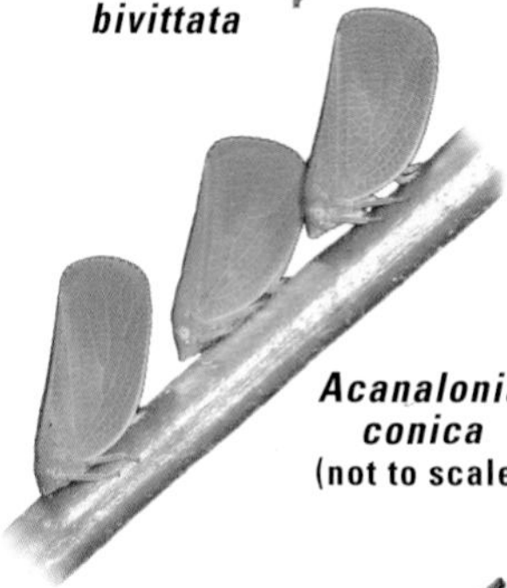
Acanalonia conica (not to scale)

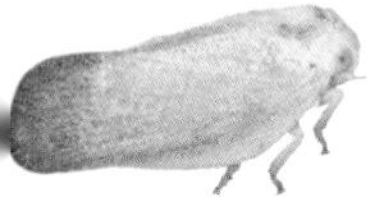
Ormenis saucia

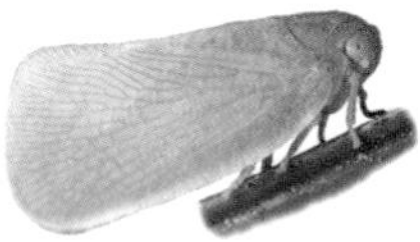
Ormenaria rufifascia

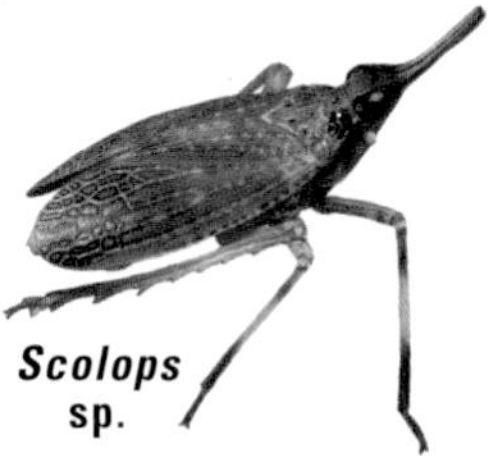
Scolops sp.

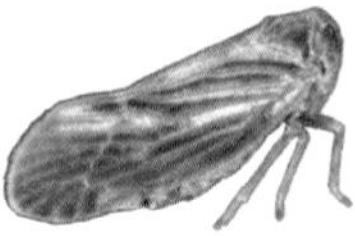
Cedusa sp.

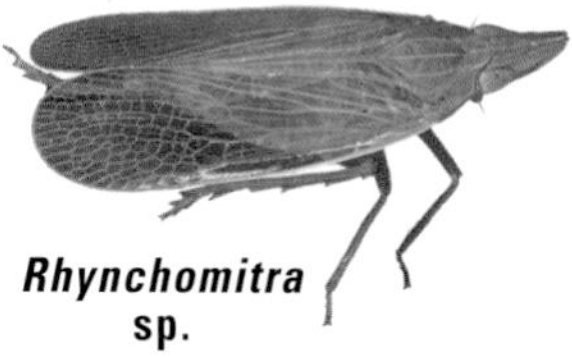
Rhynchomitra sp.

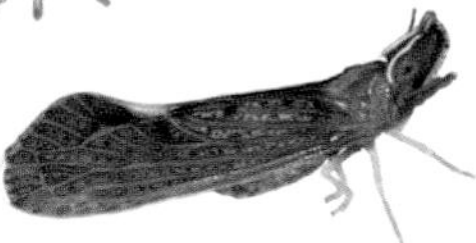
Otiocerus degeeri

APHIDS

(family Aphididae) are tiny insects that are overwhelmingly abundant and diverse. Many are best identified by association with their host plant(s). Most species alternate hosts from one season to another. Some (next page) are gall-formers. Depending on the generation, some individuals may have wings for dispersal to an alternate host plant. An amazing array of other insects are associated with aphids. Some, like lady beetles, lacewings, and braconid wasps, prey upon or parasitize the insects. Many others, especially ants, wasps, and flies, visit colonies for "honeydew," the sugary waste excreted by aphids, scales, and related insects. Most aphids have a symbiotic relationship with microbes that live inside them, essentially acting as a "sap refinery," turning nutrient-poor phloem fluids into essential compounds. Aphid colonies that do minimal or tolerable damage should be left alone. The beneficial insects they attract may decrease populations of more troublesome pests such as caterpillars.

The **Green Peach Aphid** *(Myzus persicae)* is found over most of the planet partly thanks to its ability to feed on hundreds of alternate hosts. The primary host is peach, sometimes apricot or plum, the trees upon which the egg-producing generation overwinters. Up to eight generations are produced through parthenogenesis in spring, before winged individuals are born and disperse to summer hosts. A generation may take as little as 10 to 12 days from birth to maturity. In autumn, winged aphids return to the trees, the females giving birth to wingless female offspring. These sedentary females emit a special pheromone that attracts winged males.

The **Oleander Aphid** *(Aphis nerii)* is also commonly found on milkweed. These insects sequester toxins from both plants for defense against predators. The bright yellow color indicates you should not eat them! The **Russian Wheat Aphid** *(Diuraphis noxia)* is native to temperate Eurasia and North Africa. First detected in Texas in 1986, it has spread through most of the western U.S. and three Canadian provices. Corkscrew rolling of wheat leaves is caused by toxins injected as the aphids feed. The **Poplar Leaf Aphid** *(Chaitophorus populicola)* occurs on poplar and cottonwood and does not alternate hosts. The **Giant Bark Aphid** *(Longistigma caryae)* is our largest species at about a quarter of an inch. It occurs on bark of deciduous trees over most of the eastern U.S. and west to New Mexico.

Conifer aphids (*Cinara* sp.) seldom do significant damage, although they are among our largest aphids. Colonies occur on branches, stems, and candles of the host. These aphids are often guarded by carpenter ants. In exchange for protective services, the aphids secrete honeydew the ants imbibe with relish. The **Pea Aphid** *(Acyrthosiphon pisum)* is one of the most studied of all insects. Its pest status among pea and alfalfa growers accounts for the intense scrutiny. Adults may survive mild winters, but it is usually the egg stage from which next year's aphids (all females) emerge. A single female may give birth to six or eight nymphs each day. After a number of generations, winged females are born that then migrate to alternate host legumes. Males are produced only in fall, mating with the egg-laying females. There can be up to 15 generations per year.

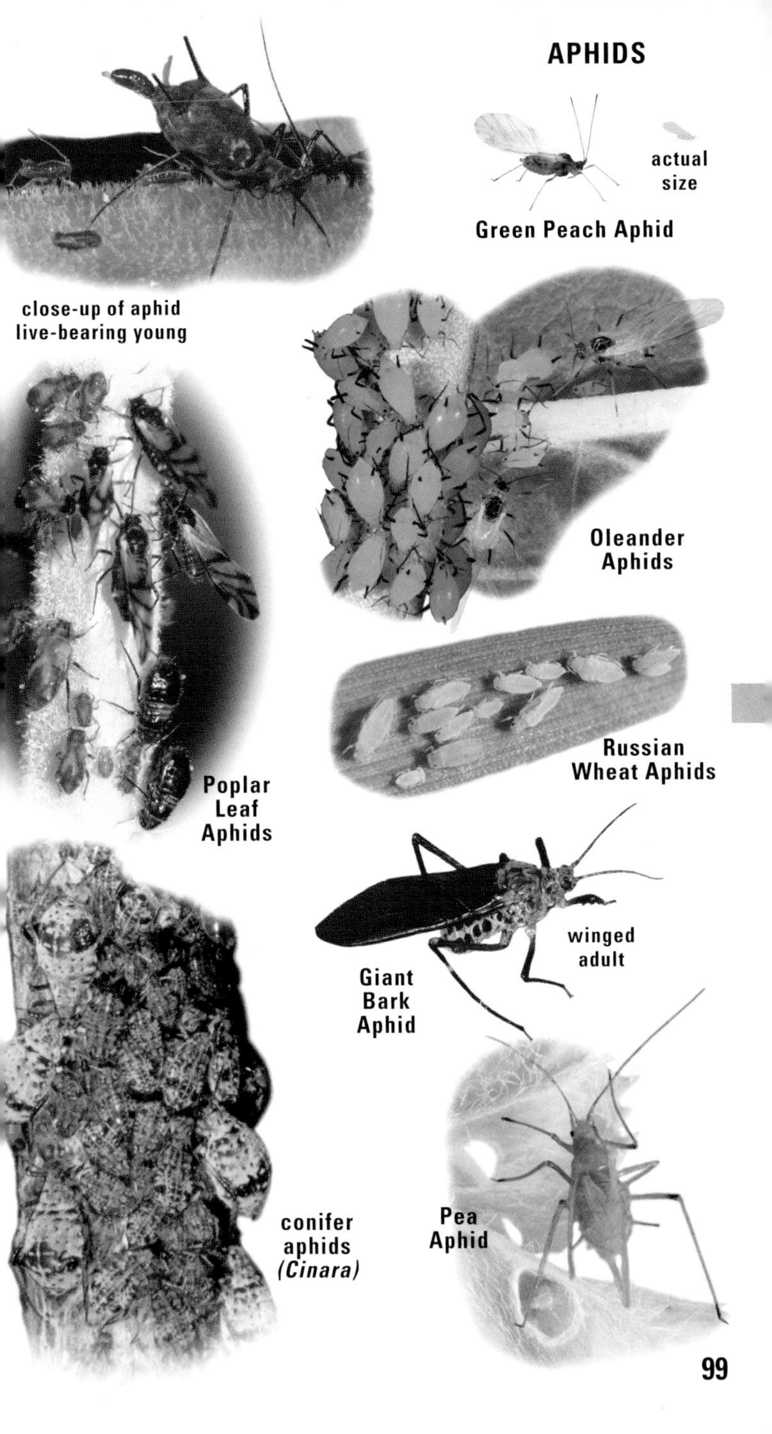
Green Peach Aphid
actual size
close-up of aphid live-bearing young
Oleander Aphids
Poplar Leaf Aphids
Russian Wheat Aphids
Giant Bark Aphid
winged adult
conifer aphids *(Cinara)*
Pea Aphid

Woolly aphids owe their appearance to filaments of wax secreted from their bodies. The coating keeps them from drying out and repels some predators that would rather not get a mouthful of wax. The **Woolly Apple Aphid** *(Eriosoma lanigerum)* feeds primarily on elm, but apple, hawthorn, and mountain ash are secondary hosts. Many other aphids have a woolly appearance; **Beech Blight Aphid** *(Grylloprociphilus imbricator),* sometimes abundant on beech twigs and leaves, is one example. The **Poplar Vagabond Aphid** *(Mordwilkoja vagabunda)* feeds at the tops of twigs, causing the formation of a large convoluted gall. Pictured is a cut-open gall showing the bloated "stem mother" aphid and a few of her offspring.

PSYLLIDS (family Psyllidae) are also called jumping plant lice. They resemble aphids, but adult psyllids are always winged and can leap away from danger as well. They suck sap from foliage and stems. Natural hosts of the **Pepper Psyllid** *(Bactericera cockerelli),* alias tomato psyllid, are confined to the Solanaceae, including nightshade. Damage to plants results in part from a toxin in the insect's saliva that causes curling and a yellow or purplish cast to the leaves. Most psyllids do not form galls, but here are a few that do. **Hackberry Nipple Galls** *(Pachypsylla celtidismamma)* each contain one developing nymph. Adults of this and other psyllids sometimes swarm on window screens in fall as they seek shelter for hibernation. **Hackberry Petiole Galls** *(Pachypsylla venusta)* form large six-chambered galls. Old galls may remain on trees for years. The small **Hairy Hackberry Galls** *(Pachypsylla pubescens)* are formed in summer.

WHITEFLIES (Aleyrodidae) are tiny and distinctive, but moth flies (p. 274) and dustywings (p. 222) might be mistaken for them. Metamorphosis in whiteflies is unique. Nymphs emerging from eggs are active "crawlers" but quickly settle down into scalelike creatures. Wings develop internally, not becoming evident until the last stage of development, an inactive, nonfeeding "pupa." The adult emerges from this stage. About 100 species are known north of Mexico. The **Greenhouse Whitefly** *(Trialeurodes vaporariorum)* includes over 200 plant species on its menu. This insect ranges across the continent but cannot overwinter outdoors in temperate climates. The **Silverleaf Whitefly** *(Bemisia argentifolii),* widespread in the southern U.S., has been found on over 500 plant species.

PHYLLOXERANS (Phylloxeridae) are about 30 species of tiny insects, some of which produce galls. ***Phylloxera subelliptica*** creates a smooth, oval gall almost an inch long on hickory leaf petioles in early summer. The **Grape Phylloxera** *(Phylloxera vitifoliae)* is a notorious pest of vineyards. Native to North America, it was accidentally exported to Europe in the mid-1800s, nearly destroying the French wine industry before resistant rootstocks were shipped from the U.S. to salvage the crops. **ADELGIDS (Adelgidae)** are associated only with conifers. Most species produce stringy wax secretions to help retard water loss and repel predators. The **Hemlock Woolly Adelgid** *(Adelges tsugae),* accidentally introduced from Asia, has had only a moderate effect on western hemlocks, but many eastern hemlocks have been killed or seriously weakened by this pest.

APHIDS, PSYLLIDS, WHITEFLIES, ETC.

(images not all to scale)

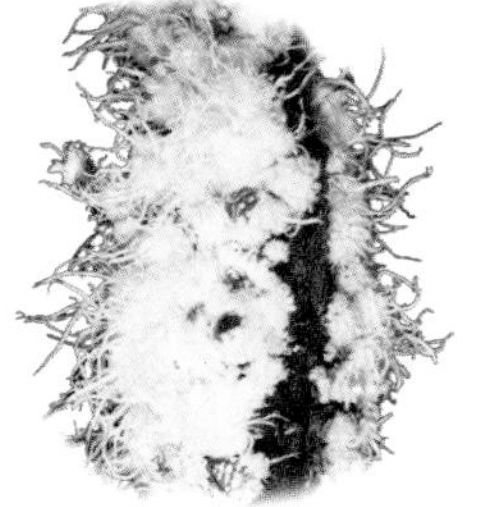

Woolly Apple Aphids

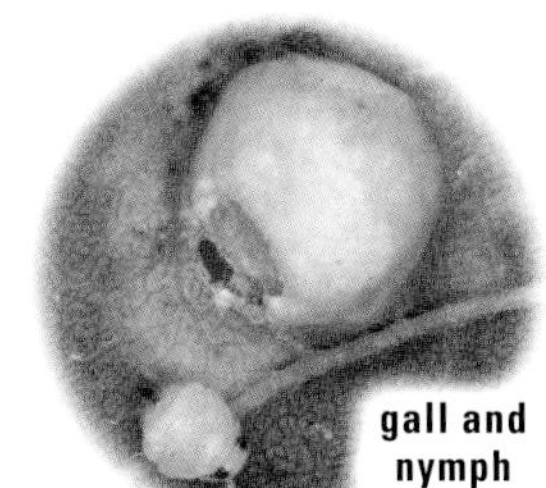

gall and nymph

actual size

adult

Hackberry Nipple Gall

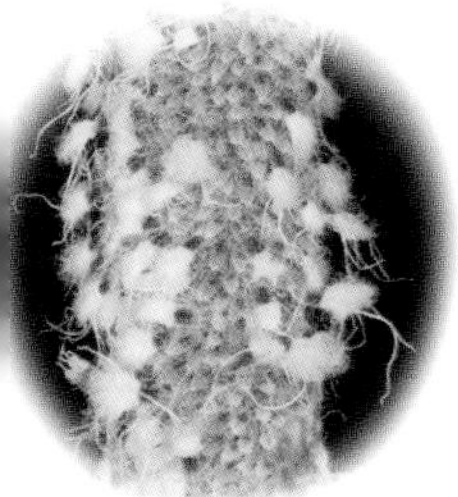

Beech Blight Aphids

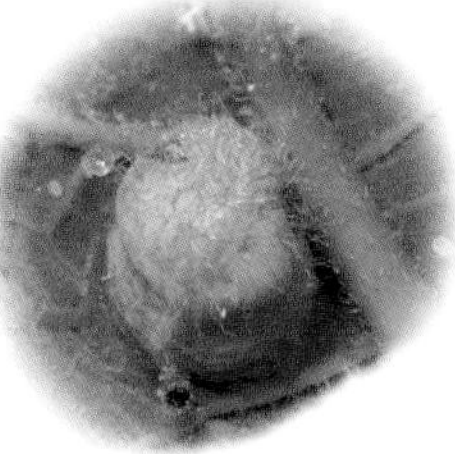

Hairy Hackberry Gall

Hackberry Petiole Gall adult

nymph

adult

Pepper Psyllid

Silverleaf Whitefly

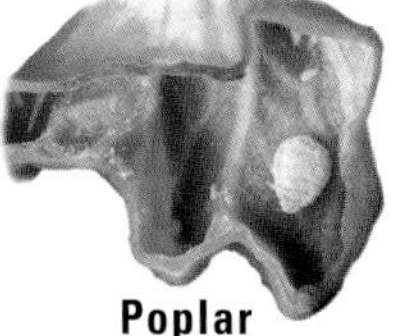

Poplar Vagabond Aphid (cut-open gall)

Phylloxera subelliptica gall

Greenhouse Whiteflies

Grape Phylloxera galls

Hemlock Woolly Adelgid

GALLS

are abnormal plant growths stimulated by another organism, such as an insect, mite, nematode worm, fungus, virus, or slime mold. Shown here are galls made by a few insect families, only one of which actually belongs in this section of the book, but they are included here because galls and scales (next page) are similar in not resembling insects at all! Galls draw a disproportionate amount of nutrients from the plant, providing the invading organism(s) with a rich supply of food and a protective home. Still, a host of predators and parasites exploit galls at the cost of the original occupant. Try "raising" galls and be amazed at the wide array of insects that emerge. A few gall flies and mites have been employed as biological controls of invasive weeds.

GALL WASPS (family Cynipidae) form many obvious galls. Dr. Alfred Kinsey, the noted sex researcher, actually began his career studying cynipids. **Oak Apple Galls**, fashioned by ***Amphibolips confluenta,*** are initially green, fading to light brown as they mature in May and June. The **Blackberry Knot Gall** is a product of ***Diastrophus nebulosus.*** **Western Speckled Oak Galls** are caused by ***Besbicus mirabilis.*** **Eastern Speckled Oak Galls** are the work of ***Loxaulus maculipennis.*** Cut into a mature gall and you will find it hollow, with a central chamber suspended from the exterior walls by a network of filaments. The **Mossy Rose Gall,** caused by ***Diplolepis rosae,*** is covered in stringy filaments. Several tiny adult wasps emerge from each gall in early spring. Bullet galls on oaks are produced by species such as the **Oak Rough Bulletgall Wasp** *(Disholcaspis quercusmamma).* **Urchin Galls** are created on oak leaves by the wasp ***Antron echinus.***

SAWFLIES (Tenthredinidae), discussed on p. 320, are usually free-living as larvae, but some species cause galls. Our 42 species of ***Pontania*** create galls, such as the ones on a willow leaf shown here.

GALL MIDGES (Cecidomyiidae) include about 1,200 species of tiny flies in North America, of which about two-thirds form galls as larvae. Legend has it that the **Hessian Fly** *(Mayetiola destructor)* arrived here in straw brought by European mercenaries to the Revolutionary War in 1776. In reality, this wheat pest probably originated in Mesopotamia, following the cultivation of its food plant to all corners of the globe. Stunted plants with bumpy stems and dark green leaves signal infestation by the stem-boring larvae. The genus ***Asphondylia*** includes 54 North American species. The **Creosotebush Gall** in the southwest is formed by ***Asphondylia auripila.*** ***Cecidomyia bedeguar*** afflicts hawthorns, creating **Tufted Thorn Galls.** **Maple Leaf Spot Galls,** formed by ***Cecidomyia ocellaris,*** are striking bull's-eyes on leaves of red maple. **Willow Cone Galls,** formed by ***Rhabdophaga strobiloides,*** look like miniature pine cones on willows.

GALL-LIKE SCALES (Kermesidae) are actually not galls at all but are related to the insects on the next page. **ERIOPHYIDS (Eriophyidae)** are not insects but tiny mites, of which many form galls on plants. **Maple Bladder Galls,** made by ***Vasates quadripedes,*** are often in thick clusters. The fingerlike or pouchlike **Linden Galls** *(Eriophyes abnormis)* are often on linden leaves, and similar species occur on maples and other trees.

GALLS AND GALL-MAKING INSECTS

(images not all to scale)

Oak Apple Gall

Blackberry Knot Gall

Eastern Speckled Oak Gall

Western Speckled Oak Gall

Mossy Rose Gall

Oak Rough Bulletgall Wasp

Pontania sp.

Urchin Gall

Hessian Fly

Creosotebush Gall

Tufted Thorn Gall

Maple Leaf Spot Gall

Willow Cone Gall

gall-like scale

Maple Bladder Gall

Linden Gall

bear little resemblance to insects, at least the adult females. Both sexes begin life as tiny "crawlers" that hatch from eggs and disperse in this phase. As they mature, females become immobile, often resorbing their legs and eyes. They usually secrete a dense protective coating of wax or similar material that hides the animal beneath. Reproduction in some cases does not involve males, females cloning themselves via parthenogenesis. Besides draining sap, some scales transmit plant viruses.

GROUND PEARLS (family Margarodidae) include our largest scales. Not all form the waxy cysts that give these insects their name. Those in the genus ***Margarodes*** feed on roots underground. The **Cottony Cushion Scale** *(Icerya purchasi)* arrived accidentally from Australia, nearly collapsing the California citrus industry. Introduction of the Vedalia lady beetle (p. 156) saved the day. **ENSIGN SCALES (Ortheziidae)** are usually covered by waxy plates. The **Greenhouse Orthezia** *(Insignorthezia insignis)* is widespread across the continent. **MEALYBUGS (Pseudococcidae)** are considered "primitive," since adult females retain legs and eyes and are more mobile than most scales. The **Large Yucca Mealybug** *(Puto yuccae)* is known from California and Texas. The **Long-tailed Mealybug** *(Pseudococcus longispinus)* is recognized by twin tail-like projections. The **Citrus Mealybug** *(Planococcus citri)*, a European native, now occurs over most of North America. **FELT SCALES (Ericoccidae)** include over 50 species here. Adult females resemble mealybugs but lack the dense wax coating.

COCHINEAL SCALES (Dactylopiidae) create a fluffy white coating on pads of prickly pear and related cacti. The insects themselves are tiny. Cochineal scales produce a defensive chemical called carmitic acid. Its brilliant red color attracted the attention of indigenous peoples, who harvested the insects and dried and crushed them to produce dye. Cochineal was a major Mexican export until roughly 1875, when synthetic dyes were developed, but is still used to dye beverages, medicines, candy, even cosmetics. **ORNATE PIT SCALES (Cerococcidae)** are represented in North America by five species, including the **Oak Wax Scale** *(Cerococcus quercus)*. Native Americans once gathered these insects and chewed them like gum.

TORTOISE SCALES (Coccidae) are unarmored, incorporating waxy excretions into their exoskeletons. Females retain their legs through adulthood but have reduced antennae. There are 92 species in North America. The **Red Wax Scale** *(Ceroplastes rubens)*, introduced in Florida, is important as a citrus pest. The **European Fruit Lecanium** *(Parthenolecanium corni)* can be on almost any tree or shrub. The **Hemispherical Scale** *(Saissetia coffeae)* secretes very little wax. Its size varies with the host plant.

ARMORED SCALES (Diaspididae) represent "advanced" scales: adult females have no legs or eyes. This is the largest family of scales, with over 300 species in North America. The **Cycad Aulacaspis Scale** *(Aulacaspis yasumatsui)* is a major introduced pest of cycads in Florida. The **Pine Needle Scale** *(Chionaspis pinifoliae)* is widespread, infesting almost all types of conifers. The **Latania Scale** *(Hemiberlesia lataniae)* is very numerous in Florida, attacking mostly Australian pine and other ornamentals.

SCALE INSECTS

Cottony Cushion Scale

Margarodes sp.
on roots

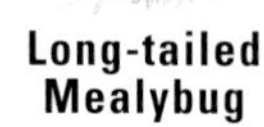

Long-tailed Mealybug

actual size

Large Yucca Mealybug

Margarodidae sp.

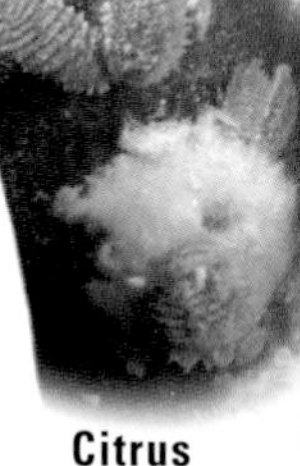

Citrus Mealybug

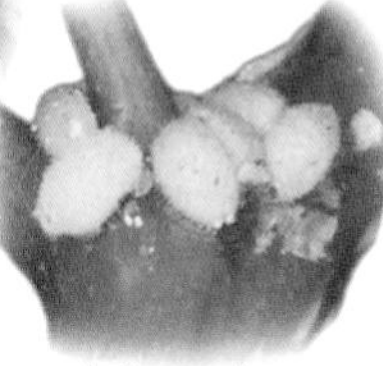

felt scale sp.

Greenhouse Orthezia

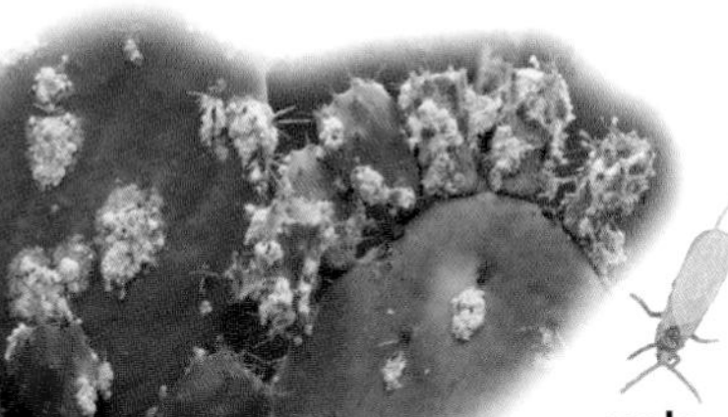

cochineal scale on prickly pear cactus

male cochineal scale

Oak Wax Scale

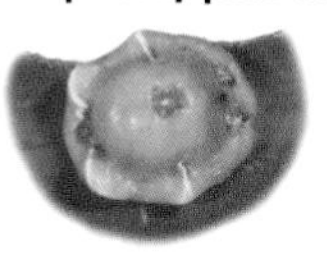

Red Wax Scale

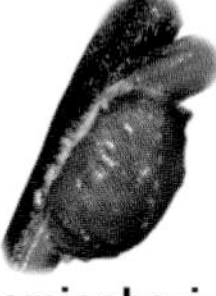

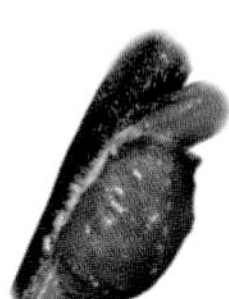

Hemispherical Scale

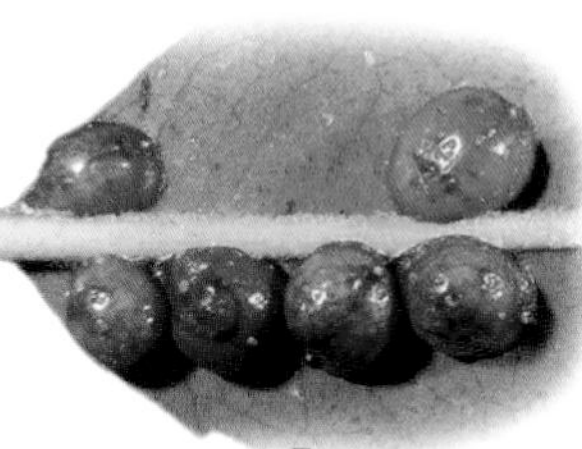

European Fruit Lecanium

Cycad Aulacaspis Scale

Pine Needle Scale

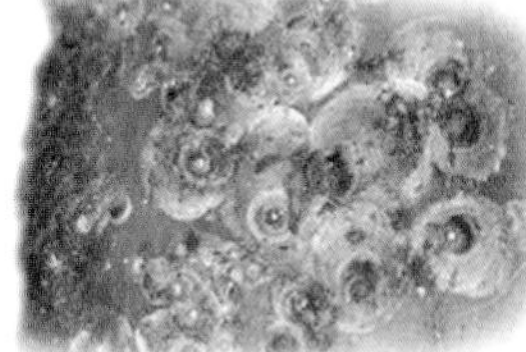

Latania Scale

WATER SCORPIONS AND GIANT WATER BUGS

This page begins the section that might be called the "truest" of the true bugs, those that have always been classified in the **order Hemiptera,** while those in the preceding section (cicadas, aphids, and others) have at various times been included in a separate order (Homoptera). See the introduction to all of the true bugs on pp. 86–87.

GIANT WATER BUGS (family Belostomatidae) are the leviathans of freshwater insects. The largest species regularly turn the tables on the usual order of things by preying on vertebrates, including small fish, frogs, and even snakes. All have the middle and hind pairs of legs somewhat flattened and fringed with fine hairs, to help them "row" swiftly underwater; the front pair of legs is modified for grasping their prey. Members of this family are generally in stagnant or slow-moving waters and may spend much of their time hiding in the mud near the bottoms of ponds. Many species live in the tropics, but only three genera and about 20 species occur in North America.

The genus ***Lethocerus*** includes five species north of Mexico. All are true giants, at 45–65 mm in length at maturity (or regularly more than 2 inches long). These are the "electric light bugs" that often fly to lights at night. Females ferociously guard their eggs, which are deposited on emergent vegetation above the waterline.

Members of the genus ***Belostoma*** are the "toe biters," so named for their supposed affinity for your tootsies. In truth they are nonaggressive, at least toward humans, but like other members of their family they may bite if handled roughly. There are nine species north of Mexico. The female lays her eggs on the back of the male. He is a devoted father, keeping the eggs free of fungal infection and guarding against predators. By doing "push-ups" at the water's surface, he keeps the ova properly oxygenated.

Abedus indentatus is one of five or six North American species in its genus, most of them found in the southwestern states, with only one eastern species. As with species of *Belostoma,* the male will "babysit" the eggs that have been laid by the female upon his back. Look for these bugs in ponds and slow-moving streams.

WATER SCORPIONS (Nepidae) are rather like underwater versions of assassin bugs (p. 116). Their front legs are modified for seizing prey, which is dispatched with a bite from the short beak. ***Ranatra fusca*** is a skinny insect easily mistaken for a stalk of grass, especially since it crawls about very slowly and often sits motionless for long periods. The stingerlike tail is actually a pair of breathing tubes used to connect the insect to the water's surface. This species ranges over most of the U.S. Eight other species in its genus are found in North America, most in the southern states. ***Nepa apiculata*** is very flat and oval, usually found in mud or dense vegetation. It occurs over the eastern half of the U.S.

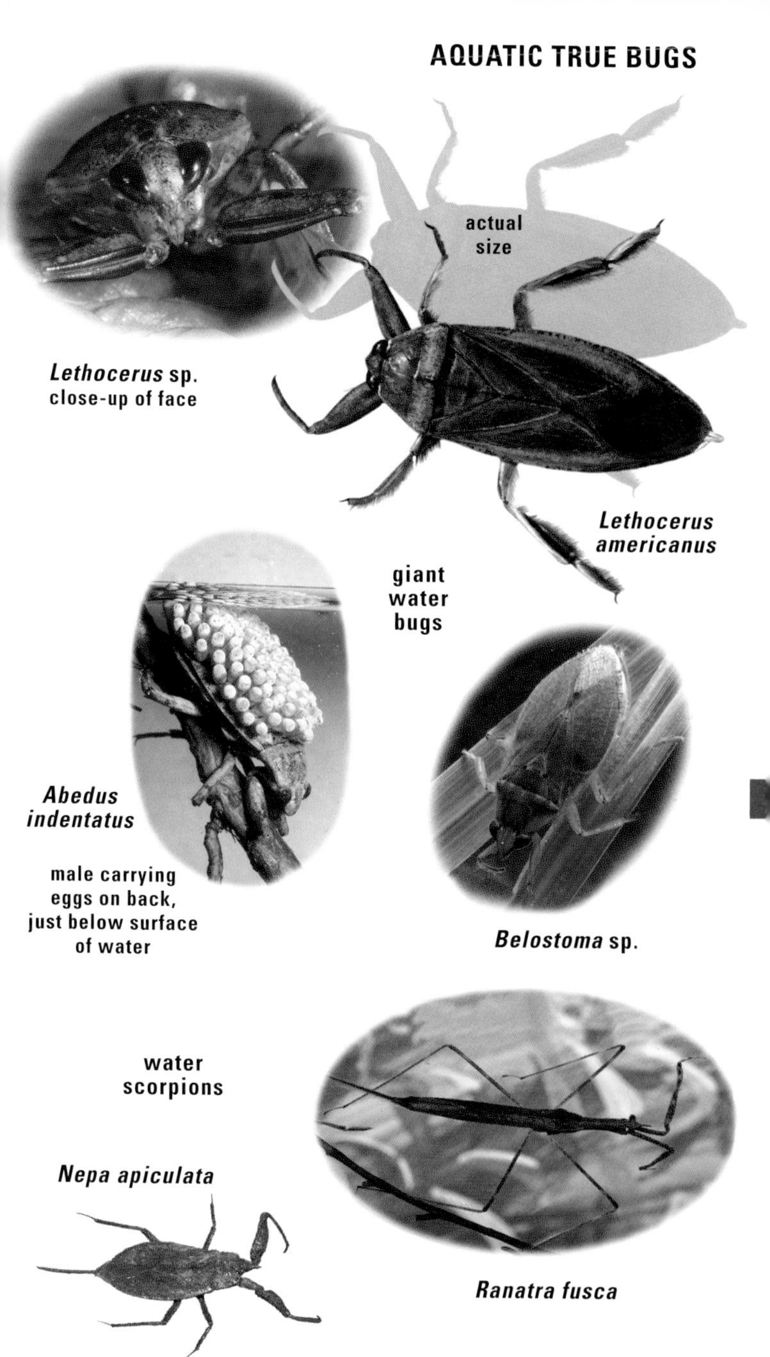

Lethocerus sp.
close-up of face

Lethocerus americanus

Abedus indentatus

male carrying eggs on back, just below surface of water

Belostoma sp.

Nepa apiculata

Ranatra fusca

BACKSWIMMERS (family Notonectidae) are common predators in ponds, lakes, even fountains and neglected swimming pools. Careful: they can give a painful bite if carelessly handled. They are frequently mistaken for water boatmen (Corixidae) but are usually larger and swim upside down. There are three genera in North America, with about 35 species. Members of the genus ***Notonecta*** are encountered most frequently. There are nearly 20 species in North America. Members of the genus ***Buenoa*** (not illustrated) are slender and small, with 14 species north of Mexico, most in the southern and southwestern states.

WATER BOATMEN (Corixidae) are a complex family of aquatic scavengers. Frequently mistaken for backswimmers, water boatmen are generally much smaller, and they swim right side up. At least 120 species occur in North America, in almost 20 genera, with two of the most important genera being ***Sigara*** (with almost 50 species) and ***Hesperocorixa*** (with almost 20 species). Most are very similar to each other, many showing a reticulated pattern of black and brown.

Most water boatmen live in ponds, not flowing streams, and their swimming is fast and erratic, interrupted by long periods of resting on submerged plants.The oarlike hind legs propel the insect through the water, while the reduced, spoonlike front legs shovel microscopic organisms into the mouth. The puncturing beak common to all Hemipterans barely qualifies as such in the Corixidae, as at least some species are able to ingest solid food, unlike other true bugs. The front wings are unique as well in that they are leathery throughout, with no membranous portion. Males of some species produce ultrasonic mating calls with their genitals. Look for water boatmen flying to lights at night, sometimes in large numbers.

CREEPING WATER BUGS (Naucoridae) are cryptic, predatory insects found in a variety of aquatic habitats. Five genera and about 20 species are known in North America. Of the three species in the genus ***Pelocoris*** (not illustrated), two extend into the east, but the other species in the family are all western, mainly in the southwest. ***Ambrysus mormon*** ranges from California and New Mexico to southern Montana and South Dakota. Look for it in streams with pebbly bottoms, or in marshes and lakes, including saline pools.

TOAD BUGS (Gelastocoridae) are common on the muddy and sandy shores of ponds, lakes, streams, rivers, and springs. Their bumpy texture makes them well camouflaged until they move. They are predatory on small insects and mites, spied with those big bulging eyes. The genus ***Gelastocoris*** includes two of the most common and widespread of the seven species known for North America.

backswimmers
(*Notonecta* sp.)

water boatmen

Graptocorixa

Corixa

Hesperocorixa

Sigara

creeping water bug
(Ambrysus mormon)

toad bugs
(*Gelastocoris* sp.)

WATER STRIDERS (family Gerridae) are also known as pond skaters. These are the unsinkable insects seen exploiting the surface tension of water on ponds, rivers, and streams. The front legs are short, built for grabbing prey, but the middle and hind legs are extremely long. These two pairs of legs are covered in very fine hairs that are not easily saturated, allowing the insect to rest and skate on top of the water without breaking the surface film. What appears to be a water strider with eight visible legs will actually be a mating pair. Males often guard their mates for long periods of time, ensuring no other male will usurp their genetic investment. Female water striders lay their eggs at or just below the surface of the water on floating objects. This family includes about 45 species in North America.

Gerris is the most familiar genus, with nearly 20 species in North America, some of them very common. Some have both winged and wingless forms. ***Gerris remigis*** is the largest and most familiar of our species, often seen in large numbers on ponds or on quiet eddies of slow-moving streams. These striders move with quick, erratic actions, darting forward and then skating in place, sometimes jumping in the air and landing again without breaking the water's surface tension. About nine species in the genus ***Trepobates*** are found north of Mexico. These are much smaller, and the wingless adults may be mistaken for nymphs of *Gerris* species.

Most remarkable are the more than 40 species of ***Halobates*** (not illustrated). The only truly marine insects, these striders live on the surface of the open ocean, often miles from shore. They are mostly tropical, but one species *(Halobates sericeus)* may occur close to our Pacific Coast.

RIFFLE BUGS (Veliidae) (not illustrated) are also known as small water striders or broad-shouldered water striders. All are very small (less than 5 mm or one-quarter inch long) and usually wingless, and their bodies are widest at the "shoulders," tapering to a narrower abdomen. They are usually dark with silvery or orange markings, and they often live in swarms on the surface of the water near ripples in fast-moving streams.

SHORE BUGS (Saldidae) are found along the banks of rivers, streams, bogs, and springs, but are most abundant in the mudflats of salt marshes. They run rapidly and, like tiger beetles (p. 130), are quick to take short flights. They prey on smaller insects. Members of the genus ***Salda,*** with 13 species in North America, seem to favor marshes and wet meadows among grasses and sedges. Look for members of the genus ***Saldula*** along the shores of rivers, streams, and lakes over most of the continent.

MARSH TREADERS or WATER MEASURERS (Hydrometridae) might suggest aquatic versions of stilt bugs (p. 118). Look for them on the surface of small ponds with lots of emergent vegetation, where they prey on other small insects. ***Hydrometra*** is the only genus in North America, with seven species. ***Hydrometra martini*** is the most widespread of these, ranging in the east, with scattered records in the west.

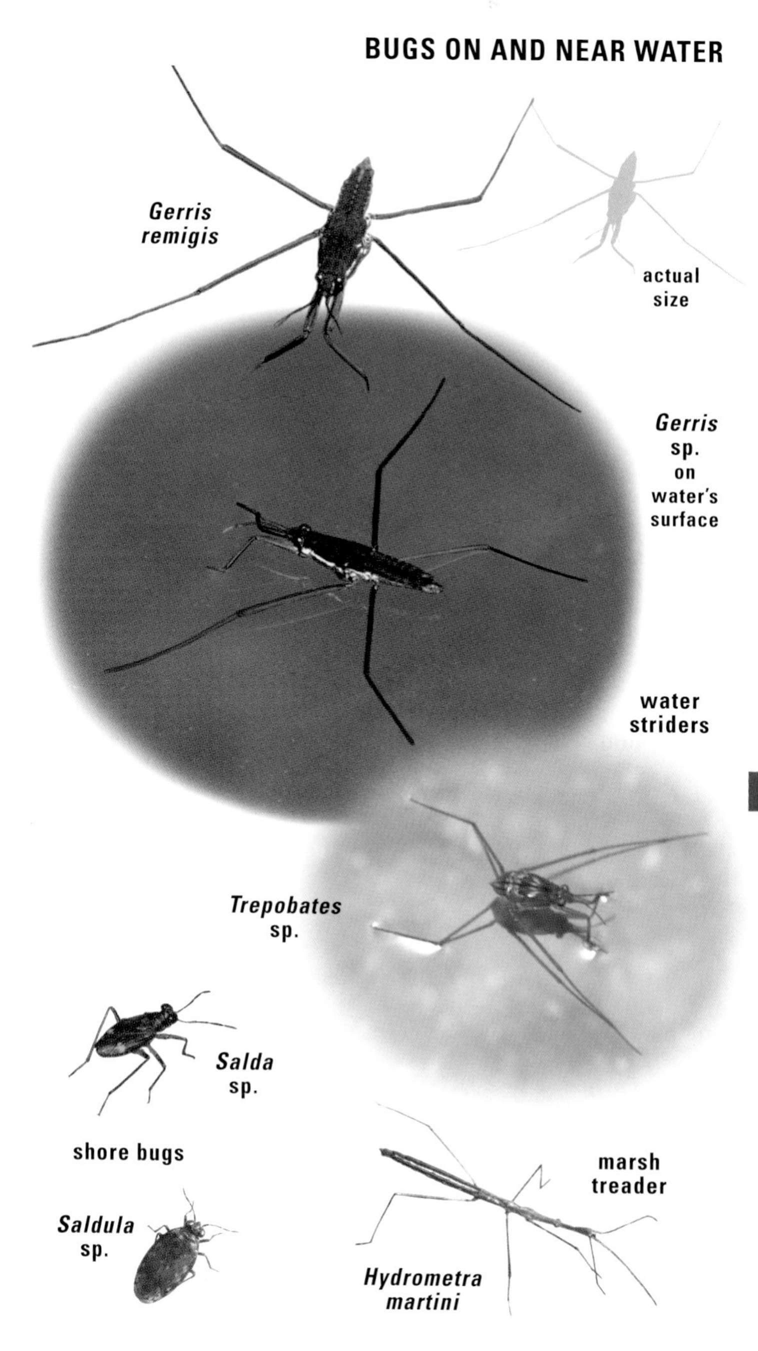
Gerris remigis
actual size
Gerris sp. on water's surface
water striders
Trepobates sp.
Salda sp.
shore bugs
Saldula sp.
marsh treader
Hydrometra martini

PLANT BUGS

(family Miridae) are incredibly diverse and often abundant. Several genera are holarctic, meaning they occur in Europe and temperate Asia as well as North America. The vast majority are plant feeders, often specific to a particular host tree or plant. A few are predatory. They exhibit a wide array of shapes, colors, and sizes, though most are small and delicate. A special crease in the front wing, called the cuneus, helps separate these insects from similar true bugs.

Lygus is a large genus of nearly identical species (at least 34 in North America). The **Tarnished Plant Bug** *(Lygus lineolaris)*, highly variable in color and markings, is abundant throughout eastern and central regions. An ivory V mark on the scutellum (back) helps identify the genus.

The genus ***Prepops*** includes 35 species in North America, many of them widely distributed. ***Prepops fraternus*** is common on sumac. Individual specimens vary widely in color pattern. ***Cylapus tenuicornis*** is a bizarre species found on the trunks of trees, where it apparently feeds on fungi. It ranges in the northeast and north-central U.S. ***Phytocoris*** is perhaps the most diverse and complicated genus of mirids, with approximately 220 species known for North America. Many species are predatory on aphids and related small insects. Some species come to lights at night.

The **Meadow Plant Bug** *(Leptopterna dolabrata)* is abundant in pastures and meadows across much of the northern U.S. Females may be wingless. This species can sometimes be injurious to bluegrass. ***Stenotus binotatus*** is a very common insect on grasses throughout the northern and central U.S. and southern Canada. It was probably introduced from Europe.

The genus ***Lopidea*** includes at least 125 species in North America. Some appear to be specific feeders on sagebrush, goldenrod, or other plants, while others are generalist feeders. The genus ***Collaria*** includes three species in North America. Look for them on grasses. ***Pseudoxenetus regalis*** is apparently an ant mimic, one of two in its genus. It is found on oak trees in the southern U.S.

The **Four-lined Plant Bug** *(Poecilocapsus lineatus)* feeds on a variety of plants from the Great Plains eastward, as well as northwest to British Columbia. Four other species in the genus occur north of Mexico. ***Oncerometopus nigriclavus*** is one of eight species in its genus in North America, most of them occurring in the western U.S. ***Metriorrhynchomiris dislocatus,*** one of three North American species in the genus, ranges in the eastern U.S. and adjacent Canada, west to Minnesota and Texas. Look for it from April to July on false Solomon's-seal and wild geranium, especially in shady habitats. There are numerous color variations.

The **Garden Fleahopper** *(Halticus bractatus)* is a tiny plant bug, at only 2 mm long, but quite common over the eastern and central U.S. Males are winged, while females are flightless with thickened forewings lacking a membranous tip. Both sexes have enlarged hind legs, making them good jumpers. Look for them on various legumes, even the white clover in your lawn. There are two other North American species in the genus.

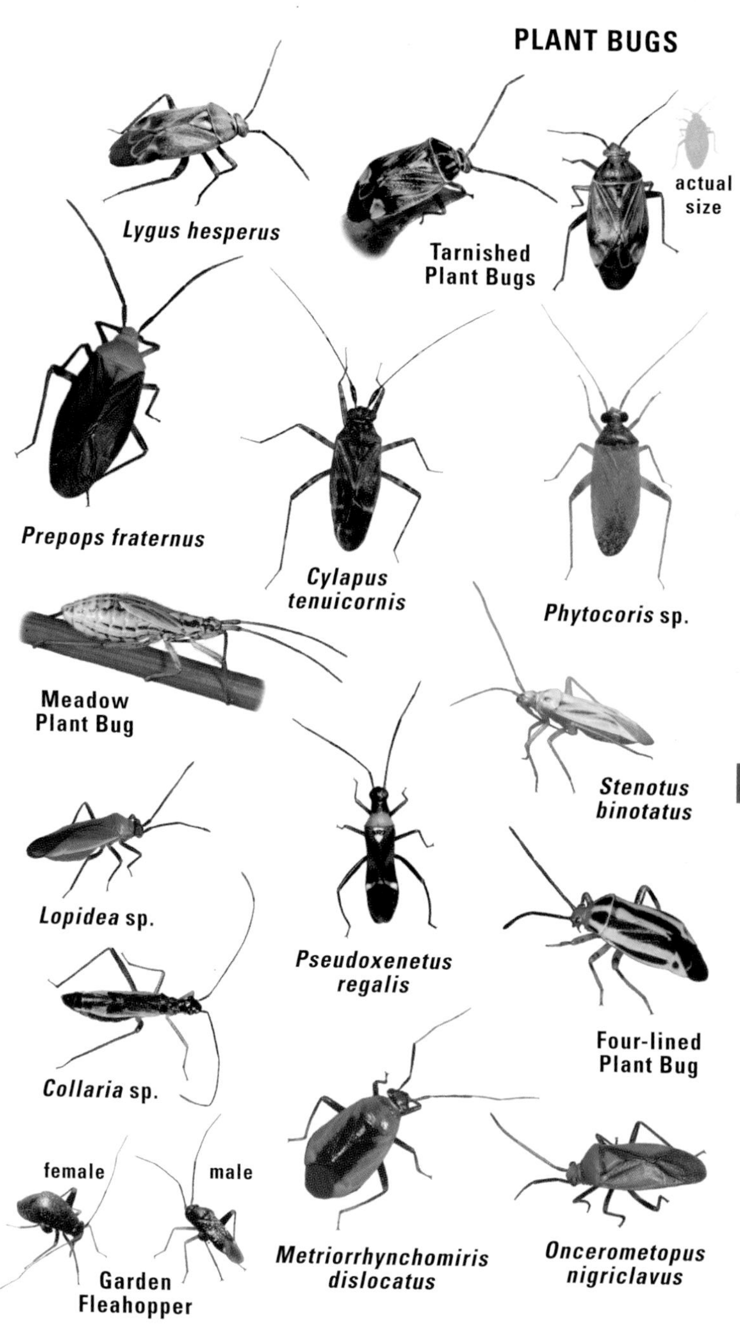

Lygus hesperus

Tarnished Plant Bugs

Prepops fraternus

Cylapus tenuicornis

Phytocoris sp.

Meadow Plant Bug

Stenotus binotatus

Lopidea sp.

Pseudoxenetus regalis

Four-lined Plant Bug

Collaria sp.

Garden Fleahopper

Metriorrhynchomiris dislocatus

Oncerometopus nigriclavus

LACE BUGS (family Tingidae), tiny but pretty, are slow-moving insects that feed on the undersides of leaves or on stems, often in groups. More than 150 species occur in North America. Their small size, lacelike structure and partially transparent wings may make them hard to spot against the foliage on which they feed. However, their feeding activities on the undersides of leaves may produce whitish spots on the upper sides, and their droppings (black fecal pellets) on the leaves may be noticeable with a close look. ***Corythuca*** is a large genus with over 50 North American species. Many of these are very host specific, meaning they feed on only one kind of plant or tree. **Oak Lace Bug** *(Corythuca arcuata),* for example, is typically on oaks, although it is sometimes found on leaves of maple, apple, or other trees. The **Eggplant Lace Bug** *(Gargaphia solani)* is one of 15 species in its genus. It is widespread but most common in the southern half of the U.S. Adult females have been observed guarding their eggs and the young nymphs that hatch from them, "herding" the youngsters to safety in the face of threats from predators.

MINUTE PIRATE BUGS or FLOWER BUGS (Anthocoridae) are small enough to be overlooked but may be noticed for their surprisingly painful bite, as they frequently alight on people. Normally, most are predatory on other insects, but a few are plant feeders. There are more than 20 genera and about 70 species in North America. ***Orius insidiosus*** is abundant east of the Rockies. Six other species in its genus occur on the continent, four of them on the Pacific Coast.

DAMSEL BUGS (Nabidae) are also predatory and are often abundant in open fields. Look for them at lights at night as well. ***Nabicula subcoleoptrata*** is a mimic of ants, usually wingless, in meadows from the Atlantic Coast to British Columbia and south to Kansas. ***Nabis*** is a genus with about 13 species in North America.

BED BUGS (Cimicidae) include blood-feeding parasites of birds and bats as well as humans. All are wingless and small. Females suffer "traumatic insemination" by males, whose genitals are essentially a blade used to stab the abdominal cuticle of the female, sperm being released directly into her body cavity. The genus ***Cimex*** includes eight species found in the U.S., but the famous one is the **Bed Bug** *(Cimex lectularius),* which afflicts humans and sometimes other mammals and birds. Bed Bugs hide away in crevices by day, creeping out at night to make small bites in the skin and suck small amounts of blood. These bites are not immediately painful, so they may not be noticed until later when they become irritating. Populations of this insect are on the rise. While some blame restrictions on pesticide use in housing structures, others cite the increase in popularity of thrift stores where infested beds are pawned. Also in this family is the **Swallow Bug** *(Oeciacus vicarius),* an often epidemic pest of Cliff Swallows that can account for up to 50 percent of the mortality of nestlings.

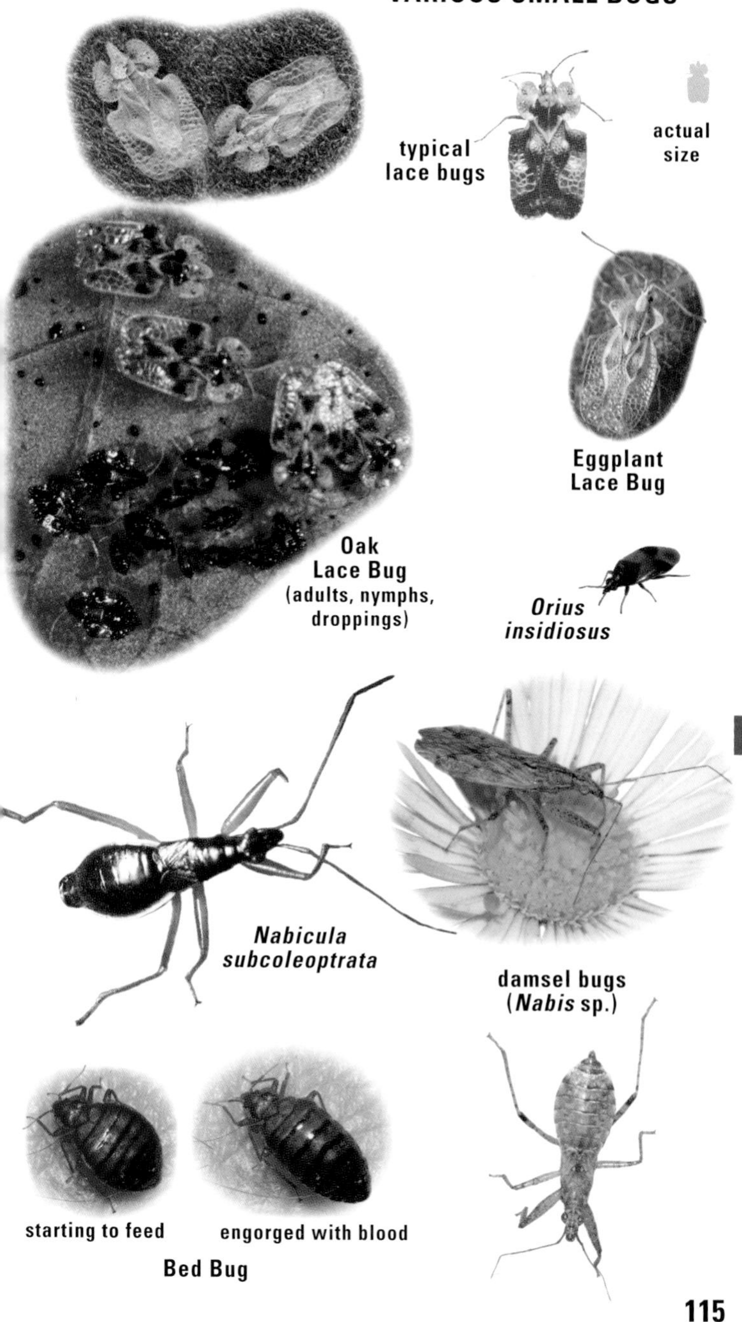
typical
lace bugs
actual
size
Eggplant
Lace Bug
Oak
Lace Bug
(adults, nymphs,
droppings)
Orius
insidiosus
Nabicula
subcoleoptrata
damsel bugs
(*Nabis* sp.)
starting to feed
engorged with blood
Bed Bug

ASSASSIN BUGS AND AMBUSH BUGS

(family Reduviidae) live up to their name by slaying other insects, often claiming victims larger than themselves. Defensive bites of assassin bugs are excruciating. The short beak rests in a ridged groove on the "chest." Most assassins make a squeaking noise by rasping the beak against those ridges. Many species come to lights at night. There are 110 species in North America.

The **Wheel Bug** *(Arilus cristatus)* prowls foliage by day and also comes to lights after dark. Nymphs lack a crest and have a bright red abdomen they hold erect. Abundant in the south-central states, Wheel Bugs range north to New York and Illinois. Members of the genus ***Apiomerus*** are known as bee assassins for their habit of lying in wait at flowers. Sticky hairs on the front legs help them grab prey. The **Yellow-bellied Bee Assassin** *(Apiomerus flaviventris)* is common in the southwest. ***Apiomerus crassipes*** occurs over much of the U.S. ***Microtomus purcis*** is a spectacular insect not uncommon at lights in the east, south to Florida and Texas. By day it hides under bark. The seven species of ***Zelus*** are slender, lanky assassins occurring on foliage, occasionally coming to lights. Like bee assassins, these bugs have sticky hairs on the front legs, the better to hold prey.

The genus ***Triatoma*** includes 12 U.S. species of kissing bugs, or bloodsucking conenoses. They normally thrive in association with wood rats but sometimes bite sleeping humans on the face, the victim awakening to a fat lip or itchy welt. South of the border, these insects are vectors of Chagas' disease, caused by a microbe that is eliminated in the conenose's feces and then scratched into the bite wound by the unwitting victim. Luckily, our species are potty trained and poop after they leave the host.

The **Masked Hunter** *(Reduvius personatus)*, native to Europe and Africa, now occurs across most of North America. Nymphs are covered in sticky hairs and mask themselves with dust, lint, and other debris. They invade homes, feeding almost solely on bed bugs (previous page), and they frequent swallow colonies, feeding on Swallow Bugs.

The two species of ***Melanolestes*** are nocturnal. Females are wingless; males are common, swift runners and strong fliers often seen at lights. The widespread **Black Corsair** *(Melanolestes picipes)* was responsible for the "kissing bug scare" of 1899 in the northeastern U.S.

The two species of ***Rhiginia*** are active, alert, and strong fliers. ***Rhiginia cruciata*** is found during the day in forested areas of the east. The genus ***Pselliopus*** includes six species of zebra-striped bugs. ***Pselliopus barberi*** is common over most of the east on foliage and flowers in spring and fall, overwintering under bark or stones. The three members of the genus ***Rasahus*** are the true corsairs. They are occasionally seen at lights, hiding under stones and debris by day. ***Narvesus carolinensis*** is common at lights, especially in the south, ranging north to New Jersey and west to Arizona.

The 22 species of **AMBUSH BUGS (subfamily Phymatinae)**, formerly placed in their own family, sit motionless on flowers, waiting to clutch other insects with their hooked front legs. Tiny and well camouflaged, they often take prey larger than themselves, including bees and large flies.

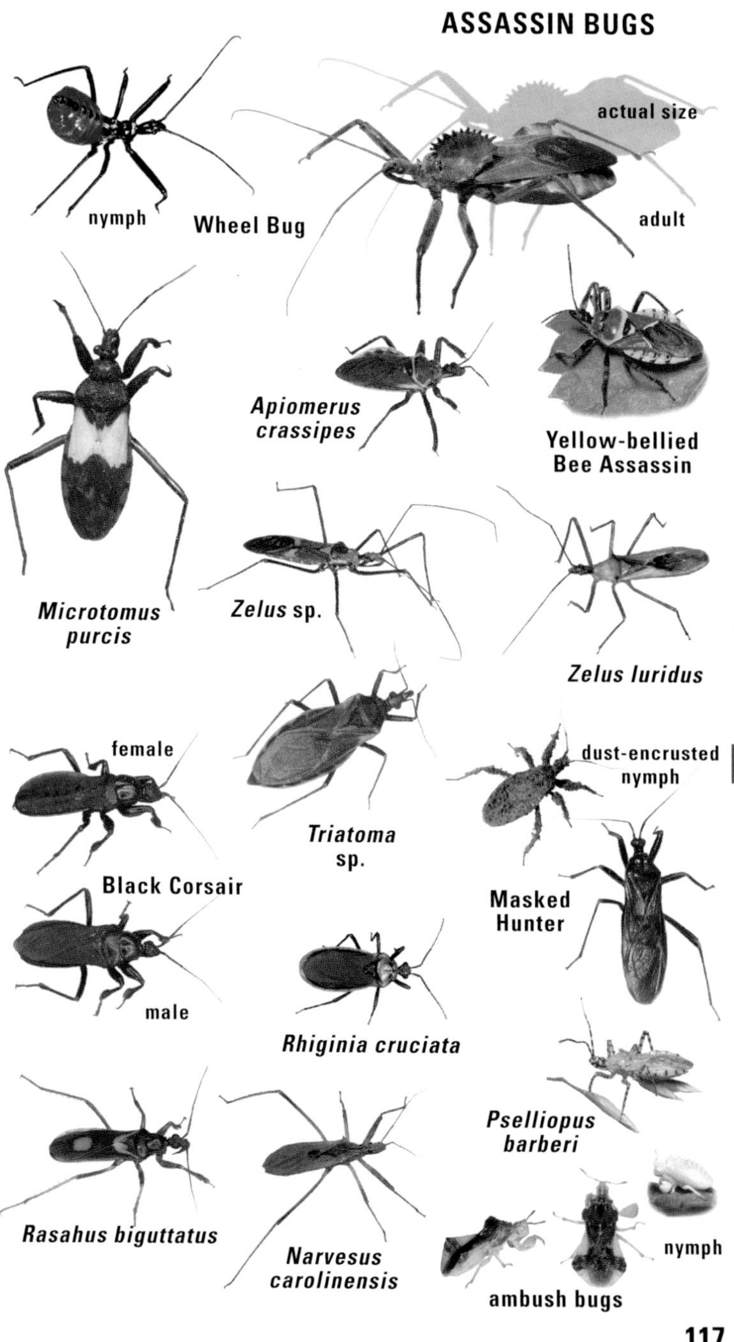

actual size
nymph
Wheel Bug
adult
Apiomerus crassipes
Yellow-bellied Bee Assassin
Microtomus purcis
Zelus sp.
Zelus luridus
female
Black Corsair
male
Triatoma sp.
dust-encrusted nymph
Masked Hunter
Rhiginia cruciata
Pselliopus barberi
Rasahus biguttatus
Narvesus carolinensis
nymph
ambush bugs

SEED BUGS were long regarded as making up one family **(Lygaeidae),** but have recently been split among ten families. For the sake of simplicity, we are mostly following the original classification here.

The **Large Milkweed Bug** *(Oncopeltus fasciatus)* is common on its host plant over much of North America. The **Small Milkweed Bug** *(Lygaeus kalmii)* is even more numerous and widespread. There are five other species in the genus ***Lygaeus,*** collectively occurring throughout most of the continent. They are often abundant and they often stray far from milkweeds. The **White-crossed Seed Bug** *(Neacoryphus bicrucis)* is a bright and conspicuous insect likely to be found over most of the U.S. The related ***Neacoryphus lateralis*** is more subdued in color and often comes to lights in the western U.S., east to the Great Plains.

Members of the genus ***Geocoris*** are big-eyed bugs with at least 19 species found north of Mexico. They are predatory on other small insects and as a result may be highly valued in agricultural systems.

The **Chinch Bug** *(Blissus leucopterus)* is a well-documented pest of wheat, rye, barley, oats, corn, and sorghum in the midwest U.S. A closely related form or subspecies, often called **Hairy Chinch Bug** *(Blissus leucopterus hirtus),* is a pest on turfgrasses of lawns, causing damage that looks like patchy drought. There are more than a dozen other species in the genus in North America, with a collectively widespread distribution.

The **False Chinch Bug** *(Nysius raphanus)* is common in the west, sometimes occurring in large numbers on low plants in open habitats. During the hottest period of summer it may seek sheltered spots, and at this time it may invade houses or other buildings. At least eight other species in the genus ***Nysius*** occur collectively over much of the continent. General feeders on broad-leaved plants, they sometimes have population outbreaks and cause damage to crops or garden plants.

Neortholomus scolopax (formerly *Ortholomus scolopax)* is very similar to *Nysius* in appearance and can be very abundant in dry weeds in open fields across most of the U.S. and southern Canada. There are four additional species in the genus in North America, found in the west and southwest. ***Myodocha serripes*** is easily recognized by its long "neck." Look for it at lights east of the Rockies in the U.S. and adjacent southern Canada.

STILT BUGS (family Berytidae) are delicate, slender, long-legged creatures found on plants. They may be confused with thread-legged assassin bugs (below) but do not have viselike front legs. ***Neides muticus*** is most common in weedy fields in the northern U.S.

THREAD-LEGGED BUGS form a distinctive subfamily **(Emesinae)** of the assassin bug family (previous page). These slender bugs suggest walkingsticks (p. 66) but have front legs modified for grasping the smaller insects on which they feed. They may be found under loose bark, in brush piles, or in abandoned buildings, sometimes stalking prey trapped in spiderwebs. Members of the genus ***Emesaya*** may be more than 30 mm long, but most (such as species of ***Empicoris)*** are smaller.

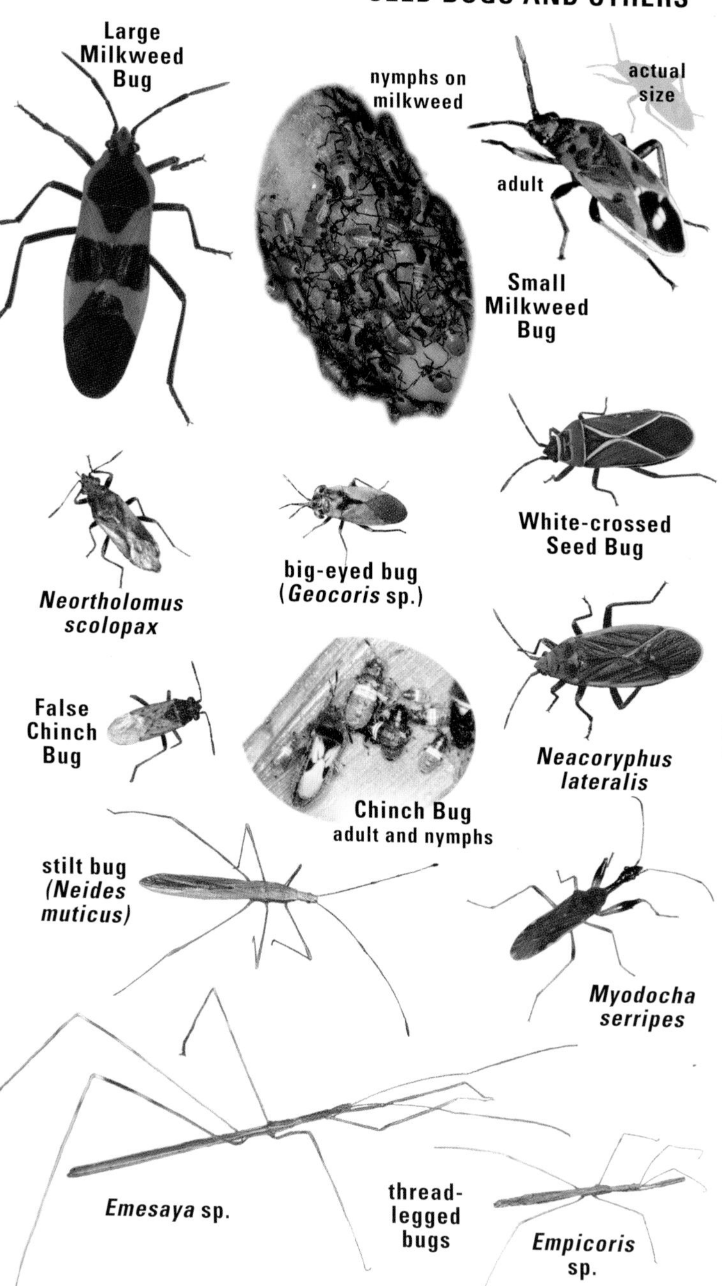
Large Milkweed Bug
nymphs on milkweed
actual size
adult
Small Milkweed Bug
White-crossed Seed Bug
Neortholomus scolopax
big-eyed bug (*Geocoris* sp.)
Neacoryphus lateralis
False Chinch Bug
Chinch Bug adult and nymphs
stilt bug (*Neides muticus*)
Myodocha serripes
Emesaya sp.
thread-legged bugs
Empicoris sp.

LARGID BUGS AND OTHERS

These four small families are not closely related to each other.

LARGID BUGS (family Largidae), despite the name, are not particularly large but can be conspicuous at times, as they often feed in groups. The genus ***Largus*** includes at least eight species, mostly occurring in the southern and southwestern states. The **Bordered Plant Bug** *(Largus cinctus)* ranges from Washington to the southwestern states. ***Largus succinctus*** ranges from New York to Florida and west to Arizona. ***Arhaphe carolina*** is a spectacular flightless mimic of tiger beetles (p. 130), found from North Carolina to Arizona. There are three or four other species in the genus, collectively ranging from Kansas to the southwest.

COTTON STAINERS or RED BUGS (Pyrrhocoridae) are mostly tropical insects, with seven species in the genus ***Dysdercus*** in the southern U.S. The **Cotton Stainer** *(Dysdercus suturellis)* feeds on citrus, hibiscus, and several other plants besides cotton in the southeast U.S. When feeding on cotton, it damages more than it eats, leaving a stain on the fibers. Other species of ***Dysdercus,*** lacking the Cotton Stainer's taste for crops, are less destructive but equally colorful.

SCENTLESS PLANT BUGS (Rhopalidae) are so called because they do not have the stink glands possessed by most other true bugs. North America has about 35 species in this family, all feeding on plants. The **Box Elder Bug** *(Boisea trivittata)* and the **Western Box Elder Bug** *(Boisea rubrolineata)* are conspicuous species. Both congregate in large numbers to hibernate, sometimes choosing human abodes in which to do so. The **Red-shouldered Bug** *(Jadera haematoloma)* is widespread and often very common in the south, including the southwest. In the southeastern states it is often associated with the introduced goldenrain tree *(Koelreuteria),* but it also feeds on various other plants. The **Hyaline Grass Bug** *(Liorhyssus hyalinus)* resembles the False Chinch Bug (previous page). It is found chiefly in the southern U.S. with scattered records north to Wyoming and Massachusetts. The genus ***Harmostes*** includes eight species, most diverse in the southern U.S. ***Harmostes reflexulus*** is common across the entire continent but highly variable in color. Look for it on low plants in weedy fields. The genus ***Arhyssus*** includes 14 species with a collectively transcontinental distribution, most diverse in the west.

FLAT BUGS (Aradidae) are paper thin and often wide-bodied, with wings that are usually not large enough to cover the whole abdomen. They feed on fungi under bark and in decaying wood. Their unique mouthparts consist of stylets coiled like a watch spring when not in use. ***Aradus*** is our most diverse genus, including over 75 species out of the nearly 100 of this family in North America. ***Neuroctenus*** is a genus of four species occurring in the east, including ***Neuroctenus simplex,*** common under the bark of oak and beech.

Arhaphe cicindeloides

Bordered Plant Bug

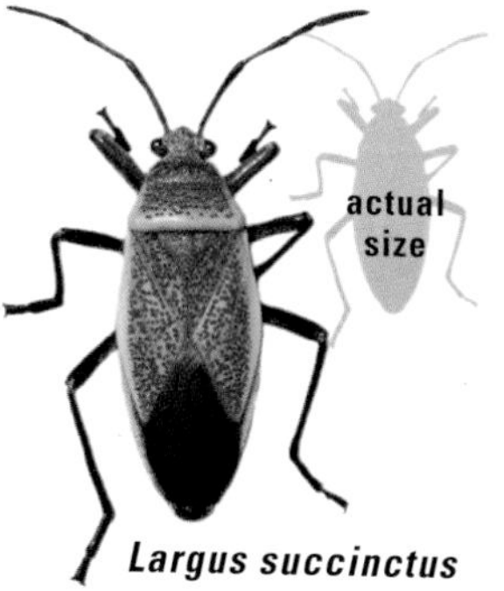

Largus succinctus

actual size

Cotton Stainer

nymph

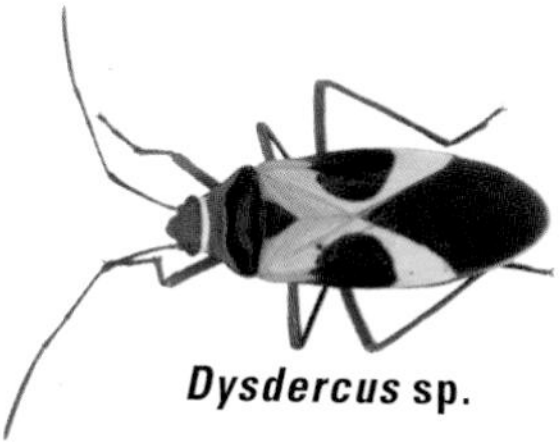

Dysdercus sp.

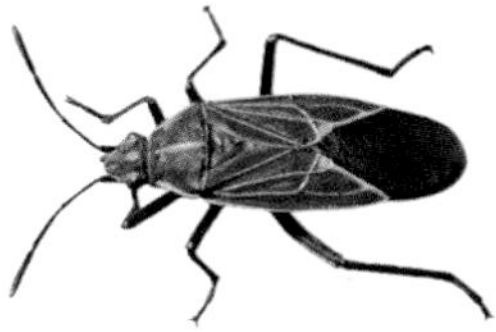

adult and nymph

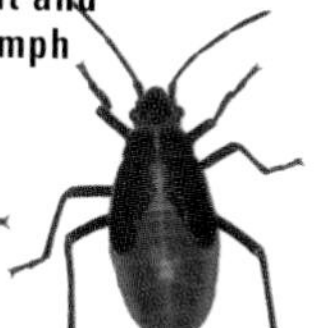

Box Elder Bug

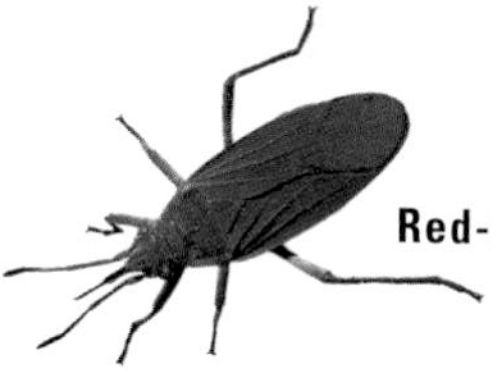

Western Box Elder Bug

Red-shouldered Bug

Harmostes reflexulus

Hyaline Grass Bug

Neuroctenus simplex

Arhyssus sp.

Aradus sp.

LEAF-FOOTED BUGS AND BROAD-HEADED BUGS

LEAF-FOOTED BUGS OR SQUASH BUGS (family Coreidae) are large plant-feeding bugs. Many of them sport leaflike flanges on the hind legs. The males of some species have thickened spiked "thighs," used in combat over females. They fly well, with a loud droning sound. When molested, they emit a sharp odor from scent glands. There are about 80 species in North America, most occurring in the south.

Members of the genus ***Leptoglossus*** are the typical leaf-footed bugs. Most of our 11 species are 18–20 mm long and feed on a variety of fruits, vegetables, and seeds. ***Leptoglossus oppositus*** is common on the pods of catalpa trees in the east, ranging southwest to Arizona. The **Western Conifer Seed Bug** *(Leptoglossus occidentalis)* is common along the West Coast, east to Maine, and appears to be expanding its range. The **Eastern Leaf-footed Bug** *(Leptoglossus phyllopus)* has a bright white band across the forewings. It is most abundant in the southeastern U.S. but has been recorded as far north as New York and west to Missouri. The three species of the genus ***Thasus,*** collectively referred to as **Giant Mesquite Bugs,** are hard to miss at 35–40 mm long. They also hang out in large groups on the food plants in Arizona and New Mexico. This social behavior helps reinforce their bright warning colors. Adults may come to lights at night.

Chariesterus antennator, one of three members of its genus in the U.S., occurs on various flowering weeds east of the Rockies, mostly in the southeast. The genus ***Mozena*** is mostly tropical, with about half a dozen species in the southern U.S., often associated with mesquites or acacias. The **Opuntia Bug** *(Chelinidea vittiger)* is widespread in the west and south, feeding on prickly-pear cactus. The genus ***Acanthocephala*** includes five species in the U.S. These large insects are often abundant in forested areas, flying noisily among the trees and shrubs. ***Acanthocephala femorata*** is a common species across the southern U.S.

Euthochtha galeator frequents hedgerows, fields, and forest edges throughout the eastern U.S., west to the Great Plains. Our six species of the genus ***Anasa*** are collectively known as squash bugs. The one that is officially "the" **Squash Bug** *(Anasa tristis)* is a pest of gourds, pumpkins, and other types of squash over much of North America, most commonly in the south. The genus ***Hypselonotus*** includes two species in North America.

BROAD-HEADED BUGS (family Alydidae) are active and alert. As nymphs, many species mimic ants. Twelve genera are known in North America. ***Alydus*** is the most frequently encountered genus, with at least seven species north of Mexico. Adults can look and behave like wasps, flicking their wings and bobbing their antennae just like the real McCoy! ***Megalotomus quinquespinosus,*** known by the white mark on its narrow dark antennae, occurs coast to coast over much of the U.S. and southern Canada. It has been found on lupines, poison oak, clover, and other plants along forest edges. The two species of ***Stenocoris*** in the southeastern U.S. are atypical and easily mistaken for large versions of stilt bugs (p. 118).

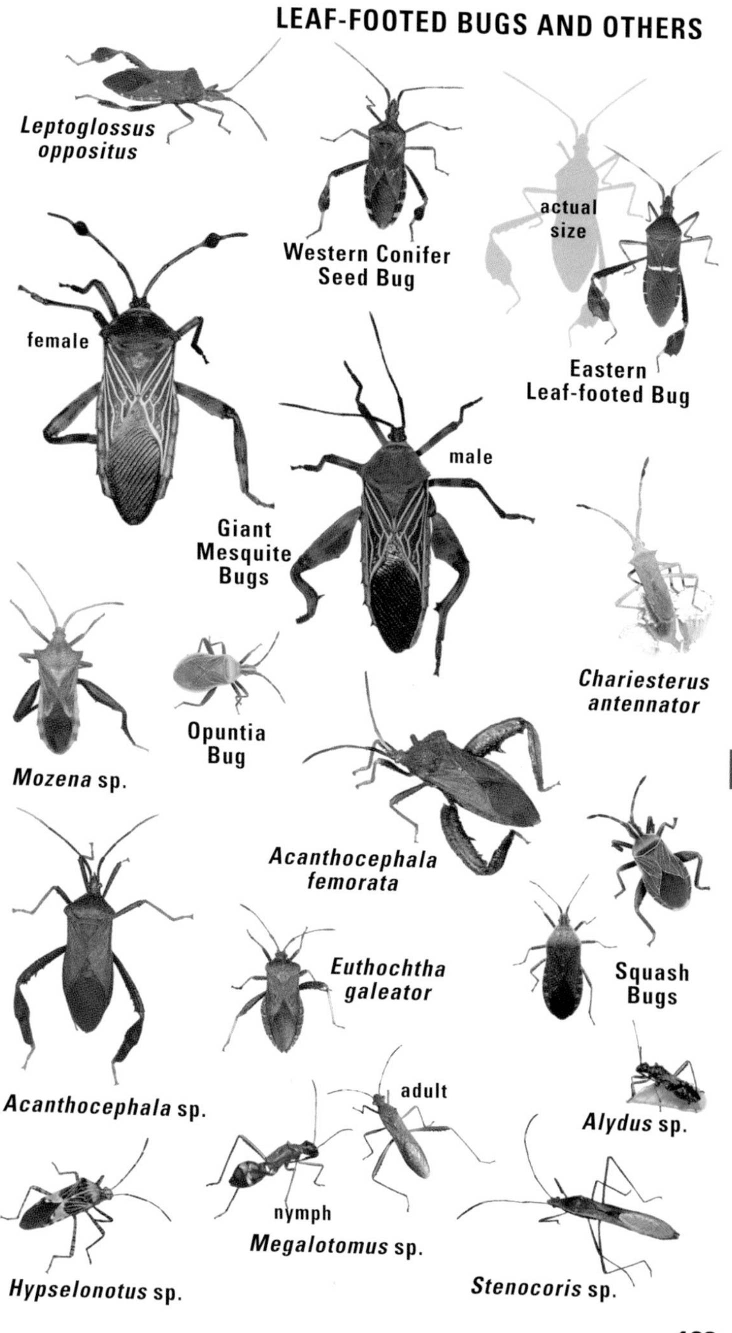

Leptoglossus oppositus
Western Conifer Seed Bug
actual size
Eastern Leaf-footed Bug
female
male
Giant Mesquite Bugs
Chariesterus antennator
Mozena sp.
Opuntia Bug
Acanthocephala femorata
Squash Bugs
Euthochtha galeator
Acanthocephala sp.
adult
Alydus sp.
nymph
Megalotomus sp.
Hypselonotus sp.
Stenocoris sp.

STINK BUGS

(family Pentatomidae) include some notorious crop and garden pests, and they smell bad, too. Thoracic glands produce volatile aromatic compounds sure to repel all but the most desperate predators. A few stink bugs are hunters themselves. There are 250 species in North America.

Euschistus includes our most abundant stink bugs, with at least 19 species. Feeding on a great variety of plants, they are occasionally pests on crops; they are sometimes attracted to lights at night. The genus ***Banasa*** includes 11 species in North America. Collectively they range over most of the continent, feeding on a variety of trees and shrubs and visiting lights at night. ***Thyanta*** is a confusing genus with at least 14 North American species. Many resemble species of *Banasa.* Those in the west and south are usually less than 8 mm long, while northern and eastern species tend to be larger. The two species of ***Mecidea*** are unusually slender stink bugs, feeding mostly on grasses but also on spinach, cotton, and wheat. ***Brochymena*** includes 22 species resembling bark in color and texture. Collectively they range across most of the continent on various trees. At least some species hibernate as adults beneath bark. ***Mormidea*** includes four species, most common in the south. ***Mormidea lugens*** ranges north to Ontario and west to Wyoming. The white triangle outlining the scutellum helps identify it.

The genus ***Holcostethus*** includes seven North American species. They can be quite common in weedy fields. ***Chlorochroa*** is a confusing genus of 19 species. The **Conchuela** *(Chlorochroa ligata),* common in the northwest, ranges to the Rockies and south to Texas. **Say's Stink Bug** *(Chlorochroa sayi)* occurs from Idaho and Kansas to California. Both are pests of cotton, alfalfa, and other crops. The **Rice Stink Bug** *(Oebalus pugnax)* is sometimes a serious pest of that crop and other cereals east of the Rockies. The forward-pointing shoulder spines are distinctive for this genus. The two species of ***Proxys*** feed on plants but may also be predators. Our two species of ***Coenus*** are identified by the large broadly rounded scutellum. Look for them on a variety of plants in weedy fields. The six small species of ***Cosmopepla*** feed on a wide variety of plants. Look for ***Cosmopepla lintneriana*** on mullein throughout North America.

Menecles insertus feeds on deciduous trees throughout the east, southwest to California. The **Harlequin Bug** *(Murgantia histrionica),* native to Mexico, now ranges to New England and Colorado. It is a major pest of cabbage and related crucifers, but it feeds on many other plants. The **Green Stink Bug** *(Acrosternum hilare)* is abundant in most of the U.S. and southern Canada. Many plant species are on its menu, and it is often attracted to lights. The **Southern Green Stink Bug** *(Nezara viridula)* is an introduced species now common in our southeast. This is a major pest of soybeans. ***Zicrona caerulea*** resembles the metallic leaf beetles it preys on. It occurs in northern and western North America and in Europe.

Elasmostethus cruciatus feeds on alder in northern regions. ***Elasmucha*** species, often on birch, are notable for the fact that adult females may tend and guard their young, a rare behavior in bugs. These two genera are sometimes placed in a separate family, **Acanthosomatidae**.

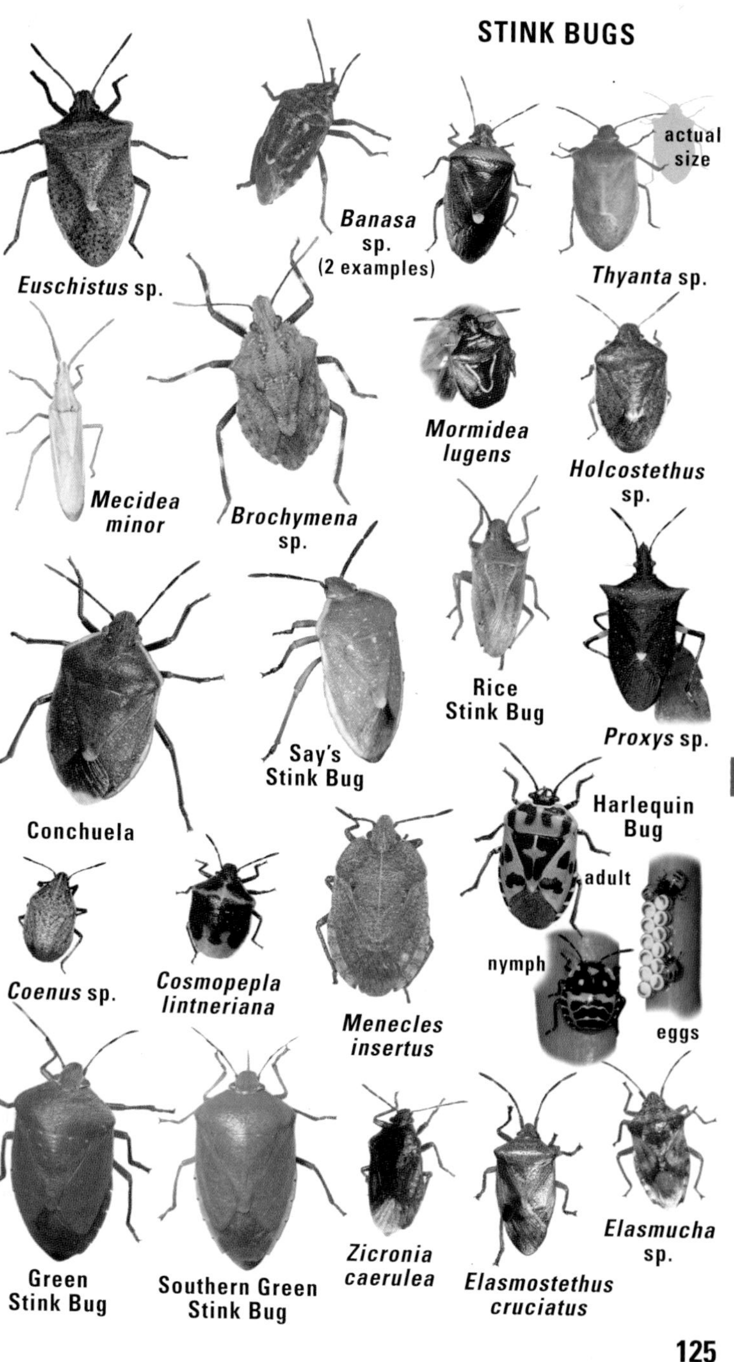
actual
size
Banasa
sp.
(2 examples)
Euschistus sp.
Thyanta sp.
Mormidea
lugens
Holcostethus
sp.
Mecidea
minor
Brochymena
sp.
Rice
Stink Bug
Proxys sp.
Say's
Stink Bug
Conchuela
Harlequin
Bug
adult
nymph
eggs
Coenus sp.
Cosmopepla
lintneriana
Menecles
insertus
Green
Stink Bug
Southern Green
Stink Bug
Zicronia
caerulea
Elasmostethus
cruciatus
Elasmucha
sp.

STINK BUGS AND OTHERS

The first few species below belong to the stink bug family, introduced on the previous page.

Podisus is a genus of stink bugs with nine North American species. There are a few records of plant-feeding by nymphs, but adults are predators. The **Spined Soldier Bug** *(Podisus maculiventris)* is armed with spikes on its "shoulders." Abundant throughout most of North America, it comes to lights. The genus ***Perillus*** includes six predatory species. The **Two-spotted Stink Bug** *(Perillus bioculatus)* is highly variable in color and pattern. It ranges over most of North America, spending the winter in the adult stage. ***Perillus circumcinctus*** is less common, occurring in the northeast quadrant of the U.S.

Stiretrus anchorago is a strikingly patterned predatory species that feeds mostly on caterpillars and on the larvae of leaf beetles. It is widespread in the east, from Ontario south. It varies in color from black and ivory to deep metallic blue and red. ***Euthyrhynchus floridanus*** is a beautiful large stink bug occurring in the southeastern U.S., north to Pennsylvania and southern Illinois. Adults and nymphs are predatory on a variety of insects, including other stink bugs.

SHIELD-BACKED BUGS (family Scutellaridae) are often lumped with stink bugs but are recognized by the enlarged scutellum. Large and triangular on most hemipterans, here it covers the entire back of the insect, concealing the wings. There are about 33 species in North America. The five in the genus ***Homaemus*** are mostly associated with grasses in a variety of habitats. ***Orsilochides guttata*** (formerly *Chelysoma guttatum)* is an eastern species, fairly common in Florida.

EBONY BUGS (Thyreocoridae or Corimelaenidae) are tiny, black spherical insects often seen on flowers in the parsley family. They can be mistaken for beetles at first glance, and it is difficult to separate the species. About 30 species, in three genera, collectively range over the entire U.S. and southern Canada. The genus ***Corimelaena*** includes 18 species in North America, some of them very common and widespread.

BURROWER BUGS (Cydnidae) resemble small spiny versions of stink bugs. As the name suggests, they are mostly subterranean, sucking sap from plant roots. They often surface, however, and are frequently attracted to lights at night. About 28 species are known in North America.

The genus ***Cyrtomenus*** includes two species, both of which are large and robust in contrast to the usually flattened bodies of most burrower bugs. They are often seen at lights in the southern states. ***Sehirus cinctus*** is actually more common aboveground, on nettles, mints, and grasses across much of Canada and the U.S. Five species of the genus ***Pangaeus*** occur here. ***Pangaeus bilineatus*** is the burrower bug most commonly encountered east of the Great Plains, and it ranges southwest to California. It can be a minor pest of cotton seedlings, pepper seed beds, spinach, and peanuts.

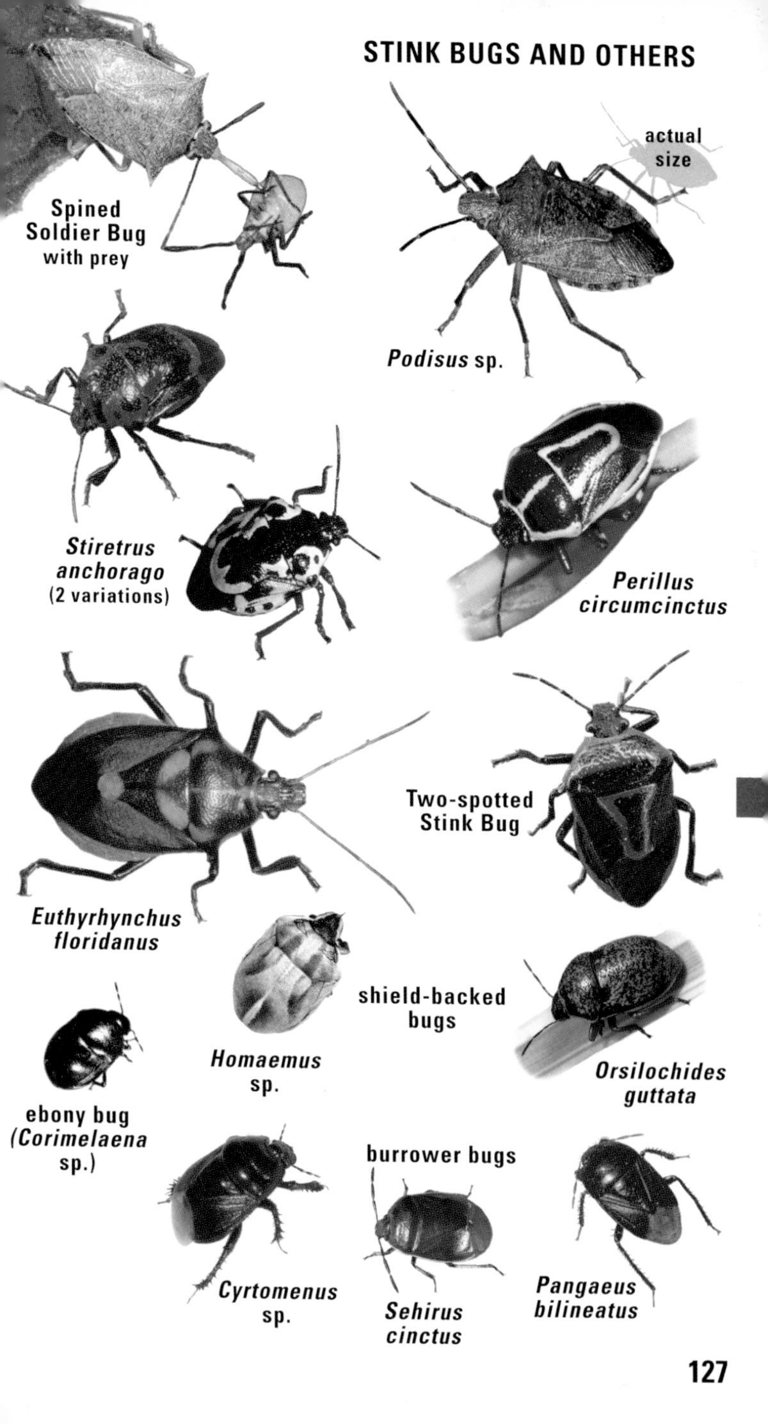

Spined Soldier Bug with prey

Podisus sp.

Stiretrus anchorago (2 variations)

Perillus circumcinctus

Euthyrhynchus floridanus

Two-spotted Stink Bug

shield-backed bugs

Homaemus sp.

Orsilochides guttata

ebony bug (*Corimelaena* sp.)

burrower bugs

Cyrtomenus sp.

Sehirus cinctus

Pangaeus bilineatus

BEETLES (order Coleoptera) are by far the most diverse organisms on the planet. In fact, out of all the species of living things (including plants), one in every five is a species of beetle. One out of every four animals is a beetle. There are over 24,000 species known in North America north of Mexico, in 113 families. Common examples include insects we may not usually think of as beetles: "lady bugs," "June bugs," "fireflies" or "lightning bugs," glowworms, and weevils.

All beetles have chewing mouthparts, and all develop through complete metamorphosis. Most have the first pair of wings hardened into stiff plates (called elytra) that protect the softer abdomen and, usually, a second pair of membranous wings used for flight. In some cases, the wing covers are abbreviated (as in rove beetles, p. 200), and in flightless species the second pair of wings is reduced or absent. When a beetle flies, typically it raises the elytra straight out to the sides and uses the hindwings for actual flight, creating a distinctive silhouette in the air.

Poised for takeoff, a firefly beetle has raised its elytra and is spreading its hindwings for flight

BEETLE, BUG, or ROACH? Beetles are frequently mistaken for other insects, especially cockroaches and true bugs. Roaches have long filamentous antennae, whereas beetles always have distinctly segmented antennae. True bugs may have clearly segmented antennae, but they always have piercing-sucking mouthparts (read "beaklike"). The hardened elytra of beetles always meet along the midline of the insect's body, with no overlap. The leathery front wings of cockroaches and true bugs nearly always have some degree of overlap at rest, especially in the rear third of the body. Additionally, rove beetles may be easily mistaken for earwigs, but rove beetles generally move much faster and have no well-developed forceps on the end of the abdomen. Several long-horned beetles, and a few scarabs, buprestids, and other beetles, mimic bees or wasps. A few kinds of beetles mimic ants.

Beetle antennae sometimes help in placing a given specimen in its correct family. For example, scarab beetles, stag beetles, and passalid beetles all have lamellate antennae, the terminal segments resembling a stack of plates. Antennae of adult weevils are sharply angled or "elbowed" (right-hand figure).

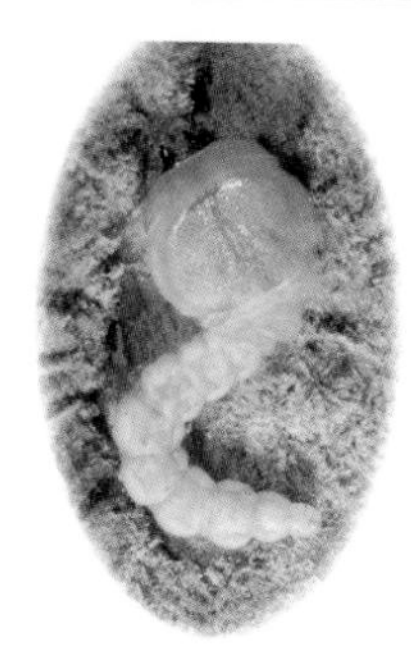

larva of a wood-boring beetle (family Buprestidae)

Larval beetles are called by several different names, depending on their habits and appearance: grubs, wireworms, rootworms, roundheaded borers, flatheaded borers, and glowworms, for example. Many species, especially woodborers, may take several years to complete the life cycle. It is in this stage that beetles can become pests, when densities reach damaging levels among crops, orchards, or forest trees. Most adult beetles live only long enough to procreate.

Beetles live in virtually every habitat imaginable. Several entire families lead a totally aquatic existence. At the other extreme are dune specialists outfitted with "sand shoes" for traction on the shifting sands. They can go without drinking, distilling all the water they need from the dry organic matter they scavenge in the bleak landscape. There are cold-hardy beetles that live above the timberline, and others, adapted to caves, that have lost the need for eyes.

Adult beetles span a wide range of sizes, too. Feather-winged beetles are nearly microscopic inhabitants of compost heaps and decaying debris on the forest floor. Enormous scarab beetles and longhorned beetles may exceed 6 inches and weigh in at several ounces. They can fly, also.

Beetles have quite a collective arsenal of chemicals with which to defend themselves. Bombardier beetles literally blast their attackers with a hot spray. Blister beetles and their relatives bleed a caustic chemical from their body joints that can raise blisters on human skin and kill some animals if they ingest the insect. Even lady beetles are not all sweetness and light. They bleed from their knees, the sticky yellowish liquid being an effective repellent. Most of these well-defended insects advertise themselves with "warning colors" of red, yellow, or orange and black, or bright metallic colors. Indeed, the physical beauty of beetles has made them favorites of many collectors, including Charles Darwin, Alfred Russel Wallace, and David Rockefeller.

Some beetles produce sounds by stridulating. That is, they rub one body part against another. The purpose of such noisemaking may be defensive, designed to startle an enemy, or communicative, helping the sexes and/or larvae to find one another.

For further information about North American beetles, please consult these resources:

Introduction to California Beetles by Arthur V. Evans and James N. Hogue (University of California Press, Berkeley, 2004). Excellent, highly readable, useful far beyond the borders of California.

A Field Guide to the Beetles of North America by Richard E. White (Houghton Mifflin, 1983). Slightly outdated, but comprehensive.

GROUND BEETLES AND TIGER BEETLES

(family Carabidae) are swift hunters and beneficial predators of pest insects. Tiger beetles work mostly days, ground beetles the graveyard shift. There are more than 2,600 species in North America. **TIGER BEETLES,** often placed in their own family (**Cicindelidae**), prowl beaches, dunes, and other exposed terrain. A few woodland species are active on sunlit paths or at night. Some are endangered by habitat destruction or off-road vehicle traffic. Wicked jaws and bulging eyes make them distinctive. They can be colorful but surprisingly cryptic and hard to spot. The larvae each live in a vertical burrow, using a hooked hump to brace their body against the inside walls. Poised with their flat heads flush with the surface of the ground, they grab insects and other invertebrates that happen by.

Cicindela are alert, sun-loving tiger beetles exhibiting shorebird behavior. Run. Stop. Run. Fly a short distance if startled. About 98 species occur in North America. Some have two-year life cycles in which newly emerged adults overwinter, breeding the following spring. Several are tied to specific habitats and/or seasons. The **Six-spotted Tiger Beetle** *(Cicindela sexguttata)* is abundant in springtime on forest paths in the eastern half of the U.S. and Canada. From the Rockies westward, look for the **Pacific Tiger Beetle** *(Cicindela oregona)* along streams and gravel beaches. The **Bronzed Tiger Beetle** *(Cicindela repanda)* is widespread except in parts of the far west. This beach-lover stalks the edges of lakes, streams, and rivers. The **Oblique Tiger Beetle** *(Cicindela tranquebarica)* is transcontinental except for the Gulf and Pacific coastlines. It favors riverbanks and has a long weaving flight when flushed. The **Backroad Tiger Beetle** *(Cicindela punctulata)* is widespread, especially in the east. Despite the name, it can turn up anywhere, even on city sidewalks! It also visits lights at night. The **Long-lipped Tiger Beetle** *(Cicindela longilabris)* is transcontinental in Canada, extending southward in the northeastern U.S. and the western mountains. Alpine areas, pine forests, and sandy habitats are its typical haunts. Across its wide range, the **Cowpath Tiger Beetle** *(Cicindela purpurea)* varies greatly in appearance. Grassy hillsides and forest clearings are places to seek it in spring. A spring and fall species, the **Splendid Tiger Beetle** *(Cicindela splendida)* is seen in barren habitats in central latitudes of the eastern U.S. The **Beautiful Tiger Beetle** *(Cicindela formosa)* is a spring and fall resident of dune edges in the midwest and northeast U.S. and adjacent Canada. Our largest species, it is highly variable. The **Ghost Tiger Beetle** *(Cicindela lepida)* inhabits dunes in the southern reaches of the prairie provinces as well as the northeastern, central, and southwestern U.S. Most of its two-year cycle is spent as a larva.

Members of the genus ***Omus*** are nocturnal and flightless. Five species live in forests of the far west. ***Omus dejeani*** has pitted wing covers and a wrinkled appearance. ***Omus audouini*** (not shown) has a granular texture. The three species in the genus ***Megacephala*** (or ***Tetracha***) are nocturnal, winged, and attracted to lights. ***Megacephala virginica*** is widespread in the east. ***Megacephala carolina,*** with ivory-tipped wing covers, occurs across the southern U.S.

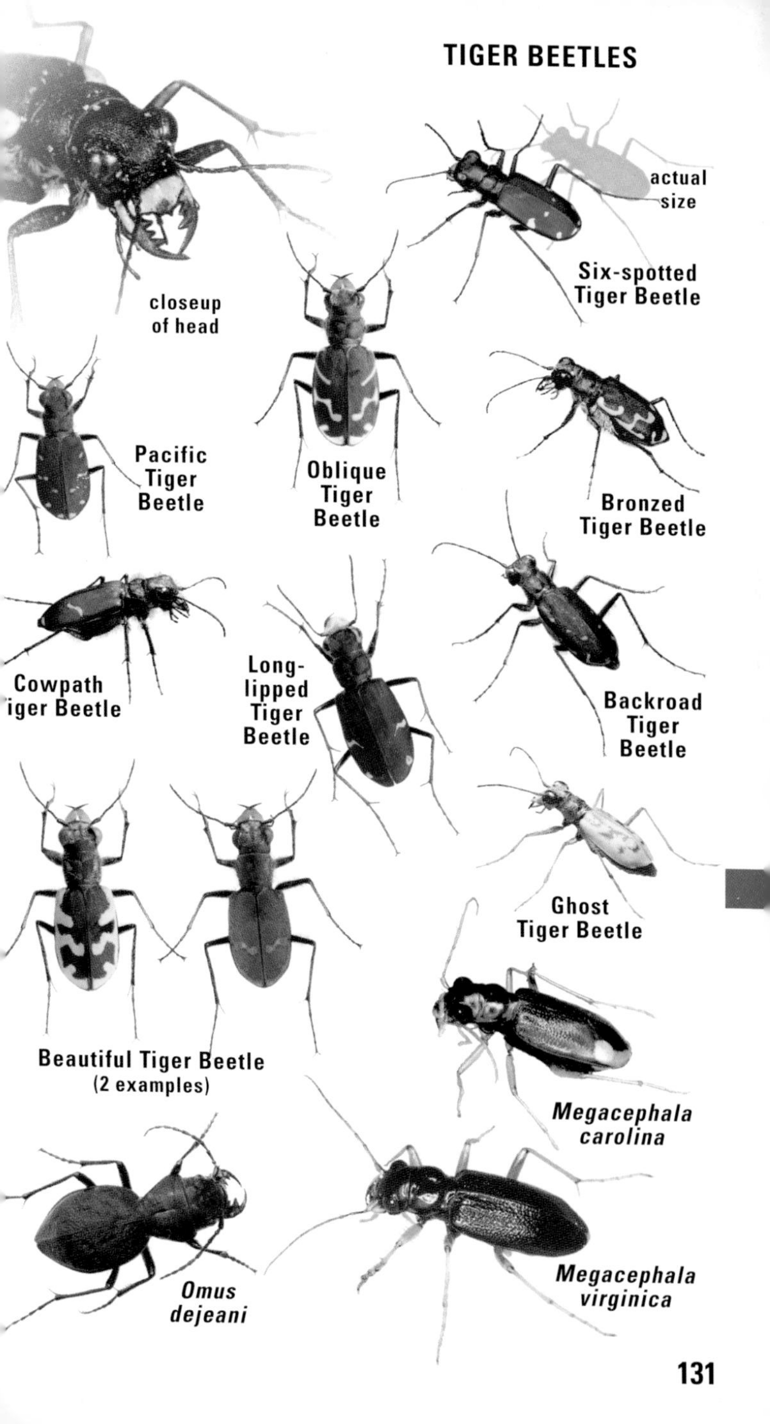

actual
size
Six-spotted
Tiger Beetle
closeup
of head
Pacific
Tiger
Beetle
Oblique
Tiger
Beetle
Bronzed
Tiger Beetle
Cowpath
iger Beetle
Long-
lipped
Tiger
Beetle
Backroad
Tiger
Beetle
Ghost
Tiger Beetle
Beautiful Tiger Beetle
(2 examples)
Megacephala carolina
Omus dejeani
Megacephala virginica

GROUND BEETLES

(family Carabidae) are abundant, diverse insects. The tiger beetle subfamily was treated on the preceding page. More typical ground beetles, below and on the next four pages, are usually seen scurrying in all directions when you turn over a stone, log, or other object. Most are black, *often shiny with grooved wing covers,* and nocturnal. Darkling beetles (pp. 192–195) are similar but less agile and without such obvious jaws. Surprisingly long-lived (two to three years as adults is not unusual), many ground beetles defend themselves with noxious secretions deployed from anal glands. Most are predatory, though a few are omnivores and a minority eat seeds. Look for them at lights at night, under debris by day. The larvae are also predatory or, rarely, parasitic.

Members of the genus ***Calosoma*** are known as "searchers" for their feverish hunting behavior. Most of the 41 species are large and found principally in the south. They are voracious predators of caterpillars and often climb trees in pursuit of such prey. They repel enemies with a foul odor. The **Caterpillar Hunter** *(Calosoma scrutator)* flies to lights at night, particularly in spring, throughout the U.S. and southern Canada but especially in the east. The **Fiery Hunter** *(Calosoma calidum)* is studded with gold or ruby spots on its wing covers. It frequents open woods and meadows, where it preys on cutworms. Common east of the Rockies, it is less so in the west. The **Black Caterpillar Hunter** *(Calosoma sayi)* is very similar in appearance but averages larger and is more common in southern regions. ***Calosoma externum*** is recognized by the metallic violet margins of the thorax and wing covers.

The genus ***Carabus*** is represented on this continent by 14 species, three of which are foreign. ***Carabus nemoralis*** is a flightless import from Europe. It is expanding its range here, from southeastern Canada to the Great Lakes, and parts of the west and northwest. Preying on slugs, it thrives in yards and gardens.

Members of the genus ***Pasimachus*** are large heroic hunters of pests in fields and grasslands, especially in the central and southern U.S. All 11 species are wingless. They can be confused with stag beetles (p. 148), but stags have elbowed antennae. ***Pasimachus viridans*** is a typical representative.

The 53 species of **boat-backed beetles** *(Scaphinotus)* are flightless, with *narrow heads and long faces.* The majority live in mountainous areas. ***Scaphinotus interruptus*** is widespread in forested country in California, including redwood forests. Active at night, both adults and larvae feed on snails and slugs, although adults will also eat berries. ***Scaphinotus angusticollis*** occurs from Alaska to northern California, west of the Cascades. Larvae feed on snails, but adults eat fruit as well, sometimes climbing to reach berries. Try attracting them at night with a trail of apple chunks.

The 19 species of ***Elaphrus*** are chiefly arctic and boreal. They *resemble miniature tiger beetles,* frequenting the muddy shores of watercourses, marshes, and mudflats by day. The **Delta Green Ground Beetle** *(Elaphrus viridis),* not illustrated, federally listed as threatened, is restricted to a few seasonal pools in the Central Valley of California.

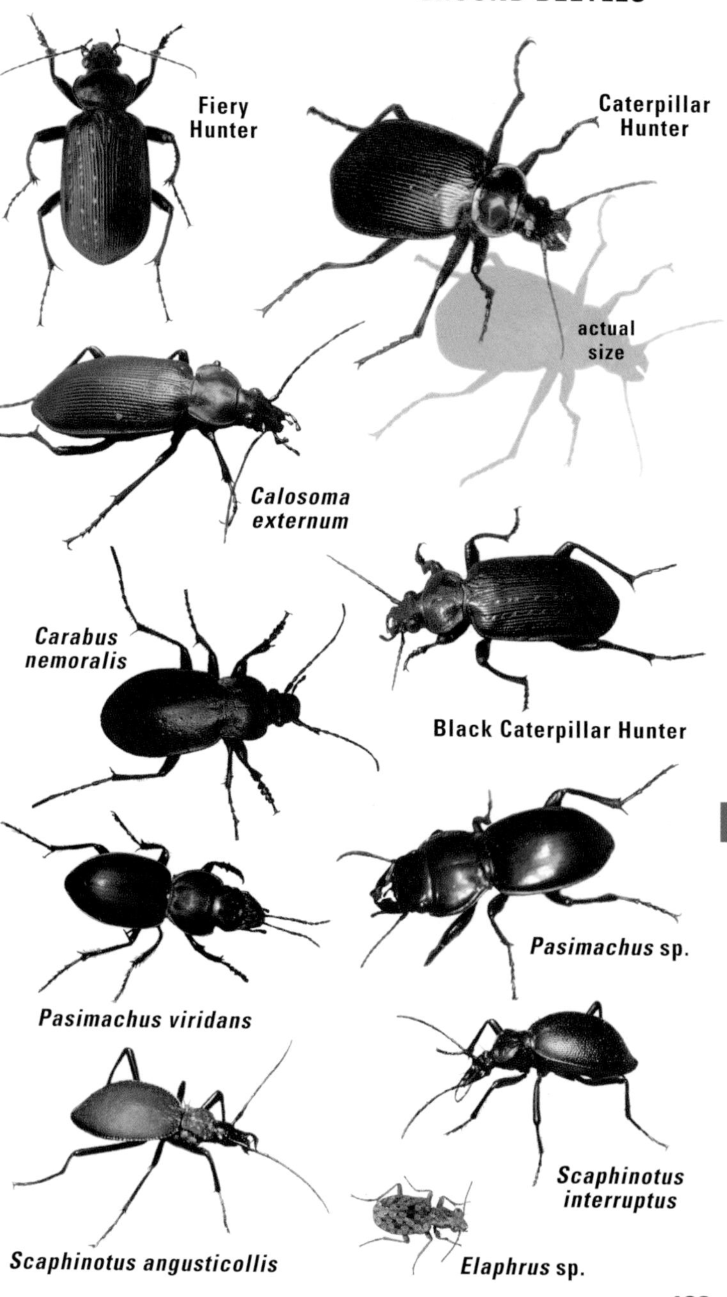
Fiery
Hunter
Caterpillar
Hunter
actual
size
Calosoma
externum
Carabus
nemoralis
Black Caterpillar Hunter
Pasimachus sp.
Pasimachus viridans
Scaphinotus
interruptus
Scaphinotus angusticollis
Elaphrus sp.

Members of the genus ***Harpalus*** are classic ground beetles in shape, color, and habits. They are widely distributed, and about 60 species are currently recognized. The **Murky Ground Beetle** *(Harpalus caliginosus)* flies to lights at night and feeds on cutworms and ragweed seeds around dry fields across North America. ***Harpalus pensylvanicus*** is abundant throughout the U.S. and much of southern Canada. It comes to lights in summer and fall, scavenging dead insects and damaged corn seed.

The 17 species of ***Dicaelus*** occur only in North America, mostly east of the Rockies. They live in woodlands, feeding on snails. Some are exquisitely metallic, but plain black species are more typical. Members of the genus ***Scarites*** have a resemblance to stag beetles (p. 148) at first glance. They often play dead when harassed. These beetles are commonly found

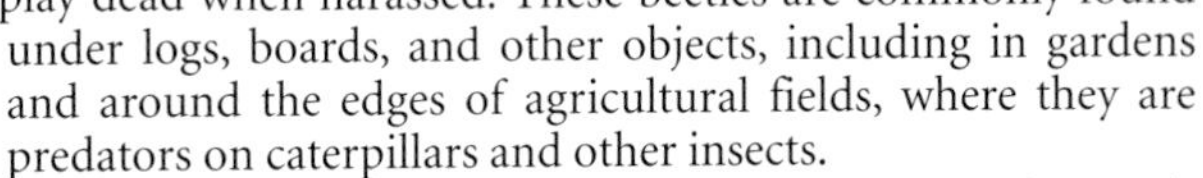

under logs, boards, and other objects, including in gardens and around the edges of agricultural fields, where they are predators on caterpillars and other insects.

larva of a ground beetle

Pterostichus is a large group with about 180 species north of Mexico, most diverse in northern forests and mountains. Some subterranean species have reduced pigment and no eyes. ***Pterostichus lama*** (not illustrated), our largest species at more than an inch long, is wingless and lives under loose bark and in rotten logs from British Columbia to southern California. The 13 species of ***Poecilus*** were formerly considered part of the genus *Pterostichus.* ***Poecilus chalcites*** has been recognized as a valuable predator on crop pests, especially early in the season, because it overwinters as an adult and becomes active in early spring. Other species in this genus are also considered valuable biological controls.

Bembidion is a complex genus of over 250 species, which are small but abundant. Most live near water, but a few live in fields. They are usually active by day, and most are good fliers. The genus covers the continent, from the Arctic to Florida.

Eleven species of **round sand beetles** *(Omophron)* are distributed over most of the continent. They live in burrows along the shores of lakes, rivers, and streams, emerging at night to hunt. You can flush them by pouring water on the beach. ***Omophron americanum*** occurs east of the Rockies and locally farther west. ***Omophron obliteratus*** is common in parts of the southwestern U.S.

The more than 100 species in the genus ***Amara*** are small compact beetles. They are active day and night in most of North America, although few species inhabit the Arctic or the extreme southeast. They feed on fruits, seeds, and small insects in mostly open habitats.

Members of the genus ***Stenolophus*** are sometimes considered pests because of their seed-eating habits. The 32 species thrive in many open areas. The **Seedcorn Beetle** *(Stenolophus lecontei)* is often seen at lights in eastern North America.

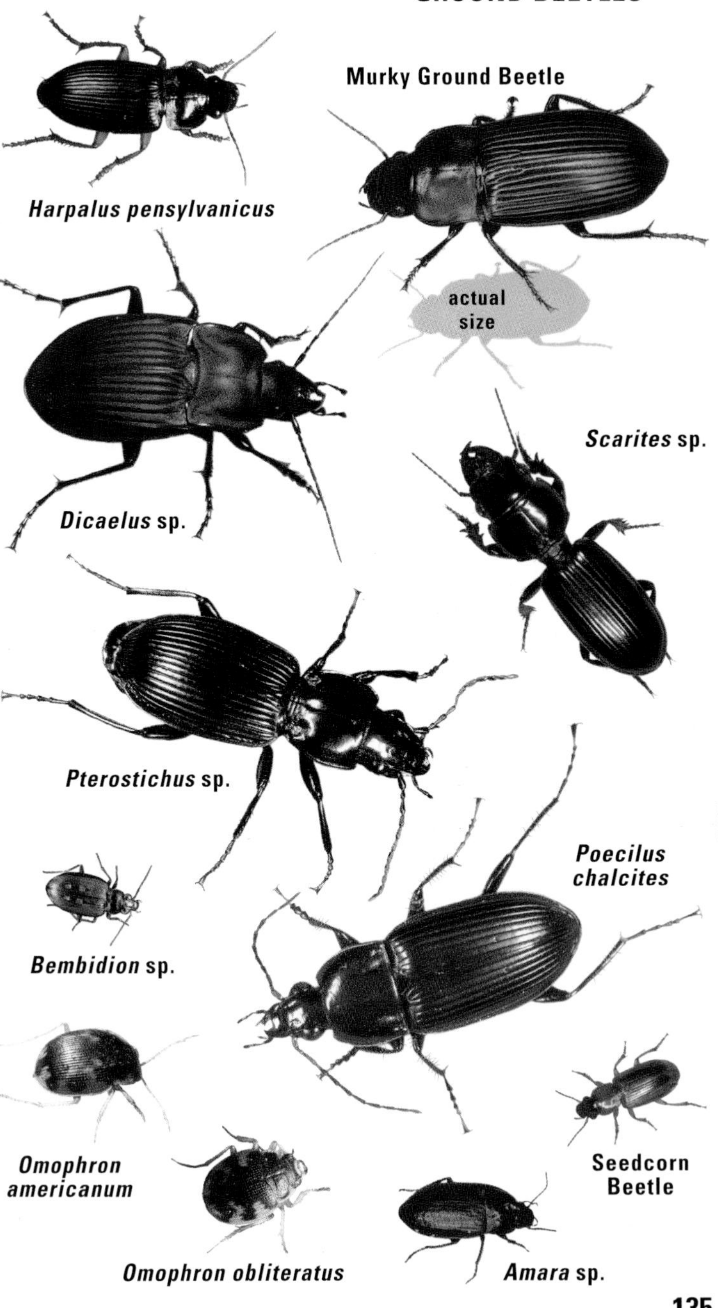

Murky Ground Beetle

Harpalus pensylvanicus

Scarites sp.

Dicaelus sp.

Pterostichus sp.

Poecilus chalcites

Bembidion sp.

Omophron americanum

Seedcorn Beetle

Omophron obliteratus

Amara sp.

Members of the genus ***Chlaenius*** have a *velvety look* due to fine hairs coating their wing covers, but you get only a glimpse as they dash for cover. If cornered, they *emit a pungent odor.* The 51 species range widely, with fewer in the far north. Favoring wet habitats such as muddy bottomlands, they hide under debris by day. Females lay eggs in mud cells attached to vegetation. ***Chlaenius sericeus*** and ***Chlaenius tricolor*** are widespread species found along the shores of rivers.

The **bombardier beetles** (genus *Brachinus*) are named for their explosive means of self-defense. About 48 species live in North America. These nocturnal insects fire rapid pulses of hot chemicals (100°C) when molested. An audible pop and a puff of smoke accompany the blast. They are 4–15 mm and not uncommon in moist areas near lakes and streams, especially in the south-central and southwestern states. The larvae of most species are parasitic on the pupae of aquatic beetles.

The three species in the genus ***Panagaeus*** are distinctly *hairy* and may be mimics of velvet ants (p. 350). ***Panagaeus fasciatus*** favors dry forests in the central and southern portions of the eastern U.S., west as far as Kansas and Arizona.

The genus ***Rhadine,*** with 40 species, includes some rare cave-dwellers. Others live in mammal burrows or under rocks or logs. ***Rhadine jejunum*** occurs east of the Cascades in the Pacific northwest.

Agonum is a genus of 72 species distributed from the Arctic to Nicaragua. Most live near water. ***Agonum placidum*** is one of the most numerous species across southern Canada and the northern U.S., occurring mainly in open fields, thickets, and woodland edges.

The genus ***Colliurus*** is represented by four species in North America, known by their oddly elongated head and thorax, as if the front of some other insect had been stuck onto a "normal" beetle. Most of these bizarre species are tropical, and little is known about them. ***Colliurus pensylvanicus*** is commonly seen at lights.

Members of the genus ***Lebia*** are attractive and distinctive beetles. The genus is most diverse in the south but ranges north into the Yukon and the Northwest Territories. The 48 species are recognized by the *squared-off hind margins of the wing covers* that expose the tip of the abdomen. Many are boldly marked and are active by day on foliage. Some visit lights at night. As larvae, they are parasites of the pupae of leaf beetles (pp. 160–165). ***Lebia viridis*** is common on goldenrod flowers and sometimes shows up at lights. ***Lebia pulchella*** is one of the most widespread and common species, found from southern Canada to southern Texas. ***Lebia grandis*** (not illustrated), typically colorful with dark blue wing covers contrasting with the reddish head and pronotum, is a parasite of pupae of the Colorado Potato Beetle (p. 160).

The eight species of ***Galerita*** are large and gangly. The genus is transcontinental in the southern U.S., farther north in the Mississippi Valley and the eastern seaboard. These beetles may be seen in open woodlands or under brush piles or old stone walls.

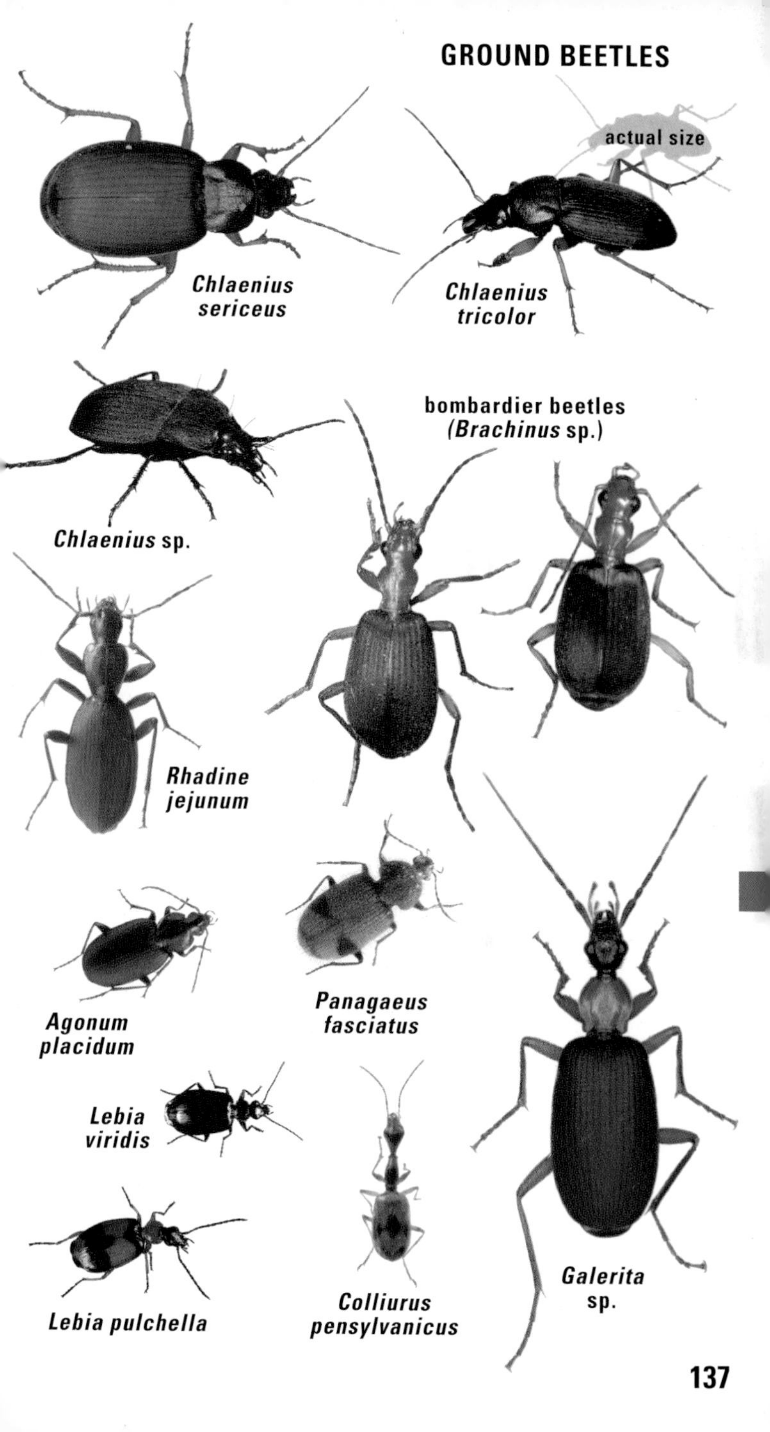
actual size
Chlaenius sericeus
Chlaenius tricolor
Chlaenius sp.
bombardier beetles (Brachinus sp.)
Rhadine jejunum
Agonum placidum
Panagaeus fasciatus
Lebia viridis
Lebia pulchella
Colliurus pensylvanicus
Galerita sp.

SCARABS

(family Scarabaeidae) are incredibly diverse and include some of our biggest insects. Certain dung-rolling species were sacred to the ancient Egyptians, but our own species are not so revered. A few are bona fide pests, but most are valuable decomposers and important to the diets of skunks, bats, and other animals. The C-shaped grubs are consumed by moles and parasitized by a variety of wasps and flies. There are about 125 genera and 1,700 species in North America.

FLOWER SCARABS (subfamily Cetoniinae) are mostly colorful, diurnal species that fly well, with their wing covers closed. The four species of ***Cotinis*** are active by day. In flight they are easily mistaken for carpenter bees. They feed on pollen, nectar, fruit, and fluid from plant wounds. The **Green June Beetle** *(Cotinis nitida)* is sometimes a pest of lawns and golf links as a root-feeding larva. Adults can damage peaches and other fruits in the southeastern U.S. The **Fig Beetle** *(Cotinis mutabilis)* is more common in the southwest.

The 20 species of ***Euphoria*** are day-flying, flower-loving scarabs. Adults mimic bees, flying with wing covers closed. Larvae feed in decomposing organic matter. The **Emerald Euphoria** *(Euphoria fulgida)* ranges throughout the northeastern U.S., southwest to Texas. Adults visit blossoms of thistle, dogwood, sumac, and other plants and also feed at sap flows. The **Bumble Flower Beetle** *(Euphoria inda)* occurs over much of the U.S. Grubs have been found in nests of ants. Adults emerge in late summer, then overwinter and resurface in early spring. The **Spangled Flower Beetle** *(Euphoria sepulcralis)* is found in the southeastern and south-central U.S. Its habits are similar to those of the preceding species. The **Olive Flower Beetle** *(Euphoria herbacea)* is another eastern member of this genus.

The genus ***Osmoderma*** includes three species found east of the Rockies. When handled, they reportedly emit the odor of Russian leather, whatever *that* is. Larvae develop in rotting wood of dead or dying hardwoods. The **Hermit Flower Beetle** *(Osmoderma eremicola)* visits lights and sap flows. It ranges in the east, west to Montana, and south to Georgia. The life cycle takes 2–3 years, longer in the north.

The genus ***Trichiotinus*** includes eight species ranging throughout the U.S. and southern Canada. Adults are active fliers found on flowers. Larvae feed in decaying stumps, logs, or branches of deciduous trees. ***Trichiotinus assimilis*** ranges coast to coast, mostly in the north. Look for adults on roses, viburnum, blackberry, and other flowers. The **Emerald Flower Scarab** *(Trichiotinus bibens)* occurs at central latitudes of the eastern U.S. The **Bee-like Flower Scarab** *(Trichiotinus piger)* can be fairly common on flowers in the east.

The genus ***Trigonopeltastes*** includes two species in the southeastern quarter of the U.S. The **Delta Flower Scarab** *(Trigonopeltastes delta)* is common on Queen Anne's lace and other flowers. Larvae feed in decaying wood. ***Valgus*** is a genus of four relatively small species. Adults are usually seen on flowers, while the larval stages are often found in association with termites in dead wood.

Fig
Beetle
Green
June Beetle
actual size
Bumble
Flower Beetle
Emerald
Euphoria
Spangled
Flower Beetle
Hermit
Flower Beetle
Olive Flower
Beetle
Emerald
Flower Scarab
Trichiotinus assimilis
Bee-like
Flower Scarab
Delta Flower
Scarab
Valgus sp.

MAY BEETLES, JUNE BEETLES, CHAFERS

(subfamily Melolonthinae) make up a diverse group of scarab beetles that are mostly drab and nocturnal.

Members of the genus ***Phyllophaga*** are the May beetles or "Junebugs," well known for their noisy, clumsy arrivals at porch lights and window screens on early summer nights. At least 400 species are collectively distributed across North America except for Alaska and far northwestern Canada. A few species have been historical pests of field crops, orchards, and turf grasses, but today most damage is negligible. Adults feed at night on foliage and can be abundant at lights. Larvae, called white grubs, feed on roots of grasses and other herbs. Depending on species, latitude, and weather, the life cycle takes one to four years. The members of this genus vary in size, and vary in color from reddish brown to gray, but most cannot be identified to species; we illustrate several here just to show the variation in this group. ***Phyllophaga cribrosa*** is notable as a mimic of certain tenebrionid beetles (pp. 192–195).

Species of ***Polyphylla*** are widely distributed except for parts of the midwest. Males have huge, platelike antennal clubs. When held, the beetles make a huffing sound by rubbing the abdomen against the inside of the wing covers. The **Ten-lined June Beetle** *(Polyphylla decemlineata)* occurs throughout the west, and east to the edge of the Great Plains. Adults eat needles of various conifers and come to lights at night. Larvae feed on roots of trees, taking three to four years to reach adulthood. There are 30 other species in the genus.

Members of the genus ***Diplotaxis,*** at 6–14 mm long, resemble miniature May beetles. They are active at night, but of our nearly 100 species, not all are attracted to lights. Look on trees and shrubs at night for the adults. The larval stages are not well known, but they probably feed on plant roots. The genus ***Hoplia*** includes 12 North American species of plant-eating scarabs. Look for them by day on flowers and leaves. Larvae feed on roots of grasses. ***Serica*** is a genus of about 100 North American species. A few are diurnal, but most are active at night and attracted to lights. Many have an iridescent sheen. Adults eat foliage of plants and trees such as oak and wild rose. Larvae feed on the roots of grasses and other plants.

There are close to 30 species in the genus ***Dichelonyx.*** Collectively they range everywhere in North America except for parts of northern Canada. Adults feed on leaves of a variety of trees and shrubs. Some species are nocturnal, others active by day. Larvae may be root feeders. ***Dichelonyx backii*** is common west of the Rockies, feeding on the needles of conifers as an adult.

Members of the genus ***Macrodactylus*** are the rose chafers. Three species occur in the south and east, as far north as Ontario and Quebec, and west to Arizona. Adults feed on foliage, flowers, and fruit of native and ornamental plants during the day. Larvae feed on roots. ***Macrodactylus subspinosus*** occurs over most of the eastern U.S. and adjacent southeastern Canada. Look for it on wild grape.

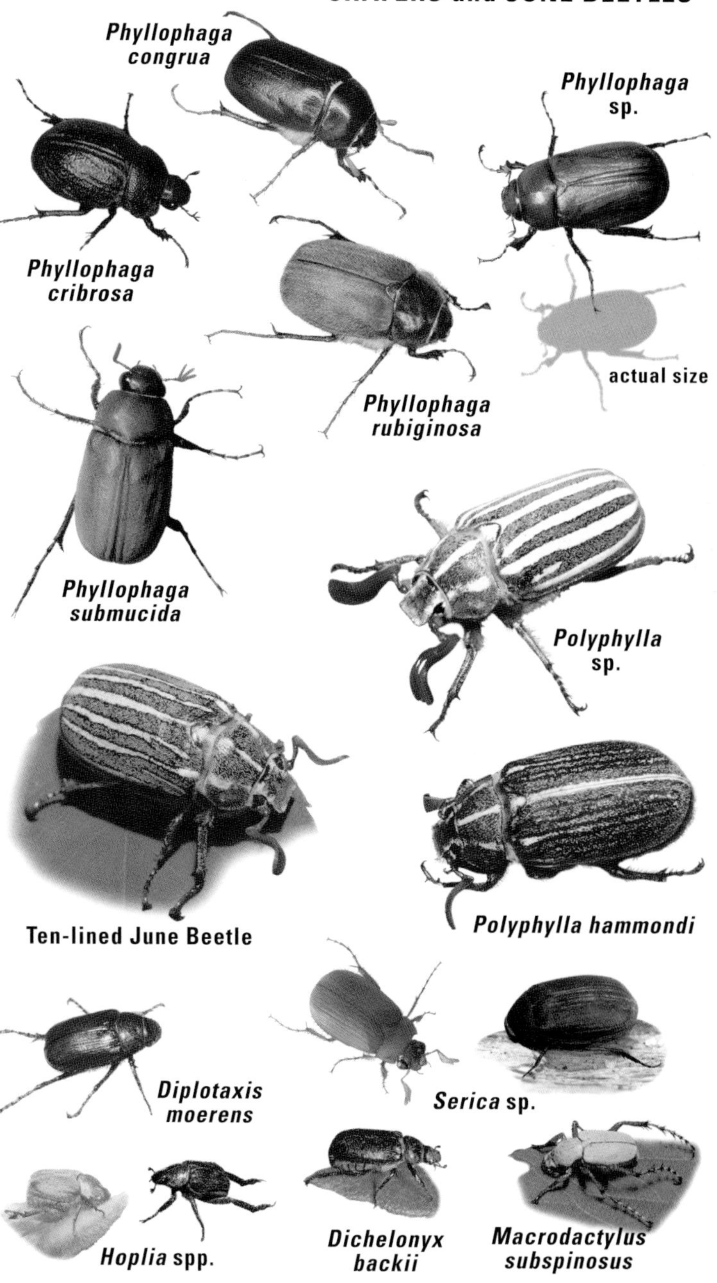
Phyllophaga congrua
Phyllophaga sp.
Phyllophaga cribrosa
actual size
Phyllophaga rubiginosa
Phyllophaga submucida
Polyphylla sp.
Ten-lined June Beetle
Polyphylla hammondi
Diplotaxis moerens
Serica sp.
Hoplia spp.
Dichelonyx backii
Macrodactylus subspinosus

LARGE DUNG BEETLES (subfamily Scarabaeinae) are often found in dung, sometimes in carrion. The genus ***Copris*** has nine nocturnal species in the U.S., including the eastern ***Copris fricator.*** Adults dig a shallow burrow beneath a manure source. Dung is hauled to the bottom and molded into two to five pear-shaped masses, inside each of which a larva grows. The female grooms the balls of mold and fungi until her offspring emerge.

The 21 species of ***Canthon*** are the comical "tumblebugs." Males roll balls of dung, while females ride atop or assist alongside. The pair eventually bury each sphere a few inches underground. The female lays an egg on the "brood ball," and a humpbacked larva develops inside. "Food balls" are rolled for consumption by the adults. The larger ***Deltochilum gibbosum*** rolls all manner of rotting wastes, from cat poop to chicken feathers to carrion. The resulting ball is bulb-shaped and may exceed 2 inches in diameter. This diurnal species favors wooded habitats in the southeast. Members of the genus ***Dichotomius*** are large spherical beetles. Both sexes may cooperate in packing dung at the end of simple or branching burrows in pastures and prairies. Look for them flying to lights at night.

The genus ***Onthophagus*** includes 40 species in North America, some introduced. They can be abundant in mammal dung, rotting fungi, and decaying vegetation. A fair number live in animal burrows, one associated only with the endangered Gopher Tortoise in the southeast. Female dung-eaters dig a burrow under a manure source, terminating in a cluster of cells they pack with dung. The first male on the scene mates with her, then guards the entrance against rivals with sheer strength and/or horns. Smaller, hornless males may squeeze past him or dig an adjacent tunnel, cutting behind him. The **Scooped Scarab** *(Onthophagus hecate)* ranges throughout most of the U.S. and southern Canada, except for parts of the far west. The **Gazelle Scarab** *(Onthophagus gazella)*, native to Africa and India, has expanded its U.S. range eastward from southern California at least as far as Missouri. Look for it at lights. ***Onthophagus taurus,*** another Old World species, occurs throughout the southern U.S.

The **Rainbow Scarab** *(Phanaeus vindex)* is a large metallic day-active dung beetle of the east and southwest. It exploits a variety of dung. Adults dig separate food storage burrows for themselves and their offspring. Brood balls are coated with clay and are not maintained by the female. Adults may live over a year but are preyed on by Burrowing Owls.

SMALL DUNG BEETLES (subfamily Aphodiinae) are small, cylindrical, and found mostly in dung. ***Aphodius*** is a genus with 210 species in North America, many abundant in cow patties or at lights. Some of our common species were introduced from Europe, such as ***Aphodius fimetarius.*** Another import, ***Aphodius distinctus,*** is common in fresh manure in spring and fall. ***Aphodius fossor,*** very common on the northern Great Plains, is our largest species (although still less than half an inch long). The genus ***Ataenius*** includes 65 species in North America. Adults sometimes fly in large numbers at sunset in early spring and are attracted to lights. Larvae of most species feed on humus, but others feed in dung.

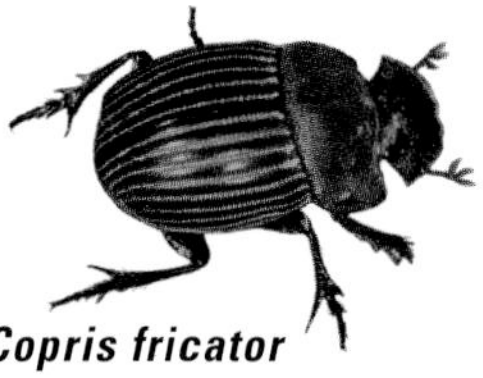

Copris fricator

actual size

Canthon sp. with dung ball

Deltochilum gibbosum

Dichotomius sp.

Scooped Scarab

Onthophagus taurus

Gazelle Scarab

Aphodius fossor

Aphodius fimetarius

female

Rainbow Scarab

male

Ataenius sp.

Aphodius distinctus

SHINING LEAF CHAFERS

(subfamily Rutelinae) are scarabs of mostly tropical distribution, with a limited number of genera in North America.

Six species of the genus ***Cotalpa*** are widespread in the U.S. One of these, ***Cotalpa lanigera,*** is the goldsmith beetle or "gold bug" of Edgar Allan Poe's short story. Found in sandy streamside habitats, adults feed on leaves of aspen, cottonwood, and willow. Larvae feed on roots of various plants.

The genus ***Anomala*** includes at least 40 North American species, plus many others around the world, as this is among the largest genera of insects. They are foliage-feeders, common in the east and southwest but absent from the Pacific Coast. Some have been introduced from abroad. Most are variable in color and difficult to tell apart, even for experts.

The spectacular beetles of the genus ***Chrysina*** (formerly *Plusiotis*) are most diverse in the New World tropics. Four species reach the southwestern U.S., where they are attracted to lights at night and where they are avidly sought by beetle aficionados. ***Chrysina gloriosa*** feeds on juniper needles at elevations above 4,000 feet from the Davis Mts. of Texas to southeastern Arizona. ***Chrysina beyeri*** feeds on oak in the mountains and foothills of southeastern Arizona. ***Chrysina lecontei*** is another mountain species of southern Arizona and New Mexico, while ***Chrysina woodi*** (not illustrated), pale green with metallic blue on the tarsus (lower leg), occurs in the mountains of western Texas.

The two species of ***Pelidnota*** occur east of the Rockies. ***Pelidnota punctata*** commonly flies to lights at night in midsummer. Adults eat mostly the leaves of grape, while larvae feed in decaying stumps or the rich soil in proximity to rotting wood. The life cycle takes two years.

Members of the genus ***Paracotalpa*** are "little bears," named for the adults' woolly appearance. The four species range west of the Rockies. ***Paracotalpa granicollis*** occurs in the Pacific northwest, feeding on tree buds and blossoms by day. The larvae of a related species are known to feed on the roots of sagebrush.

The **Japanese Beetle** *(Popillia japonica),* recognized by five white tufts of hair on the edge of the abdomen, is native to Japan and northern China. Introduced to New Jersey in 1916 in nursery stock, it is now established in southeastern Canada and all states east of the Mississippi. Adults are active by day, feeding in groups on nearly 300 kinds of plants, the leaves of which they skeletonize. Larvae eat roots of herbaceous plants and can damage turf grasses. Despite attempts at control with imported pathogens, parasites, and predators, the Japanese Beetle is still expanding its range.

BUMBLE BEE SCARABS (family Glaphyridae) are hairy, fast-flying mimics of bees. The one North American genus, ***Lichnanthe,*** includes eight species. ***Lichnanthe rathvoni*** is common from southwestern British Columbia to California, Nevada, and Idaho. Adults fly by day in sandy habitats but may also appear at lights at night. Larvae feed in debris 8 to 15 inches underground, taking three years to mature. ***Lichnanthe apina*** is another member of this genus common in California.

Cotalpa sp.

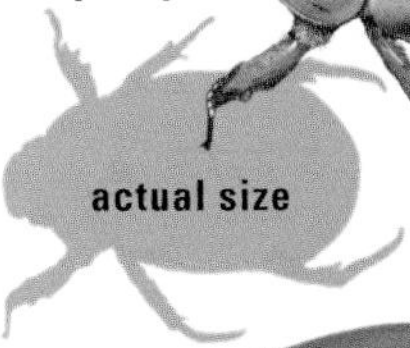

actual size

Anomala sp.

Chrysina beyeri

Chrysina gloriosa

Chrysina lecontei

Pelidnota punctata

Paracotalpa granicollis

Lichnanthe rathvoni

Japanese Beetles

Lichnanthe apina

RHINOCEROS BEETLES AND OTHERS

The first group of scarabs below **(subfamily Dynastinae)** includes some real giants. The other two groups treated here are now considered to be separate families related to the true scarab beetles.

In the genus ***Strategus*** are five species in the southern U.S. The **Ox Beetle** *(Strategus antaeus)* occurs in pine forests along the southern Atlantic Coast and west along the Gulf. Larvae feed initially on leaf litter dragged into a burrow by the female parent, then graduate to eating decaying tree roots.

The **Rhinoceros Beetle** *(Xyloryctes jamaicensis)* is widespread in deciduous woodlands from southern New England southwest to Arizona. Larvae feed in decomposing leaf litter. Adults feed underground on decaying roots but may come to lights at night.

Members of the genus ***Phileurus*** resemble stag or bess beetles (pp. 148–149) at first glance. Two species occur in the southern U.S. Adults are carnivorous on other insects and may survive two years. Larvae live in rotting wood. The genus ***Tomarus*** (formerly *Ligyrus)* includes four species. One, the **Carrot Beetle** *(Tomarus gibbosus),* is common throughout the U.S., especially in the south. Both larvae and adults are subterranean root and tuber feeders, but the beetles frequently appear at lights at night.

The massive ***Dynastes*** beetles never fail to draw attention. The **Eastern Hercules Beetle** *(Dynastes tityus)* occurs in mixed forests as far north as Indiana and New Jersey. Larvae feed in decaying heartwood of many tree species, taking two years to mature. Adults are attracted to fermenting sap and fruits and fly to lights at night. The **Western Hercules Beetle** *(Dynastes granti)* occurs at higher elevations in Utah and Arizona.

EARTH-BORING SCARABS (family Geotrupidae) are called dor beetles in Europe. Our species excavate burrows, some as deep as 3 meters, in which decaying leaves, fungi, or humus is stored as food for the larvae, each in its own cell. Adults are attracted to dung and, rarely, lights. ***Bolboceras*** includes 10 species, most in the northeastern U.S. Some males have a horn; others do not. ***Bolboceras obesus*** occurs in the Pacific northwest. Species of ***Bolbocerasoma*** are widespread, especially in the east, while ***Bolbocerastes imperialis*** is found in the southwest. The genus ***Geotrupes*** includes 10 species of secretive beetles. The widespread ***Geotrupes splendidus*** varies from black to metallic purple or bronze. Adults are attracted to carrion, dung, and other decomposing material. Males dig branching tunnels, packing each cell with dead leaves or decaying fungi, and then await a receptive female that will lay an egg in each chamber.

HIDE BEETLES (Trogidae) look like living dirt clods, often caked with debris, making them cryptic among the remains of carrion. Adults and larvae eat feathers, fur, and skin on dry carcasses, the grubs living in shallow burrows beneath the corpse. Some species feed on debris in nests of mammals or birds. There are 41 species in two genera in North America. The 16 species in ***Omorgus*** include the widespread ***Omorgus suberosus.*** It comes to old, dry carrion and visits lights at night. ***Trox*** is a genus of 25 species. They have been recorded in the nests of various birds and mammals and also in carrion, owl pellets, bone meal, and even a wool coat.

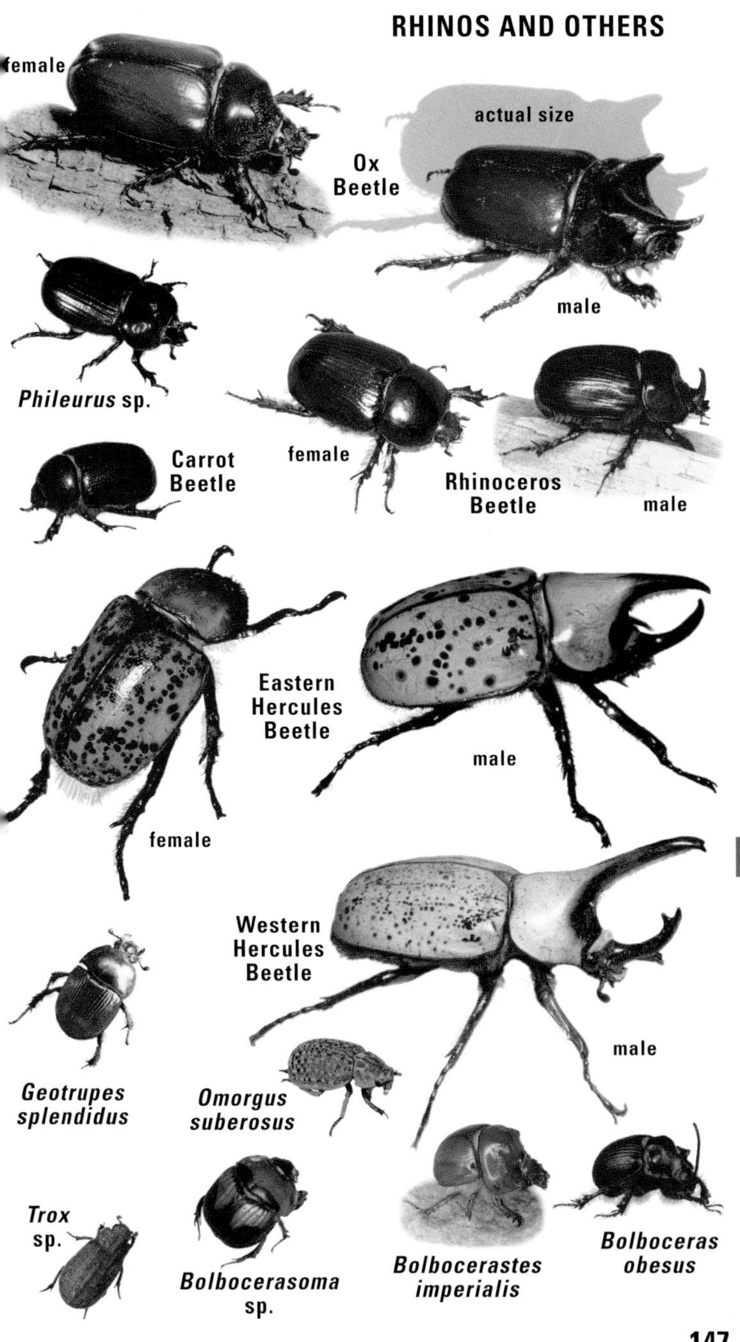

female
actual size
Ox Beetle
male
Phileurus sp.
female
Carrot Beetle
Rhinoceros Beetle
male
Eastern Hercules Beetle
male
female
Western Hercules Beetle
male
Geotrupes splendidus
Omorgus suberosus
Trox sp.
Bolbocerasoma sp.
Bolbocerastes imperialis
Bolboceras obesus

STAG BEETLES, BESS BEETLES, RAIN BEETLES

The **Stag Beetle family (Lucanidae)** is named for the fact that males of some species have extended, antlerlike mandibles, used in combat over females. Many tropical varieties of stag beetles are gigantic. Most of our 24 species are small, but a few are large enough (and large-mandibled enough) that they are sure to draw attention. Larvae develop in rotting wood, while adults are attracted to lights and tree sap.

Our five largest species of stag beetles all belong to the genus ***Lucanus.*** Within some of these species, individual males vary from manly ("major") to modest ("minor"), in reference to body size and jaw length. The **Giant Stag Beetle** *(Lucanus elaphus)* occurs in the eastern U.S., from Virginia and North Carolina west to Oklahoma, and might be found around rotting oak stumps. The **Pinching Beetle** *(Lucanus capreolus,* formerly *Pseudolucanus capreolus),* often called pinching bug, is fairly common in the northeastern U.S. Look for these two species visiting lights at night.

The genus ***Dorcus*** includes many Asian species and two in North America. The **Antelope Beetle** *(Dorcus parallelus)* of the northeastern U.S. might be mistaken for a *Pasimachus* ground beetle (p. 132) or a *Parandra* longhorn (p. 184). ***Platycerus*** is a genus of five small species. The **Oregon Stag Beetle** *(Platycerus oregonensis)* lives in decaying maples and other hardwoods in western British Columbia, Washington, and Oregon.

The **Rugose Stag Beetle** *(Sinodendron rugosum)* is our only member of its genus. Males have a horn instead of enlarged jaws. This species occurs in rotting logs and stumps of alder, ash, willow, and other trees in the Pacific northwest. The genus ***Ceruchus*** includes three small species found mostly in the northwest and northeast. ***Ceruchus piceus*** occurs in dead hardwoods in the northeastern U.S.

BESS BEETLES (family Passalidae) are also called patent leather beetles for their shiny bodies. A few rare species occur in Florida and Arizona, but only the **Horned Passalus** *(Odontotaenius disjunctus)* is common and widespread in the eastern U.S. All life stages live together in colonies in well-rotted wood. Adults and larvae communicate with sounds made by rubbing body segments together. Fourteen different calls have been recorded. Look for these beetles in forests by day as they clamber on logs.

RAIN BEETLES (Pleocomidae) are so unique that one species was nominated for state insect of Oregon. "Rufus" even had a campaign button. There are about 30 species of ***Pleocoma,*** collectively found from southwestern Washington to northern Baja California, but many of these have very limited ranges. Larvae feed on roots deep underground and may take 8–13 years to reach adulthood; adults do no feeding at all, as they have vestigial mouthparts and no digestive tract. Adult females have reduced wings and remain in or near their larval burrows, attracting mates by releasing a strong scent. Males fly in search of females, operating on stored fat that may fuel only a couple of hours of flight. Most species emerge in fall or winter, males flying mainly at dawn or dusk at times of high humidity, such as rainfall or snowmelt. Males die shortly after mating, but females burrow back down into the soil and lay their eggs within a few weeks.

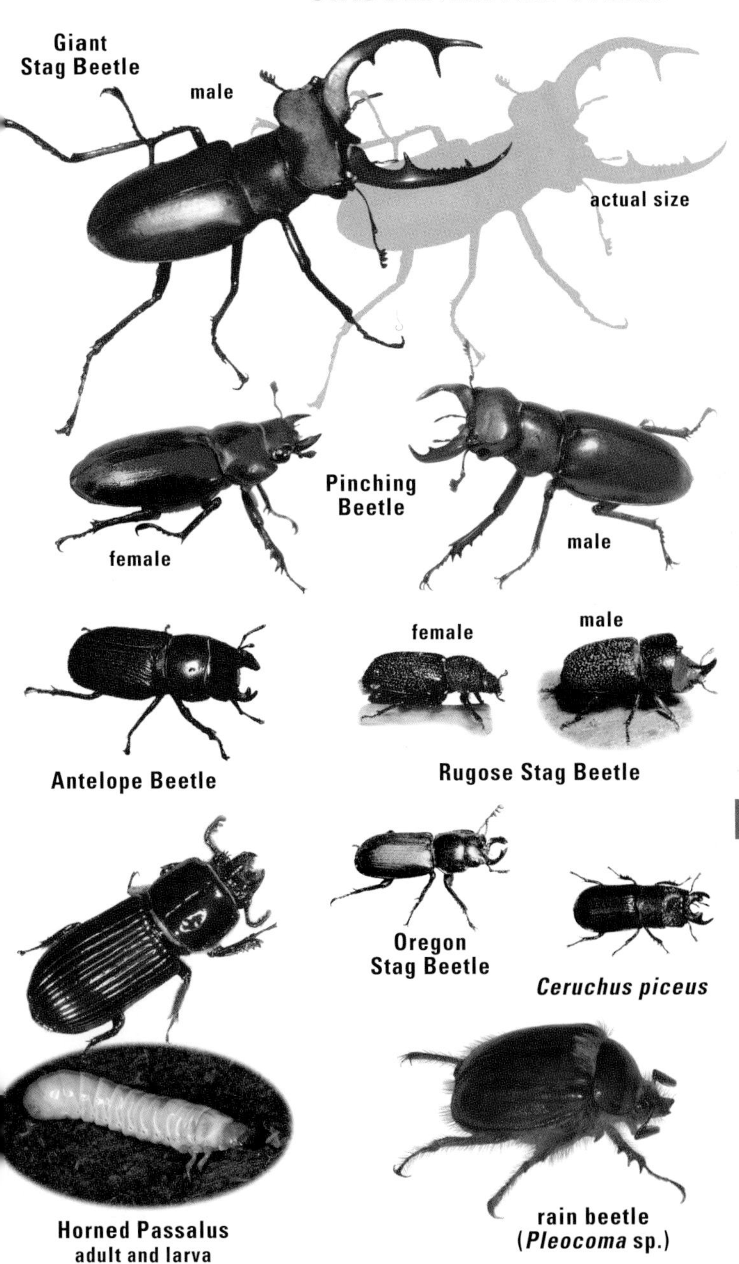
Giant
Stag Beetle
male
actual size
Pinching
Beetle
female
male
Antelope Beetle
female
male
Rugose Stag Beetle
Oregon
Stag Beetle
Ceruchus piceus
Horned Passalus
adult and larva
rain beetle
(*Pleocoma* sp.)

Predaceous Diving Beetles (family Dytiscidae) are known as water tigers in the larval stage. Larger species can kill small vertebrates such as tadpoles. Adults swim by moving their flattened hindlegs in unison, like oars. They collect air at the surface in a "bottoms-up" pose, trapping a bubble beneath their wing covers. Males have disklike front feet. Look for them in any body of fresh water, especially ponds with emergent vegetation. They fly well and can exploit temporary pools and potholes. There are roughly 475 species known in North America.

Members of the genus ***Dytiscus*** are monstrous, ranging up to 40 mm long. Twelve species are found in North America. They often fly to lights at night. Females may differ in appearance from males, with deeply grooved wing covers. Newly mated females sport an obvious "plug" designed to discourage competing males. Both sexes can excrete potent steroids in self-defense. ***Dytiscus marginicollis*** is widespread in the west, from Alaska to Baja, and east to Alberta and Colorado.

The six species in the genus ***Acilius*** are oval, medium-sized, and found over a wide area of Canada and the northern half of the U.S. They are typically seen in woodland ponds, pools, beaver ponds, or ditches. The genus ***Rhantus*** includes 10 species, all similar in appearance and size. They are often numerous in a given body of water, even potholes and other temporary pools. ***Rhantus binotatus*** occurs throughout much of North America. ***Thermonectus*** includes six species north of Mexico, with more in the Neotropics. ***Thermonectus marmoratus*** is a striking insect occurring in the southwestern U.S. It is not uncommon in pools and slower stretches of intermittent desert streams. ***Agabus*** is a huge genus (about 100 species) of medium-sized insects. Most common in small bodies of still water, they will exploit a diversity of habitats, including swimming pools.

Members of the genus ***Cybister*** are very large and live in the deeper waters of large ponds and lakes. They frequently come to lights at night. Three species occur in the eastern U.S., two in the southwest. ***Cybister fimbriolatus*** is one of the larger eastern varieties.

WATER SCAVENGER BEETLES (family Hydrophilidae) are not all aquatic, nor all scavengers. Easily mistaken for diving beetles, most hydrophilids have a keel-like spine running down the middle of their undersides. They swim by moving their legs alternately and surface "heads-up," using their short antennae to channel air to a bubble on their belly. Some species squeak by rubbing the abdomen against the inside of the wing covers. Eggs are laid in a silken sac (or individually wrapped). The larvae are nearly all predatory. There are 258 species in 35 genera in North America.

The 17 species in the genus ***Berosus*** are aquatic but frequently seen at lights at night. Members of the genus ***Tropisternus*** are the most commonly encountered species (14 in North America). ***Tropisternus lateralis*** is widely distributed, with several subspecies across the continent. The three species in the genus ***Hydrophilus*** are enormous predators. ***Hydrophilus triangularis,*** which ranges throughout North America, is an occasional pest in fish hatcheries. Look for these beetles visiting lights at night.

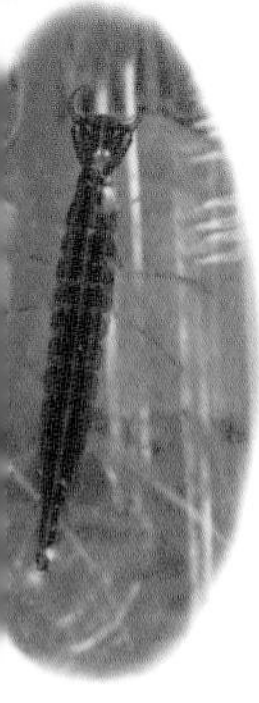

Dytiscus marginicollis

Dytiscus sp.
larva

Acilius sp.

Rhantus binotatus

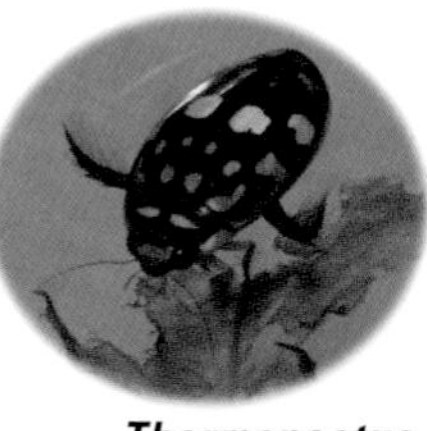

Thermonectus marmoratus

Agabus sp.

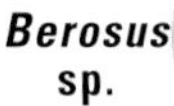

Berosus sp.

Cybister fimbriolatus

Tropisternus lateralis

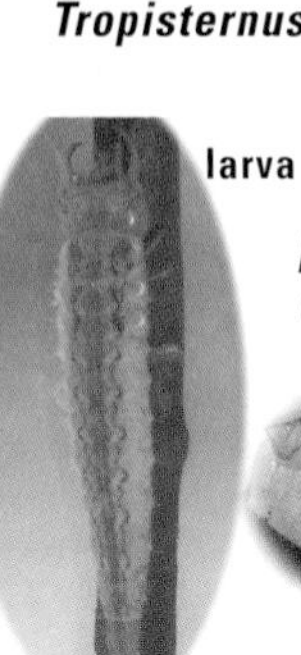

larva

Hydrophilus triangularis

adult

egg case

AQUATIC BEETLES

Compare these to the beetles on the previous page and also to some aquatic true bugs (order Hemiptera), especially those on pp. 106–109.

WHIRLIGIG BEETLES (family Gyrinidae) are the bumper cars of the beetle world. Their comical antics make them entertaining to watch as they skim the surface of ponds, lakes, and streams. Flipperlike middle and hind legs propel them at great speed. They can also dive, carrying an air bubble beneath them. With divided eyes, essentially one pair of eyes on top of the head and one pair on the underside, they see both above and below the water. Adult whirligig beetles feed on insects that fall onto the surface film. The larvae are underwater predators.

larva of whirligig beetle

Dineutus is a genus of 11 large species that often assemble in rafts of hundreds of individuals, especially in late summer and fall. When captured, they secrete a milky substance that smells like apples. They generally favor streams and small rivers.

Members of the genus ***Gyrinus*** are small but diverse, with more than 40 species. Although they are usually seen in small groups, they may form aggregations of amazing proportions. One flotilla observed on a Florida lake contained an estimated one million beetles. Different species prefer different habitats, but many in this genus favor lakes and ponds rather than flowing streams. Their defensive secretions have an unpleasant odor.

CRAWLING WATER BEETLES (Haliplidae) are small and mostly aquatic, but they are not particularly skilled swimmers. The coxal (basal) segments of the hind legs are expanded into plates covering much of the abdomen. Most are encountered along the edges of ponds, lakes, and slow-flowing streams, clambering about in algal mats. Adults feed on insect eggs, small invertebrates, and algae. The larvae eat algae exclusively. There are more than 60 species in North America, most belonging to just two genera, ***Haliplus*** and ***Peltodytes.***

WATER PENNIES (Psephenidae) are named for the round, disklike larval stage, found adhering to stones and submerged wood in fast-flowing streams and rivers. There they graze on algae or detritus. Adults are soft-bodied and terrestrial, found mostly on foliage near the larval habitat. There are six genera, and 16 species, in North America. The genus ***Psephenus*** includes seven species, collectively found continentwide except the Great Plains and Rockies. ***Eubrianax edwardsii,*** found only in California and Oregon, is the only North American species in its genus.

TOED-WINGED BEETLES (Ptilodactylidae) visit lights at night. They are easily confused with other beetles like adult water pennies but are generally more common. There are 28 North American species in six genera, but only ***Ptilodactyla,*** with nine species, is usually seen. Adults graze on mold spores on foliage, using brushlike modifications of their mouthparts. Larvae eat decaying vegetation or wood in moist habitats.

AQUATIC BEETLES

actual size

whirligig beetles (Gyrinidae)

Dineutus sp.

Gyrinus sp.

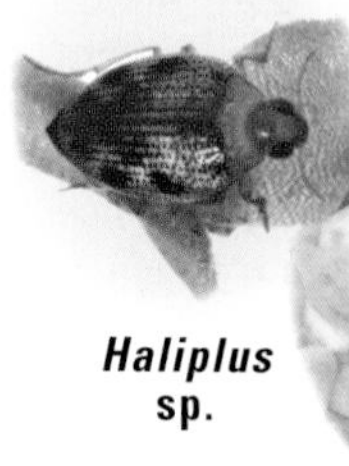

Haliplus sp.

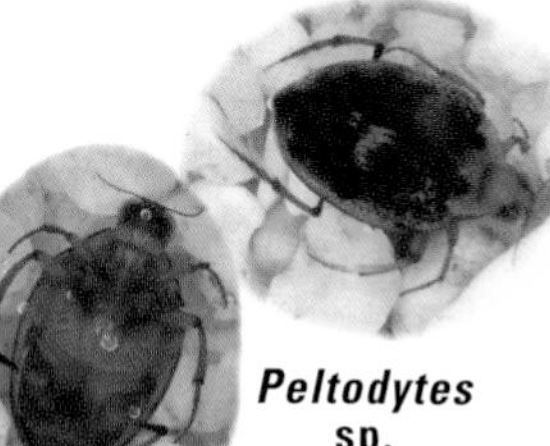

Peltodytes sp.

from below

crawling water beetles (Haliplidae)

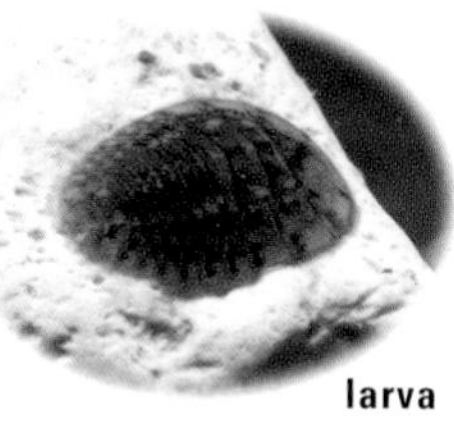

larva

Ptilodactyla serricollis

toed-winged beetle (Ptilodactylidae)

water pennies (Psephenidae)

Psephenus sp.

Eubrianax edwardsii larva

adult

LADY BEETLES

(family Coccinellidae) are also called ladybird beetles or ladybugs. Their polka-dot patterns and appetite for aphids make them among our favorite insects, and they have been chosen as the official state insect in Delaware, Massachusetts, New Hampshire, New York, Ohio, and Tennessee. Among our more than 480 species, most are tiny (1–3 mm). Many defend themselves by "reflex bleeding," secreting a sticky cocktail of irritating chemicals from their knee joints when molested. A few species migrate vertically to avoid hot, food-scarce summers in dry valleys, and/or hibernate in masses of hundreds or thousands. Some species have been introduced here from elsewhere. Other beetles can be mistaken for lady beetles, such as leaf and tortoise beetles (pp. 160–165), fungus beetles (p. 166), soft-winged flower beetles (p. 172), and carpet beetles (p. 210). Don't count on identifying all lady beetles by their color and pattern, as there is often great individual variation.

The genus ***Hippodamia*** includes 18 species in North America north of Mexico. The **Convergent Lady Beetle** *(Hippodamia convergens)*, an abundant transcontinental insect, is the one sold at gardening shops. The beetles will disperse upon release, likely benefiting someone else's yard. In many areas this species estivates, and hibernates, by the thousands, the masses from which commercial vendors make their harvest. This is the lady beetle found in huge concentrations among rocks on mountaintops in the west. Some individuals are devoid of spots. All lady beetles lack spots when freshly emerged from the pupa, gradually attaining full pigmentation as the cuticle hardens. The **Parenthesis Lady Beetle** *(Hippodamia parenthesis)* is widespread in the north, southward to the Carolinas, and in the Rockies to New Mexico. The **Thirteen-spotted Lady Beetle** *(Hippodamia tredecimpunctata)* is transcontinental in the northern half of North America.

The **Two-spotted Lady Beetle** *(Adalia bipunctata)* is another abundant species, and it sports many different color forms besides the typical pattern. It feeds on aphids and adelgids. ***Scymnus*** (not illustrated) is a genus with 92 North American species, collectively transcontinental in distribution. They are tiny (1–3 mm in length), and *many are dark-colored with a pale band across the tips of the wing covers.*

The genus ***Hyperaspis*** is represented by at least 94 species in North America, many of them blackish with yellow or orange spots. One species that is fairly common, especially in the west, is ***Hyperaspis quadrioculata.*** Another typical species that is more widespread is ***Hyperaspis proba.*** The genus ***Brachiacantha*** includes about 25 North American species, widely distributed. Larvae feed on scale insects protected inside ant nests. ***Brachiacantha ursina*** is a common eastern species found from the Carolinas and Missouri to Canada.

A species with a long history of use in integrated pest management, the **Mealybug Destroyer** *(Cryptolaemus montrouzieri)* was brought from Australia in 1892 to control mealybugs infesting California citrus groves. It is still propagated by insectaries in Ventura County for sale everywhere. The larvae resemble large mealybugs themselves.

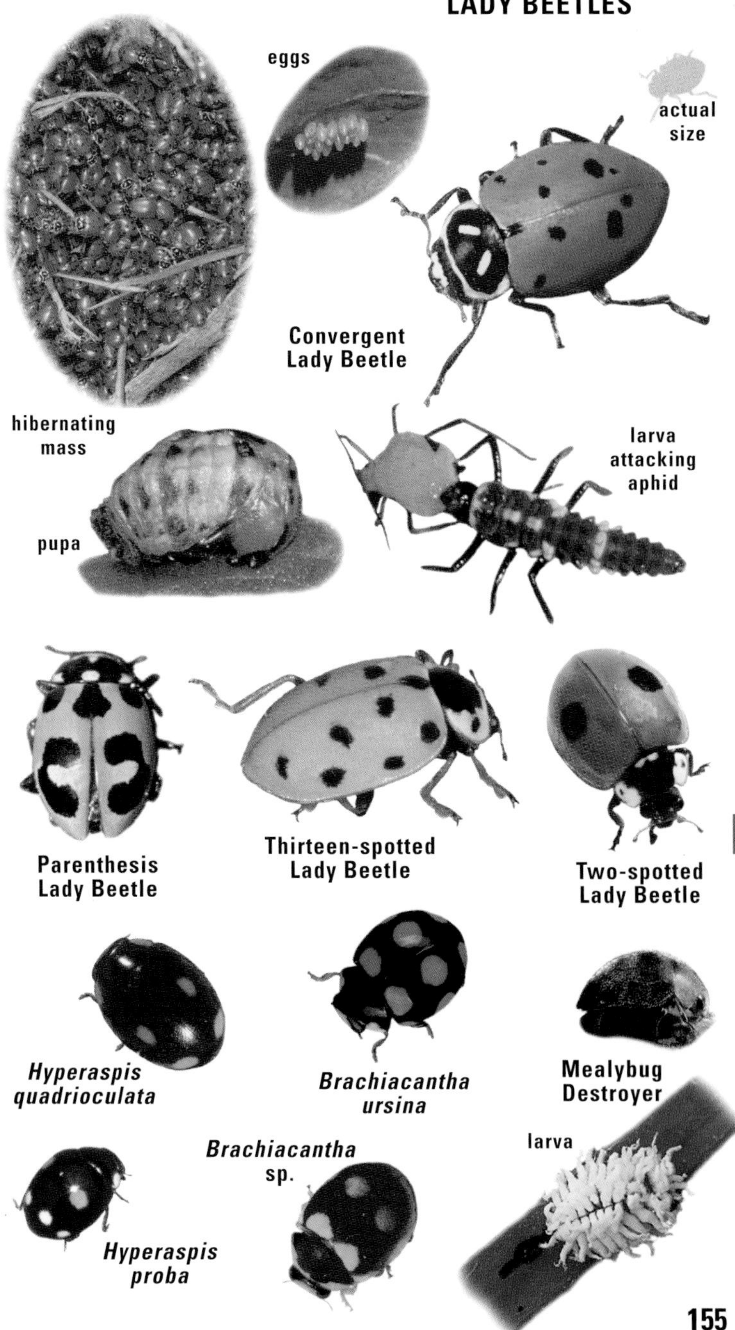
eggs
actual size
Convergent Lady Beetle
hibernating mass
larva attacking aphid
pupa
Parenthesis Lady Beetle
Thirteen-spotted Lady Beetle
Two-spotted Lady Beetle
Hyperaspis quadrioculata
Brachiacantha ursina
Mealybug Destroyer
larva
Brachiacantha sp.
Hyperaspis proba

Lady beetles in the genus ***Coccinella*** are oval convex insects with 12 species in North America. They are primarily aphid predators. The **Three-banded Lady Beetle** *(Coccinella trifasciata)* ranges across the northern U.S. and southern Canada, north to Alaska. The **Transverse Lady Beetle** *(Coccinella transversoguttata)* is common in the northern half of the continent, and south as far as Arizona and New Mexico in the west. The **Nine-spotted Lady Beetle** *(Coccinella novemnotata)* was formerly abundant here, occurring across the U.S. and southern Canada. It is the state insect of New York. Recently it has become scarce, perhaps displaced by the next species. The **Seven-spotted Lady Beetle** *(Coccinella septempunctata),* also known as the "C-7," has been spreading since its accidental introduction to Bergen County, New Jersey, in 1973. Ironically, a number of previous intentional attempts to bring it from Europe between 1956 and 1971 had failed. It is now common across southern Canada and the northern states.The **California Lady Beetle** *(Coccinella californica)* is common along the Pacific Coast. The **Eleven-spotted Lady Beetle** *(Coccinella undecimpunctata),* not illustrated, was accidentally introduced from its native Eurasia. It has been recorded from most southern Canadian provinces and parts of the northern and central U.S.

The **Spotted Lady Beetle** *(Coleomegilla maculata)* ranges throughout the eastern U.S. and across the southwest to California. The diet of this common beetle is evenly divided between aphids and pollen. Look for adults on flowers in open areas.

Chilocorus is a worldwide genus of effective predators on scale insects, with eight species here. The **Twice-stabbed Lady Beetle** *(Chilocorus stigma)* is eastern, ranging west to Alberta and Arizona. ***Chilocorus orbus*** (not illustrated) occurs along the Pacific Coast. Several species in other genera share the pattern of two red spots on a black ground color.

The **Vedalia** *(Rodolia cardinalis)* is widely hailed as the first example of biological control attempted in the U.S. It was introduced from Australia in 1888 and 1889 to save the California citrus industry from the cottony cushion scale, also of Australia. The experiment was a triumph, and the beetle is now found in a number of states from California to Florida.

The genus ***Harmonia*** includes three introduced species. The **Asian Multicolored Lady Beetle** *(Harmonia axyridis)* is an arboreal (tree-dwelling) aphid predator, coveted as a biological control in pecan and apple orchards. It was repeatedly introduced by the U.S. Department of Agriculture to California in 1916, 1964, and 1965, and to various eastern states between 1978 and 1982. Finally, an accidental introduction took hold near New Orleans in 1988. Today it is abundant north to Quebec and west to Texas and Missouri, with scattered occurrences elsewhere. In agricultural systems, this species may be displacing native lady beetles. They hibernate en masse, frequently selecting manmade structures in which to cluster. Their occupation results in a permeating odor, and the beetles bite on occasion. These habits have turned them into a "nuisance pest" on the order of Box Elder Bugs (p. 120).

actual
size
Three-banded
Lady Beetle
Transverse
Lady Beetle
Nine-spotted
Lady Beetle
California
Lady Beetle
Spotted
Lady
Beetle
Seven-spotted
Lady Beetle
Twice-
stabbed
Lady
Beetle
Vedalia
Asian
Multicolored
Lady
Beetle
larva
pupa
pattern
and color
variations

The genus ***Cycloneda*** includes three species here. The **Red Lady Beetle** *(Cycloneda munda)* is widespread in the eastern half of the U.S. and southern Canada, west to Colorado and Wyoming. The **Western Red Lady Beetle** *(Cycloneda polita)* occurs in the Pacific northwest, south to southern California and down the Rockies to northern New Mexico. ***Cycloneda sanguinea*** (not illustrated) is a southern species.

The genus ***Anatis*** includes our largest (7–10 mm) lady beetles, with four North American species usually found in trees. Adults darken in color with age. ***Anatis rathvoni*** ranges from western Canada to northern California. The **Eye-spotted Lady Beetle** *(Anatis mali),* which was featured on a U.S. postage stamp in 1999, ranges from southern Alaska to Tennessee and Virginia. The **Giant Lady Beetle** *(Anatis lecontei),* a giant at 10 mm long, is widespread and fairly common in the west.

The **Cream-spotted Lady Beetle** *(Calvia quatuordecimguttata)* comes in a bewildering array of patterns, to the extent that its English name often seems inappropriate. Widespread in northern and central regions of the continent, it preys on psyllids and aphids. The **Fourteen-spotted Lady Beetle** *(Propylea quatuordecimpunctata)* was introduced from the Old World, probably accidentally via shipping on the St. Lawrence Seaway in the late 1960s. It is now known from several areas of the northeast.

Another species with a misleading common name, the **Ashy Gray Lady Beetle** *(Olla v-nigrum)* actually occurs in various color forms. An important predator on aphids in California, it occurs across most of the U.S. except for most of Oregon and Washington.

The genus ***Mulsantina*** includes four species of pine lady beetles in North America. The **Painted Lady Beetle** *(Mulsantina picta),* found over much of the continent, is an aphid and adelgid predator. ***Psyllobora*** is a mostly Neotropical genus with six species in North America. They feed on powdery mildews, using specially adapted mouthparts to rake up spores. The **Twenty-spotted Lady Beetle** *(Psyllobora vigintimaculata)* occurs from Alaska and southern Canada to Mexico but is absent from Florida and the southern Atlantic Coast.

Members of the genus ***Epilachna*** are vegetarians. Our three U.S. species are widespread but are absent from the Pacific and north-central states. The **Mexican Bean Beetle** *(Epilachna varivestis)* has two populations in our area, one in the mountain west and one in the east. First found in Colorado in 1853, it may have stowed away in hay destined for cavalry horses used in the Mexican-American War. The hay scenario was repeated in 1918 when the beetle arrived in Alabama. The **Squash Lady Beetle** *(Epilachna borealis)* has a scattered distribution over the eastern half of the U.S. south of Massachusetts.

HANDSOME FUNGUS BEETLES (family Endomychidae) may be confused with lady beetles but are generally uncommon. The "dimples" on the thorax are distinctive. Forty-five species, in 22 genera, are known in North America. Only ***Endomychus biguttatus*** of eastern U.S. forests is likely to be noticed by the casual observer.

Anatis rathvoni

Western Red Lady Beetle

actual size

Red Lady Beetle

Eye-spotted Lady Beetle

Giant Lady Beetle

Cream-spotted Lady Beetle

Ashy Gray Lady Beetle
(2 color forms)

Fourteen-spotted Lady Beetle

Painted Lady Beetle

Twenty-spotted Lady Beetle

Endomychus biguttatus

Mexican Bean Beetle

Squash Lady Beetle

LEAF BEETLES

(family Chrysomelidae) are diverse, colorful vegetarians. Many species are host-specific, feeding only on one particular group of plants (willows, for example). Several species cause significant damage to crops, while others have been imported to control noxious weeds. Adults of some, known as flea beetles, jump when disturbed. Larvae are generally grublike and feed on foliage, but larvae of some species feed on roots, and a few are leaf miners, tracing patterns within the leaf as they feed. Spotted adults are often mistaken for lady beetles (pp. 154–159) or fungus beetles (p. 166). There are over 1,700 species of leaf beetles known in North America, with many new species awaiting formal description.

Beetles of the genus ***Calligrapha*** are well-named for their hieroglyphic markings. North America has 37 species. ***Calligrapha multipunctata*** is common on willows throughout much of the U.S. and Canada. ***Calligrapha dislocata*** is widespread in the southwestern U.S. and Mexico. Look for other representatives on elm, alder, birch, hazel, hollyhocks, tickseed *(Coreopsis),* and goldenrod, to name but a few known hosts.

The genus ***Chrysomela*** includes at least 17 species in North America, ranging as far north as the Arctic. Willows, alders, and poplars are the typical host plants. Adults of most species are small and heavily spotted. One of the most familiar is the **Cottonwood Leaf Beetle** *(Chrysomela scripta),* abundant over most of North America. ***Chrysolina*** is an enormous genus, but only 15 species are known from North America. Some of those are introduced, including the **Klamathweed Beetle** *(Chrysolina quadrigemina),* native to southern Europe, brought here to combat an exotic St. John's-wort that causes sun sensitivity in cattle. The **Milkweed Leaf Beetle** *(Labidomera clivicollis)* is widespread in the east. Adults nip milkweed leaf veins to drain the tips of toxic latex before feeding.

The poorly named **Colorado Potato Beetle** *(Leptinotarsa decemlineata)* is native to Mexico, where it originally fed on burweed. As cattle were grazed farther north, they spread burweed and the beetles followed, reaching Colorado about 1822. Previously, Pizarro had found the Incas cultivating potatoes in South America and had shipped some to Spain. From there the crop made it to England and on to the states. Beetle and potato met in the midwest, the insect developing a preference for this new foodplant.

The **Larger Elm Leaf Beetle** *(Monocesta coryli)* ranges widely in the east and midwest. It varies from dull yellow to orange, with or without black spots or bands. The **Elm Leaf Beetle** *(Xanthogaleruca luteola)* was introduced from Europe and is now common over most of North America. ***Trirhabda*** is a genus with 23 species in North America. Most species are striped and small (about a quarter inch long). Host plants include sagebrush, desert broom, rabbitbrush, sunflowers, thistles, and goldenrod.

The genus ***Diabrotica*** includes eight species in North America. The **Spotted Cucumber Beetle** *(Diabrotica undecimpunctata)* is abundant on flowers and foliage of many plants. The related **Western Corn Rootworm** *(Diabrotica virgifera)* is also common, and not just in the west. Larvae of both are bona fide pests in the roots of corn and related cereal crops.

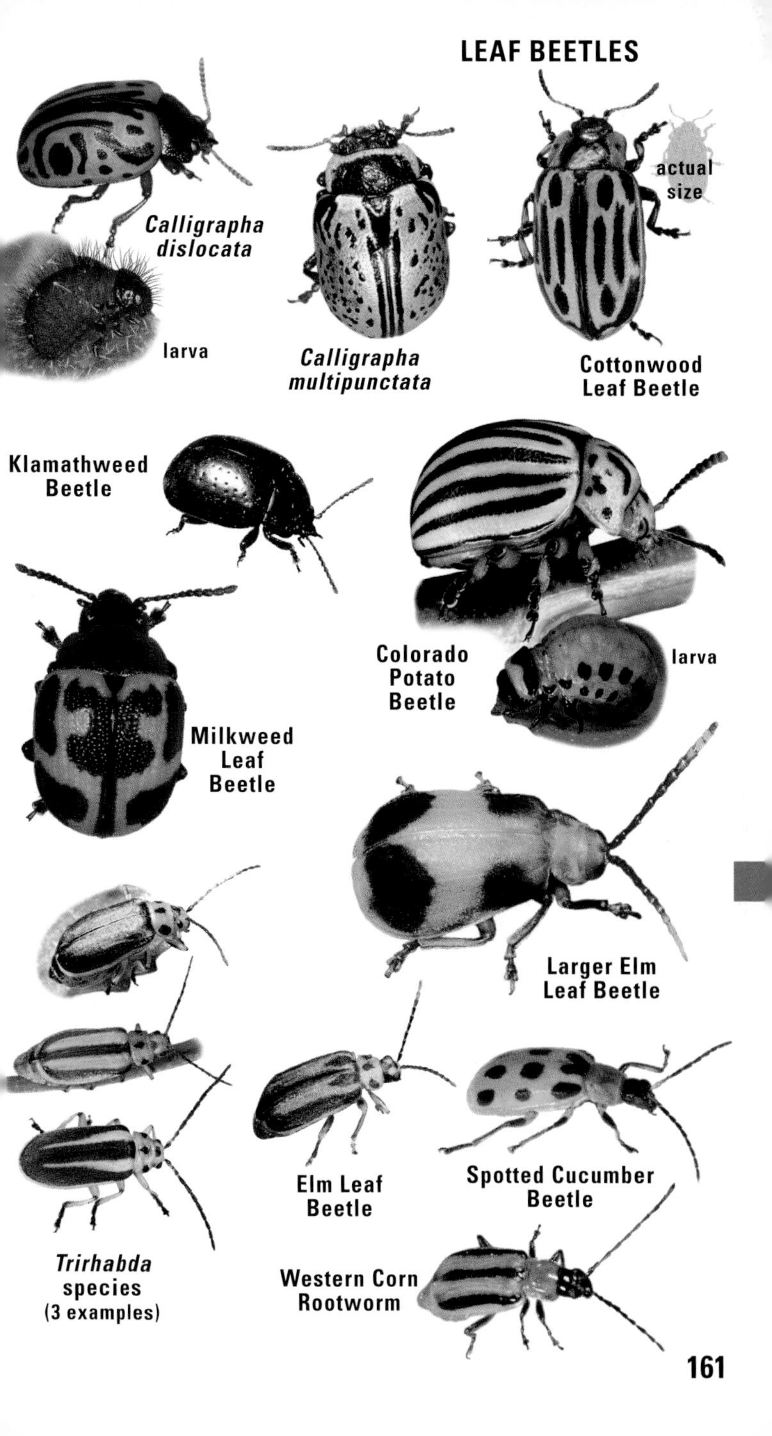
actual
size
Callligrapha
dislocata
larva
Calligrapha
multipunctata
Cottonwood
Leaf Beetle
Klamathweed
Beetle
Colorado
Potato
Beetle
larva
Milkweed
Leaf
Beetle
Larger Elm
Leaf Beetle
Elm Leaf
Beetle
Spotted Cucumber
Beetle
Trirhabda
species
(3 examples)
Western Corn
Rootworm

Adults of the **Bean Leaf Beetle** *(Cerotoma trifurcata),* one of three North American species in its genus, may be yellow, olive, or red. As the name suggests, they feed on various members of the legume (bean) family, and they sometimes become superabundant in soybean fields.

Immortalized on a U.S. postage stamp in 1999, the **Dogbane Beetle** *(Chrysochus auratus)* is widespread in the east, extending into the interior of the west. It feeds mainly on dogbane (Apocynaceae). Its western counterpart, the **Blue Milkweed Beetle** *(Chrysochus cobaltinus),* has a wider variety in its diet, feeding on milkweeds as well as dogbane. These two beetles meet and hybridize in southern Washington state, and their interactions there have been the subject of some important scientific studies. The genus ***Fidia*** is represented by at least six species in North America, associated mostly with grape. Adults are gangly, fast-moving, and covered with pale gray or brown scales. The **Grape Rootworm Beetle** *(Fidia viticida)* is common on that host plant in the eastern half of the continent.

Altica is a genus of "flea beetles" with about 70 species in North America. Many are solid metallic black, blue, or green, and most are very small (often a mere eighth of an inch). As their name implies, they jump when danger threatens. The genus ***Disonycha*** includes about three dozen species in the U.S. and southern Canada. Most are small and marked with a colorful pattern of stripes, and most of the species feed on weeds. The **Alligatorweed Flea Beetle** *(Agasicles hygrophila)* was brought from South America to control the equally exotic alligatorweed, a plant that was clogging Florida waterways.

The genus ***Pachybrachis*** (formerly spelled *Pachybrachys)* includes 150 North American species, found mostly in the southern U.S. and drier habitats elsewhere. Adults are very small and can vary greatly in color pattern from one individual to the next, even within species. ***Cryptocephalus*** is a genus with more than 70 species in the U.S. and southern Canada. The larvae apparently feed in leaf litter, adults on foliage of various plants.

The genus ***Saxinis*** includes 10 species in North America, eight of those restricted to the southwestern U.S. ***Saxinis saucia*** is more widespread in the west, extending north into British Columbia. Larvae live with ants, protected by a case constructed from fecal material and detritus. Look for adults feeding on foliage of wild rose and other plants.

The **Clay-colored Leaf Beetle** *(Anomoea laticlavia)* is widespread in the U.S. and southeastern Canada. Both adults and larvae feed on the foliage of legumes and many other plants. ***Chlamisus*** is a mostly neotropical genus with eight species in North America, most occurring in the southwest. ***Chlamisus foveolatus*** is widespread in the east. Adults are found on various oaks and other plants; larvae probably feed on the same plants as adults, but their habits are poorly known. Members of the genus ***Exema*** look like animated caterpillar droppings. Our nine species collectively range south of southern Canada. Most are very small, no more than 2–3 mm in length. Adults and larvae are largely associated with plants in the aster family. When disturbed, the beetles feign death and roll off the leaf.

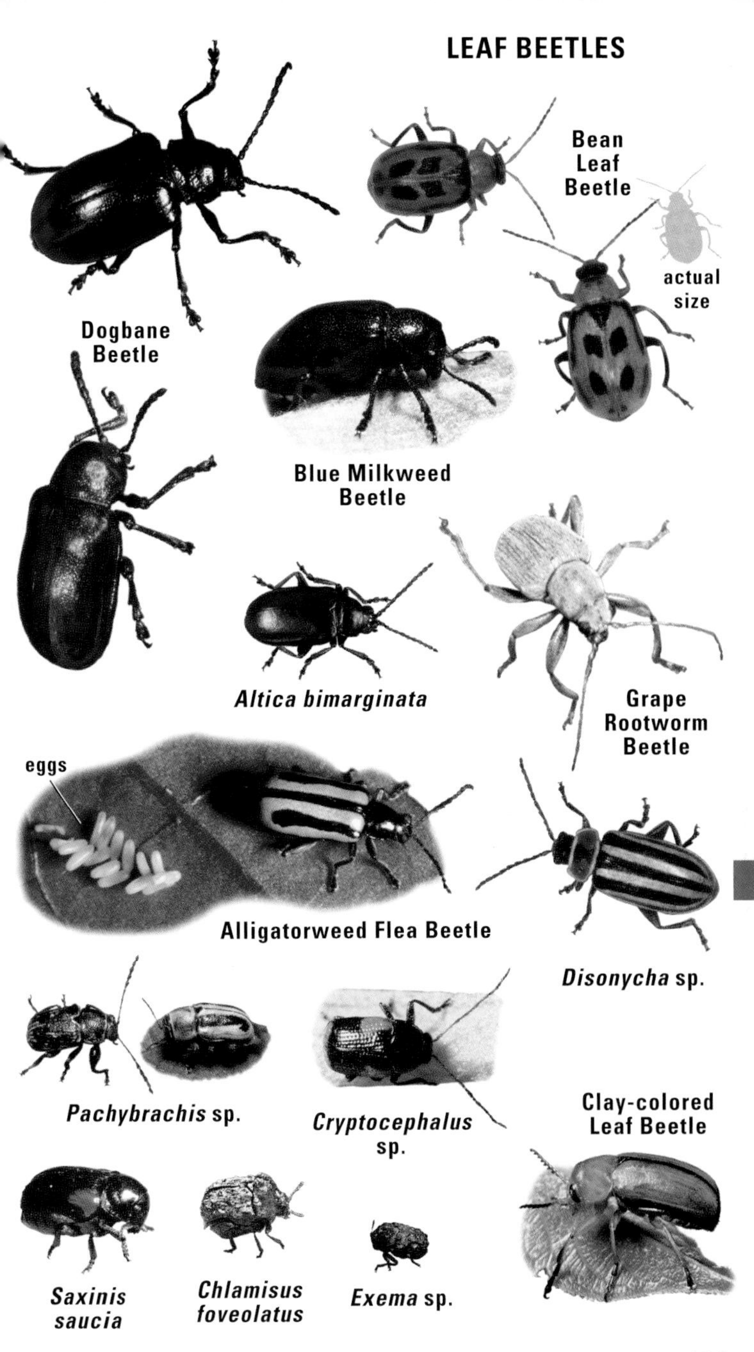

Bean Leaf Beetle

Dogbane Beetle

Blue Milkweed Beetle

Altica bimarginata

Grape Rootworm Beetle

Alligatorweed Flea Beetle

Disonycha sp.

Pachybrachis sp.

Cryptocephalus sp.

Clay-colored Leaf Beetle

Saxinis saucia

Chlamisus foveolatus

Exema sp.

The **Golden Tortoise Beetle** *(Charidotella sexpunctata bicolor),* one of six members of its genus here, has brilliant colors formed by the microscopic structure of the cuticle. The live beetle can change from metallic to non-metallic "at will" by emptying or filling microscopic cavities with fluid; the change in size of the cavities causes light to refract differently. Since the cavities collapse in dry beetles, preserved specimens almost always lack the metallic color. This is a common, widespread insect found on morning glory. Larvae hide beneath an umbrella of dried fecal matter that they pile onto forked tail-like appendages that curl over their backs.

Deloyala is a genus of tortoise beetles with three representatives here, generally distributed except for the Pacific northwest. The **Mottled Tortoise Beetle** *(Deloyala guttata)* is common on morning glory and related plants. The **Argus Tortoise Beetle** *(Chelymorpha cassidea)* is one of three North American species in its genus. Adults are found on morning glory and milkweed. As with other tortoise beetles, it overwinters in the adult stage and usually has only one generation per year.

The genus ***Syneta*** includes eight species occurring over much of North America. The **Western Fruit Beetle** *(Syneta albida),* not illustrated, can be a pest in orchards and fields where it feeds on the blossoms and new leaves of apple, pear, currant, cherry, peach, and strawberry. Willow may be one of its native hosts. Others in this genus feed on various other trees, including both deciduous trees and conifers.

Crioceris is a genus of Old World beetles, with two introduced to North America from Europe: the **Asparagus Beetle** *(Crioceris asparagi)* and the **Spotted Asparagus Beetle** *(Crioceris duodecimpunctata).* Both are occasional pests of asparagus across southern Canada and the northern U.S.

Lema is a mostly tropical and subtropical genus, with only 13 native species (and one introduced species) in North America. Adults usually drop or fly away when approached. Many are associated with plants in the nightshade family (Solenaceae); for example, ***Lema daturaphila*** has been found feeding on ground cherry *(Physalis).* ***Oulema collaris,*** on the other hand, has been found on spiderwort *(Tradescantia)* in the iris family.

Odontota is a genus with six species in the eastern U.S. and one in the southwest. The **Locust Leafminer** *(Odontota dorsalis)* can be abundant on black locust, but larvae also feed on foliage of some other leguminous trees, and adults feed on a wider variety of plants. Larvae form blotch mines between the layers of leaflets, creating large patches that give the foliage a reddish brown appearance. The genus ***Microrhopala*** includes nine species in the U.S. and southern Canada. As with the preceding genus, these beetles spend the larval stage mining within leaves. One of the more widespread species in this group is the **Goldenrod Leafminer** *(Microrhopala vittata).* ***Donacia*** is a genus of aquatic plant-eaters with 31 species here. Adults are recognized by their long antennae and often thickened "thighs." Adults feed on emergent vegetation, but larvae feed on submerged stems and roots. They tap air from the plant itself, sticking their knifelike terminal segments (and associated breathing holes) into the stem.

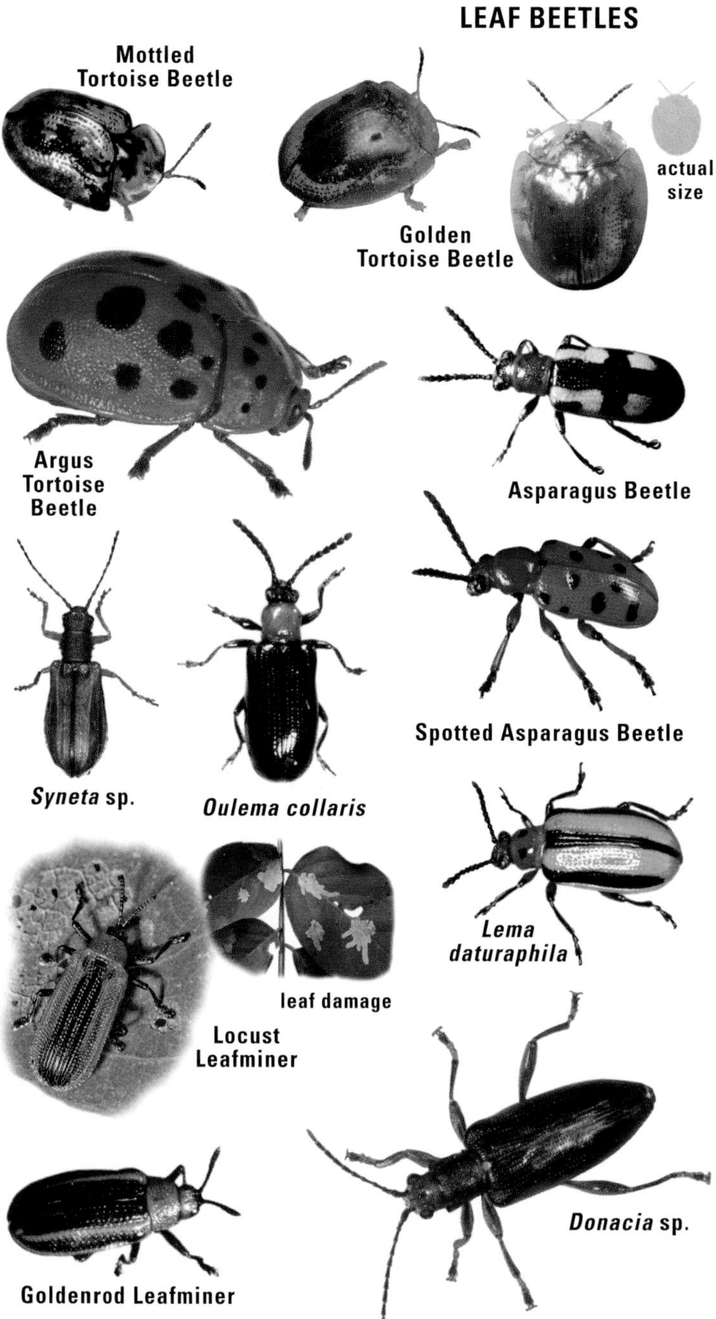
Mottled
Tortoise Beetle
Golden
Tortoise Beetle
actual
size
Argus
Tortoise
Beetle
Asparagus Beetle
Spotted Asparagus Beetle
Syneta sp.
Oulema collaris
Lema
daturaphila
leaf damage
Locust
Leafminer
Goldenrod Leafminer
Donacia sp.

FUNGUS BEETLES

The four families included on this page are not all closely related to each other, but all include species regularly found around fungi. Note that beetles from some other families are also common around fungi: for example, some of the darkling beetles (see p. 192).

PLEASING FUNGUS BEETLES (family Erotylidae) are often present in large numbers on the fruiting bodies of various fungi, especially bracket fungi on logs. There are about 50 species in North America.

Megalodacne heros and ***Megalodacne fasciata*** are the only two North American species in their genus, both found in the U.S. east of the Rockies. Larvae bore in woody bracket fungi. Adults are active at night, hiding by day. The beetles also hibernate in groups under bark or logs.

Ischyrus is a mostly tropical genus with two species in the eastern U.S. and one in Arizona. ***Ischyrus quadripunctatus*** is the common eastern species, variable in its color pattern. Adults are nocturnal and sometimes come to lights.

The genus ***Triplax*** includes 18 species in North America, generally distributed. Adults are active by day. Some species can be abundant in fleshy bracket fungi on dead, standing deciduous trees in the eastern U.S. ***Triplax californicus*** is common from British Columbia to California, especially in *Pleurotus* fungi.

Gibbifer californicus is the only North American representative of this neotropical genus (formerly placed in *Cypherotylus*). It is a fairly common insect on fungi or rotted wood in moist habitats at higher elevations in the southwestern U.S. Larvae are nocturnal and feed in groups.

HAIRY FUNGUS BEETLES (Mycetophagidae) are not uncommon in fleshy fungi, escaping rapidly if disturbed. They also occur under bark and in moldy plant matter. There are five genera, with 26 species, in North America. The 15 members of the genus ***Mycetophagus*** are the most often encountered. ***Typhaea stercorea*** (not illustrated) is sometimes considered a pest in stored grain, but it favors grain that is already moldy.

FALSE DARKLING BEETLES (Melandryidae) include 24 genera and 60 species north of Mexico. Most common species are found under bark or in fungi. The five species of ***Eustrophinus*** (not illustrated) occur in fleshy fungi. They are difficult to catch, being quick and slippery. ***Dircaea liturata*** (not illustrated), elongated and black with zigzag orange markings above, sometimes comes to lights at night in the eastern U.S. and southern Canada. Look for it on trunks of trees at night as well.

POLYPORE FUNGUS BEETLES (Tetratomidae) were once lumped in with the preceding family, the false darkling beetles. There are 10 genera and 26 species in North America, of which the two species of ***Penthe*** are the most commonly seen. ***Penthe obliquata*** has an orange spot between its wing covers. It ranges in the eastern half of the U.S. and adjacent Canada. ***Penthe pimelia*** lacks the orange spot and has a similar range, southwest to Texas. Both occur under bark, in fungi, and in decaying wood.

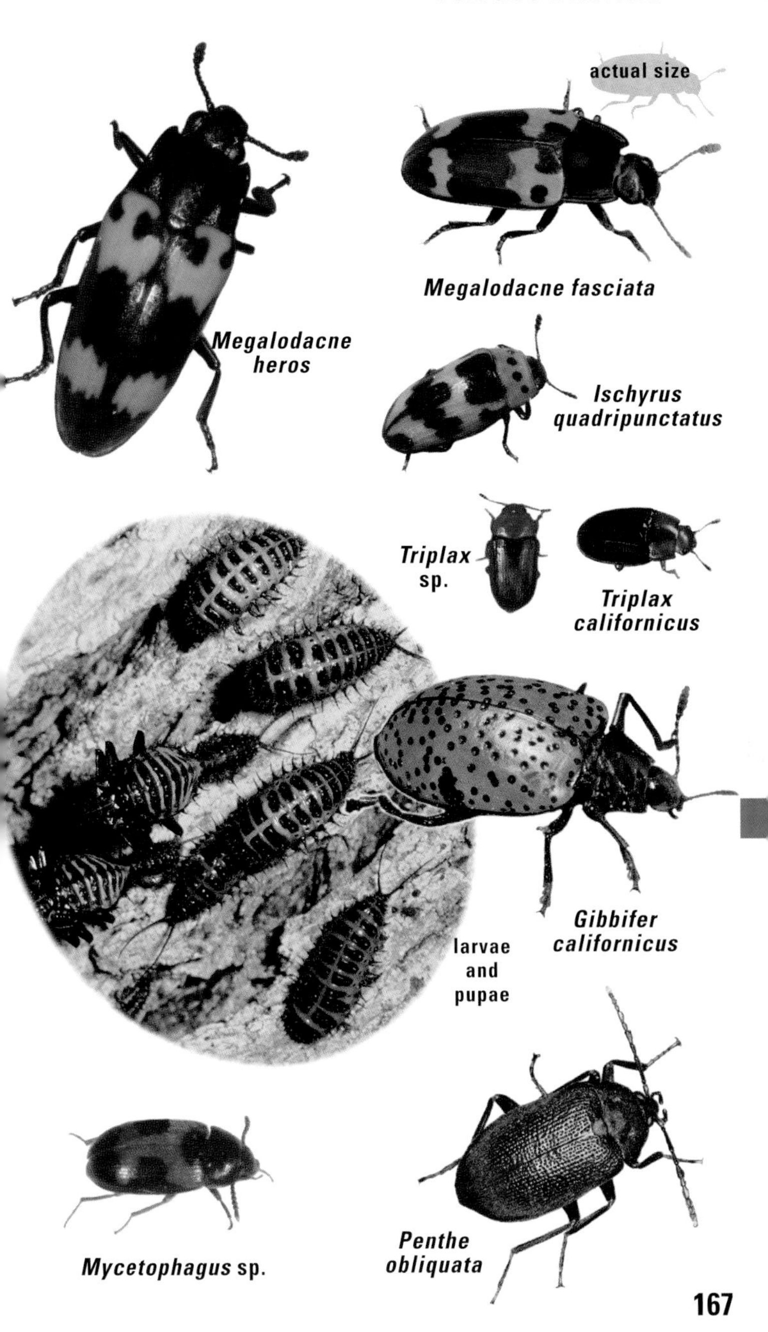
actual size
Megalodacne heros
Megalodacne fasciata
Ischyrus quadripunctatus
Triplax sp.
Triplax californicus
Gibbifer californicus
larvae and pupae
Mycetophagus sp.
Penthe obliquata

NET-WINGED BEETLES (family Lycidae) resemble fireflies (next page), but usually have a *raised, netlike pattern of veins* on their soft wing covers. Many are brightly colored, suggesting a distasteful or poisonous nature. Most are active by day, or at dawn and dusk, occurring on foliage or flowers, or feeding on honeydew at aphid colonies. Larvae feed on fungi. There are more than 75 species in North America.

The genus ***Calopteron*** includes five common eastern and northern species of net-winged beetles, usually seen sitting on leaves or flying slowly through forests at dusk. They may raise their wing covers in an odd defensive display when threatened. Look for larvae in decaying wood or crawling on the forest floor. ***Lycus*** is a common genus in the southwest. The 11 species have elongated heads, the better to probe blossoms for nectar, and they are usually found on flowers. ***Dictyopterus*** is a genus with four North American species ranging in northern and coastal areas. ***Dictyopterus hematus*** is a stunning scarlet creature of the coniferous forests of the northwest. The genus ***Plateros*** includes 31 species of small beetles, easily mistaken for *Pyropyga* fireflies (p. 170), widespread across southern Canada and throughout the U.S.

SAP BEETLES (Nitidulidae) are a diverse lot, most of which are attracted to fermenting fruit, grains, sap, or decaying fungi. A few are found in flowers, or live in nests of ants or bees. Larvae of some species are predatory, and a few breed in carrion. There are 165 species, in 30 genera, north of Mexico. The four species of ***Conotelus*** resemble rove beetles and are commonly found in flowers of morning glory, hollyhock, and dogwood over much of North America excepting the Pacific northwest. The genus ***Carpophilus*** includes 32 species in North America. Some are common in flowers of yucca. One species, the **Dusky Sap Beetle** *(Carpophilus lugubris)*, not illustrated, is found in rotting fruit but also in ripening corn, where it can be an occasional pest. Members of the genus ***Glischrochilus*** are the picnic beetles, with nine species in North America. All are glossy black marked with yellow or red. They may fly in fair numbers to sap and decaying fruit.

GLOWWORMS (Phengodidae) are generally uncommon. Females reach maturity while retaining a larval appearance. Males are soft-bodied beetles with feathery antennae, and they sometimes fly to lights at night. Females and larvae glow between their body segments, like a worm ringed in halos of light. They feed exclusively on millipedes. There are 23 species in the U.S. The genus ***Phengodes*** is represented here by 10 species found in the eastern and central U.S., southwest to southern California. Males have shortened wing covers and plumelike antennae. ***Phengodes plumosa*** is the most frequently seen species in the east. The genus ***Zarhipis*** includes three species in the Pacific states and southwest.

actual size

Calopteron discrepens

Calopteron reticulatum

Calopteron terminale

Lycus lecontei

Plateros sp.

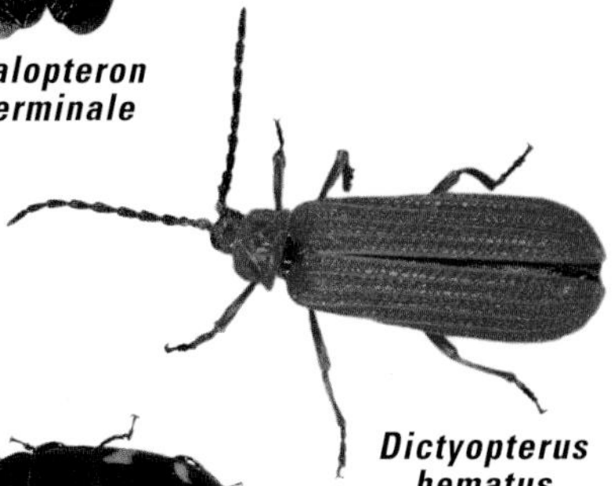

Dictyopterus hematus

Conotelus sp.

Glischrochilus quadrisignatus

Carpophilus sp.

Phengodes plumosa

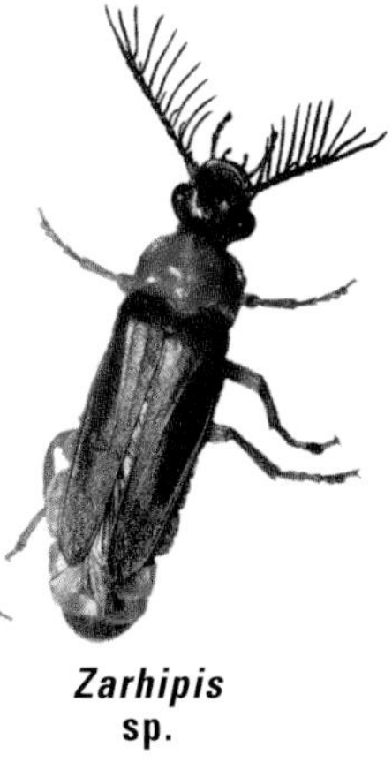

Phengodidae larva or female

Zarhipis sp.

FIREFLIES

(family Lampyridae), also popularly known as lightning bugs, are neither flies nor bugs, but beetles. They are not all luminescent in the adult stage, but all known larvae possess light-bearing cells. The glow is produced in a chemical reaction catalyzed by an enzyme, a remarkable process in which virtually no energy is lost as heat. These soft-bodied beetles usually have the head hidden under the thoracic shield. Larvae are predatory, some specializing on mollusks or earthworms. Larvae reach 12–25 mm (half an inch to an inch) before pupating. In some species, females are wingless, and a few mature sexually while retaining a larval form. Lampyrids have poisonous blood, exuding droplets from the base of their wing covers in defense. There are 20 genera in North America, with at least 150–200 species. New species have recently been identified based on differences in light-flashing patterns. Other beetles mimic fireflies, including soldier beetles (next page), net-winged beetles (previous page), and some checkered beetles (p. 206).

Members of the genus ***Photinus*** are among our most familiar fireflies, with 34 species in our fauna, most occurring in the east. In several species the female is wingless. Both sexes signal with yellow flashes. Larvae are subterranean and may hunt earthworms in "packs." ***Photinus pyralis*** is one common species. Watch for males in flight just after sundown, blinking over fields and lawns where the females wait.

Photuris is a genus with 22 known species in North America (at least 28 more await formal description), mainly in the east. They are most common in low-lying fields and in marshes of the northeast and northern midwest. Males begin flying well after dark, and some species fly above the tree canopy. Their flashes are green or yellow-green. Females sometimes mimic the flashes of female *Photinus* fireflies. A male *Photinus* that falls for the ploy will be devoured! Female *Photuris* accumulate defensive steroids from these meals. Larvae of *Photuris* are omnivores, eating dead insects and fallen berries. ***Photuris pennsylvanica*** is the state insect of Pennsylvania, but despite the name, there is some question whether it actually occurs in that state.

The genus ***Pyractomena*** includes 16 species in the U.S., mostly in the east. Their flashes are orange-yellow or amber. Larvae are snail predators. They sometimes shine from several feet above the ground, especially on damp nights. Most species occur along forest edges or lake shores.

Our dozen species of ***Ellychnia*** are day-active nonluminescent beetles. ***Ellychnia hatchi*** is the common firefly in the Pacific northwest. Adults are found on tree trunks or foliage. Members of the genus ***Lucidota*** are also diurnal fireflies that do not light up as adults. Our three species are widespread. Adults have flattened antennae. The common eastern species, ***Lucidota atra,*** is found in open woodlands.

The genus ***Pyropyga*** includes four species. Their light organs are feeble or nonfunctional, and courtship is initiated with odors (pheromones) liberated by females. Females of some species have short wings and wing covers. Adults are often found on flowers or at aphid colonies.

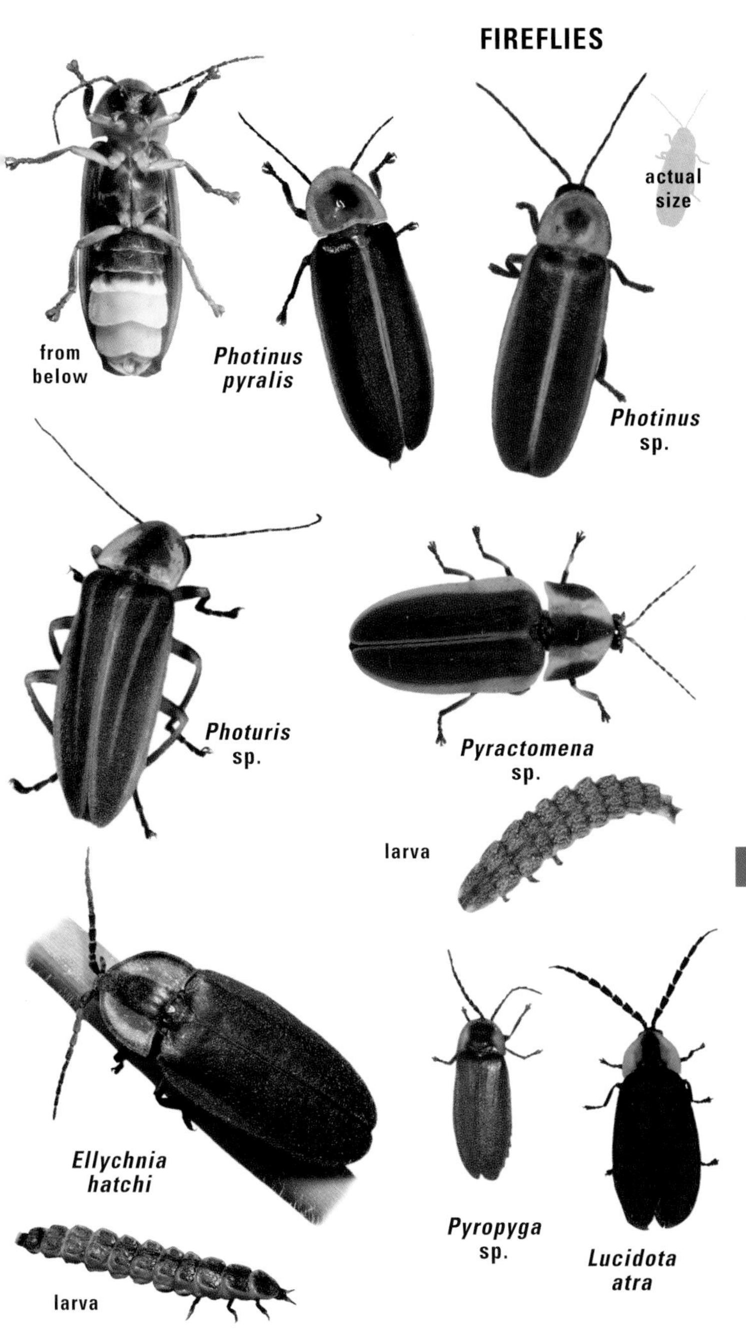

from below

Photinus pyralis

Photinus sp.

Photuris sp.

Pyractomena sp.

larva

Ellychnia hatchi

larva

Pyropyga sp.

Lucidota atra

SOLDIER BEETLES (family Cantharidae) are also known as leatherwings. Many of our common species emerge in spring or fall and can be abundant by day on flowers and foliage. They fly well and are valuable pollinators, and many eat aphids and other insects. Adults can be confused with many other kinds of beetles, some of which mimic them. Both adults and larvae have glands at the rear of the abdomen that secrete defensive chemicals. Larvae live mostly in leaf-litter; under stones, logs, and other debris; or in decaying wood. They prey on eggs and larvae of other insects. There are more than 470 species in North America.

Members of the genus ***Chauliognathus*** are abundant, especially on spring and fall flowers. There are 19 species in our fauna, all of them valuable pollinators and agents of pest control. The **Pennsylvania Leatherwing** *(Chauliognathus pennsylvanicus)* occurs on goldenrod and thoroughworts in the east in autumn. ***Chauliognathus profundus*** is often common on flowers in the southwest in late summer. ***Chauliognathus marginatus*** is another eastern species, seen in spring and early summer on flowers of wild hydrangea, tree of heaven, linden, and New Jersey tea.

The genus ***Cantharis*** includes 22 species in our fauna, two of them introduced from Europe, collectively distributed over most of the continent in forested and open habitats. The **Brown Leatherwing** *(Cantharis consors)* is common in parts of the west. ***Cantharis rufa*** is a very widespread species, common over most of Europe and northern Asia as well as in North America. ***Ancistronycha bilineata*** (formerly *Cantharis bilineatus*) is especially common on flowers of red haw in the eastern U.S.

Podabrus is a widely distributed genus in North America, with more than 100 species. Adults are 7–14 mm in length and feed on aphids, the honeydew they exude, or both. They sometimes come to lights at night. ***Podabrus pruinosus*** is a common western species typical of the genus. A few species are more colorful, with metallic colors above.

SOFT-WINGED FLOWER BEETLES (Melyridae) are small, active omnivores, predators, and pollen-feeders. Many are brightly colored and resemble lady beetles (pp. 154–159). There are at least 520 species, in 58 genera, in North America. In about one-third of our fauna, adults have "inflatable" sacs on the sides of their bodies, apparently deployed in defense. Larvae are probably scavengers, predators, or both, and live in leaf litter or under bark. We have 27 species in the genus ***Malachius,*** including the **Scarlet Malachite Beetle** *(Malachius aeneus)*. Widespread in Europe and the Middle East as well as North America, this species has become rare and possibly endangered in England. We have six species in the genus ***Anthocomus,*** including one introduced from Europe. Look for them feeding on aphids in trees. The genus ***Collops*** includes close to 50 species, common on flowers over most of North America. Adults are mostly less than a quarter inch long. ***Collops vittatus*** and ***Collops bipunctatus*** are two widespread and common species that have been credited with helping to control various crop pests in agricultural fields.

actual size

Pennsylvania Leatherwing

Chauliognathus profundus

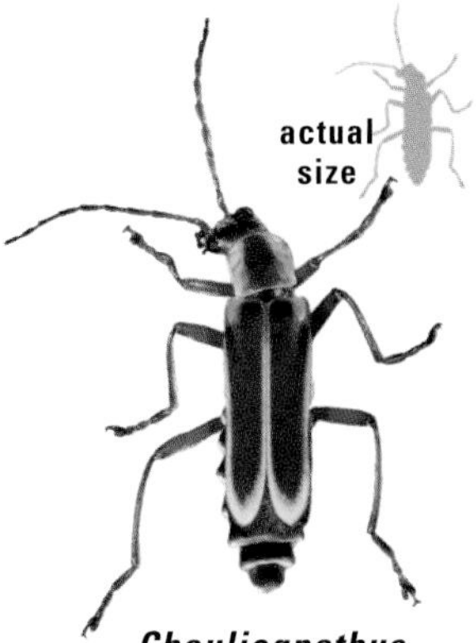

Chauliognathus marginatus

Ancistronycha bilineata

Brown Leatherwing

Podabrus pruinosus

Cantharis rufa

Scarlet Malachite Beetle

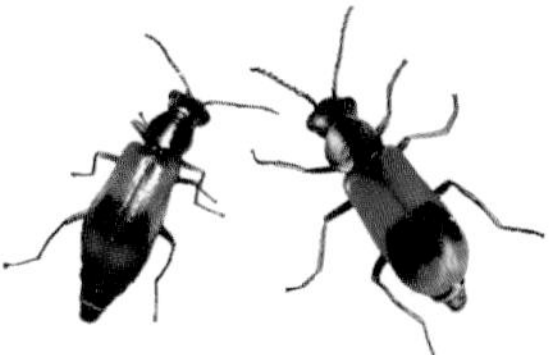

Anthocomus sp.

Podabrus sp.

Collops vittatus

Collops bipunctatus

LONG-HORNED BEETLES

(family Cerambycidae) include "sawyers," "pruners," "girdlers," and a variety of borers. More than 900 species are known in North America. Larvae are "round-headed borers," most found in dead or dying wood. Others mine live plants. Serpentine paths etched in wood, packed with concentric arcs of fibrous "frass" (dried poop), are the work of the larvae. Adults have exceedingly long antennae, longer in males. When held, some squeak by rocking their head, rubbing minute ridges against the inside surface of the thorax. Most fly well and can be found on logs, tree trunks, flowers, or at lights at night. A few are occasional pests, but some literally shape the forest canopy or help recycle dead wood into soil.

On this page, the **FLAT-FACED LONGHORNS (subfamily Lamiinae)** have *vertical faces* and often a single spike on each side of the thorax.

The genus ***Saperda*** counts 15 species in North America, collectively widely distributed. Larvae of some species form galls on the host tree, and adults often feed on foliage of the trees that are the larval hosts. The **Round-headed Appletree Borer** *(Saperda candida)* also bores in serviceberry and other members of the rose family. It occurs in the eastern U.S. and southeast Canada. The life cycle takes two years in the south, three to four years in the north. The **Elm Borer** *(Saperda tridentata)* is common at lights in the east. Larvae bore in live elms, just under the bark and into the sapwood, for two to three years. The **Small Poplar Borer** *(Saperda populnea)* ranges across southern Canada and the northern U.S., south along mountain ranges. The **Woodbine Borer** *(Saperda puncticollis)* ranges across the east. Larvae bore in dead vines of Virginia creeper, grape, and poison ivy. Adults are active from May to August. Their coloration suggests they may be firefly mimics.

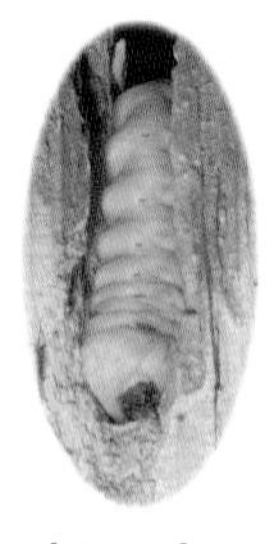

larva of a long-horned borer (*Ergates* sp.)

The eight species of ***Monochamus*** are known as "sawyers." Adults are mostly active by day, feeding on leaf petioles and twig bark. Antennae of males can be three times the body length. The **White-spotted Sawyer** *(Monochamus scutellatus)* is highly variable in color and pattern and is widespread except for the central U.S. (The **Oregon Fir Sawyer** is but one subspecies, *M. s. oregonensis,* occurring in the Pacific states and British Columbia.) Larvae mine dead and dying conifers, particularly pine. Should their paths cross, one larva usually cannibalizes the other. The life cycle takes one year in southern parts of the range, two years in the north. The **Northeastern Sawyer** *(Monochamus notatus)* ranges over central and eastern North America, with isolated populations in the Pacific northwest. Conifers, especially pine, are the larval hosts.

The **Cottonwood Borer** *(Plectrodera scalator)* is widespread east of the Rockies in the U.S. Larvae bore in the base of cottonwood and other poplars. Their galleries often extend into taproots. Adults are active by day, feeding on leaf petioles, new twigs, and soft bark of host trees. They fly well and never fail to draw attention with their spectacular pattern.

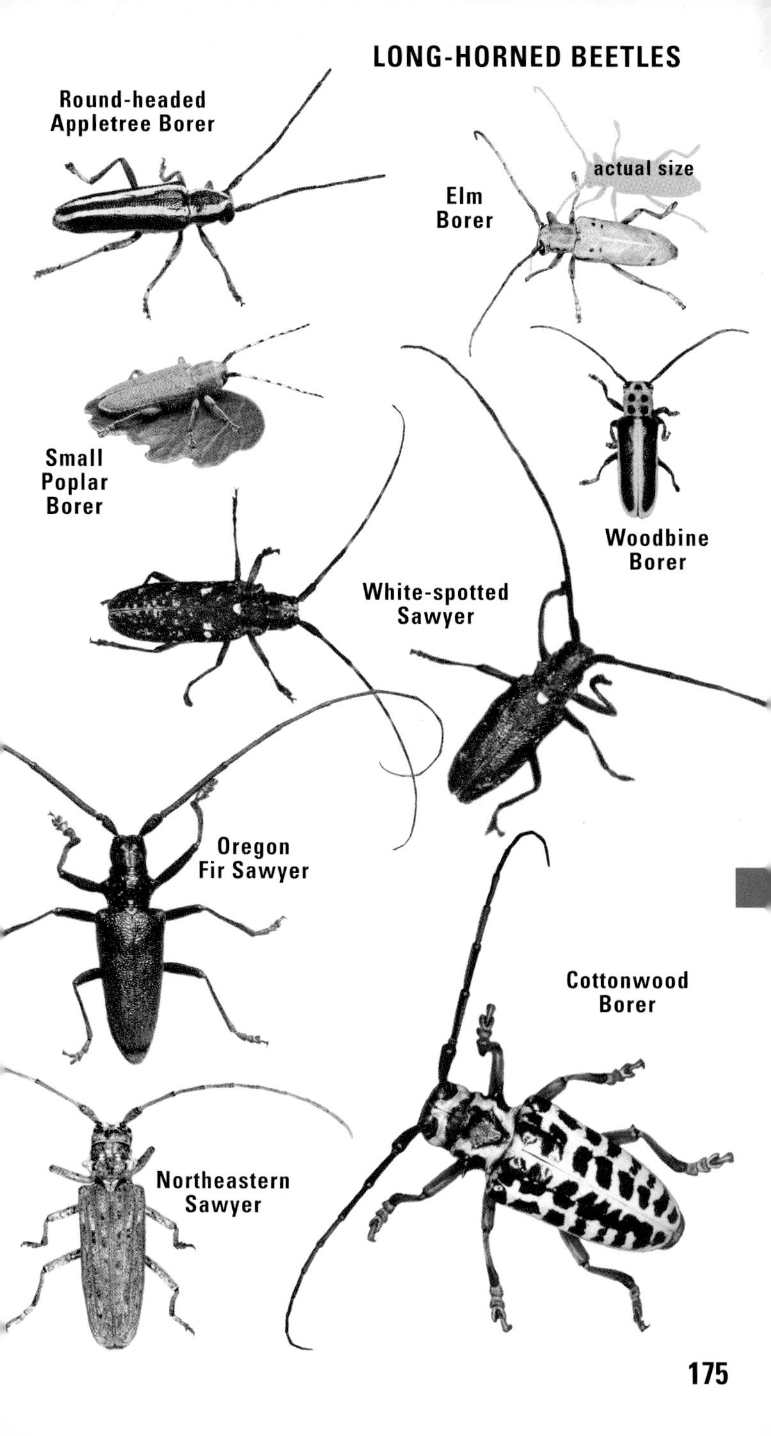
Round-headed
Appletree Borer
Elm
Borer
actual size
Small
Poplar
Borer
Woodbine
Borer
White-spotted
Sawyer
Oregon
Fir Sawyer
Cottonwood
Borer
Northeastern
Sawyer

The nine species in the genus ***Goes*** are stocky beetles confined to the eastern and central U.S. Larvae of the **White Oak Borer** *(Goes tigrinus)* bore in the sapwood of young, living white oak trees, occasionally becoming serious pests. They generally take three to four years to complete development. Adults are relatively long-lived, active from April to August, varying with latitude.

The genus ***Oncideres*** includes four species of girdlers ranging from the eastern U.S. to Texas and Arizona. The common name describes the female's habit of biting a deep groove around a twig or small branch where she lays an egg, to furnish her larva with dying wood beyond the scar. The **Mesquite Girdler** *(Oncideres rhodosticta)* is common throughout the southwest. Look for adults at lights at night.

The **Asian Long-horned Beetle** *(Anoplophora glabripennis)* is native to Japan, Korea, and southeastern China, but since 1996 it has invaded the U.S. in solid wood packing material shipped from China. Now established in Chicago and on Long Island, New York, it infests healthy hardwoods of all kinds. Larvae can kill trees in the course of their life cycle, which spans one to two years. Preemptive removal of trees is quickly changing the urban landscape of affected cities.

Sternidius variegatus is very common east of the Continental Divide. Adults are recognized by the pale oval spot at the bottom of each wing cover. Larvae feed in branches of hardwoods, conifers, shrubs, and vines.

The genus ***Graphisurus*** (formerly *Urographis)* includes three species, ranging in eastern North America, west to Minnesota and Texas. Females have a long sticklike ovipositor that they use to lay eggs. ***Graphisurus fasciatus*** is widespread in eastern North America, boring in various hardwoods as a larva. Adults come to lights at night.

Oberea is a genus of 15 elongate species found east of the Rockies. The **Sassafras Borer** *(Oberea ruficollis)* is one of the largest species. It ranges over most of eastern North America, where its host plants occur. Larvae are stem borers in live sassafras and spicebush. The life cycle takes one to two years. Adults emerge from late April to early August, varying with latitude. Woodpeckers often prey on the larvae and pupae.

The genus ***Tetraopes*** includes 13 species of milkweed beetles in North America (plus others in Mexico and Central America). Most of the species studied so far specialize on one or a few species of milkweeds, the larvae mining in the roots, the adults feeding on the upper foliage and blooms. Adults feed boldly in the open by day, their bright colors advertising their toxic nature. Each eye is divided by the antennal socket, hence they are often called four-eyed beetles. The **Red Milkweed Beetle** *(Tetraopes tetrophthalmus)*, the most numerous eastern and central species, feeds mainly on common milkweed *(Asclepias syriaca)*, although it has been found on at least two other species of milkweeds. ***Tetraopes femoratus*** is associated with showy milkweed *(A. speciosa)* throughout the western U.S. but has been found in low numbers on common milkweed at the eastern edge of its range.

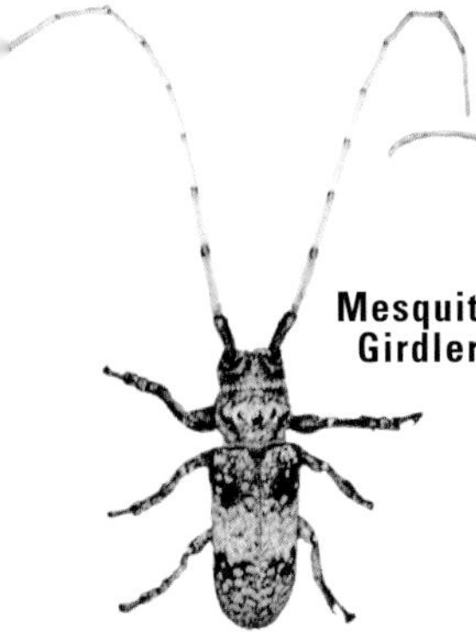

actual size

Mesquite Girdler

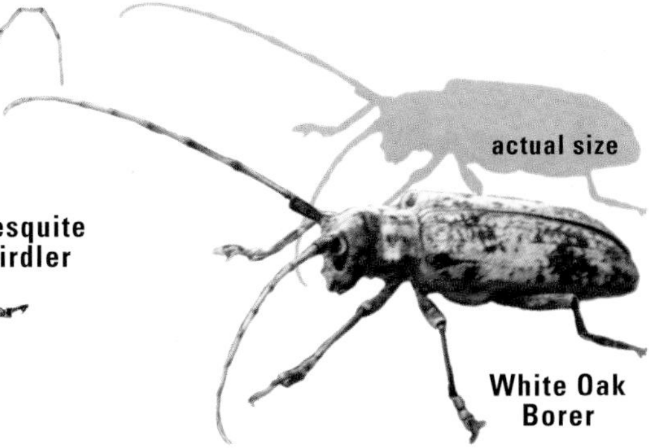

White Oak Borer

twig attacked by Mesquite Girdler

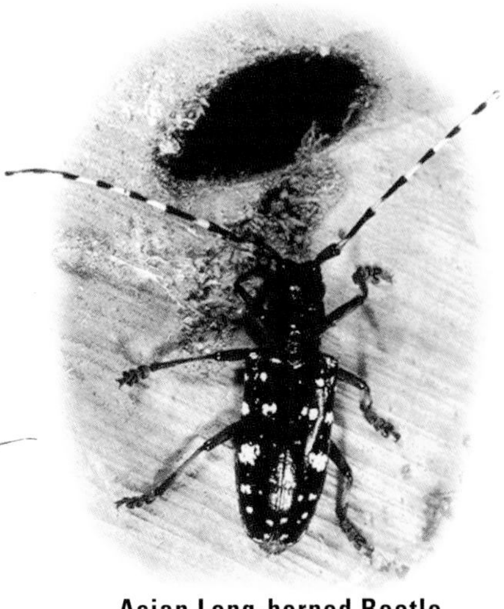

Sternidius variegatus

Asian Long-horned Beetle

male

Graphisurus fasciatus

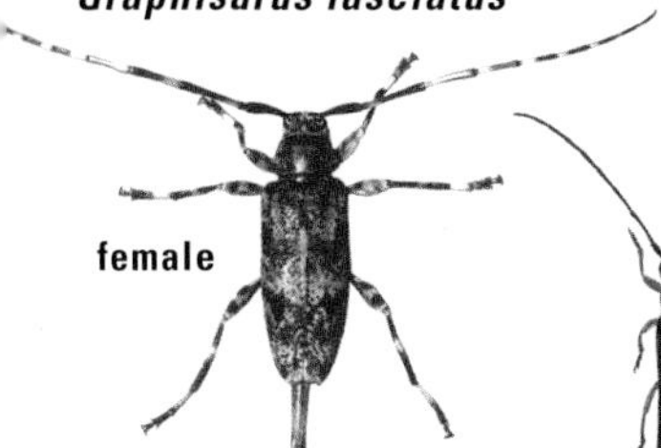

female

Sassafras Borer

Red Milkweed Beetle

Tetraopes femoratus

(subfamily Lepturinae) are mostly active by day, visiting flowers to feed on pollen. Many have a distinctive shape, broad at the "shoulders," tapering toward the neck and toward the rear of the abdomen.

Leptura is a genus with 14 North American species. Adults are often seen on umbelliferous flowers. ***Leptura obliterata*** is common in the Pacific northwest, south to California and east to Montana. It infests fire-killed spruce, fir, Douglas-fir, hemlock, and pine. ***Leptura emarginata*** is an eastern U.S. species. Adults do not visit flowers but circle dead, standing beech and other hardwoods, 20–30 feet up. Two species of ***Lepturobosca*** (formerly *Cosmosalia)* live in boreal and western North America. ***Lepturobosca chrysocoma*** owes its velvety sheen to fine pubescence on its body. Look for this species from June to September on flowers in mountain meadows across Canada and throughout the western U.S. Larvae mine dead pine, spruce, poplar, and alder.

The genus ***Typocerus*** includes 15 similar species across the continent. ***Typocerus velutinus,*** common in the east, is variable in color and pattern. Larvae live in decaying oak, hickory, and other hardwoods. Look for adults from May to August on flowers in the parsley family. ***Typocerus zebra*** occurs in the Atlantic and Gulf Coast states. Larvae bore in pines. Markings on the adults are generally uniform and consistent. They are active March to August, seen on a wide variety of flowers.

The **Elderberry Borer** *(Desmocerus palliatus),* widespread in the east, is one of three species in its genus. Larvae bore in stems of the living host, migrating to the roots. The life cycle takes two years. Adults occur on flowers and foliage of the host.

The **Ribbed Pine Borer** *(Rhagium inquisitor)* is unusual for this family in having *short* antennae. It occurs throughout much of North America as well as in Eurasia. Hosts include a variety of conifers. Larvae bore between the wood and bark of dead trees. Pupae are in cells ringed with coarse wood fibers.

Centrodera is a genus of 11 species in North America. The **Yellow Douglas-fir Borer** *(Centrodera spurca)* occurs from British Columbia to Utah. Besides Douglas-fir, it also uses oaks, madrone, and woody shrubs as hosts. Larvae live in soil, feeding on stumps and dead roots, taking two years to mature. Adults may fly to lights at night.

The 10 species of the widespread genus ***Strangalia*** have a distinctive slender-bodied appearance. ***Strangalia luteicornis*** is one of several nearly identical eastern species. Adults actively tumble amid flowers in deciduous woods from May to August. Larvae bore in various hardwoods. ***Strangalia strigosa*** occurs only in Florida. Adults are on the wing from April to July. Look for them on flowers of buckthorns, redbay, fleabane, holly, and various other plants.

The genus ***Necydalis*** includes convincing wasp mimics with seven North American species. The short wing covers and elongate body are distinctive. Look for adults on foliage or around woodpiles.

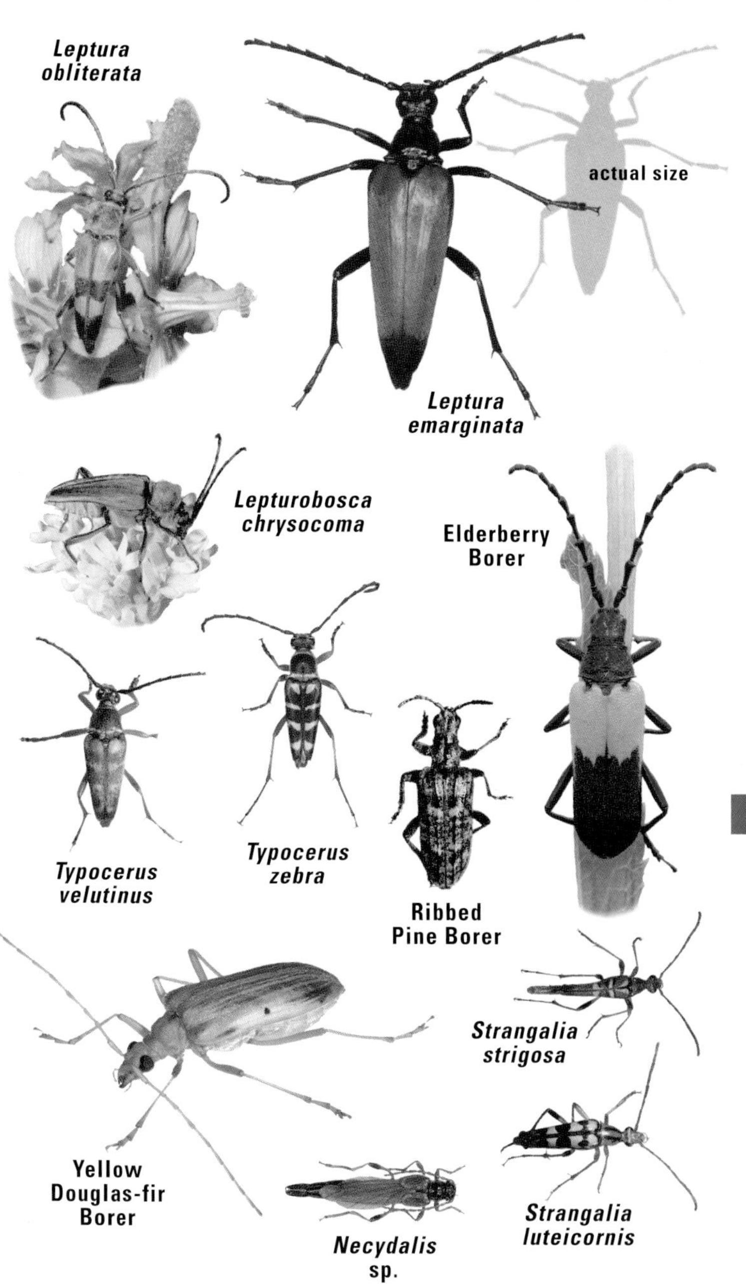
Leptura obliterata
actual size
Leptura emarginata
Lepturobosca chrysocoma
Elderberry Borer
Typocerus velutinus
Typocerus zebra
Ribbed Pine Borer
Strangalia strigosa
Yellow Douglas-fir Borer
Necydalis sp.
Strangalia luteicornis

ROUND-NECKED LONGHORNS

Members of this subfamily **(Cerambycinae)** have protruding faces and lack spines on the thorax. This is a very diverse group.

The genus ***Megacyllene*** includes 10 species, found mostly in the eastern and central U.S. The **Locust Borer** *(Megacyllene robiniae)* is abundant in the fall on goldenrod. Larvae bore in the heartwood of living locust trees. This species is expanding its range as locust trees become more widely used as ornamentals. The **Painted Hickory Borer** *(Megacyllene caryae)*, not illustrated, is identical to the Locust Borer but emerges in spring and visits lights at night. Adults also emerge indoors from firewood.

Eburia is a genus of 12 species ranging from eastern North America to the southwest. The **Ivory-marked Borer** *(Eburia quadrigeminata)* occasionally emerges from flooring, sills, or furniture after 10–40 years! Normally development takes a shorter time, larvae boring in the solid heartwood of dead deciduous trees. Adults are on the wing in the east during June and July, flying to lights at night.

The genus ***Elaphidion*** includes eight species in the eastern and southeastern U.S. The **Spined Bark Borer** *(Elaphidion mucronatum)* is an abundant eastern species often seen at lights. Larvae tunnel beneath bark, and into the sapwood, of most deciduous trees and shrubs.

The two species of ***Plinthocoelium*** are spectacular beetles. ***Plinthocoelium suaveolens*** bores as a larva in the roots of tupelo and bumelia in the southern U.S. Adults are strong diurnal fliers fond of fermenting sap. Look for them from May to August on the blossoms of bumelia or at the base of the tree trunks. They give off a strong, musky odor when handled.

The genus ***Enaphalodes*** includes eight species of large beetles. ***Enaphalodes hispicornis*** occurs throughout the U.S. Larvae bore in oaks. Adults are active July to October and come to lights at night. The **Red Oak Borer** *(Enaphalodes rufulus)*, not illustrated, is found throughout the eastern U.S. and southeastern Canada. Where timber management has resulted in stands of older, even-aged trees, this beetle has become a serious pest. Larvae bore in living, but weakened, red and white oaks.

The **Banded Alder Borer** *(Rosalia funebris)* ranges from Alaska to southern California and east to the Rockies. Adults are active in daylight and are often attracted to fresh paint. Larvae bore in the trunks of dead alder, maple, ash, oak, willow, and California laurel.

The genus ***Stenaspis*** includes two species in the desert southwest. ***Stenaspis solitaria*** adults are active from May to September. Look for them on flowers of mesquite or acacia.They give off a musty smell. Larvae bore in small living branches of acacia and mesquite.

The genus ***Semanotus*** includes five species. ***Semanotus ligneus*** is a variable species, found coast to coast wherever its preferred cedar and cypress hosts occur. Larvae mine just under the bark, entering the wood only to pupate. Adults are active May to August. Look for them on log piles. Members of the genus ***Aneflus*** are mostly large dull-colored beetles. They range from the southern U.S. south into Central America and are often found in association with mesquite trees.

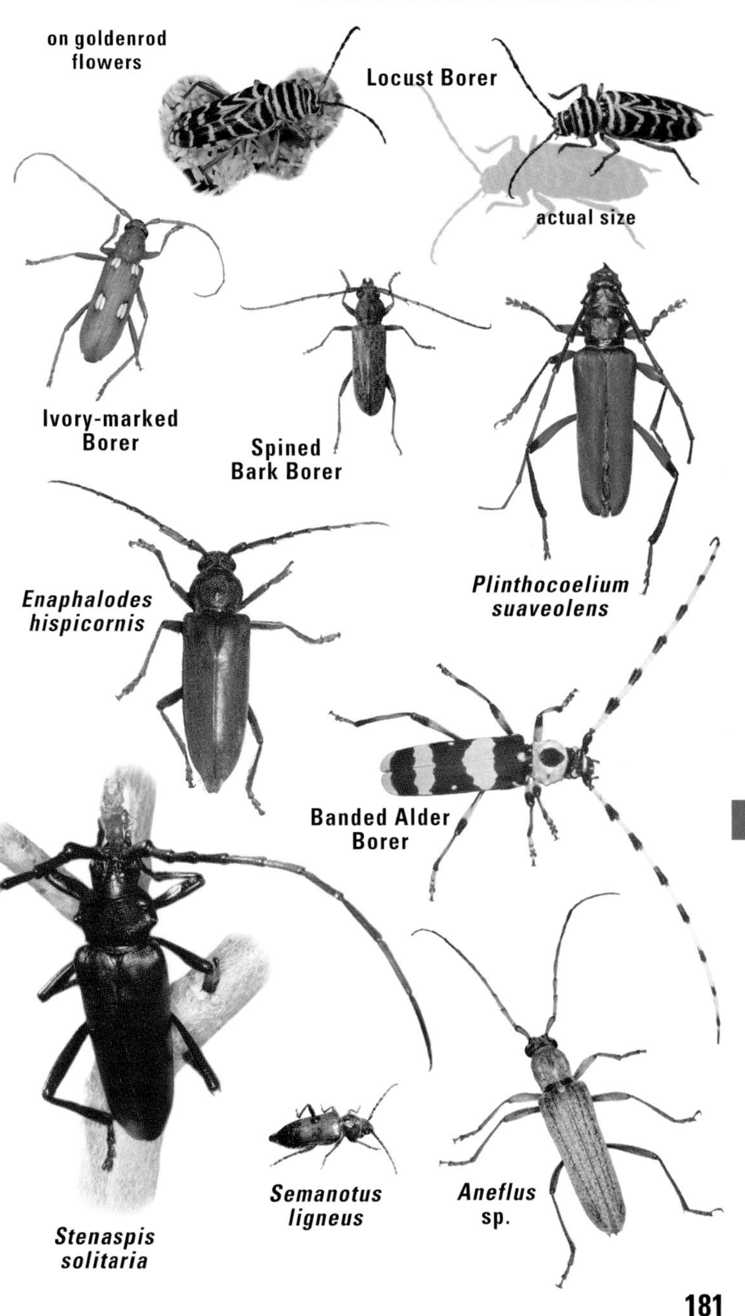
on goldenrod flowers
Locust Borer
actual size
Ivory-marked Borer
Spined Bark Borer
Plinthocoelium suaveolens
Enaphalodes hispicornis
Banded Alder Borer
Stenaspis solitaria
Semanotus ligneus
Aneflus sp.

The 22 species in the genus ***Xylotrechus*** collectively range across North America. The **Rustic Borer** *(Xylotrechus colonus)* is abundant east of the Rockies. Adults come to lights but are also active by day on dead or dying hardwoods. Larvae tunnel just beneath the bark. ***Xylotrechus undulatus*** is transcontinental in Canada and Alaska. Larvae bore under the bark of dead conifers, especially spruce. Adults are active from June to September.

The genus ***Callidium*** includes 19 species, collectively distributed across most of North America, especially the west. The **Black-horned Pine Borer** *(Callidium antennatum)* ranges from British Columbia and California east to the Rockies. Larvae bore in dead pine, Douglas-fir, and hemlock.

The genus ***Clytus*** includes eight species in North America, mostly in boreal regions. ***Clytus ruricola*** breeds in decaying deciduous trees, particularly maples, in eastern Canada and the northeastern U.S. Adults are wasp mimics, active May to July. ***Neoclytus*** is a genus of 26 wasp mimics collectively found over the entire continent. The **Red-headed Ash Borer** *(Neoclytus acuminatus)* occurs east of the Rockies. Adults run rapidly over freshly cut wood of many hardwoods, shrubs, and vines. Larvae bore in the sapwood of weakened, dead, or dying hosts. The **Western Ash Borer** *(Neoclytus conjunctus)* is a common West Coast species. Larvae honeycomb the sapwood and heartwood of ash, oak, madrone, eucalyptus, and wild crabapple. Watch for adults emerging from firewood. The clerid beetle *Chariessa elegans* (p. 206) is a chief predator of this species.

Cyrtophorus verrucosus is an ant mimic, common on spring flowers in the east. Its knobby "shoulders" are distinctive. Larvae mine dead, solid hardwoods and pine. The genus ***Batyle*** is represented by three species, collectively found everywhere but the Pacific Coast. Look for adults of ***Batyle suturalis*** on composite flowers in eastern and central North America from May to August. Larval hosts include oak and hickory. ***Batyle ignicollis*** is also common on flowers in the central and southeast states.

Crossidius is a genus with at least a dozen species in the western and central U.S. Adults are common in late summer on flowers of rabbitbrush. Larvae mine the roots of rabbitbrush, sagebrush, and burroweed *(Haplopappus)*, host plants varying with the species of beetle.

The big **Horse-bean Longhorn** *(Trachyderes mandibularis)* ranges from Florida to southern California. Some males have enlarged jaws and may use them to defend food resources like ripe saguaro fruits from competing males. Larvae bore in palo verde, horse-bean, willow, and citrus trees. These beetles sometimes fly to lights at night, and can be active any time between March and November.

Two species of eucalyptus borers, introduced from Australia, are established in California and may be spreading into Arizona. They can do extensive damage to ornamental eucalyptus trees, especially in drought years. ***Phoracantha semipunctata*** reached California in the 1980s, while ***Phorocantha recurva*** was first detected there in 1995. The latter species may have a longer season of activity, and some of the biological controls developed for *P. semipunctata* may not be as effective for *P. recurva.*

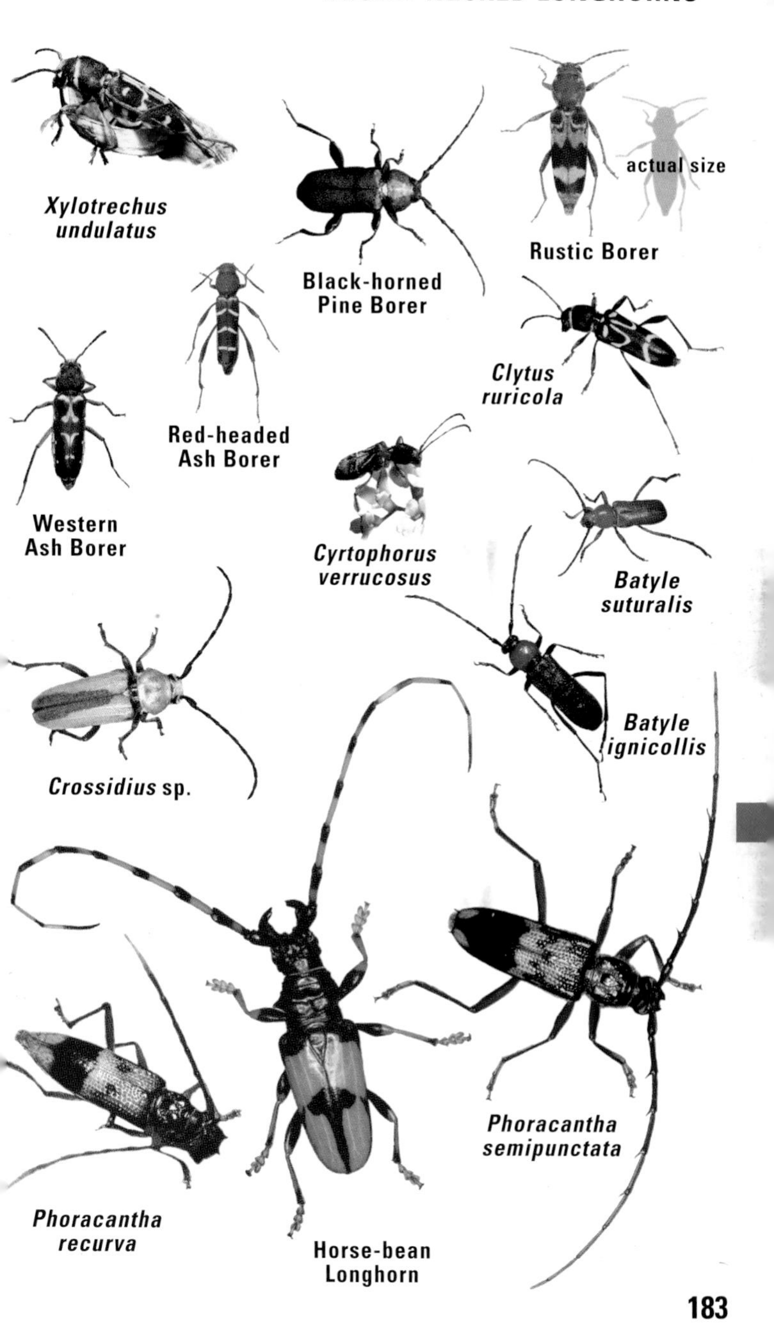

Xylotrechus undulatus

Black-horned Pine Borer

Rustic Borer

Clytus ruricola

Red-headed Ash Borer

Western Ash Borer

Cyrtophorus verrucosus

Batyle suturalis

Batyle ignicollis

Crossidius sp.

Phoracantha semipunctata

Phoracantha recurva

Horse-bean Longhorn

TOOTH-NECKED AND ABERRANT LONGHORNS

Tooth-necked Longhorns **(subfamily Prioninae)** are huge and dark.

The **Hardwood Stump Borer** *(Mallodon dasytomus)* ranges widely in the southern U.S. Larvae feed gregariously in a variety of hardwoods, usually in stumps and structural wood, but also in live trees. The life cycle takes three to four years. Males have enlarged jaws.

The genus ***Ergates*** includes two western species, one ranging to South Dakota. Fine teeth along the edge of the thorax distinguish this genus. The **Ponderous Borer** *(Ergates spiculatus)* bores in Douglas-fir and ponderosa pine. The life cycle takes several years. Mature larvae may reach 70 mm (almost 3 inches), and the gnawing action of their jaws inspired the invention of the chain saw.

Prionus is a genus of 16 species ranging across most of the continent. Larvae feed in roots of living trees and shrubs. The **Broad-necked Root Borer** *(Prionus laticollis)* occurs in eastern North America. Adults are active in July and August. The enormous females do not fly. Look for them scrambling across the forest floor or crawling on logs.

The four species of ***Derobrachus*** range from the southeastern U.S. to southern California. ***Derobrachus geminatus*** occurs in the southwest, the larvae boring in the roots of palo verde, poplar, and oak. Adults begin flying at dusk and are attracted to lights. They are on the wing from July to September.

The genus ***Tragosoma*** includes three species in coniferous forests across the northern hemisphere. The hairy thorax of the adult insects is distinctive. ***Tragosoma chiricahuae*** occurs only in Arizona's Chiricahua Mountains. The **Hairy Pine Borer** *(Tragosoma depsarius)*, not shown but essentially identical, is common in coniferous forests from British Columbia to California, east to Montana, and also in the northeastern U.S. Larvae mine the sapwood of dead, downed pines and Douglas-fir.

ABERRANT LONGHORNS (subfamily Parandrinae) have short antennae. Members of the genus ***Parandra*** are easily mistaken for stag beetles (p. 148) or darkling beetles (p. 192). Two species together range throughout the eastern U.S. and west to southern California. The **Pole Borer** *(Parandra brunnea)* develops in structural wood in contact with soil. The life cycle typically takes three to four years. Look for adults at lights.

The remaining species, despite their odd appearance, are classified among the **subfamily Cerambycinae** like those on pp. 180–183.

The **Lion Beetle** *(Ulochaetes leoninus)* ranges from British Columbia to California and Idaho. Adults resemble bumble bees in flight. Larvae bore in the roots or base of dead, standing pines, Douglas-fir, fir, hemlock, and spruce in mountainous areas.

The genus ***Moneilema*** includes six species of cactus-eating longhorns in the western U.S., with at least another 10 in Mexico. Adults are flightless and mimic *Eleodes* darkling beetles (p. 192). One of the largest is the **Opuntia Borer** *(Moneilema gigas)* of the southwest; both adults and larvae often feed on prickly-pear and cholla cactus.

ODD LONG-HORNED BEETLES

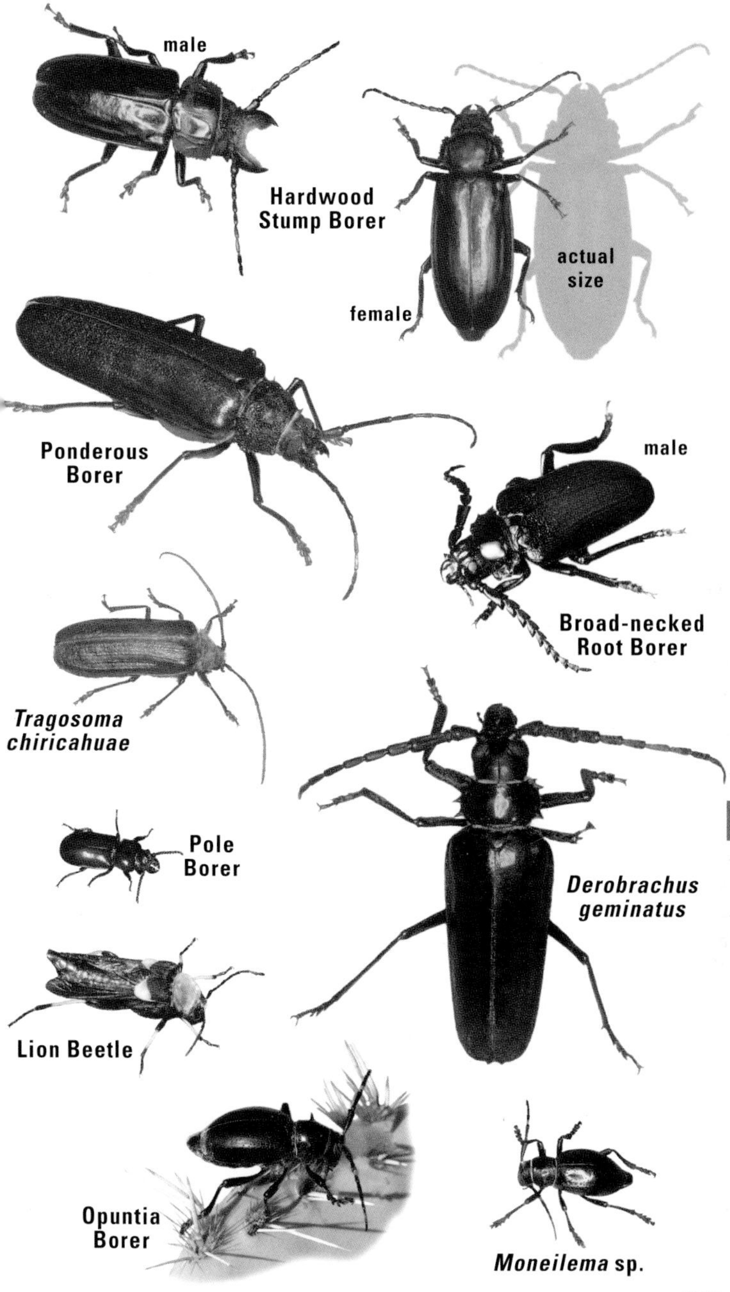
male
Hardwood Stump Borer
actual size
female
Ponderous Borer
male
Broad-necked Root Borer
Tragosoma chiricahuae
Pole Borer
Derobrachus geminatus
Lion Beetle
Opuntia Borer
Moneilema sp.

METALLIC WOODBORERS

(family Buprestidae) are wary and active, usually seen on tree trunks, logs, foliage, or flowers in bright sunshine. Our 750-plus species include many brilliant, bullet-shaped insects. Larvae are called flatheaded borers for their pancake-like thoracic section. Some are forest pests, though they seldom attack healthy trees. They can, however, interfere with timber salvage after fires and blowdowns. Meandering etchings in wood, just under the bark, packed with powdery "frass" (dried poop) are the work of the larvae. A few are leaf miners, cone borers, or gall makers.

Buprestid larval galleries under bark of dead tree

Buprestis is a genus of 18 species, widely distributed. ***Buprestis laeviventris*** is common in western forests. Larvae bore in pines and Douglas-fir. ***Buprestis confluenta*** ranges from British Columbia to Ontario, south to Texas and Illinois. Adults feed on foliage of injured, dead, and dying trees. Larvae bore in the wood of aspen and cottonwood. ***Buprestis gibbsii*** occurs from Washington to southern California. Oregon white oak and California black oak are the larval hosts. ***Buprestis lineata,*** a variable species, is common in the east, the larvae boring in pines.

The genus ***Cypriacis*** includes eight species formerly included in *Buprestis,* collectively widespread. The **Golden Buprestid** *(Cypriacis aurulenta)* ranges from British Columbia to Manitoba and south throughout the western U.S. Larvae mine lightning-struck (or otherwise wounded or dead) pines, Douglas-firs, and western redcedars. The life cycle takes two to four years, but larvae trapped in milled lumber may bore for 20–30 years or longer before emerging as adults. ***Cypriacis fasciata*** is fairly widespread in the east, but its hosts and habits remain undiscovered.

Members of the genus ***Chalcophora*** are large, roughly textured beetles. Five species occur in coniferous forests. ***Chalcophora virginiensis*** (formerly called *Chalcophora angulicollis* in the west) is transcontinental wherever its host trees are found. The top of the abdomen is metallic blue-green, revealed as the beetles take flight noisily from tree trunks. Larvae bore in pines, Douglas-fir, and true firs and may require several years to mature.

The four species of ***Hippomelas*** in the U.S. are limited to the southwest. ***Hippomelas sphenicus*** is typical, found on mesquite and other trees. The larvae bore in the heartwood of a variety of trees. ***Gyascutus caelatus,*** one of eight species in its genus, ranges from Arizona to Texas.

Members of the genus ***Anthaxia*** are small, usually 3–7 mm long. The 36 species range collectively across the continent, and many can be found on flowers. ***Anthaxia inornata*** (formerly called *A. expansa)* is common from Alaska to Labrador and over much of the western U.S. Larvae are suspected to bore in branches of conifers.

Buprestis laeviventris

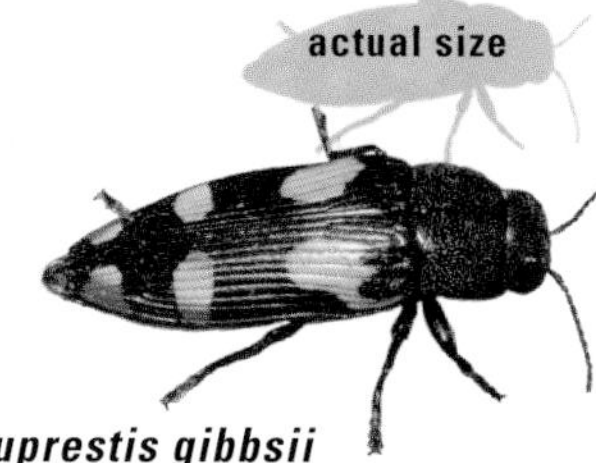

Buprestis gibbsii

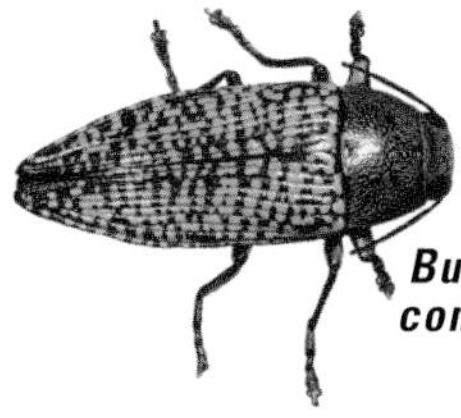

Buprestis confluenta

Buprestis lineata (2 variations)

Cypriacis fasciata

Golden Buprestid

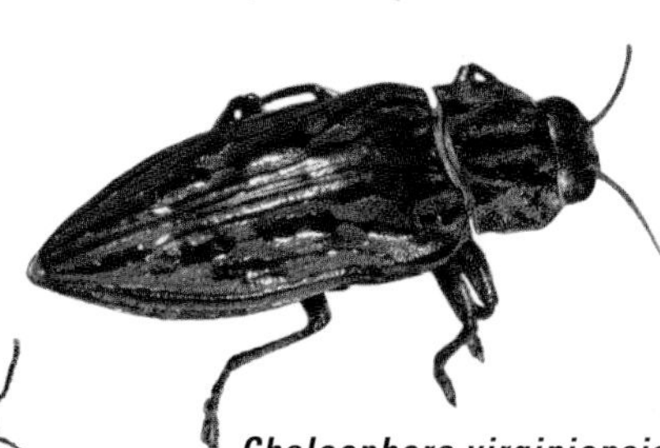

Chalcophora virginiensis

Hippomelas sphenicus

Gyascutus caelatus

Anthaxia inornata

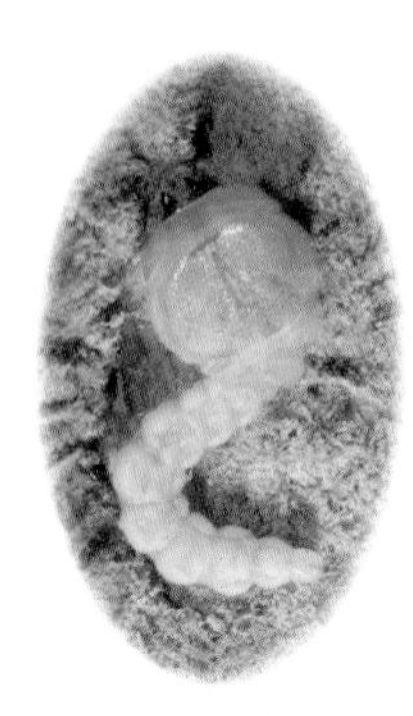

Buprestid larva

Members of the genus ***Chrysobothris*** are among the most frequently encountered Buprestids. The front legs have toothed or swollen "biceps." At least 134 species occur in North America, most 7–14 mm in length. They may be numerous on recently felled and storm-damaged trees. Males bang their abdomens rapidly against the boles to court nearby females. Both sexes are cryptic but in flight expose a metallic blue or green abdomen. ***Chrysobothris dentipes*** usually feeds on conifers, especially pines while ***Chrysobothris rugosiceps*** is associated with hardwoods. ***Chrysobothris basalis*** is common in the southwest, usually feeding on species of acacia.

Members of the genus ***Agrilus*** are abundant, diverse (170-plus species), and widespread. Several are serious pests. The **Two-lined Chestnut Borer** *(Agrilus bilineatus)* now feeds on beech and oak, since blight wiped out most chestnuts. Adults nibble foliage, while larvae bore between the bark and wood. This species ranges east of the Rockies in the U.S. and southeastern Canada. The **Red-necked Cane Borer** *(Agrilus ruficollis)* bores in canes of blackberry and raspberry in the eastern U.S. and southern Canada. Its coloration varies individually. The **Emerald Ash Borer** *(Agrilus planipennis)* is an Asian species discovered in Ontario and southeastern Michigan in 2002. In infested areas, thousands of ash trees have been cut down and burned in the attempt to limit the beetle's spread, but by 2005 it had already expanded its range into northwestern Ohio.

We have 24 species in the genus ***Dicerca,*** seven associated with conifers, the remainder with various deciduous trees and shrubs. Members of this genus are recognized by the flared or tapered tips of the wing covers. ***Dicerca tenebrosa*** is transcontinental in Canada, occurring south locally to California and Florida. It feeds on a wide variety of conifers.

Beetles in the genus ***Melanophila*** are called fire bugs, owing to the affinity of some species for forest blazes. Heat-sensing pits on the thorax help them seek smoldering trees. Females eager to lay eggs on burning boles may singe their feet off. ***Melanophila acuminata*** attacks conifers throughout the continent. The **Charcoal Beetle** *(Melanophila consputa)* flocks to heavy smoke, sometimes biting firefighters. Its hosts are lodgepole and ponderosa pine, from eastern Washington and northern Idaho to Arizona. Similar to *Melanophila* but classified in a different genus, ***Phaenops lecontei*** is widespread in the western U.S.

Acmaeodera is a genus of more than 140 species. Many of them are hairy bee and wasp mimics, and many are marked with contrasting variegated patterns of yellow and black. Look for adults throughout the warmer months on flowers in deserts, chaparral, and woodlands, especially in the west and southwest.

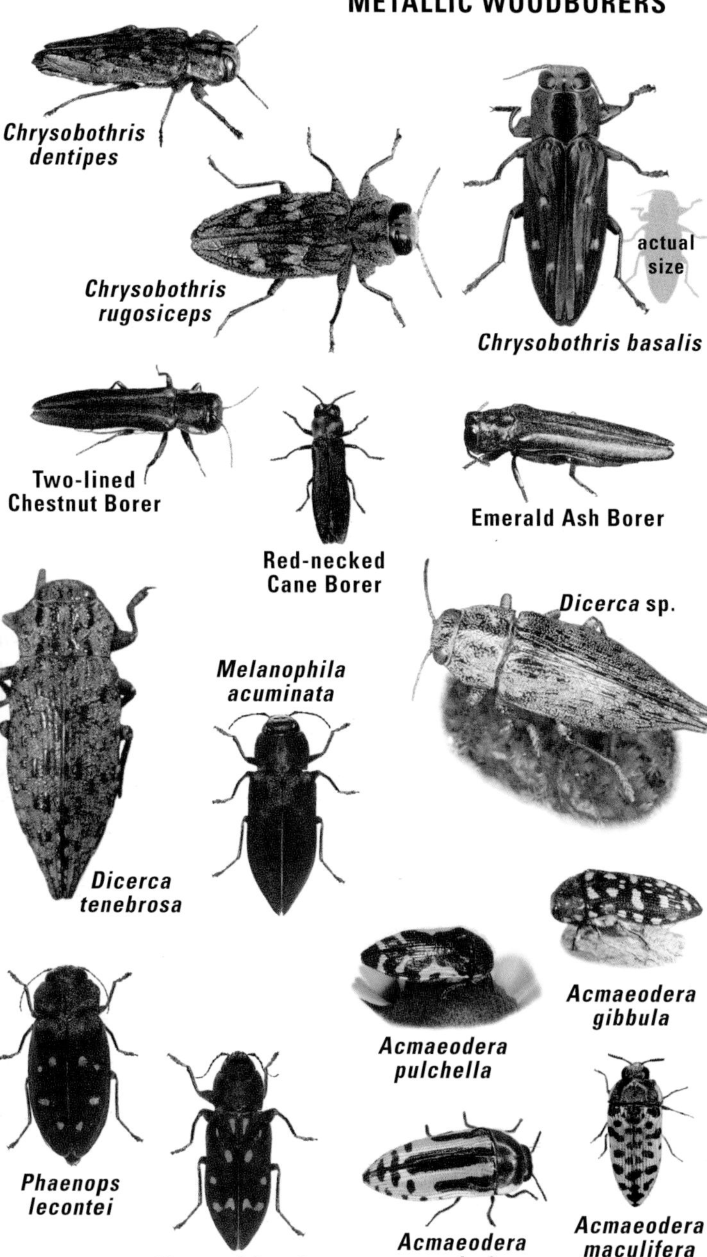
Chrysobothris dentipes
Chrysobothris rugosiceps
actual size
Chrysobothris basalis
Two-lined Chestnut Borer
Red-necked Cane Borer
Emerald Ash Borer
Dicerca sp.
Melanophila acuminata
Dicerca tenebrosa
Acmaeodera gibbula
Acmaeodera pulchella
Phaenops lecontei
Charcoal Beetle
Acmaeodera scalaris
Acmaeodera maculifera

TUMBLING FLOWER BEETLES AND OTHERS

TUMBLING FLOWER BEETLES (family Mordellidae) are lively little insects common on flowers, especially those in the parsley and composite families. Their wedgelike shape is distinctive. They fly well and jump away when disturbed. Depending on the species, larvae feed in stems of herbaceous plants or in fungi or decaying wood. There are over 205 species in 17 genera north of Mexico.

Mordellistena is by far the largest genus in this family in North America, with more than 130 species, collectively distributed over most of the continent. Larvae typically live in the stems of plants. Some species occur in goldenrod stem galls, taking the place of the rightful occupant (p. 268). ***Mordellistena trifasciata*** is one widespread species.

The genus ***Mordella*** includes 25 species in North America, seen most often on flowers. ***Mordella atrata*** is one of the most abundant and widely distributed species.

Hoshihananomia octopunctata is one of three U.S. species in its genus (formerly placed in the genus *Mordella)*. It is common on Queen Anne's lace, mainly in the northeast. ***Glipa oculata*** is found bouncing around on understory foliage at forest edges in the eastern U.S. and Canada.

Yakuhananomia bidentata (formerly placed in the genus *Tomoxia)* is an active, wary, and cryptic species. Look for it on the trunks of dead beech, oak, and hickory trees in the northeastern U.S., west to Missouri.

WEDGE-SHAPED BEETLES (Ripiphoridae) are bizarre parasites. Adults typically visit flowers where females deposit eggs. Each larva that hatches climbs aboard a female solitary bee, hitching a ride back to her nest. Once inside a cell, the larval beetle feeds as an internal parasite of the larval bee. Eventually it molts into a sedentary grub that feeds externally on its host. The larva molts four more times before becoming a pupa. An adult beetle emerges two weeks later but lives only one or two days. There are about 50 species in North America, in six genera. The genus ***Macrosiagon*** includes 11 species that resemble large versions (4–12 mm) of tumbling flower beetles (see above). They are parasitic on solitary bees and on mason wasps and solitary wasps. The genus ***Rhipiphorus*** is represented in North America by close to 30 species. Most are 4–11 mm, with short wing covers, easily mistaken for sweat bees. Look for them on composite flowers.

IRONCLAD BEETLES (Zopheridae) are named for their dense exoskeletons. North America has about 30 species that feed mostly on dense, woody fungi. ***Phellopsis obcordata*** occurs in coniferous forests from Alaska and Canada well southward along the Cascades, Sierra, and Appalachians. It is found beneath bark and in rotten logs, playing dead when disturbed. The genus ***Phloeodes*** includes 14 species in the southwest U.S. ***Zopherus haldemani*** is one of 10 species in its genus, ranging widely in the southwestern U.S. Live specimens of one Mexican species of *Zopherus* are decorated by jewelers and sold as living brooches.

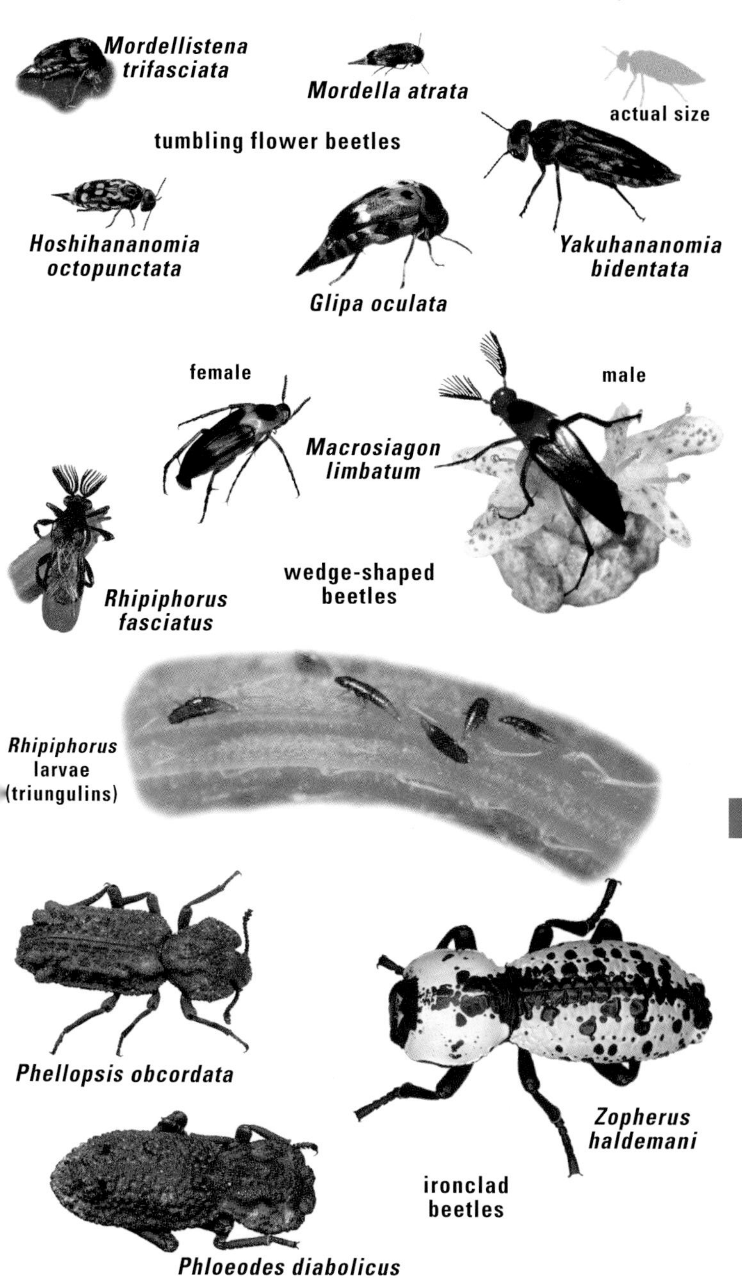
Mordellistena trifasciata
Mordella atrata
actual size
tumbling flower beetles
Hoshihananomia octopunctata
Glipa oculata
Yakuhananomia bidentata
female
male
Macrosiagon limbatum
Rhipiphorus fasciatus
wedge-shaped beetles
Rhipiphorus larvae (triungulins)
Phellopsis obcordata
Zopherus haldemani
ironclad beetles
Phloeodes diabolicus

DARKLING BEETLES

(family Tenebrionidae) represent perhaps the fifth-largest family of beetles, with at least 1,180 species in North America alone. The former families Alleculidae (comb-clawed beetles) and Lagriidae (long-jointed bark beetles) are now included here. No one character easily identifies this group, though the front of the head is usually broad and round, hiding the jaws, and the antennae are often beadlike. Many discharge repulsive aromatic chemicals when threatened or handled. They vary greatly in size, shape, color, habitat, and behavior. The majority are scavengers.

The genus ***Eleodes*** includes the critters popularly called skunk beetles, pinacate beetles, or head-stander beetles in the arid west. There are at least 121 species north of Mexico. Adults are flightless, the fused wing covers helping to prevent dehydration. Mostly nocturnal, they search open ground for windblown organic debris. Try attracting them by laying an oatmeal trail. Look under boards and rocks or in rodent burrows by day. When disturbed, some species tilt their bodies, secreting or spraying volatile chemicals from the tip of the abdomen. Grasshopper mice foil this maneuver by jamming the beetle's rear end into the ground and eating it from the head down. Larvae of some are known as false wireworms and feed on the roots of grasses, sometimes damaging wheat or other crops. Larvae of some in drier regions feed on plant detritus underground.

The **Desert Ironclad Beetles** *(Asbolus laevis* and *Asbolus verrucosus)* are related to the preceding species but lack their superb chemical defenses. If threatened, they flop over sideways and play dead, an act that may last for several minutes or even for hours, hence the alternate name of death-feigning beetles.

The genus ***Coniontis*** includes 48 species in the western and central U.S. Adults are flightless and fairly small, usually 8–16 mm in length. The oval, cylindrical shape is distinctive. Look for them in arid habitats under rocks, logs, dried cow patties, and other debris. They can be minor pests at times in California, nibbling on sugar beets, young tomatoes, and lima beans.

Eusattus is a genus represented by 14 species in the western and central states. Adults are round in shape, wingless, and average 6–14 mm in length. ***Eusattus muricatus*** (not illustrated) inhabits isolated dune systems throughout the west, burrowing in the shifting sands. The 29 species of ***Blapstinus*** are collectively widespread in the U.S. Adults are small (3–8 mm), sometimes abundant, and are attracted to lights at night. They may girdle plants at the soil surface, sometimes causing damage to cotton, peppers, strawberries, and other crops in the western states.

Bolitotherus cornutus is our only member of its genus. Males have thoracic horns that they use to battle over females. Both adults and larvae inhabit dense, woody bracket fungi on beech, maple, and other hardwoods in the eastern U.S. The rough texture of these beetles, and their talent for playing dead, make them difficult to spot.

The genus ***Stenomorpha*** includes 61 species, collectively ranging from southwestern Canada to Mexico. Most diverse in the southwestern U.S., they occur in the central states as well.

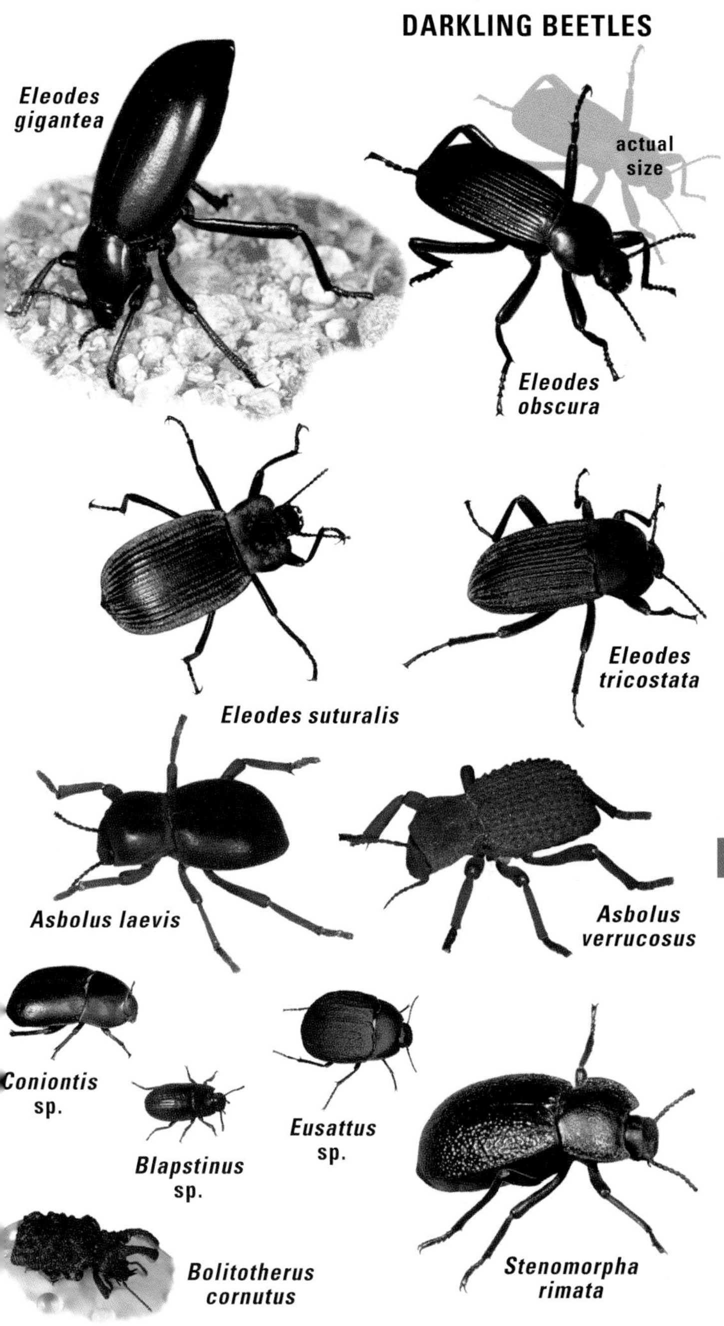
Eleodes gigantea
actual size
Eleodes obscura
Eleodes suturalis
Eleodes tricostata
Asbolus laevis
Asbolus verrucosus
Coniontis sp.
Blapstinus sp.
Eusattus sp.
Bolitotherus cornutus
Stenomorpha rimata

Members of the genus ***Tenebrio*** are the "mealworms" commonly sold i
pet stores. The two cosmopolitan species, **Yellow Mealworm** *(Tenebri
molitor)* and **Dark Mealworm** *(Tenebrio obscurus,* not shown), are world
wide pests of stored grain. Depending on conditions, their life cycle ma
take anywhere from four months to two years.

Alobates pennsylvanicus is a common, flightless beetle found in group
under loose bark in the eastern U.S. The genus ***Helops*** is represented b
40 species in North America, collectively distributed across the continen
Adults vary in size from 4–40 mm in length, depending on the specie
Some species congregate beneath loose bark.

Embaphion is a genus that includes eight species in North America, col
lectively ranging from the southwest to South Dakota. ***Meracantha con
tracta*** occurs in eastern forests. Look for this flightless insect on the bas
of tree trunks, especially under clinging foliage or under bark on deac
standing trees. The genus ***Lobopoda*** (formerly included in the family Al
leculidae) includes seven species in North America, collectively ranging i
the eastern U.S. to west Texas. Adults are mostly found in rotting wood o
bracket fungi. They also come to lights at night.

The seven species of ***Coelocnemis*** are large flightless western beetle
Adults live up to three years and hibernate in small groups in dry, shel
tered situations. They eat a variety of plant matter, including whole grain
lichens, and fungi. ***Coelocnemis californicus*** is common under loose bar
in dry forests of southwestern Canada and the western U.S. ***Iphthiminu***
is represented by six species here, collectively found in the western an
northeastern U.S. ***Iphthiminus serratus*** is common in the west, especiall
on the east slope of the Cascades. Look for the flightless adults under loos
bark in pine forests.

Strongylium is a genus of 10 mostly slender, lanky, and agile specie
collectively found in the eastern and southern U.S. ***Strongylium tenuicoll***
is the common eastern species, ranging into extreme southern Quebe
and Ontario. Adults fly to lights at night. Larvae develop in decaying o
wounded deciduous trees. The genus ***Statira*** (formerly included in th
family Lagriidae) includes 15 North American species, mostly in the south
eastern U.S. The antennae have a long terminal segment. Adults are activ
by day on foliage. Larvae occur in leaf litter or under bark on stumps.

The **Confused Flour Beetle** *(Tribolium confusum)* and the **Red Flou
Beetle** *(Tribolium castaneum,* not shown) are only 3–5 mm long. They ar
global pests in processed grain, pasta, nuts, beans, and other dry foods, bu
also valuable subjects in genetic research. Hardy (they need no water), an
long-lived (up to a year for adult females), these beetles apparently hav
been associated with humans for over 4,000 years.

The genus ***Uloma*** includes five forest species. ***Uloma longula*** is com
monly found in rotten logs along the Pacific Coast. The genus ***Diaperi***
includes three species in the eastern and southern U.S. ***Diaperis maculat***
can be abundant in certain polypore fungi throughout the eastern U.S. an
is frequently mistaken for a lady beetle.

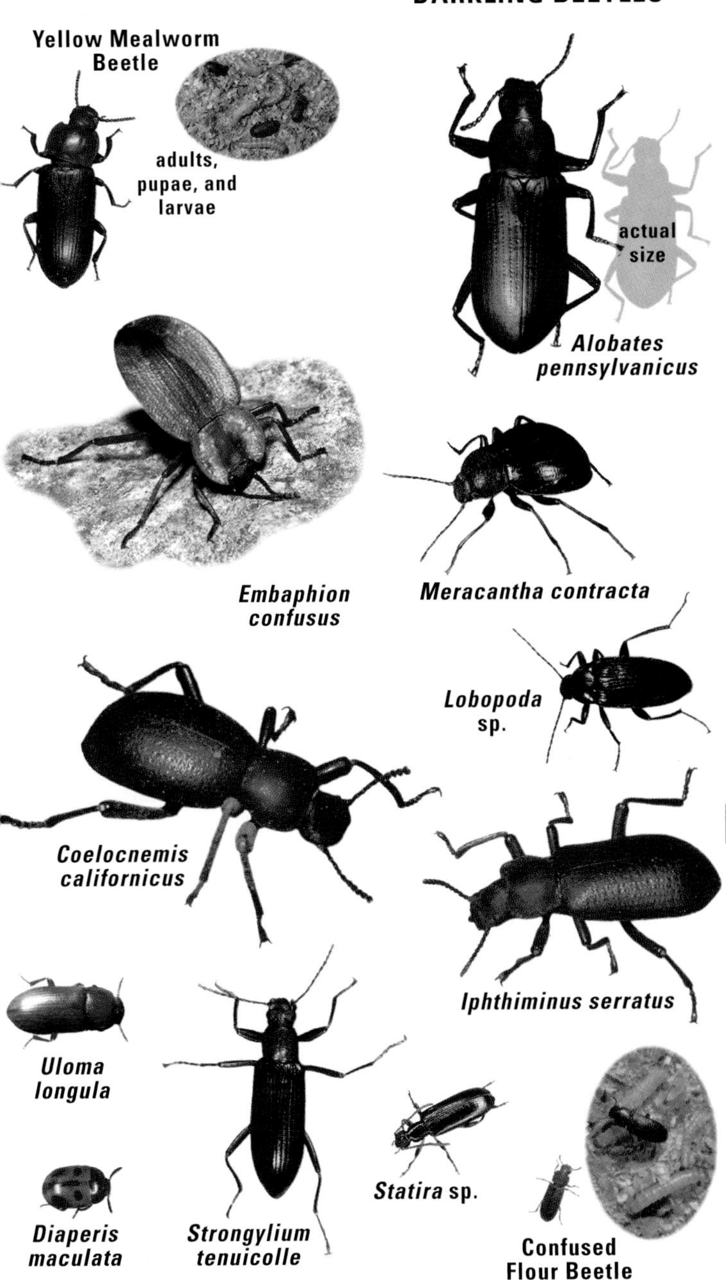

Yellow Mealworm Beetle

Alobates pennsylvanicus

Embaphion confusus

Meracantha contracta

Lobopoda sp.

Coelocnemis californicus

Iphthiminus serratus

Uloma longula

Statira sp.

Diaperis maculata

Strongylium tenuicolle

Confused Flour Beetle

CLICK BEETLES AND CEDAR BEETLES

(family Elateridae) are also known as skipjacks or jacknife beetles, referring to their acrobatic behavior. An individual can snap a spinelike process into a groove on its "chest" with enough force to propel the beetle into the air. This maneuver may startle a predator into dropping the beetle. Often they are recognizable by their long narrow shape, but if in doubt, you can gently flip the beetle onto its back and see if it does the click trick. Adults are active in the afternoon and evening, often coming to lights at night. Look under bark on logs or on foliage as well. Larvae are "wireworms" that occur in soil or rotting wood. Most feed on roots, tubers, or seeds, and a few species are occasional pests of sprouting grass seeds or roots. Some species are predatory on other larval insects or feed on fungi. The life cycle varies from one to three years, depending on the species. There are at least 965 species known in North America, with nearly 100 more awaiting formal description.

larva of eyed elater *(Alaus)*

The genus ***Alaus*** includes the eyed elaters, named for the prominent eyespots on the thorax. North America has at least five species, generally distributed. Their larvae live in decaying wood, especially decaying roots, preying on larvae of wood-boring beetles and other insects. ***Alaus oculatus*** is widespread in the east and as far west as Texas. While not abundant, it is sure to draw attention whenever found. ***Alaus melanops*** replaces this species in the west, although the two may overlap in the Black Hills of South Dakota. ***Alaus zunianus*** is southwestern. The genus ***Chalcolepidius*** includes about a dozen large species, mostly in the southern states. Some are notable for bright patterns or colors, such as ***Chalcolepidius webbi*** and ***Chalcolepidius smaragdinus.***

The genus ***Melanotus*** is represented by about 50 species in North America. Larvae of some species are harmful to corn, potatoes, and other crops. Larvae of the various species of ***Conoderus,*** found chiefly in the eastern and southern U.S., may also have a significant impact on crops such as tobacco, cotton, and potatoes. Members of the genus ***Deilelater*** (formerly in *Pyrophorus)* are luminous insects with glowing spots on the thorax. ***Ampedus*** includes more than 70 North American species, distributed widely across the continent. ***Limonius*** is a genus with more than 50 North American representatives, collectively widespread. Some do damage to the buds and flowers of various fruit trees. The genus ***Ctenicera*** currently includes more than 120 North American species, although some authorities would reclassify most of these into other genera.

CEDAR BEETLES (family Rhipiceridae) are, as larvae, parasites of cicada nymphs. All five of our species are in the genus ***Sandalus.*** Adults are active in September and early October. One female on a tree trunk will attract large numbers of strong-flying males. Eggs are laid in bark crevices, and the larvae that hatch crawl underground in search of cicada nymphs. Once attached, they become sedentary grubs.

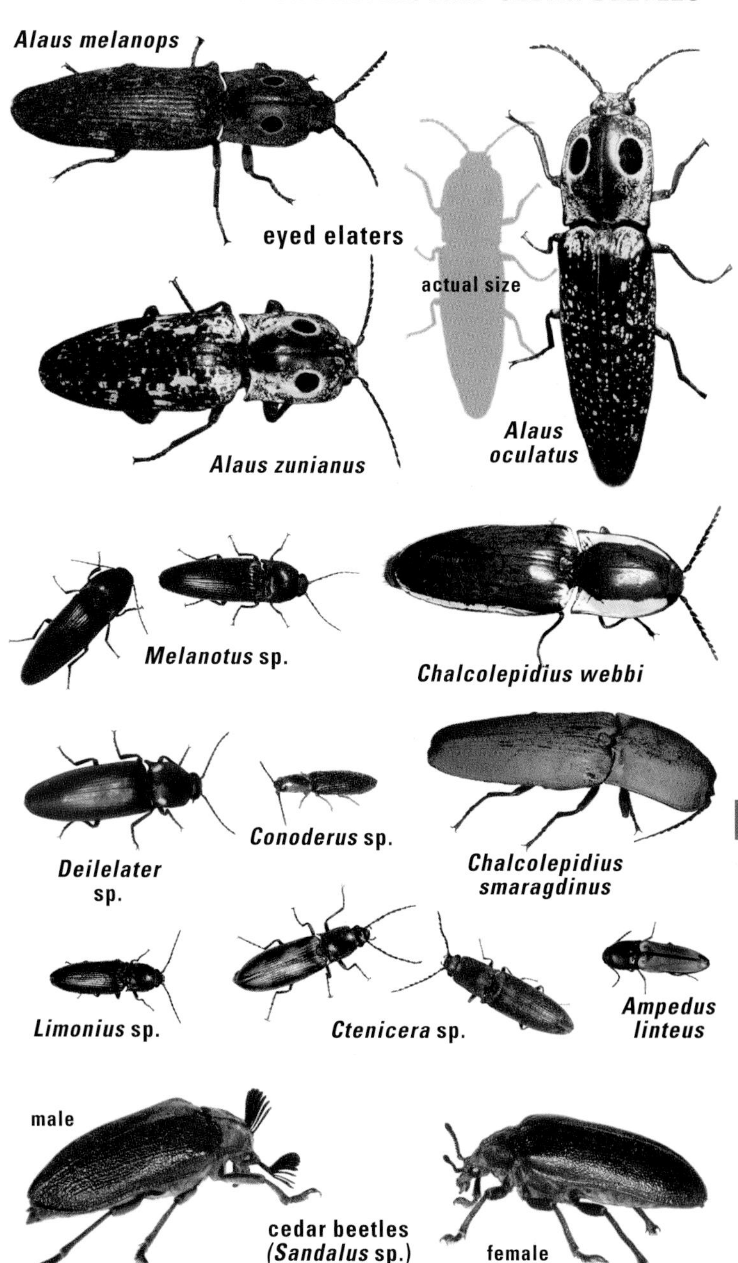
Alaus melanops
eyed elaters
actual size
Alaus zunianus
Alaus oculatus
Melanotus sp.
Chalcolepidius webbi
Deilelater sp.
Conoderus sp.
Chalcolepidius smaragdinus
Limonius sp.
Ctenicera sp.
Ampedus linteus
male
cedar beetles (Sandalus sp.)
female

CARRION BEETLES (family Silphidae) are seldom seen apart from freshly deceased vertebrates. Adults feed mostly on fly maggots, but the larvae eat the carcass itself. There are about 30 species in our area.

Our 15 members of the genus ***Nicrophorus*** are the burying beetles. They bury the bodies of small vertebrates, removing fur or feathers and fashioning the corpse into a meatball. The larvae are initially fed morsels by one or both parents. Adults squeak by rubbing the abdomen against the wing covers. Many carry mites that disembark to eliminate the beetles' competition: fly eggs that would hatch into voracious maggots. The **American Burying Beetle** *(Nicrophorus americanus)* is our largest species at 20–35 mm. Formerly common over much of the eastern U.S., it is now endangered, known only in a few sites in Oklahoma, Arkansas, Kansas, and Rhode Island. ***Nicrophorus orbicollis,*** abundant in the eastern half of the U.S. and adjacent Canada, forages at night for carrion, dung, maggots, even rotting fruit. It exhibits extensive parental care. ***Nicrophorus marginatus*** ranges over most of the U.S. and southern Canada. It is active by day in meadows and grasslands. ***Nicrophorus tomentosus*** occurs over almost the entire U.S., plus Canada east of the Rockies. Adults fly during the day, resembling bumble bees. They do not completely bury a carcass but sink it in a shallow pit and cover the body with debris.

Necrodes surinamensis flies to lights at night but otherwise occurs on carrion, where it feeds on maggots. Males have enlarged "thighs" and bowed "shins" on their hindlegs. The species ranges east of the Rockies in the U.S., and, sparingly, in the Pacific northwest.

The remaining carrion beetles shown here were all formerly classified in the genus *Silpha.* ***Heterosilpha ramosa,*** widespread in the west, is found on carrion and in decaying vegetation but was a noted pest of strawberries in Naches, Washington, in 1937–1938. The eastern ***Necrophila americana*** resembles a carpenter bee or bumble bee in flight. Adults feed on maggots, the larvae on carrion "jerky," sinew, and skin. ***Oiceoptoma inaequale*** is common in forests from southeastern Canada to Texas. It flies in daylight, exposing the metallic blue underside of its wing covers. ***Oiceoptoma noveboracense*** is another eastern forest dweller. ***Thanatophilus lapponicus,*** widespread in the north, is one of six species in its genus. Common on fish and other carrion, it sometimes damages dried, stored meats and hides.

CLOWN BEETLES (family Histeridae) might be amusing in shape if not behavior. Most of our 435 species are small, spherical, black, and highly polished. Their legs and antennae are short, retracting close to the body when a threat causes the beetle to feign death. They are common in carrion, dung, or under bark, preying on eggs and larvae of flies and beetles. Others live in burrows of rodents or reptiles, or in ant nests. ***Hister*** is a genus of small beetles found in carrion and dung. Our 10 species of ***Hololepta*** are bizarre, paper-thin beetles. Look for them under the skintight bark of recently felled hardwoods. Members of the genus ***Saprinus*** are found in carrion in eastern and central North America. Species of ***Platysoma*** occur under bark and are valued predators of bark beetles.

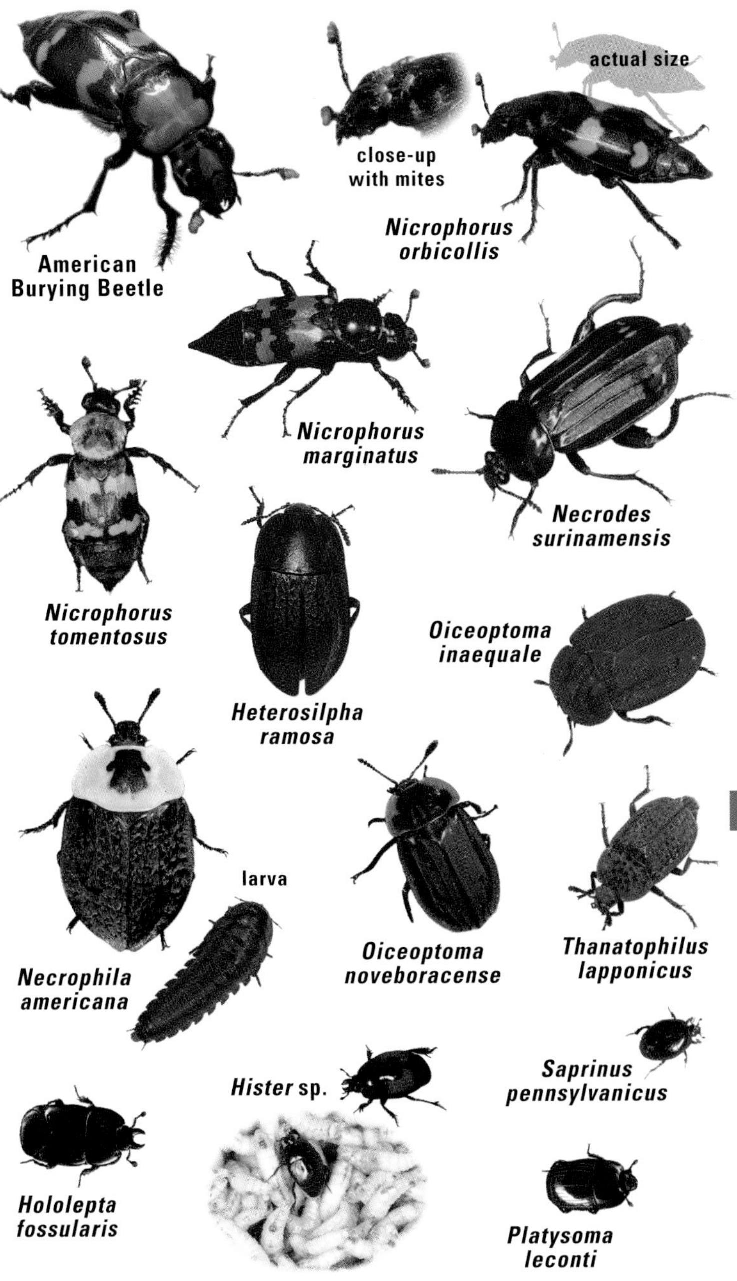
actual size
close-up
with mites
American
Burying Beetle
Nicrophorus
orbicollis
Nicrophorus
marginatus
Necrodes
surinamensis
Nicrophorus
tomentosus
Heterosilpha
ramosa
Oiceoptoma
inaequale
larva
Necrophila
americana
Oiceoptoma
noveboracense
Thanatophilus
lapponicus
Hister sp.
Saprinus
pennsylvanicus
Hololepta
fossularis
Platysoma
leconti

ROVE BEETLES

(family Staphylinidae) are sleek and serpentine, often mistaken for earwigs. They generally can be recognized by their short wing covers, which leave much of the abdomen exposed. This is one of the largest families of beetles, with over 4,100 species in North America, but the majority are tiny creatures that tend to escape the notice of the casual observer. Most rove beetles are predators as adults and larvae, living in a variety of niches, including intertidal zones, logs, fungi, carrion, dung, and ant nests. None are harmful to human interests, and many are beneficial as predators on pest insects. Although their short wing covers might suggest that they are also short-winged, most have well-developed hindwings that are folded under the wing coverts at rest, and most can fly well. Some carry their abdomens upright when alarmed and secrete defensive chemicals. Only a few examples are shown here, and you should not expect to be able to identify most rove beetles to species.

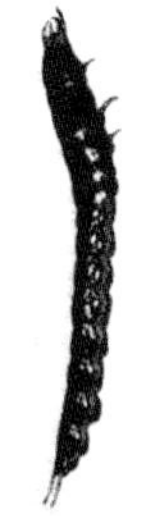

rove beetle larva

Platydracus (formerly *Staphylinus*) is a genus with 27 species in North America. ***Platydracus maculosus*** occurs in carrion, dung, and decaying fungi, but in flight it resembles a wasp. It is widely distributed, especially in the east. ***Platydracus cinnamopterus*** is common in rotting logs and fungi in eastern North America. The **Violet Rove Beetle** *(Platydracus violaceus)* is notable for its metallic reflections.

Ontholestes cingulatus, of the east and northwest, is very agile, and you only glimpse that bright gold tail as it takes cover or flies from beneath carrion or rotting fungi.

The **Hairy Rove Beetle** *(Creophilus maxillosus)* is found in and under carrion, munching on maggots. It ranges over most of North America. The **Pictured Rove Beetle** *(Thinopinus pictus)* is a unique, wingless beach species living along the Pacific Coast from Alaska to California. It feeds mostly on sand fleas at night.

The genus ***Philonthus*** contains well over 100 species. Many are found in fleshy fungi, others on manure or carrion. ***Philonthus cyanipennis,*** named for its stunning metallic blue (or sometimes green) wing covers, occurs in North America and Europe.

The genus ***Homaeotarsus*** includes 36 species. They frequent leaf-litter on the forest floor or in marshes or prairies. Some species come to lights at night. The genus ***Paederus,*** with about 15 species, is difficult to distinguish from other related genera.

Tachinus fimbriatus, which has a very roachlike appearance, is found in fleshy fungi, dung, and occasionally carrion over much of the continent. There are about 40 other North American species in the genus.

The so-called **Devil's Coach Horse** *(Ocypus olens),* native to Europe, was discovered in California in 1931. It is now common in the Bay Area and in urban southern California. At up to 32 mm long and with a menacing defense posture (jaws wide open, abdomen upraised, stink glands deployed), it never fails to attract attention. It earns its keep by eating snails, slugs, and other pests.

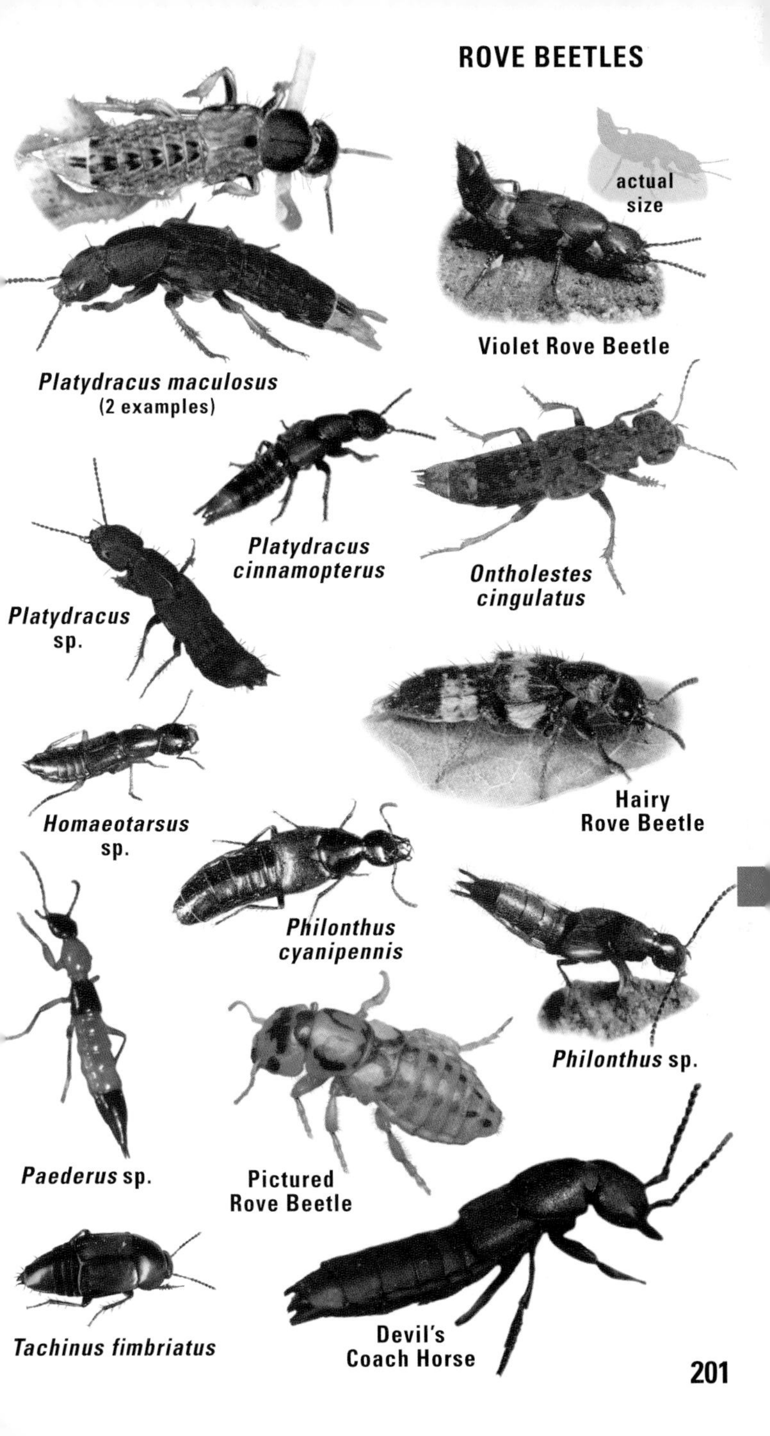

Platydracus maculosus
(2 examples)

Violet Rove Beetle

Platydracus cinnamopterus

Ontholestes cingulatus

Platydracus sp.

Hairy Rove Beetle

Homaeotarsus sp.

Philonthus cyanipennis

Philonthus sp.

Paederus sp.

Pictured Rove Beetle

Tachinus fimbriatus

Devil's Coach Horse

BLISTER BEETLES

(family Meloidae) are often common locally for a short time. Soft-bodied, with round heads and long legs, adults of most species feed on nectar or pollen or consume the flower itself. A few foliage feeders sometimes damage crops. When squeezed, adults exude skin-blistering irritants in a defensive tactic called reflex bleeding. Cantharidin is the harmful chemical agent. Larvae are parasitic on larvae of solitary bees or on grasshopper egg pods. They go through "hypermetamorphosis," beginning life as a sleek, host-seeking missile, becoming a plump couch potato once they locate a host. In North America there are about 410 species in 22 genera, most diverse in the west. "Spanish Fly" is a blister beetle native to Europe.

Our 21 species of oil beetles ***(Meloe)*** are collectively widespread, from Alaska to Florida and Mexico. Adults are flightless ground-dwellers with short wing covers and bloated abdomens. Males of some species use strange kinks in their antennae to grip females during mating. Larvae hatch from the eggs as active "triungulins" that scale flower stalks and climb aboard a female bee that visits the blossom. Once inside the bee's burrow, the larva molts into a sedentary grub that consumes the pollen and nectar stored for the larval bee, as well as the larva itself.

About 30 species of ***Pyrota*** are collectively widespread in the U.S. Look for members of this genus at lights at night or on composite flowers by day. ***Pyrota akhurstiana*** is common in southern Arizona and New Mexico. ***Pyrota invita*** is particularly numerous in Texas.

Epicauta includes 173 species north of Mexico. Adults are velvety gray, tan, black, or a combination of these. Larvae feed on egg capsules of grasshoppers, but adults are occasional crop pests. Beetles baled with hay have caused the death of horses that unwittingly consumed them. The **Black Blister Beetle** *(Epicauta pennsylvanica)* is common on goldenrod in fall in the eastern U.S., west to Montana. The **Margined Blister Beetle** *(Epicauta pestifera)* can be all gray. It ranges throughout the eastern and north-central U.S. The **Striped Blister Beetle** *(Epicauta vittata)* is one in a complex of nearly identical species found in the eastern U.S. ***Epicauta andersoni,*** one of several dotted species, is common in southwestern mountains.

The genus ***Lytta*** includes 69 species recognized in North America, most occurring in the west. Larvae are parasitic on solitary bees. ***Lytta aenea*** can be found in early spring in eastern North America on the flowers of apple and related trees. **Nuttall's Blister Beetle** *(Lytta nuttalli)* is widespread in the west, east to Saskatchewan and Minnesota. Look for it on legumes such as lupine. The genus ***Nemognatha*** includes 28 North American species, with a collectively wide distribution. Adults are dull orange, black, or both, with extremely long mouthparts. Look for them on thistles and other composite flowers. Females lay their eggs there, and the larvae that hatch hitch rides to the nests of solitary bees where they parasitize the larval bees. ***Zonitis*** is a genus with 16 species in North America, generally distributed. Adults are usually shiny yellow, half an inch long or less, and similar in habits to ***Nemognatha*** species.

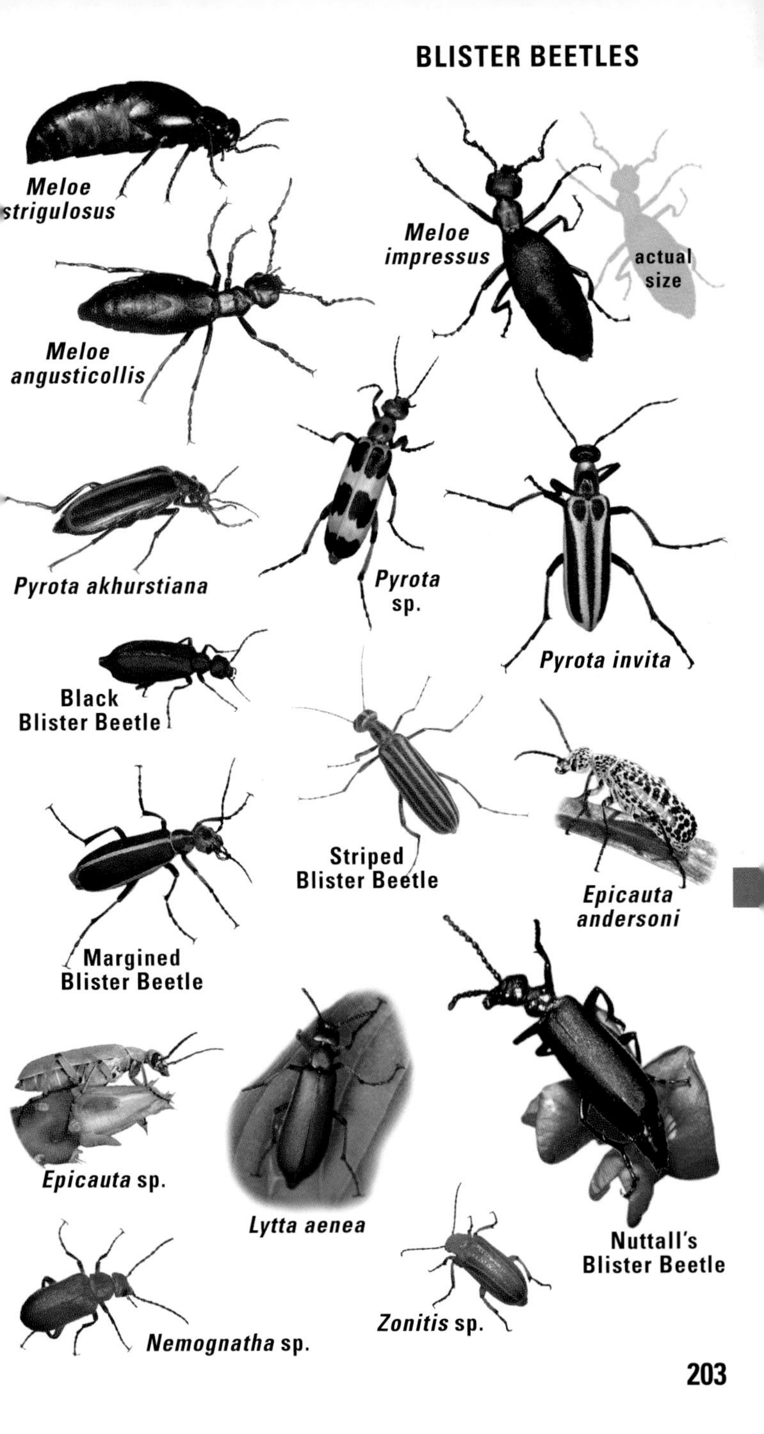
Meloe strigulosus
Meloe impressus
actual size
Meloe angusticollis
Pyrota akhurstiana
Pyrota sp.
Pyrota invita
Black Blister Beetle
Striped Blister Beetle
Epicauta andersoni
Margined Blister Beetle
Epicauta sp.
Lytta aenea
Nuttall's Blister Beetle
Zonitis sp.
Nemognatha sp.

FALSE BLISTER BEETLES (family Oedemeridae) are most common along coastlines on flowers, foliage, under driftwood, or in wet, rotten wood (where the larvae of most species develop). They resemble blister beetles (previous page) and soldier beetles (p. 172) in shape and soft-bodied texture. When crushed they secrete potent skin irritants that may raise blisters. There are about 87 species in North America.

The **Wharf Borer** *(Nacerdes melanura)* is native to Europe but is now cosmopolitan. It sometimes does severe damage as a larva boring in old, moist wood pilings, boardwalk timbers, and other structural lumber. Look for adults at lights from April to July.

The genus ***Ditylus*** includes three species, two on the West Coast and one in the northeastern U.S. ***Ditylus quadricollis*** is common in British Columbia and the northwestern U.S., where adults may be found under debris on the forest floor or on flowers. Larvae feed in logs of western red-cedar, Douglas-fir, and other conifers.

The genus ***Asclera*** is represented by six species, three in the western third of the continent, two in the eastern half, and one widespread. Look for adults on early spring wildflowers such as willow catkins, buttercups, violets, and wild plum. The genus ***Oxacis*** includes 29 species in North America, with a collectively wide distribution but most diverse in the southeastern U.S. Adults are usually found on flowers.

FIRE-COLORED BEETLES (family Pyrochroidae) resemble soldier beetles (p. 172). Larvae live mostly under bark on rotting logs, feeding on decaying wood and fungi. Contrasting colors of red and black in the adults of many species warn of their toxic nature. Cantharidin is a poison common to blister beetles and false blister beetles, but how it is acquired by fire-colored beetles is a mystery. There are about 50 species in seven genera in North America.

Pedilus is a genus with 30 species in North America, formerly placed in their own family, the Pedilidae. Most are found in the western and northeastern U.S. In cases in which larvae are known, they feed in decaying vegetation in or adjacent to the soil. Look for adults on foliage of shrubs or on flowers in spring. ***Pedilus terminalis*** is common in the east.

The genus ***Neopyrochroa*** includes four North American species, two restricted to California. ***Neopyrochroa flabellata*** ranges widely in eastern forests, occurring on understory foliage or at lights at night. The male has comblike antennae and a cleft on his face. From this groove he dispenses a cantharadin-laden secretion to a prospective mate. She allows him to mate if convinced he has a good store of the chemical. She uses the substance to protect her eggs from predators.

The genus ***Dendroides*** includes six North American species with a collectively transcontinental distribution and with elaborate feathery antennae in males. The dark bulging eyes (larger on males) are distinctive. ***Dendroides cyanipennis*** (the specific name means "blue winged") sometimes comes to lights at night in the east.

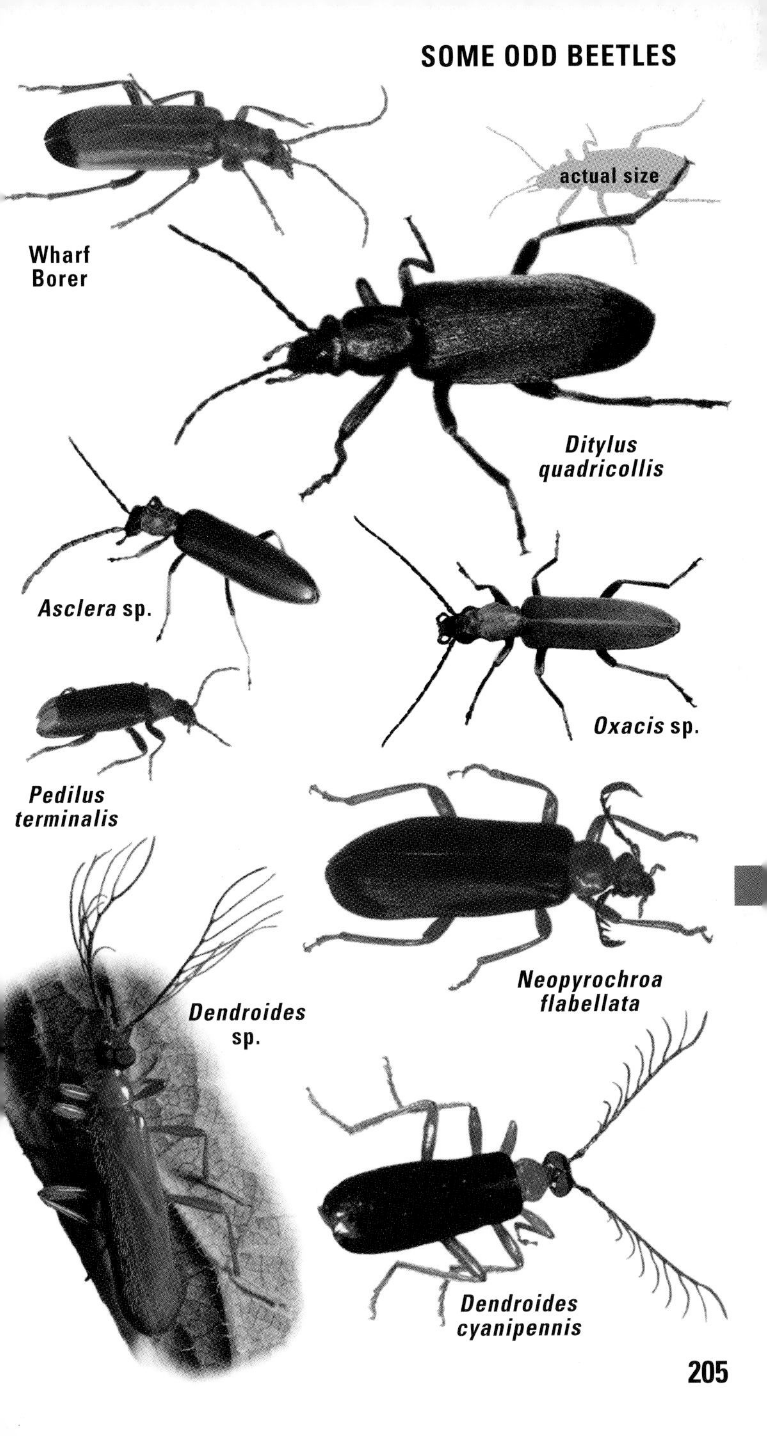
actual size
Wharf Borer
Ditylus quadricollis
Asclera sp.
Oxacis sp.
Pedilus terminalis
Neopyrochroa flabellata
Dendroides sp.
Dendroides cyanipennis

The Checkered Beetles **(family Cleridae)** are mostly colorful, agile predators of other insects. Some are important pollinators, and a minority are pests of stored products. Some species mimic velvet ants (p. 350), other wasps, or fireflies. Most are active by day, but a few come to lights at night. There are nearly 300 species in North America.

Beetles of the genus ***Enoclerus*** are usually banded in combinations of black with orange, red, gray, brown, white, or yellow. There are 36 North American species. Adults and larvae are active predators of bark beetles, weevils, and other borers. ***Enoclerus ichneumoneus*** occurs in most eastern states. Watch for them on trunks of dead, standing oak trees. ***Enoclerus quadrisignatus*** comes to lights at night in the eastern U.S. ***Enoclerus nigripes*** is another common day-active eastern species. ***Enoclerus rosmarus*** can be abundant on flowers and weeds in late spring and early summer over most of the eastern U.S.

Trichodes is represented by 10 species in North America, widely distributed, especially in the west. The **Ornate Checkered Beetle** *(Trichodes ornatus)* is common on yarrow and other flowers in western states. Larvae are parasitic in nests of leafcutter bees (p. 342), making this species a pest to alfalfa farmers who depend on the bees to pollinate their crops.

The genus ***Cymatodera*** includes 60 species in North America, widely distributed but most diverse in the southwest. Adults eat a variety of other insects, including caterpillars, the grubs of gall wasps, and the larvae of various wood-boring beetles. They come to lights at night. Our four wide-ranging species of ***Chariessa*** resemble fireflies and are also nocturnal, but the branching antennae of males are distinctive. The genus ***Necrobia*** is represented by three cosmopolitan species, all about 5 mm long. They prey on a variety of stored-product pests but also scavenge dried animal and plant matter. The **Red-legged Ham Beetle** *(Necrobia rufipes)* is sometimes a pest of grain, silk, and copra (dried coconut).

TROGOSSITID BEETLES (family Trogossitidae) are generally flattened, sometimes cylindrical insects. Most are predators of wood-boring beetles and can be found under bark on dead, standing trees or on logs. A few come to lights at night. There are about 59 species in the U.S. and Canada. ***Calitys scabra*** is one of two North American species in its genus. Look for it under bark on decaying logs over much of eastern Canada and the northeastern U.S. ***Ostoma*** is a genus with three North American representatives, collectively widespread. ***Ostoma pippinskoeldi*** occurs under bark in the western states and British Columbia. The genus ***Temnoscheila*** includes 10 species found in coniferous forests. They feed on wood-boring insects, sometimes pursuing them inside their tunnels. ***Temnoscheila chlorodia*** ranges through much of the western U.S. and Canada. Look for it under bark on pines. The 21 North American species of the genus ***Tenebroides*** are widespread and not uncommon at lights at night. Most prey on bark beetle adults and larvae (p. 216), but the **Cadelle** *(Tenebroides mauritanicus)* feeds on pests of stored products in granaries. It may eat grain itself but is seldom a pest.

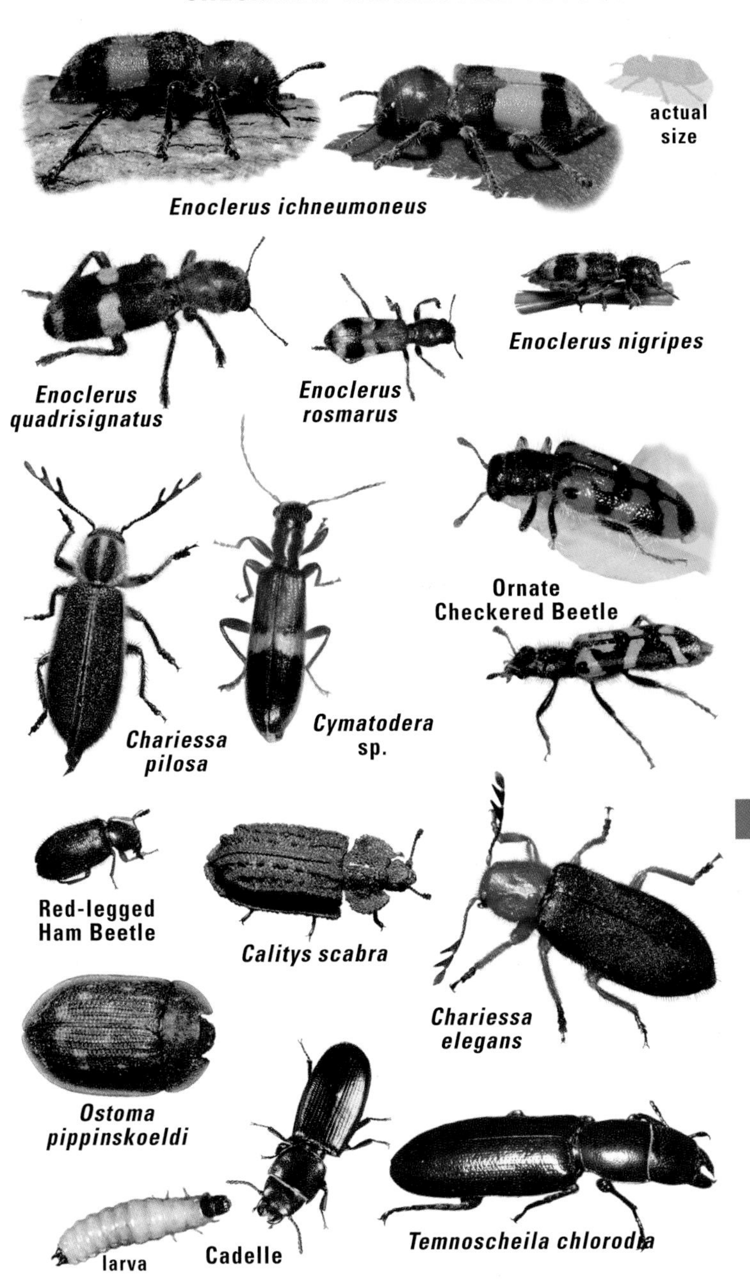
actual
size
Enoclerus ichneumoneus
Enoclerus quadrisignatus
Enoclerus rosmarus
Enoclerus nigripes
Ornate
Checkered Beetle
Chariessa
pilosa
Cymatodera
sp.
Red-legged
Ham Beetle
Calitys scabra
Chariessa
elegans
Ostoma
pippinskoeldi
larva
Cadelle
Temnoscheila chlorodia

DEATH-WATCH AND BOSTRICHID BEETLES

DEATH-WATCH BEETLES (family Anobiidae) now include the former family Ptinidae (spider beetles). These tiny insects include some of our "most unwanted" stored-product and structural pests. Death-watch beetles are largely wood borers, although a good percentage infest woody fungi, puffballs, dried herbs, spices, or tobacco. Most spider beetles live as scavengers in bird or mammal nests. Others invade nests of solitary bees and eat the pollen stored for the larval bees. A few eat flour, wool, and other goods. Roughly 360 species occur in North America.

The original **Death-watch Beetle** *(Xestobium rufovillosum)*, not illustrated, is native to Europe. It bores in wood, infesting furniture and old houses. Males and females call to each other by smacking their faces against the floor of their tunnels. The audible ticking sound was thought to portend a death in the household. More likely to be seen is the **Drugstore Beetle** *(Stegobium paniceum)*, a native that has spread far and wide through commerce. Virtually nothing organic is off-limits to this insect, including leather, flour, dried beans, and cayenne pepper. It will even bore through plastic vials to get to a meal. The genus ***Lasioderma*** includes five species in North America. The **Cigarette Beetle** *(Lasioderma serricorne)* infests a variety of dried vegetable matter but appears most addicted to tobacco at all stages of production. So far, it has not been observed smoking.

The genus ***Ptinus*** includes 38 spider beetle species in North America. ***Ptinus fur*** is typical, eating almost any dry organic material but seldom reaching pest status. ***Gibbium aequinoctiale*** is a common pest in seeds, grain, pepper, and insect collections.

BOSTRICHID BEETLES (family Bostrichidae) are mostly branch and twig borers that thrive on the starch content of cellulose. Like termites, they harbor microscopic gut fauna that aid in digestion. A few are pests of trees, wild and cultivated vines, tubers, dried herbs, seasoned wood, or stored grain. Our native fauna numbers more than 70 species. Another four are exotic in origin, and at least 34 other nonnative species have been intercepted at various ports. ***Apatides fortis*** occurs from Texas to southern California and north to Oklahoma. The short hooked horns on the thorax are distinctive. Look for it at lights. Its hosts include mesquite and palo verde. The genus ***Lichenophanes*** includes eight species, mostly in the east and the southwest. ***Lichenophanes bicornis*** is a common eastern species found at lights at night, or under bark or in dry fungi.

Stout's Hardwood Borer *(Polycaon stouti)* lives mostly in the far west, locally in western Texas. It bores in a variety of hardwoods, and adults may emerge from furniture, pallets, and other materials constructed from infested wood. The genus ***Amphicerus*** includes four species in North America, absent from the Pacific northwest. The **Apple Twig Borer** *(Amphicerus bicaudatus)* is a common eastern species that flies to lights at night in early spring. The **Giant Palm Borer** *(Dinapate wrightii)*, which can be up to 2 inches long, is native to southern California. Adult females lay their eggs in the crown of palms, and the larvae bore down, then up, inside the trunk, taking 3–9 years to complete the life cycle.

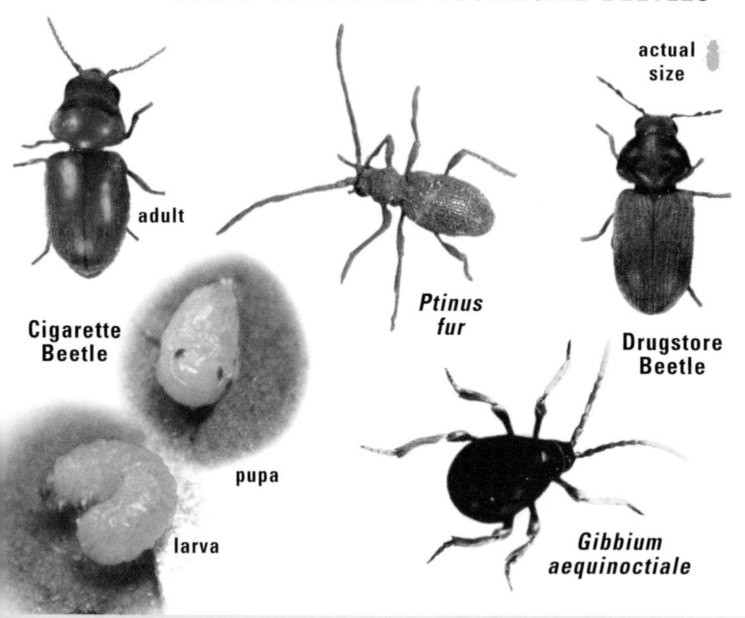

bostrichid beetles (different scale)

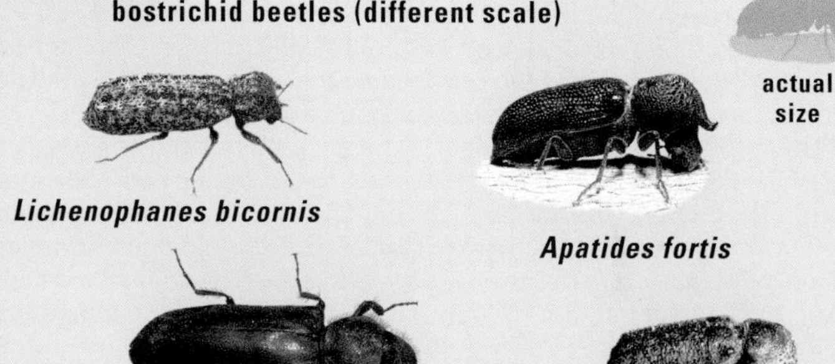

Lichenophanes bicornis

Apatides fortis

Stout's Hardwood Borer

Apple Twig Borer

Giant Palm Borer

DERMESTID BEETLES AND OTHERS

SKIN BEETLES AND CARPET BEETLES (family Dermestidae) are small dark beetles, patterned with light-colored scales or hairs. There are more than 120 species in North America.They are notorious pests of stored products of many kinds. Most are scavengers on dried animal parts or on plant matter with a high protein content. In the wild, dermestid larvae and adults generally occur in dry carrion, in bird and mammal nests, or in abandoned nests of wasps and bees. Adults of some species feed on pollen and may be found visiting flowers. Closer to home, many an entomologist has had the unsettling experience of discovering dermestid beetles in their collection, not as preserved specimens but as larvae that are rapidly devouring the preserved specimens.

The genus ***Dermestes*** includes 15 species, many introduced from other parts of the world, collectively distributed across the continent. Museums employ colonies of these insects to clean fragments of flesh and other tissue off of bones that are to be added to their collections. The **Larder Beetle** *(Dermestes lardarius)* is a cosmopolitan pest of cured ham, bacon, dried fish, dog biscuits, and other animal products. ***Dermestes frischii*** (not illustrated) is common over much of North America on dry carrion. Scientist F. W. Hope reported it from Egyptian mummies in 1834.

Members of the genus ***Anthrenus,*** with 18 species in North America, are the most familiar carpet beetles. Larvae can reduce a prized insect collection to piles of powder in no time. They also damage woolen garments, furs, and other animal products. The **Varied Carpet Beetle** *(Anthrenus verbasci)* is introduced on this continent and now widespread. Adults and larvae are often found inside houses. Outdoors, adults may be seen on flowers. The **Buffalo Carpet Beetle** *(Anthrenus scrophulariae)* has a worldwide distribution. Adults may be seen on spring flowers.

FLAT BARK BEETLES (Cucujidae), as their common name implies, have extremely flattened bodies and are usually found under bark. This once constituted a larger group of insects, but with the current classification, this family now includes only three species of beetles in North America. ***Cucujus clavipes*** is striking as an adult, but it is most often observed in the larval stage, under bark on recently felled maple, beech, elm, ash, poplar, and other hardwoods.

SILVANID FLAT BARK BEETLES (Silvanidae) were recently split from the family Cucujidae. These are tiny insects, many of which infest stored grain, nuts, or spices. Others feed on fungi under bark, or on wilting vegetation. Some of our 32 species (in 14 genera) are immigrants from the Old World. The **Sawtooth Grain Beetle** *(Oryzaephilus surinamensis)* is named for the "teeth" along the edges of its thorax. It is an introduced pest. The **Square-necked Grain Beetle** *(Cathartus quadricollis),* native to this continent, is noted as a pest of corn, both in the field and in the silo. The best way to prevent household infestations of these beetles is to seal all foods in airtight containers.

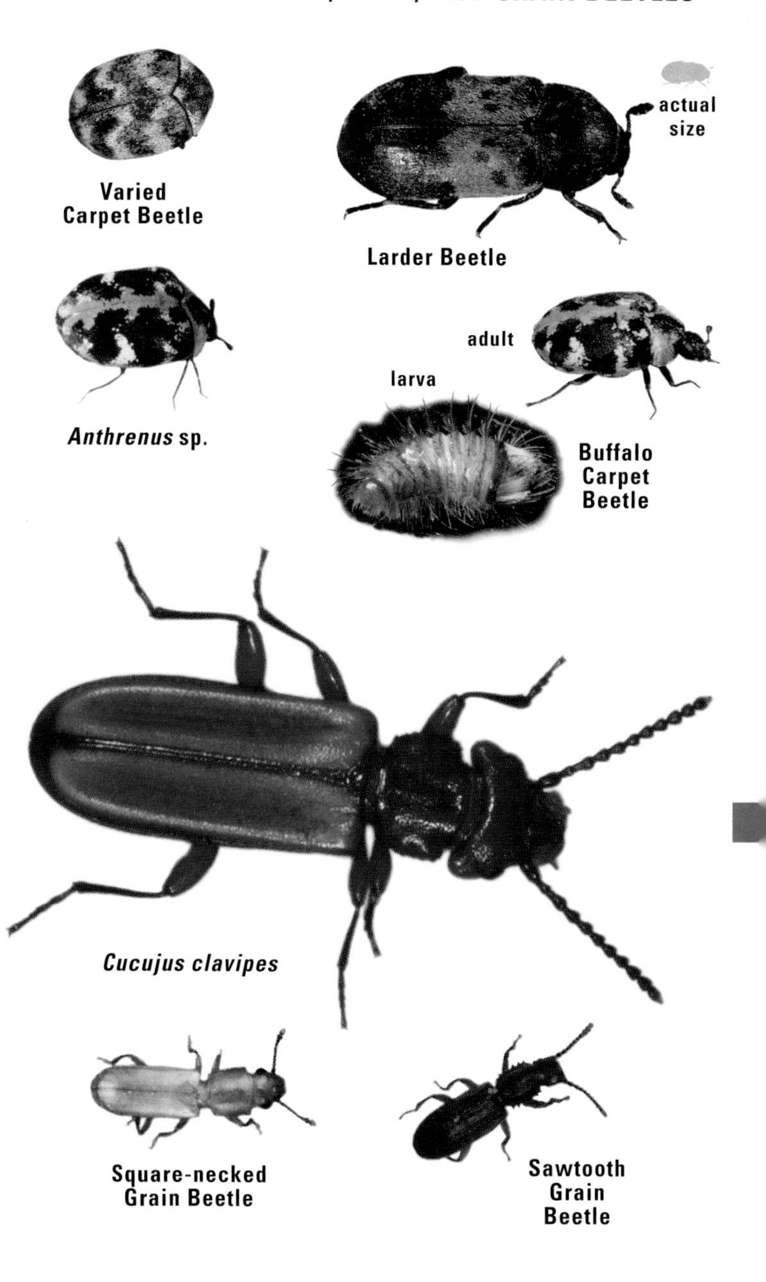

Varied
Carpet Beetle

Larder Beetle

Anthrenus sp.

Buffalo
Carpet
Beetle

Cucujus clavipes

Square-necked
Grain Beetle

Sawtooth
Grain
Beetle

TRUE WEEVILS

(family Curculionidae) are among the most diverse insect groups, with over 3,000 species in North America. They are recognized by the elbowed antennae. Other beetles may have a prolonged "snout" but not the distinctive antennae. At the tip of the snout (properly called a rostrum) are chewing mouthparts. A great number of species play dead when disturbed. Many are coated with scales that may rub off as the insect ages, so individuals of a species can vary greatly. Larvae resemble plump maggots, but their large head capsule identifies them as coleopterans. Nearly all known weevils are vegetarian in larval and adult stages. Hardly any plant is not affected by at least one species of Curculionidae.

Most in the **subfamily Curculioninae** have the classic long "nose." Most develop in fruits or seeds of the host plant. Our 27 species of ***Curculio*** are known as acorn or nut weevils. The **Larger Chestnut Weevil** *(Curculio caryatrypes)* persisted while its native chestnut host tree largely perished: it has adapted to the ornamental trees that brought chestnut blight from Asia in the first place. Of more than 100 species of ***Anthonomus,*** best known is the **Boll Weevil** *(Anthonomus grandis)*. Native to Mexico, where it fed on wild cotton, it crossed the Rio Grande around 1892. It does hundreds of millions of dollars' worth of damage to cotton crops annually.

The **subfamily Baridinae** includes about 500 species in our area, including about 50 in the genus ***Odontocorynus.*** Look for adults on flowers. **Subfamily Cryptorhynchinae** includes nearly 200 species in North America. Many have a groove in their "chest" where the beak rests when not in use. The **Poplar-and-willow Borer** *(Cryptorhynchus lapathi)*, introduced from Eurasia, is now widespread in the north. Eighteen species of ***Gerstaeckeria*** occur in the south and west. The nocturnal adults feed on cacti.

Members of the **subfamily Entiminae** are "broad-nosed weevils" with short rostrums. The **Little Leaf Notcher** *(Artipus floridanus)* is a citrus pest in Florida. The flightless adults often disperse by riding on transplanted trees. As usual, the larvae do the most damage, those juvenile delinquents, feeding on roots. The genus ***Naupactus*** includes four introduced species. The **Fuller Rose Beetle** *(Naupactus godmanni)* is a flightless species known from females only. ***Compsus auricephalus*** occurs in the southeast. Our 14 species of ***Otiorhynchus*** are all introduced from elsewhere. The **Black Vine Weevil** *(Otiorhynchus sulcatus)* is a nocturnal pest of many plants. The **Clay-colored Weevil** *(Otiorhynchus singularis)* is native to Europe. First reported in the U.S. in 1872, it now occurs over much of the north. The 35 species of ***Ophryastes*** are western. The flightless adults are associated with shrubs in arid habitats. The **Imported Long-horned Weevil** *(Calomycterus setarius)*, native to Asia, has spread from the northeast U.S. to Kansas. Only females are known, and they are flightless. ***Epicaerus*** includes 11 species, ranging over much of the U.S. Adults feed on foliage. ***Pachyrhinus*** (formerly *Scythropus*) includes eight species ranging from the western U.S. to Nova Scotia. ***Pachyrhinus elegans*** occurs mostly on white pine. ***Polydrusus*** includes about seven species here. ***Polydrusus impressifrons,*** native to Europe, was first discovered here in New York in 1906.

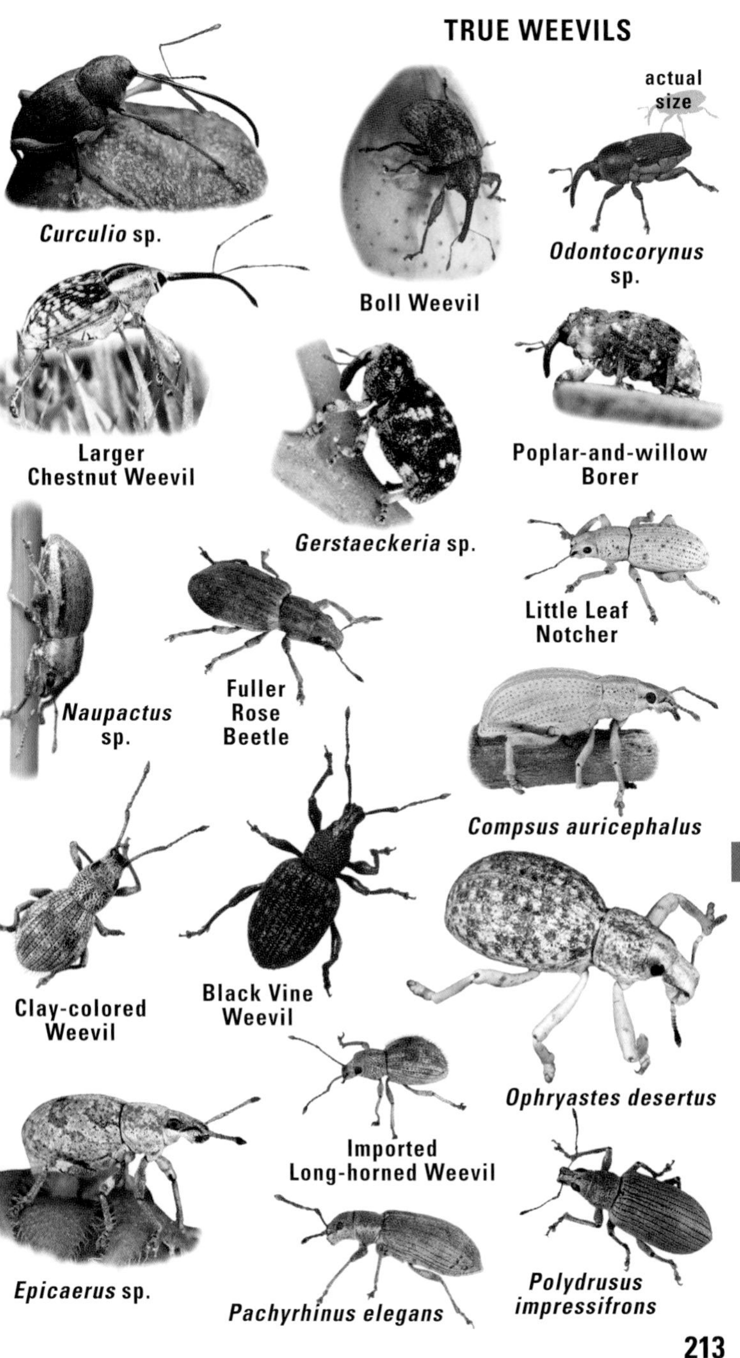

Curculio sp.

Boll Weevil

Odontocorynus sp.

Larger Chestnut Weevil

Gerstaeckeria sp.

Poplar-and-willow Borer

Little Leaf Notcher

Naupactus sp.

Fuller Rose Beetle

Compsus auricephalus

Clay-colored Weevil

Black Vine Weevil

Ophryastes desertus

Imported Long-horned Weevil

Epicaerus sp.

Pachyrhinus elegans

Polydrusus impressifrons

Members of the **subfamily Dryopthorinae** have football-shaped bodies. Some significant stored-product pests, the tiny granary weevils, are in this subfamily. Outdoors, look for the following "billbugs." ***Sphenophorus*** includes 65 North American species. Host plants are grasses, sedges, and other monocots, the larvae boring in stems or roots. ***Sphenophorus australis*** breeds primarily in cattails. The **Clay-colored Billbug** *(Sphenophorus aequalis)* occurs from southern Canada to northern Mexico. The usual host plants are bulrushes and reeds, but this species has also been known to damage young corn plants in bottomland fields. ***Scyphophorus yuccae*** is found in California and less commonly in Arizona and Texas. The larvae bore inside roots and stems of only two species of yucca. The **Cocklebur Weevil** *(Rhodobaenus quinquepunctatus)* ranges over the east. Adults are typically red with an oval black mark on the top of the thorax. Larvae are known to bore in the stems or roots of cocklebur, ragweed, and other composite plants. ***Rhodobaenus tredecimpunctatus*** can be separated from its cousin by the round black dot on the top of the thorax. Adults of both species are very active, fly well, and hibernate during the winter. The **Palmetto Weevil** *(Rhyncophorus cruentatus)* is our largest weevil. Its preferred host is the native cabbage palmetto.

The **subfamily Cyclominae** includes weevils commonly found in wetlands. Our 81 species of ***Listronotus*** are chiefly associated with arrowheads, sedges, and other aquatic plants. The **subfamily Hyperinae** includes only the 17 species of ***Hypera*** north of Mexico. The **Alfalfa Weevil** *(Hypera postica)* is originally from Europe. Its life cycle begins when a female chews a hole in an alfalfa stem and deposits one or more eggs. She can lay more than 400 in her brief life. The **Lesser Clover Weevil** *(Hypera nigrirostris)* is another pest of alfalfa and clover, introduced from the Old World.

The **subfamily Lixinae** includes over 100 species, most associated with watersides. Several species have been imported to control invasive weeds. ***Lixus*** is represented by 69 species over most of the U.S. and Canada. They are generally large cylindrical beetles, covered initially in pale scales that rub off easily. The **subfamily Mesoptiliinae** includes minor forest pests that, as larvae, bore under bark or in twigs. ***Magdalis*** includes 25 species here, the larvae mining beneath the bark of dead or dying trees.

The **subfamily Molytinae** includes some important pests of conifers, orchard fruits, and other plants, but the majority of genera are of no economic consequence. The **Pitch-eating Weevil** *(Pachylobius picivorus)* lives in pines in the eastern U.S. and southern Canada. Species of ***Rhyssomatus*** have a general distribution except the Pacific northwest, associating with plants in several families. The **Plum Curculio** *(Conotrachelus nenuphar)* is found over much of the continent. Apples, plums, and apricots are the principal hosts. The genus ***Hylobius*** includes both heroes and villains, from the human perspective. ***Hylobius transversovittatus*** was introduced to help control purple loosestrife, an invasive wetland weed. The **Pales Weevil** *(Hylobius pales)* is like its evil twin, literally grinchlike in being a serious pest of Christmas tree farms in the eastern U.S. and Ontario.

Sphenophorus dicolor

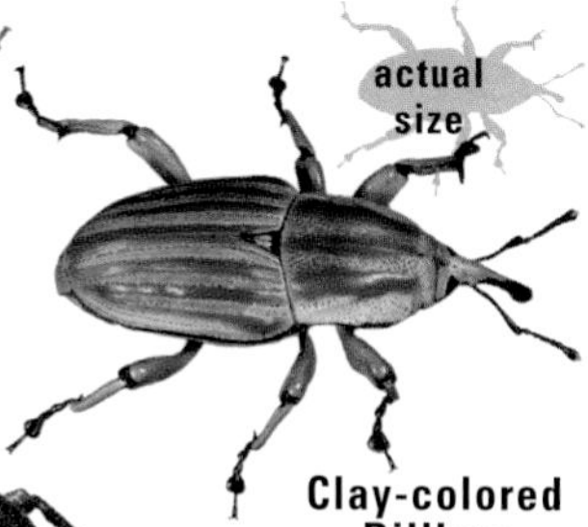

actual size

Clay-colored Billbug

Sphenophorus australis

Scyphophorus yuccae

Cocklebur Weevil

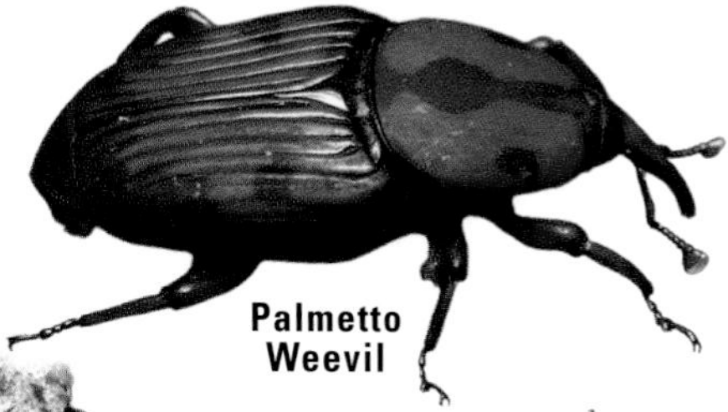

Palmetto Weevil

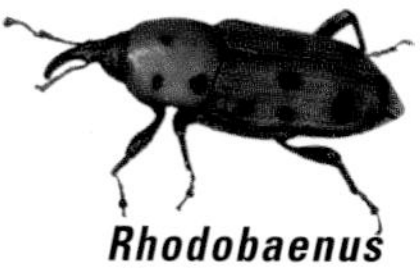

Rhodobaenus tredecimpunctatus

Listronotus sp.

Alfalfa Weevil

Lesser Clover Weevil

Pitch-eating Weevil

Lixus sp.

Magdalis lecontei

Rhyssomatus sp.

Pales Weevil

Plum Curculio

Hylobius transversovittatus

BARK AND AMBROSIA BEETLES, BEAN WEEVILS

The **subfamily Scolytinae,** now classified in the true weevil family, includes the ambrosia beetles and the notorious bark beetles. The latter, often portrayed as the silvicultural equivalent of locusts, may be simply scapegoats for poor forest management. Indeed, some evidence suggests that bark beetles are vital to ecological processes such as succession and decomposition. Their economic impact has led to fascinating discoveries about their biology and life histories. Most species use an array of chemicals (collectively called pheromones) to communicate with each other. Many genera of bark beetles, even some species, can be identified by the patterns left by their tunneling activities. Such etchings are amazingly artistic. The beetles themselves are small and may be confused with anobiid and bostrichid beetles (p. 208). All scolytids have clubbed antennae, which generally separates them from lookalikes.

The 13 species of ***Dendroctonus*** are "pine beetles" that make headlines. They can reach epidemic status, usually aggravated by drought or fire that weakens their host trees. All appear to have a symbiotic relationship with fungi that greatly neutralize the trees' natural defenses. The **Spruce Beetle** *(Dendroctonus rufipennis)* is by far the most destructive pest of spruce, accounting for an average of over one-third billion board feet of standing timber lost each year. A recent outbreak in Alaska totaled far more than that. The **Southern Pine Beetle** *(Dendroctonus frontalis)* attacks pines exclusively. Unlike many bark beetles, this species focuses on vigorous, standing trees. The **Red Turpentine Beetle** *(Dendroctonus valens)* attacks pines and spruce over most of the continent except the southeastern U.S.

Seven species of ***Hylesinus*** occur in ash trees. The egg galleries are distinctive, broadly T-shaped. A fungus usually follows the beetles, staining the wood brown. ***Scolytus*** includes 20 North American species, three introduced. In profile, the abdomen is distinctly concave. In the west, these insects attack conifers; in the east they mine in hardwoods. ***Scolytus multistriatus,*** introduced from Eurasia, has spread Dutch elm disease far and wide. ***Phloeosinus*** includes 25 species of tiny beetles, mostly western. Most mine in cedars, junipers, and related conifers. Adults in the genus ***Ips*** are recognized by the concave hind portion of the wing covers, ringed with jagged teeth. Each of the 17 shothole borers in the genus ***Xyleborus*** has an association with its own ambrosia fungus that it impregnates into the walls of its tunnels. The **European Shothole Borer** *(Xyleborus dispar)*, introduced here prior to 1816, is now found coast to coast in the north.

BEAN WEEVILS (Bruchidae) are oval and compact. There are about 150 species in North America, most associated with legumes. A few, found globally, are capable of developing in stored dried beans and other seeds. Many entomologists lump these insects with the leaf beetles (pp. 160–165). The genus ***Amblycerus*** includes seven species here. ***Amblycerus robiniae*** is common at lights in the eastern U.S. Larvae develop in the seeds of honey locust. Our 54 species of ***Acanthoscelides*** include the **Bean Weevil** *(Acanthoscelides obtectus)*, a common pest of growing and stored beans. Native to this continent, it has spread via commerce throughout the world.

BARK BEETLES, ETC.

Hylesinus
tunneling in green ash

Spruce Beetle

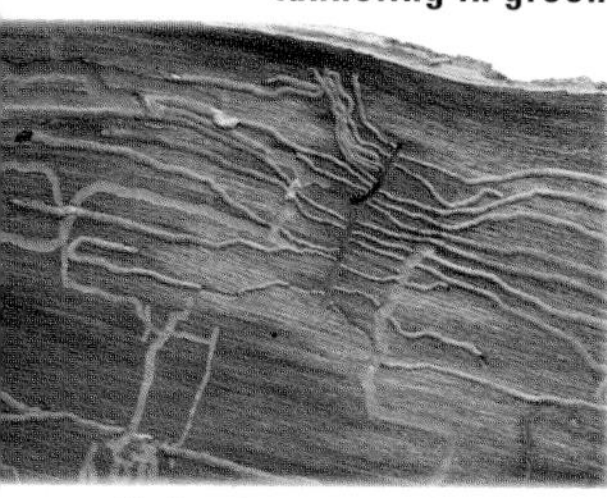

Hylesinus
galleries in green ash

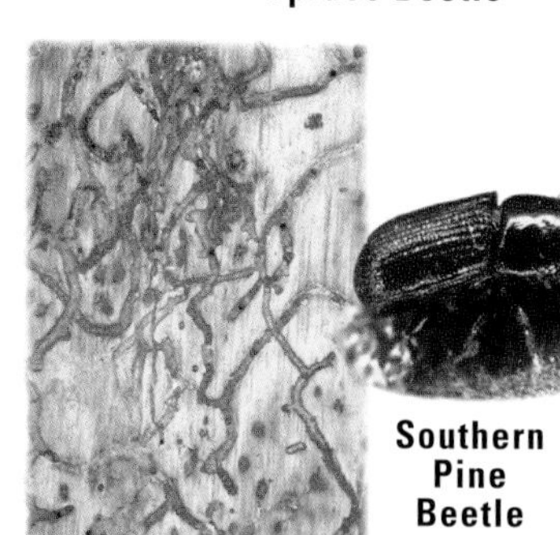

galleries of
Southern Pine Beetle

Southern
Pine
Beetle

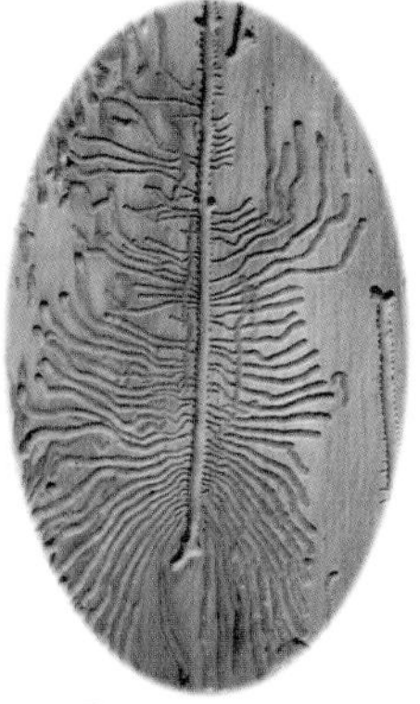

Phloeosinus
galleries in cypress

Red
Turpentine
Beetle

Scolytus
multistriatus

European
Shothole
Borer

Ips sp.

Amblycerus
robiniae

Bean
Weevil

MISCELLANEOUS WEEVILS

These beetles suggest the appearance of true weevils (Curculionidae), pp. 212–217, but are currently placed in four distinct families.

The **New York Weevil** *(Ithycerus noveboracensis)* is an eastern species in a family of its own, **Ithyceridae.** Adults are found most often on white oak and American beech. They feed on the bark of leaf petioles and shoots and acorn buds. Eggs are laid in the soil and the larvae feed on tree roots.

LEAF ROLLING WEEVILS (family Attelabidae) make origami look like child's play. Females of most species cut and fold a leaf into a "nidus," essentially an incubator in which the larva feeds and develops. In one genus, females commandeer the nidus created by another species, hence the term "thief weevil." Larvae pupate in the soil. There are about 50 species in North America.

The genus ***Attelabus*** includes two species. ***Attelabus bipustulatus*** occurs on oak and *Carpinus* in southeast Canada and the eastern U.S., as far west as Oklahoma and Texas. ***Attelabus nigripes*** (not illustrated) is reddish overall and is found on poison ivy and related plants in the eastern half of the U.S. The genus ***Merhynchites*** (formerly called *Rhynchites*) includes four species of "rose curculios." Larvae develop in the buds (hips) of various rose species. The **Eastern Rose Curculio** *(Merhynchites bicolor)* is found on the blossoms of its hosts across the northern U.S. and southern Canada. The **Western Rose Curculio** *(Merhynchites wickhami)* ranges in the western half of the U.S. and southwestern Canada.

FUNGUS WEEVILS (Anthribidae) include a few common species found on fungi or dead twigs or under bark. The antennae are not elbowed as they are in true weevils. The snout is short and broad. There are about 30 genera in North America, with 88 known species; more than 30 additional species await formal description. The genus ***Eurymycter*** includes three species, found over most of the continent except the southwest. ***Eurymycter fasciatus*** is eastern, occurring on dead twigs and on fungi growing on beech trees. The five species of ***Euparius*** range from the eastern U.S. to Montana and Arizona. ***Euparius marmoreus*** is common on fungi in eastern deciduous forests, sometimes flying to lights at night.

STRAIGHT-SNOUTED WEEVILS (Brentidae) now include some weevils formerly placed in the Curculionidae. Many genera are host specific, feeding on members of only one or two plant families. Some have been deliberately introduced here to control exotic weeds. The typical, elongated species found under bark are mostly tropical. There are more than 150 species in North America.

The **Hollyhock Weevil** *(Apion longirostre)*, an introduced species, is as specialized as its name implies. Adults feed on buds and foliage, while larvae feed on seeds, all on hollyhock. The **Oak Timberworm** *(Arrhenodes minutus)* is found beneath bark of injured, dying, or recently felled oak, poplar, and beech in the eastern U.S. and extreme southern Canada. Larvae bore in the wood. The territorial males use their broad snouts and large jaws to defend their mates from rival males during egg-laying.

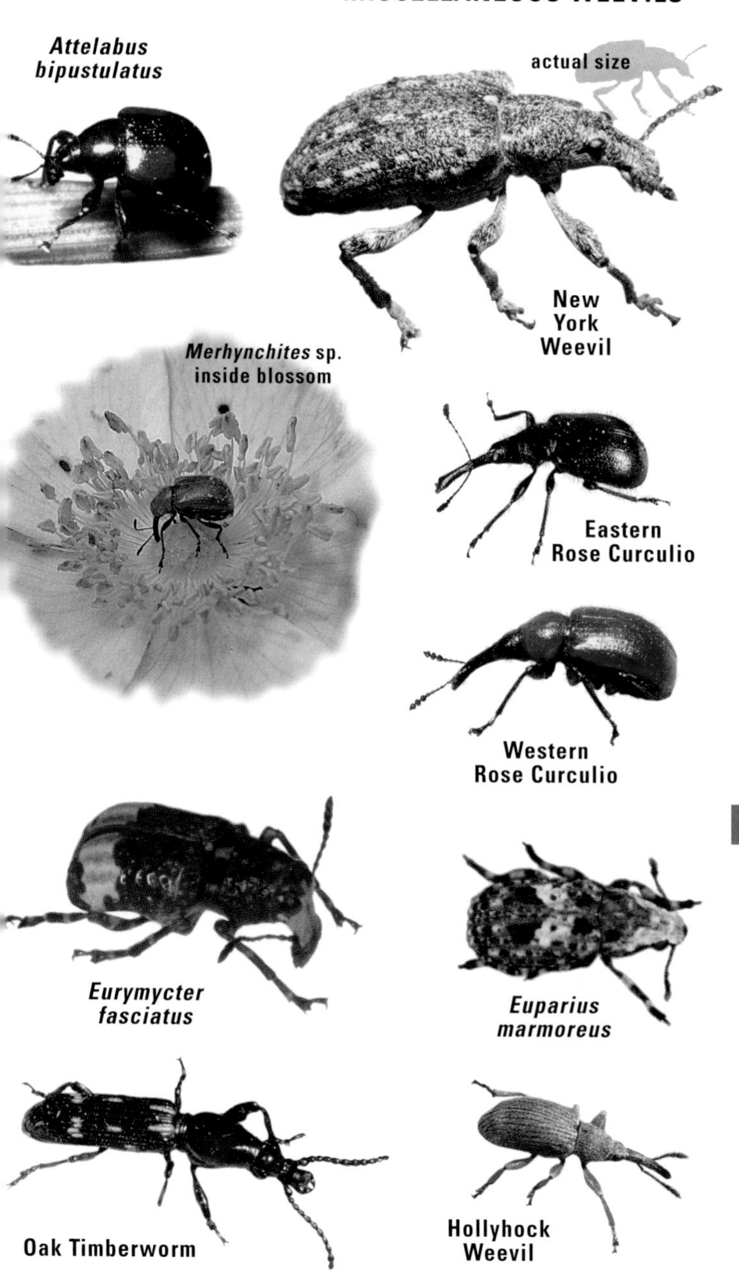

Attelabus bipustulatus

New York Weevil

Merhynchites sp. inside blossom

Eastern Rose Curculio

Western Rose Curculio

Eurymycter fasciatus

Euparius marmoreus

Oak Timberworm

Hollyhock Weevil

LACEWINGS AND THEIR RELATIVES

(order Neuroptera) are the so-called nerve-winged insects, but the name describes the netlike pattern of the wing veins, not the degree of sensitivity of the membranes. These are among our most familiar and valued insects, as they are insatiable predators of many pest insects, especially in the larval stage. While the immatures may be downright ugly, the mature creature is delicate and beautiful, like an angel born of a beast. There are 15 families known on our continent, including three small families of rare lacewings not illustrated in this guide.

GREEN LACEWINGS (family Chrysopidae) are also known as golden-eyed lacewings, the two names describing the adults of most members of the family, although some western species are pale brown. These delicate friends of the farmer and gardener are very common over most of North America, often seen fluttering through gardens or visiting lights at night. The larvae are "aphid lions" that prey heavily on pest insects. With their tonglike jaws, aphid lions skewer their victims, then suck them dry. Some larvae pile the dry carcasses atop their bristly backs or decorate themselves with other debris. This effectively disguises them against the ants that vigorously protect aphids in exchange for the aphids' honeydew secretions. In two obscure genera, the larvae actually live inside ant nests, presumably preying on the ant larvae. The distinctive eggs are laid in clusters on thin, wirelike stalks, and the pupa is formed inside an opaque, silken cocoon. North America has 15 genera and 84 species of green lacewings, but identifying them to species or even to genus is often impossible without painstaking study of male specimens. Two of the most frequently observed genera are ***Chrysopa,*** with 19 species, and ***Chrysoperla***, with 16 species in our area.

larva of green lacewing

BROWN LACEWINGS (Hemerobiidae) are known informally as aphid wolves in the larval stage. They prey on aphids and other pest insects just as voraciously as green lacewing larvae, but they are usually far less numerous. Adults are smaller than green lacewings and generally some shade of brown (varying from beige to black). There are six genera and 61 species found in North America. ***Hemerobius,*** with 12 species, is the genus often seen at lights at night.

SPONGILLAFLIES (Sisyridae) have a unique life history. Larvae develop as predators of freshwater sponges. They use long, hollow needlelike jaws to extract fluids from the cells of the sponge. We have six species, in two genera, found mainly in the east and the Pacific northwest.

BEADED LACEWINGS (Berothidae) are named for beadlike liquid-secreting scales on the wings of females. Adults resemble brown lacewings with hooked wingtips. Larvae are predators of drywood termites, living in their colonies and dispatching individual termites with a gas or chemical shot from their hind end. ***Lomamyia,*** with 10 species, is our only genus.

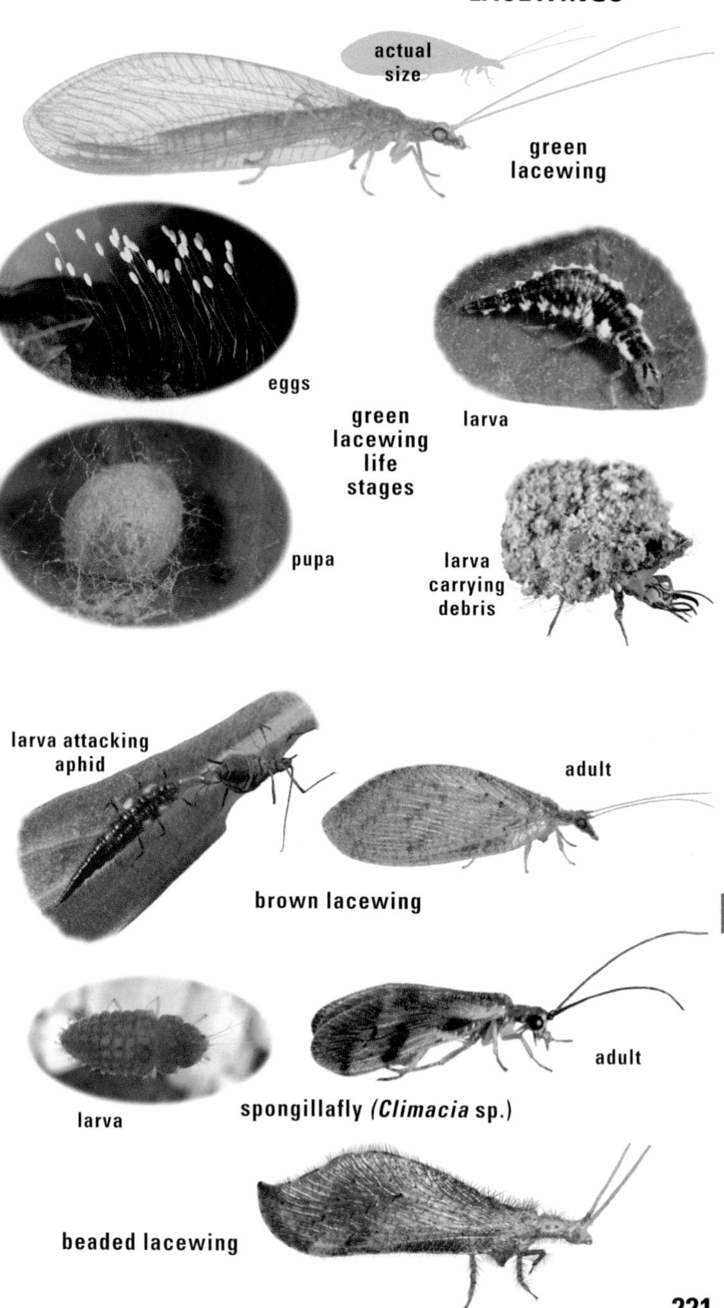
actual
size
green
lacewing
eggs
larva
green
lacewing
life
stages
pupa
larva
carrying
debris
larva attacking
aphid
adult
brown lacewing
adult
larva
spongillafly (*Climacia* sp.)
beaded lacewing

MANTISFLIES, SNAKEFLIES, DUSTYWINGS

Mantisflies and snakeflies are very distinctive relatives of the lacewings, generally uncommon but easily recognized when seen. Dustywings are so much smaller than most members of the order Neuroptera that they are probably overlooked at times.

MANTISFLIES or MANTISPIDS (family Mantispidae) resemble a science experiment gone horribly wrong. Imagine shrinking a praying mantis, then attaching its front end to the hind end of a lacewing, and you have a mantispid. Adults are predatory, using their highly modified front legs to capture smaller insects in much the same manner as a praying mantis. Many mantispids are predators of spider egg sacs in the larval stage. A larva hatches from its egg and may actively seek a spider upon which it rides until its host spins an egg sac—or, if the spider is a male, until it mates, in which case the larva maneuvers to the female spider. There are other scenarios, no less amazing, that facilitate the larva's meal ticket. Once inside the egg sac, the larva metamorphoses into a sedentary "couch potato" grub until it finishes feeding and finally pupates. The pupa is active as well, but eventually an adult emerges. Individual adults vary greatly in size depending on the quantity of their larval banquet. The genus ***Mantispa*** includes several common species, of which ***Mantispa sayi*** is most widespread in the east. Look for it around lights at night, preying on other insects. ***Mantispa interrupta*** is another widespread brownish species. The bright green ***Zeugomantispa minuta*** might be passed off as a green lacewing at first glance. ***Climaciella brunnea*** is active by day and is a striking mimic of certain paper wasps in the genus *Polistes* (see p. 354). It ranges over most of the continent. The members of the genus ***Plega*** occur in the southwestern U.S. In this genus, larvae are apparently predatory on the larvae of solitary bees, or certain paper wasps, but they may simply be opportunistic. Adults are often found around lights at night.

SNAKEFLIES (suborder Raphidioptera) are named for the elongated "neck" of the adults, actually a portion of the thoracic segment. The larvae are actually quite serpentine themselves, being active predators of small insects under bark on trees, under logs, or in leaf litter. There are two families in North America, with three genera and 22 species. Most of our species belong to the family **Raphidiidae** and the genus ***Agulla,*** which is widespread in southwestern Canada and the western U.S., east as far as Texas. The other family, **Inocelliidae,** includes only three species of the far west, from California to British Columbia. They average larger than the Raphidiidae, with thicker antennae.

DUSTYWINGS (family Coniopterygidae) are tiny lacewing relatives, averaging only 3 mm long or smaller, and are easily mistaken for whiteflies (p. 100). Larvae prey on aphids, scale insects, mites, and the eggs of various insects. There are eight genera and about 55 species known in North America. Although they are usually considered rare, they may be at least partially overlooked.

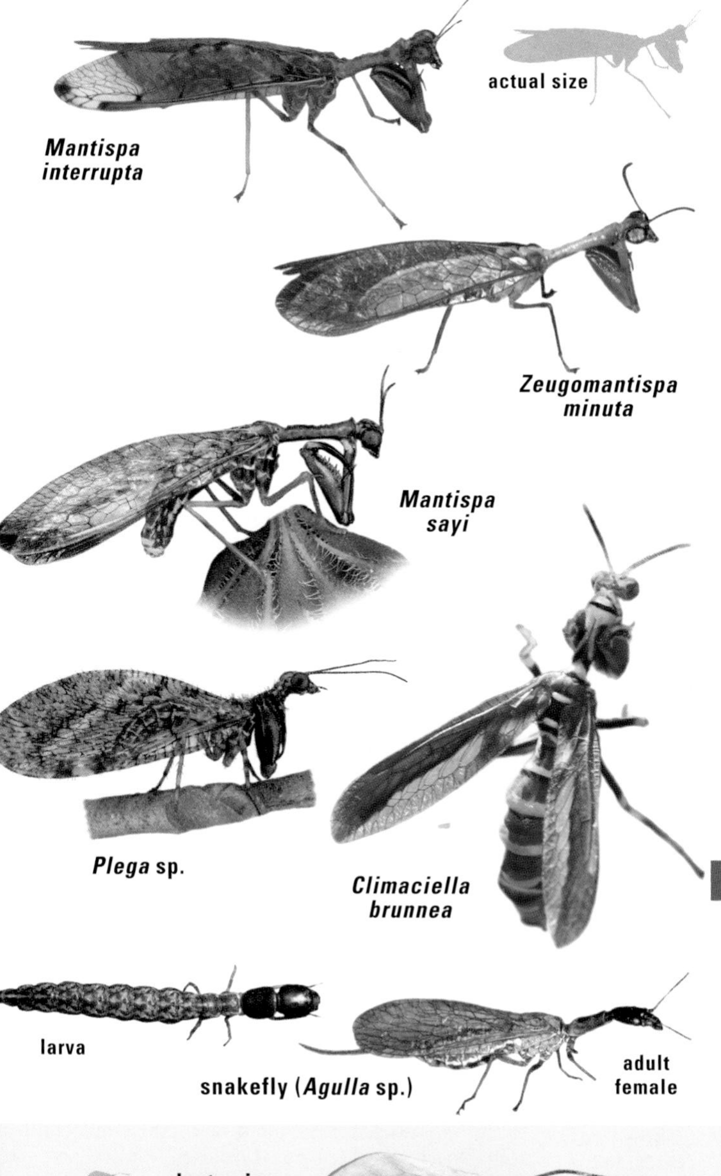

Mantispa interrupta

Zeugomantispa minuta

Mantispa sayi

Plega sp.

Climaciella brunnea

larva

snakefly (*Agulla* sp.)

adult female

actual size

dustywing (not to scale)

AQUATIC NEUROPTERA AND OWLFLIES

Fishflies, dobsonflies, and alderflies are often considered to make up a separate order, the Megaloptera, but in this guide we treat them as a suborder of the nerve-winged insects. All known species are aquatic in the larval stage, usually occurring in flowing waters.

ALDERFLIES (family Sialidae) never venture far from water, even as winged adults. Look for them perched on vegetation along streambanks or the edges of rivers and lakes. Larvae live in the beds of watercourses or lakes and are remarkably tolerant of even high levels of pollution. ***Sialis*** is our only North American genus, with 24 species. The dark, smoky wings of the adults are distinctive, but the color of the head varies from orange to black depending on the species.

DOBSONFLIES AND FISHFLIES (Corydalidae) include some real giants. Larvae are aggressive predators, but adults either do not feed or imbibe only small amounts of nectar and fruit juices. Many species are attracted to lights at night. The family is represented by seven genera and about 24 species in North America.

The **Eastern Dobsonfly** *(Corydalus cornutus)* is an enormous insect with a wingspan that may exceed 5 inches. Males have huge sicklelike jaws but are completely harmless. Females can inflict a painful bite if molested. Neither gender feeds as an adult. Larvae are known as hellgrammites and prey on aquatic invertebrates, tadpoles, and fish in fast-flowing rivers and streams. Look for adults at lights at night in midsummer across the eastern U.S. and southeastern Canada. The **Western Dobsonfly** *(Corydalus cognata),* not illustrated, is the replacement species in the western U.S. and is similar in appearance and habits.

Among the fishflies in the genus ***Chauliodes,*** males have distinctive feathery antennae, while females have beadlike antennae. Their larvae live mainly in still waters such as swamps or ponds. Most genera of fishflies have pale wings, but those in the genus ***Nigronia*** are recognized by their blackish wings with white markings. Their larvae live in flowing waters, often in narrow clear streams or small rivers.

Owlfly larvae hatching from cluster of eggs on twig

OWLFLIES (Ascalaphidae) are about six species of strong-flying predators, the near equal of dragonflies. Long clubbed antennae and large eyes help distinguish them. Adults are crepuscular, with some species flying for only 30 minutes prior to total darkness. The female lays eggs on the tips of twigs, often near streams. She also lays "trophic eggs" that furnish the larvae with their first meal. Larvae lie in ambush on the surface of the ground, or on leaves, with jaws open at 180 degrees. ***Ascaloptynx appendiculatus*** is a great golden creature with an odd thumblike process near the base of the front wing. In some species of the genus ***Ululodes,*** females have dark spots on the wings. Look for most owlflies at lights at night.

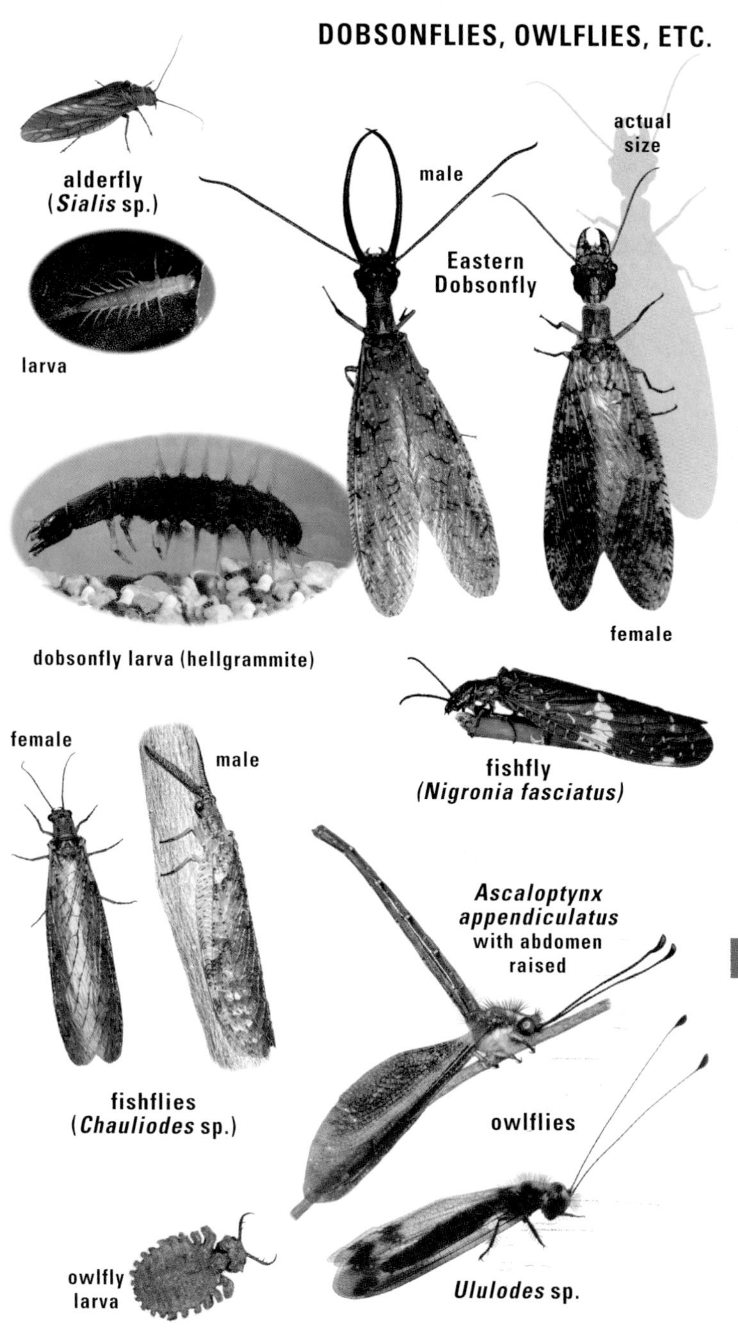

alderfly
(*Sialis* sp.)

dobsonfly larva (hellgrammite)

fishfly
(*Nigronia fasciatus*)

fishflies
(*Chauliodes* sp.)

Ascaloptynx appendiculatus with abdomen raised

owlflies

Ululodes sp.

ANTLIONS

(family Myrmeleontidae) are named for their larval stage. In North America, only the larvae of the genus *Myrmeleon* excavate the well-known funnel-shaped pits to trap ants and other insects upon which the antlion feeds. The larvae of most species lie just under the surface of the soil with jaws open in anticipation of a hapless victim. Adults, often active at night, are rather weak fliers with wings either spotted or clear. They resemble damselflies (p. 52), but the short, thick, clubbed antennae are distinctive. There are more than 90 species north of Mexico.

Our seven species in the genus ***Myrmeleon*** do not draw much attention as adults, and the larvae are usually out of sight, but the small pit traps excavated by these larvae are familiar sights for alert naturalists. Look for groups of pits, each about an inch across, in very fine, dry soil under overhanging rock ledges, beneath bridges, at the base of trees, or in the dirt floor of abandoned barns and sheds. When an ant or another tiny insect strays over the lip of one of these pits and begins to slide downward on the fine loose grains of silt, the larval antlion kicks up little fountains of earth to shower its victim and hasten its descent into the jaws of doom. On the surface of the soil, the stubby-legged larval antlion (also known as a doodlebug) can only walk backward, leaving cursive trails across the surface when changing the location of its trap. Adults of this genus do not come to lights at night and are most often seen flying weakly after being flushed from tall grass.

The genus ***Vella*** includes our largest antlions. Look for adults at lights at night or resting on tree trunks or posts by day. ***Vella fallax texana*** occurs from California to Texas and Utah, and the genus includes one other species in the southeastern states.

Glenurus gratus is a spectacular insect in the adult stage. Larvae live in dry tree holes, at least in Florida, where they also have been found living in the burrows of gopher tortoises. This species ranges north to Indiana and New Jersey, and the genus includes two other species north of Mexico.

The **Spotted-winged Antlion** *(Dendroleon obsoletus)*, one of two North American species in its genus, ranges throughout the eastern U.S. Larvae have been recorded in dry tree holes in Florida.

antlion larva (*Brachynemurus* sp.)

Brachynemurus is a diverse genus of antlions in North America, with more than 20 species. In this genus the males have a very long abdomen, far exceeding the wingtips when the insect is at rest. ***Brachynemurus abdominalis*** has been recorded over much of the eastern U.S. and in parts of southern Canada. ***Brachynemurus hubbardii*** is another common species. Members of the genus ***Scotoleon,*** formerly classified as part of *Brachynemurus,* are common in parts of the southwestern U.S. One species of *Scotoleon* has been found visiting and pollinating *Gaura* flowers in western Texas, a behavior that went undiscovered until recently because they visit the flowers at night.

pit traps

Myrmeleon sp.

actual size

inset: larva from bottom of pit

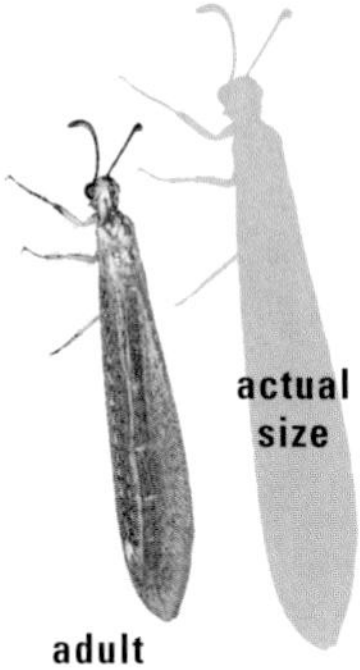

adult

Vella fallax texana

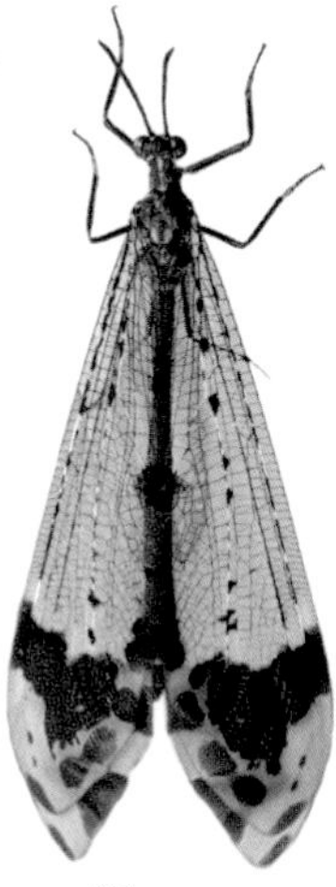

Spotted-winged Antlion

Glenurus gratus

Brachynemurus hubbardii

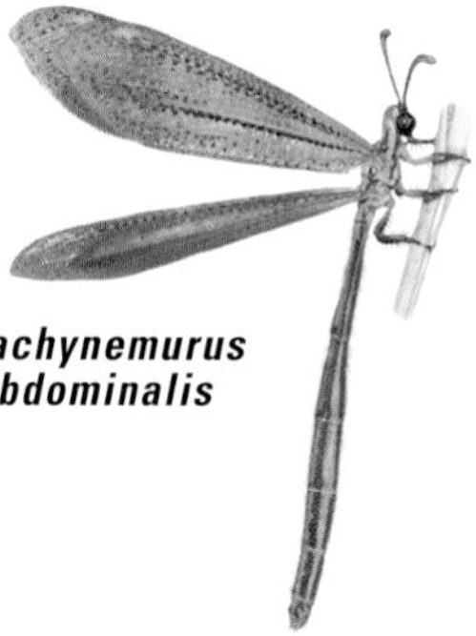

Brachynemurus abdominalis

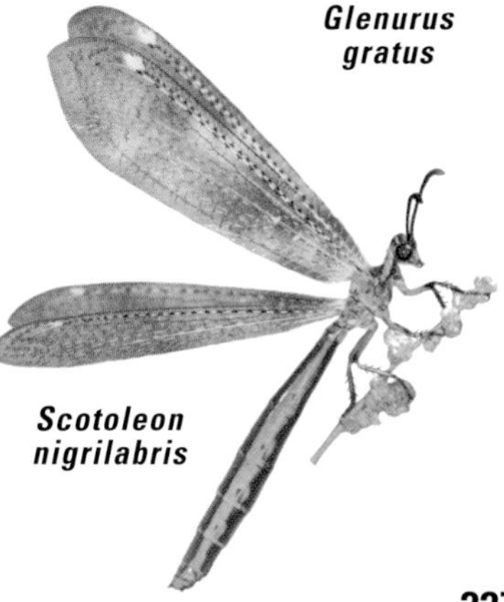

Scotoleon nigrilabris

BUTTERFLIES AND MOTHS

make up the **order Lepidoptera,** which means "scaled wings." Adults of almost all species have four wings covered with colored scales, often creating beautiful patterns. Butterflies are undoubtedly the most popular insects in the world, but moths have their admirers as well, and moths have the advantages of variety. In North America north of Mexico there are about 700 species of butterflies (including the skippers, p. 236, a very distinct group), but the number of moth species is well over 10,000.

What's the difference between butterflies and moths? Very little. Butterflies and skippers are active by day, while most moths are active at night, but hundreds of moth species are day-fliers also. Butterflies usually sit with the wings spread out to the side or raised above their backs, while many moths hold their wings folded rooflike over their bodies, but there are many exceptions. The most consistent visible difference is in the structure of the antennae. Butterflies have thickened "knobs" at the tip of each antenna (with a slight extension on most skippers), while the antennae of moths (at least in North America) are threadlike or fernlike, without a thickened tip.

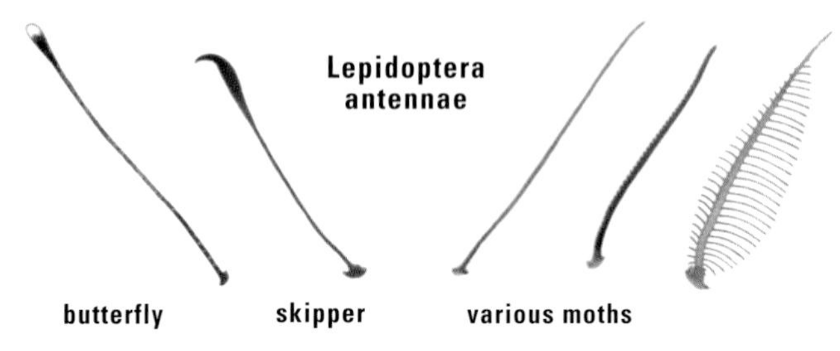

Butterfly watching has become hugely popular in recent years, and moth watching is now starting to catch up. These creatures are usually quite easy to photograph in the wild, and—unlike most other groups of insects—they often *can* be identified to species from photos. At the same time, for those with a more scientific interest, there is still a need for collecting of specimens. Species new to science are still being discovered, even in the U.S. and Canada, and they cannot be documented without collecting. Despite some misguided claims to the contrary, collectors essentially never pose a threat to butterfly or moth populations. By determining the presence of rare species and helping to protect their habitats, collectors are working for both science and conservation.

Most users of this book probably will be more interested in observing than collecting. Fortunately, observing butterflies and moths is quite easy. You can attract them by planting a garden that includes native flowers for the adults and foodplants for the caterpillars. Some butterflies and moths that seldom visit flowers may be attracted to overripe fruit, such as bananas. Males of some butterfly species will congregate at the edges of puddles, sucking up salts from the mud. Leaving the porch light on will draw many species of moths, particularly on warm, humid, moonless nights. Moths are thought to normally orient to the moon and so are fooled by our electric lights. Moths can see in the ultraviolet spectrum, so black lights will lure hard-to-find species. Mercury vapor lights also draw large numbers by broadcasting light a great distance.

Our fascination with the beautiful patterns of Lepidoptera should not blind us to their interesting behaviors. Caterpillars often take advantage of their camouflage by acting the part of a twig, leaf edge, or other inedible object. Adults may defend territories, migrate long distances, or home in on the pheromones of the opposite sex from a mile away.

All butterflies and moths go through complete metamorphosis, beginning life as an egg. From the egg hatches a caterpillar, also called a larva. The larva has six legs (it is an insect, after all) but appears to have more, as it usually bears several pairs of prolegs down the length of the body to make active crawling possible. The larva sheds its exoskeleton (or "skin") several times as it grows, passing through a series of stages called instars (the intervals between these molts). After it is full-grown, it pupates. The pupa of a butterfly is often referred to as a chrysalis, while a moth pupa is often enclosed in a cocoon. Finally, the winged adult emerges from the pupa, to begin the cycle over again.

One of the most rewarding ways to learn about nature is to raise some of your local native moths or butterflies from the egg or caterpillar stage. The female usually lays her eggs on the caterpillar's foodplant, so by providing a steady supply of fresh foliage, you should be able to get the larva through to adulthood. Cleaning the container frequently and thoroughly will prevent mold and fungal infections. Documenting the details of a caterpillar's life cycle, with photography and written notes, can be very valuable. You might record a new host plant, for example, that was not previously known for the species. Just how many species of butterflies and moths could you expect to find in your own back yard? One Connecticut entomologist has documented more than 1,000 species at his residence.

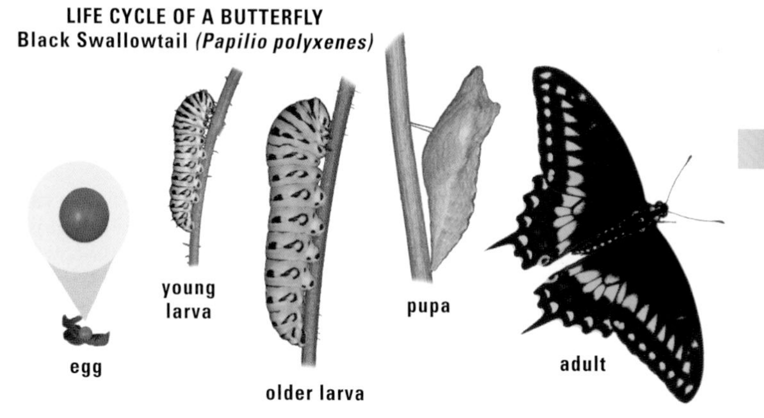

SWALLOWTAILS (family Papilionidae) are among our largest and most conspicuous butterflies. About two dozen species occur regularly in North America, with several more straying in from the tropics. The **Eastern Tiger Swallowtail** *(Papilio glaucus)* is very common east of the Rockies and south of Canada, in woodlands, open country, and gardens. Its caterpillars feed on leaves of various trees, including cottonwood and tulip tree. Very similar species replace it farther west and north. Widespread and numerous in the west is the **Anise Swallowtail** *(Papilio zelicaon)*. Its caterpillars feed on plants in the parsley family, including the introduced sweet fennel, or anise. The **Giant Swallowtail** *(Papilio cresphontes)* is most abundant in the south, where its caterpillars feed on plants in the citrus family. Caterpillars of the **Pipevine Swallowtail** *(Battus philenor)* feed upon aristolochias, or pipevines, accumulating chemicals from these plants that make even the adults noxious to predators. Several other swallowtails (and some unrelated butterflies) are surprisingly similar to this species, and they may gain some protection from predators by mimicking this bad-tasting butterfly. Classified with swallowtails but looking quite different are the parnassians, pale creatures that live mostly in the high mountains and the far north. On this continent they are strictly western, from the mountains of California and New Mexico north to Alaska. The **Rocky Mountain Parnassian** *(Parnassius smintheus)* is fairly common from open woods to tundra.

Whites, marbles, orangetips, and sulphurs **(Pieridae)**, often collectively referred to as pierids, are mostly medium-sized, but some very large and very small butterflies are in this family. About 60 species occur regularly in our area. Orangetips and marbles, such as the western **Sara Orangetip** *(Anthocharis sara)*, are among the earliest butterflies of spring in many areas. Their larvae feed on native plants in the mustard family. The **Cabbage White** *(Pieris rapae)*, one of our most abundant butterflies, is native to Europe and was introduced to North America in the 19th century. Among the most numerous native whites is **Checkered White** *(Pontia protodice)*, found coast to coast but more common in the west. Its larvae feed on many native and exotic mustards. The genus ***Colias*** includes about 18 North American species, including some in the high Arctic and high western mountains. More widespread and common is the **Clouded Sulphur** *(Colias philodice)*. The big sulphurs of the genus ***Phoebis*** are mostly southern and tropical, but the **Cloudless Sulphur** *(Phoebis sennae)* wanders far to the north, especially in fall. Oranges and yellows in ***Eurema*** and related genera are also most diverse in the tropics, but the **Sleepy Orange** *(Abaeis nicippe)* is very common in our southern and central states. Larval foodplants for the three preceding genera are mostly in the legume family, but larvae of the **Dainty Sulphur** *(Nathalis iole)*, our smallest sulphur, feed on plants in the aster family.

larva of a sulphur (Southern Dogface, *Zerene cesonia*)

SWALLOWTAILS AND PIERIDS

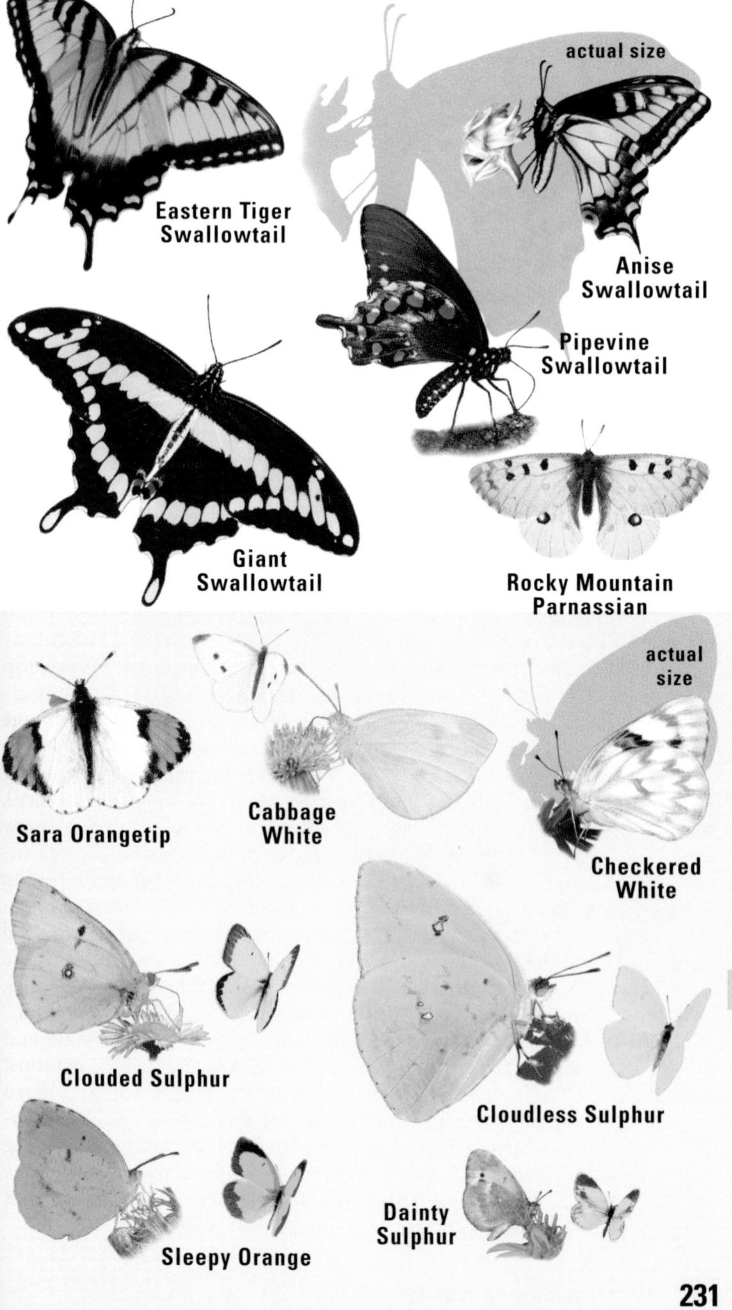
Eastern Tiger Swallowtail
actual size
Anise Swallowtail
Pipevine Swallowtail
Giant Swallowtail
Rocky Mountain Parnassian
Sara Orangetip
Cabbage White
actual size
Checkered White
Clouded Sulphur
Cloudless Sulphur
Sleepy Orange
Dainty Sulphur

GOSSAMER-WINGS AND METALMARKS

The **GOSSAMER-WINGS (family Lycaenidae)** are mostly small to very small butterflies. Included here are four distinctive groups that are sometimes treated as separate subfamilies or even families.

We have about 18 species of **COPPERS (tribe Lycaenini),** mostly in the north and west. Often seen visiting flowers, they bask with their wings spread out to the sides. The **Ruddy Copper** *(Lycaena rubidus)* is widespread in western mountain meadows, while the **American Copper** *(Lycaena phlaeas)* has populations in the northeast, the west, and the far north.

HAIRSTREAKS (tribes Theclini and Eumaeini) often have short "tails" on the hindwings. About 60 species occur regularly north of Mexico. Most perch with the wings folded above their backs. An exception is **Gray Hairstreak** *(Strymon melinus),* one of our most common hairstreaks, which may bask with its wings spread. The 18 species of ***Satyrium*** are mostly hard to identify, brown or gray with subtle pattern differences; **Banded Hairstreak** *(Satyrium calanus)* is the most common eastern species. **Red-banded Hairstreak** *(Calycopis cecrops),* small but colorful, is very common in the southeast. Its larvae feed on rotting dead leaves on the ground. Elfins are brown hairstreaks, usually without hindwing tails. Most are active in spring, but the widespread **Brown Elfin** *(Callophrys augustinus)* may fly in spring or summer. Related to the elfins are several hairstreaks that are green below, including the widespread but variable **Juniper Hairstreak** *(Callophrys gryneus),* always found close to the junipers or other evergreens on which its larvae feed. Flashy and unique is the **Colorado Hairstreak** *(Hypaurotis crysalus),* found around Gambel oak in the southwest.

BLUES (tribe Polyommatini) are mostly blue above, especially the males. They may perch with their wings open, but the most distinctive field marks are on the undersides of the wings. Larvae of many feed on the flowers and buds of legumes. The **Eastern Tailed-Blue** *(Cupido comyntas,* formerly *Everes comyntas),* one of the few blues with hindwing tails, is abundant in parts of the east. The **Melissa Blue** *(Plebejus melissa)* is very common in the west, but its eastern subspecies ("Karner Blue") is endangered. The **Marine Blue** *(Leptotes marina)* is very common in the southwest and sometimes expands its range far to the north and east. The **Spring Azure** complex (*Celastrina* sp.) may include nine or more species in North America. Classification of many groups of blues (a subject that fascinated the novelist Vladimir Nabokov) is still a topic of active research.

The **Harvester** *(Feniseca tarquinius),* our only member of the **subfamily Miletinae,** is also our only carnivorous butterfly: its larva feeds on aphids.

METALMARKS (family Riodinidae) are often treated as a subfamily of the gossamer-wings. There are hundreds of species in the tropics, many of them brilliantly colored. Fewer than 20 species are regular north of Mexico, and these are mostly rather drab except for tiny metallic markings on the wings of some. Metalmarks usually sit with the wings spread out to the side. **Little Metalmark** *(Calephelis virginiensis)* is the only species in the southeastern U.S. The **Mormon Metalmark** *(Apodemia mormo)* is one of a complex of closely related species widespread in the west.

Ruddy Copper
American Copper
actual size
Gray Hairstreak
Banded Hairstreak
Red-banded Hairstreak
Brown Elfin
Juniper Hairstreak
Colorado Hairstreak
Marine Blue
Melissa Blue
Spring Azure
Eastern Tailed-Blue
Harvester
Little Metalmark
Mormon Metalmark

BRUSH-FOOTED BUTTERFLIES

(family Nymphalidae) make up a very diverse group, including some of our most common and well-known butterflies. Now included within this family are several very distinct groups that were formerly treated as separate families. All are similar in having their front pair of legs strongly reduced and covered with short bristles, hence the name "brush-footed." Since these front legs are seldom visible, these butterflies often look four-legged, but few other generalizations are possible.

The **Gulf Fritillary** *(Agraulis vanillae)*, common in the south, is the northernmost of a tropical group known as longwings or heliconians. Larvae feed on passion vines. Among true fritillaries, those in the genus ***Boloria*** are smaller and live mostly in the north, including the Arctic, although the **Meadow Fritillary** *(Boloria bellona)* is also widespread south of Canada. The **Great Spangled Fritillary** *(Speyeria cybele)* is found coast to coast, but most members of the genus ***Speyeria*** live in western mountains. The small crescents and checkerspots are diverse, especially in the west and south; **Pearl Crescent** *(Phyciodes tharos)* is among the most common.

The **Question Mark** *(Polygonia interrogationis)* is one of the anglewings or commas, with silvery "punctuation marks" under the hindwings. Unlike most butterflies, they overwinter as adults and may fly about during midwinter thaws. Another species that hibernates as an adult, and may fly in midwinter or early spring, is the **Mourning Cloak** *(Nymphalis antiopa)*. Its larvae feed on willows and other plants across North America, Europe, and Asia. Also with a wide distribution across the northern hemisphere is the **Red Admiral** *(Vanessa atalanta)*, which uses nettles as larval foodplants. The most widespread of all butterflies is the **Painted Lady** *(Vanessa cardui)*, found practically worldwide and subject to mass "migrations" with tens of thousands of individuals all flying in the same direction.

larva of American Snout

The **Common Buckeye** *(Junonia coenia)* is most numerous in the south, its larvae feeding on plants in the acanthus and snapdragon families. Famed for mimicry is the **Viceroy** *(Limenitis archippus)*, unrelated to the Monarch (below) but strikingly similar. The **Arizona Sister** *(Adelpha eulalia)* and the very similar **California Sister** *(Adelpha californica)* are both numerous, their larvae feeding on oaks. The **American Snout** *(Libytheana carinenta)*, which sometimes swarms in the southwest, is our only member of an odd group in which adults have elongated palps (the "snout"). Larvae feed on hackberry leaves. The most famous butterfly, the **Monarch** *(Danaus plexippus)*, migrates to wintering sites in Mexico and California. Its larvae feed on milkweeds. Chemicals from those plants make both larva and adult extremely distasteful to predators, as advertised by their bright "warning colors."

Most satyrs **(subfamily Satyrinae)** are brown. Larvae of many feed on grasses. Some are common on northern or montane tundra. Others live in forests, such as the **Little Wood-Satyr** *(Megisto cymela)* of the east and the more widespread **Common Wood-Nymph** *(Cercyonis pegala)*.

Gulf Fritillary
Meadow Fritillary
Great Spangled Fritillary
actual size
Pearl Crescent
Question Mark
Red Admiral
Mourning Cloak
Common Buckeye
Painted Lady
Arizona Sister
Viceroy
American Snout
Common Wood-Nymph
Little Wood-Satyr
larva
Monarch

SKIPPERS

(family Hesperiidae) are mostly small, with relatively short wings, thick bodies, and fast flight. Almost a third of all North American butterflies, more than 200 species, are skippers. The species found in our area are divided among four major subfamilies; three are illustrated below.

The **GRASS SKIPPERS (subfamily Hesperiinae)** are mostly small and usually orange or brown. More than 125 species occur regularly north of Mexico. Their larvae feed mostly on grasses. Adults often bask in a "jet-plane" position, with hindwings spread more widely than forewings. Some grass skippers are truly tiny, such as the **European Skipper** *(Thymelicus lineola),* introduced to this continent, now abundant in the northeast and still spreading. The **Fiery Skipper** *(Hylephilus phyleus),* very common in the south and often straying northward, adapts well to suburbs because its larvae feed on Bermuda grass. Members of the genus ***Hesperia,*** such as the **Juba Skipper** *(Hesperia juba),* often have strong patterns on the underside of the hindwing, but so do many other grass skippers — for example, members of the genus ***Polites,*** such as the **Long Dash** *(Polites mystic),* common around meadows in the north and northeast. **Roadside-skippers** (genus *Ambyscirtes)* are small, dark, hard to identify, and often seen along trails, dry streambeds, or roadsides. **Panoquins** (genus *Panoquina),* such as the **Ocola Skipper** *(Panoquina ocola),* often have long forewings and a dark stripe along the abdomen. Most are southern or coastal. **Dusted-skippers** (genus *Atrytonopsis),* such as **Moon-marked Skipper** *(Atrytonopsis lunus),* are relatively large and dark. Most species are southwestern.

SKIPPERLINGS (subfamily Heteropterinae) include four species of the genus ***Piruna,*** in the southwest, and the **Arctic Skipper** *(Carterocephalus palaemon),* widespread in the north. All are small and rather weak fliers.

Unlike grass skippers, many **SPREAD-WING SKIPPERS (subfamily Pyrginae)** do rest with the wings spread flat, and their colors tend more toward black and white. One of the most easily recognized is the widespread **Silver-spotted Skipper** *(Epargyreus clarus).* However, many groups of spread-wings pose serious identification problems. The many species of **duskywings** (genus *Erynnis*) are extremely difficult to tell apart, as are the several species of **cloudywings** (genus *Thorybes*). Our eight species of **checkered-skippers** (genus *Pyrgus*) are also difficult to separate; the **Common Checkered-Skipper** *(Pyrgus communis)* is by far the most frequently seen away from the southernmost states. There are several species of **white-skippers** (genus *Heliopetes*) in the tropics and southern Texas; the **Northern White-Skipper** *(Heliopetes ericetorum)* is the only one widespread in the western U.S. Many tropical spread-wings have "tails" on the hindwings, but the **Long-tailed Skipper** *(Urbanus proteus)* is the most common in the U.S., especially in the southeast.

GIANT-SKIPPERS, formerly considered a separate subfamily, are big, heavy-bodied, fast-flying, and usually uncommon. Their larvae bore into the stems and roots of yuccas, agaves, and related plants. The **Yucca Giant-Skipper** *(Megathymus yuccae)* of the southern states flies in spring, but most agave-feeding species (genus ***Agathymus)*** fly in fall.

European Skipper
Long Dash
actual size
Fiery Skipper
Bronze Roadside-Skipper
Moon-marked Skipper
Juba Skipper
Ocola Skipper
Arctic Skipper
Northern Cloudywing
Silver-spotted Skipper
Funereal Duskywing
Northern White-Skipper
Common Checkered-Skipper
Long-tailed Skipper
Yucca Giant-Skipper

GIANT SILKMOTHS

(family Saturniidae) are not always enormous, but our largest moths are included here. The common name comes from the thick, silken cocoons woven by the larvae. About 70 species live in North America. Most have large furry bodies and small heads with only vestigial mouthparts; adults live for only a few days and do not feed. Populations of these spectacular moths are vulnerable to light pollution and to control measures aimed at Gypsy Moths, including introduced parasitic flies and wasps. The commercial silkmoth of Asia *(Bombyx mori)* belongs to a different family.

The gorgeous **Luna Moth** *(Actias luna)* is common in hardwood forests of the eastern U.S. and southeastern Canada. In our area it is unique, but several relatives range in Asia and Mexico. Adults fly in early summer in the north, sometimes appearing at suburban porch lights; in the south they have several broods per year. Foodplants of the larvae vary, including birches in the north, sweetgums in the south, and hickories, walnuts, and other trees.

Royal Walnut Moth larva ("Hickory Horned Devil")

Royal Moths (genus ***Citheronia***) include two species in the east and one in Arizona. The adult **Royal Walnut Moth** *(Citheronia regalis)* is upstaged by its own larva, the "Hickory Horned Devil," one of the most spectacular of caterpillars. More than 5 inches long when full-grown, it feeds at night on various trees and shrubs, including hickory, walnut, sweetgum, and sumac. Adults fly in summer in eastern forests.

Imperial Moths *(Eacles imperialis)* are common in forests of the eastern U.S. and southeast Canada, the adults flying at night in summer. Larvae feed on foliage of trees and shrubs, including oak, maple, sweetgum, and pine. A close relative lives in Arizona.

Our most widespread silkmoth, the **Polyphemus Moth** *(Antheraea polyphemus)*, is found almost throughout southern Canada and the lower 48 states except some desert regions. Adults often fly to lights late at night. If startled they will spread their forewings, flashing the eyespots on the hindwings. Their larvae feed on various trees and shrubs and spin cocoons among the leaves of their foodplant or on the ground.

The **Promethea Moth** *(Callosamia promethea)* is dramatically dimorphic, females being reddish brown, males nearly black. This insect is common in the eastern U.S. and southeastern Canada. Larvae feed on many trees and shrubs, but local populations often favor just one host plant.

Robin Moths (genus ***Hyalophora***) include at least three species in North America (debate still rages as to which populations constitute a species). The **Cecropia Moth** *(Hyalophora cecropia)* ranges west to the foothills of the Rockies. Look for adults around lights at night (or even the next morning) in the vicinity of forest edges, city parks, and suburban woodlots.

Native to Asia, **Ailanthus Silkmoths** *(Samia cynthia)* were introduced to the U.S. many years ago in a failed attempt to establish a commercial silk industry here. Populations persist around some northeastern cities.

GIANT SILKMOTHS

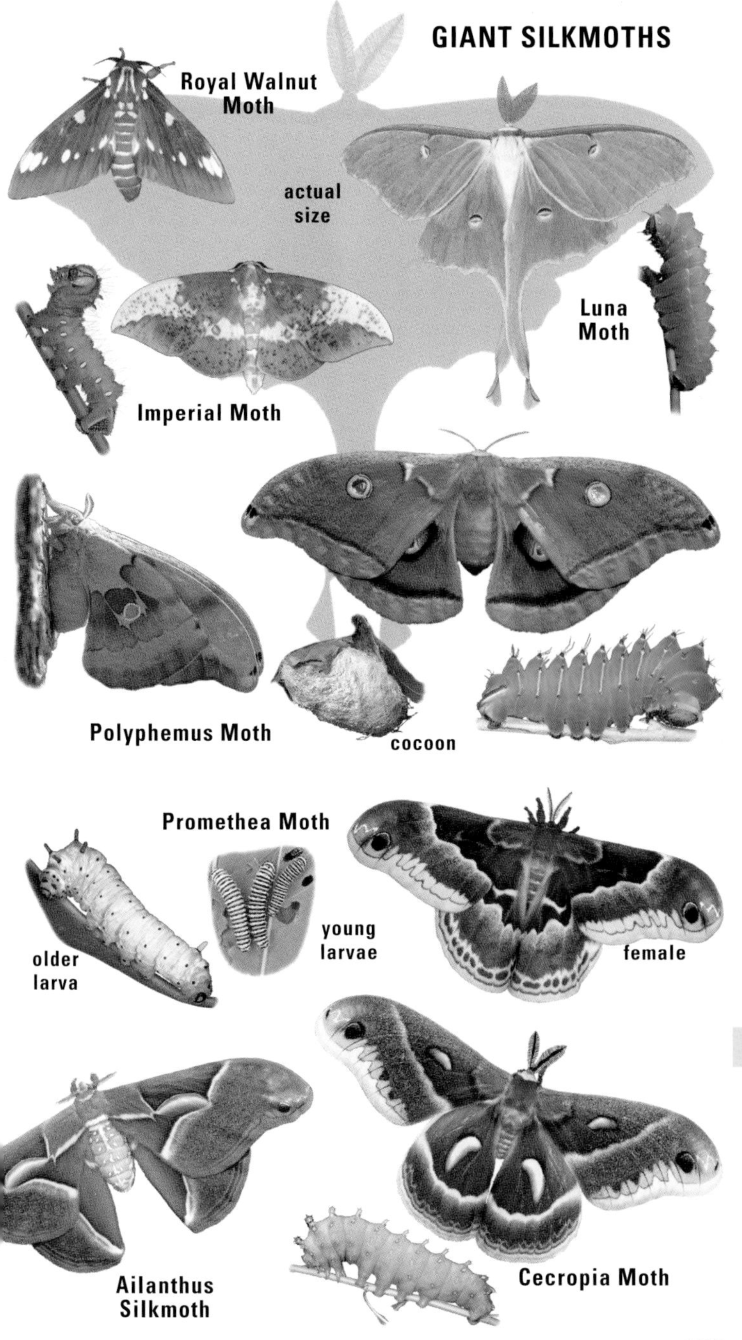

The species shown here are smaller than their spectacular relatives on the preceding page, but still larger than the average moth. Silkmoths in general are noted for the fact that adult females "call" to males by releasing a pheromone. The feathery antennae of males may detect this subtle fragrance on the breeze from half a mile away, allowing the males to home in on the location of the female.

larva of Io Moth

Members of the genus ***Automeris*** have large round spots on the hindwings. Usually hidden when the moth is at rest, these spots may flash boldly when the moth is disturbed, and their resemblance to the eyes of a much larger creature may scare away predators. Most of our species have limited southern ranges, but the **Io Moth** *(Automeris io)* is common east of the Rockies, its larvae feeding on a remarkably wide variety of plants.

Widespread and common in eastern forests is the **Rosy Maple Moth** *(Dryocampa rubicunda).* The typical pink-and-yellow adults are easily recognized, but they vary, and some are almost white. Larvae usually feed on maple leaves. Members of the genus ***Anisota*** are among the smallest silkmoths. The males are active mostly by day, when their colors and very rapid flight make them look like bees. Most species are eastern, and their larvae feed mainly on oaks. The **Pink-striped Oakworm Moth** *(Anisota virginiensis)* is one of the most numerous in this group.

The genus ***Hemileuca*** includes more than 15 species, most of them western, most active by day, and all strikingly patterned. The **Western Sheepmoth** *(Hemileuca eglanterina)* is widespread and variable in the west, the adults usually flying in summer or fall. It uses a wide variety of shrubs and small trees as larval foodplants. Young larvae are black and (as with many silkmoth larvae) feed in tight groups, but the more colorful older larvae disperse and feed singly. The **Buck Moth** *(Hemileuca maia)* is one of the few eastern members of this genus. Its larvae feed mainly on oaks, and adults usually fly in fall or early winter.

Members of the genus ***Agapema,*** including the **Mexican Agapema** *(Agapema anona),* are southwestern. Larvae feed on condalia and related shrubs, and adults fly at night, mostly in fall. Active by day in spring is ***Saturnia mendocino,*** a fast-flying moth of scrubby chaparral in California and Oregon, where its larvae feed on manzanita. Larvae of the ***Sphingicampa*** moths feed mostly on trees and shrubs in the legume family. Adults fly at night. We have two species in the east and six in the southwest, where **Hubbard's Small Silkmoth** *(Sphingicampa hubbardi)* is the most common. Four species of ***Coloradia*** live in mountain forests of the west, where their larvae feed on pine needles. The most widespread species, the **Pandora Pinemoth** *(Coloradia pandora),* flies mostly at night in late summer and fall. During occasional population explosions, the adults may fly by night or day, and the larvae may temporarily strip the foliage from large tracts of pines. Native Americans once exploited these larvae as a food source.

SMALLER SILKMOTHS

male

Pink-striped Oakworm Moth

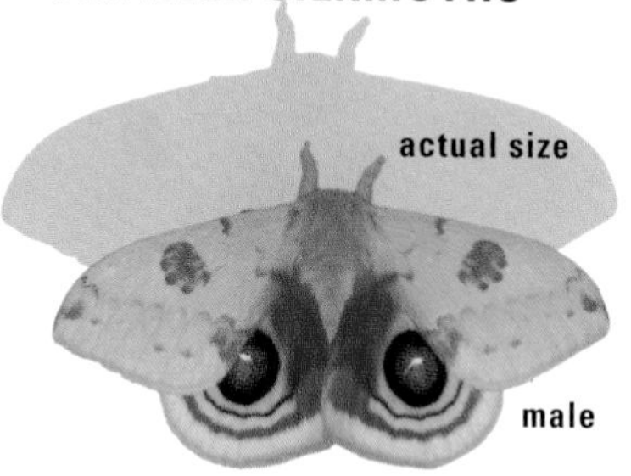

female

actual size

male

Io Moth

Rosy Maple Moth

Buck Moth

Western Sheepmoth

Saturnia mendocino

Mexican Agapema

Pandora Pinemoth

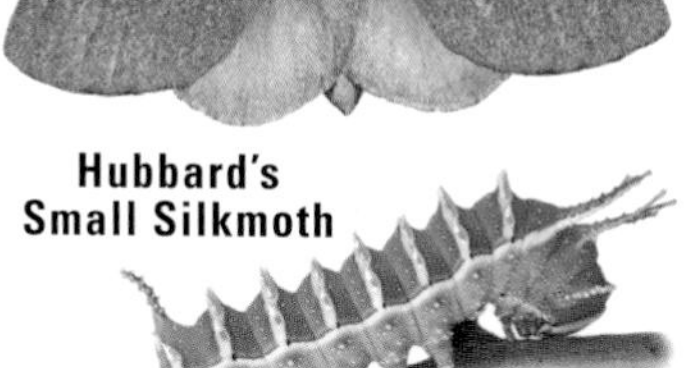

Hubbard's Small Silkmoth

SPHINX MOTHS

(family Sphingidae), also known as hawkmoths, are fast, powerful fliers, often seen hovering at flowers to sip nectar. Most are active at night. Those that forage in daylight are often mistaken for hummingbirds, though some smaller ones resemble bumble bees. Their caterpillars typically have a short posterior "horn" and are commonly called hornworms. Most do not spin cocoons but pupate in an earthen cell just below the soil surface. More than 125 species are known from North America, but some of these are only strays from the tropics.

Over much of the continent, the **White-lined Sphinx** *(Hyles lineata)* is the most commonly seen sphingid, flying by day or at night. Its larvae are remarkably varied in color pattern, and they feed on a remarkably wide variety of plants; perhaps as a result of this, the species sometimes undergoes explosive outbreaks and reaches high population densities, especially in dry country of the west. Two well-known members of the genus ***Manduca*** are often referred to by names of their larvae; thus, the **Five-spotted Sphinx** *(Manduca quinquemaculata)* is also called **Tomato Hornworm,** and the **Carolina Sphinx** *(Manduca sexta)* is also called **Tobacco Hornworm.** These caterpillars can't necessarily be identified by the plants on which they're found: both feed on each other's namesake and on potato plants and others as well. Larvae of the related **Rustic Sphinx** *(Manduca rustica)* feed on various plants, including some in the bignonia family.

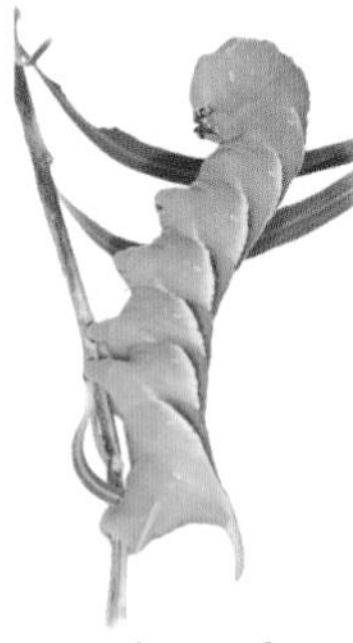

larva of Rustic Sphinx

More than 20 members of the genus ***Sphinx*** have occurred in our area, though some are only rare strays from Mexico. Adults mostly have cryptic patterns of brown or gray on the forewing and bands of black and white on the hindwing. The **Great Ash Sphinx** *(Sphinx chersis)* is found almost throughout the U.S. and southern Canada, but is rare in the southeast. Larvae feed not only on ash but also lilac, privet, and many other plants, particularly those in the olive family. The **Fawn Sphinx** *(Sphinx kalmiae),* or Laurel Sphinx, is another that favors the olive family for larval foodplants. It is widespread east of the Great Plains.

Our five members of the genus ***Ceratomia*** all have wavy forewing patterns. The **Waved Sphinx** *(Ceratomia undulosa)* is common east of the Rockies, its larvae feeding on ash and other members of the olive family. The **Big Poplar Sphinx** *(Pachysphinx occidentalis)* of the west and southwest, and the similar but more widespread **Modest Sphinx** *(Pachysphinx modesta),* not illustrated, are pale, heavy-bodied moths that often come to lights at night. As in some other sphinx moths, adults are not seen visiting flowers since they do not feed at all. The larvae more than make up for that, consuming large quantities of poplar and willow leaves.

Species in the genus ***Eumorpha*** have strong patterns, and most live in warm or tropical climates. The **Achemon Sphinx** *(Eumorpha achemon)* is more widespread, regularly reaching New England and the Dakotas.

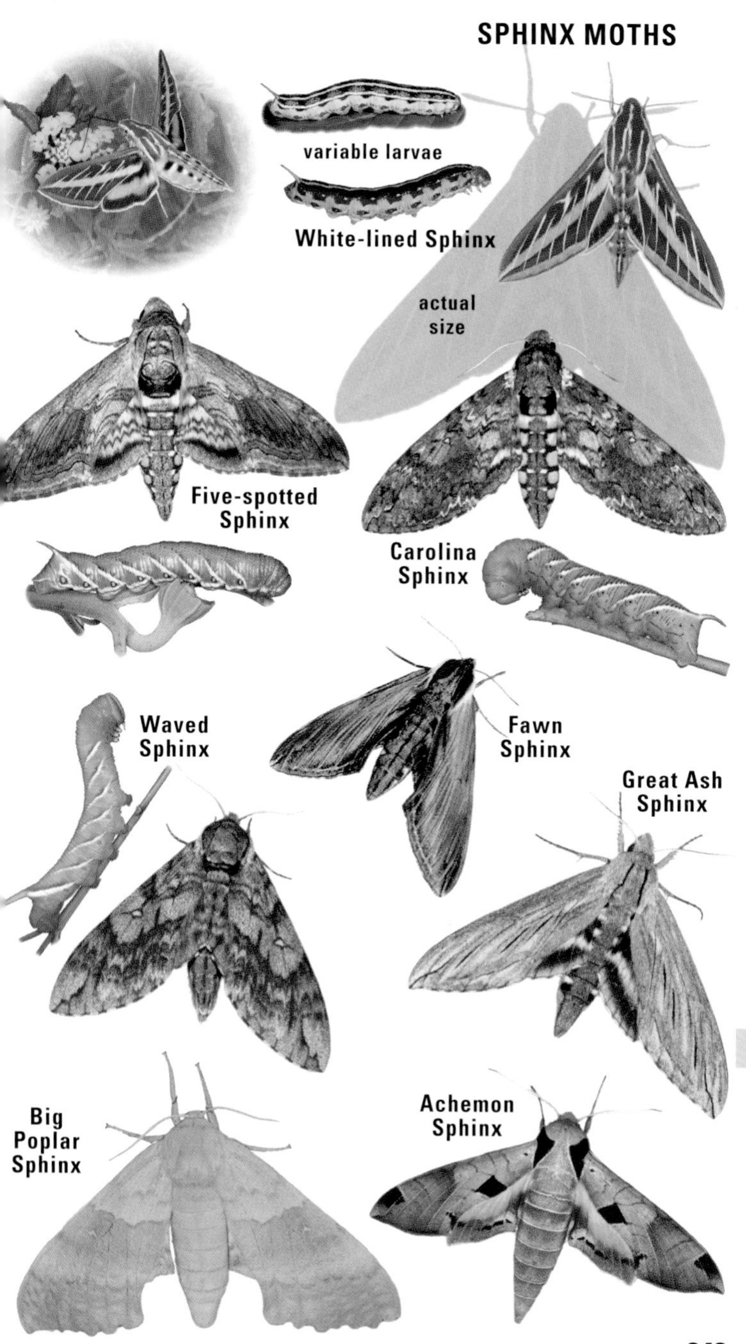
variable larvae
White-lined Sphinx
actual size
Five-spotted Sphinx
Carolina Sphinx
Waved Sphinx
Fawn Sphinx
Great Ash Sphinx
Big Poplar Sphinx
Achemon Sphinx

Several sphinx moths have English names that mention their "eyes," but the reference is actually to eyelike spots on their hindwings. Thus, the **Small-eyed Sphinx** *(Paonias myops)* really has normal eyes, and only its hindwing spots could be considered small. This species is common across much of the U.S. as well as southern Canada and northern Mexico, its larvae feeding on wild cherry, hawthorn, and other trees. Even more fanciful is the name of the **Blinded Sphinx** *(Paonias excaecatus)*, perhaps so called because it lacks a "pupil" in the center of its hindwing eyespot. Its larvae feed on a wide variety of trees and shrubs. The species is widespread in southern Canada and the U.S. and is often abundant in the northeast. The **One-eyed Sphinx** *(Smerinthus cerisyi)* is found from coast to coast, more commonly in the north — especially the northwest, where it is among the most numerous of sphinx moths. Its larvae feed on willow and poplar.

larva of One-eyed Sphinx

Moths of the genus ***Erinnyis*** live mostly in our southernmost states. The **Ello Sphinx** *(Erinnyis ello)* is typical in having colorful hindwings and camouflaged forewings. Common in Florida and the southwest, it sometimes strays far to the north. Its larvae feed on members of the spurge family, such as poinsettias and cassavas. The **Obscure Sphinx** *(Erinnyis obscura)* is another common tropical and southern species that wanders northward. Larvae feed on papaya, climbing milkweed, and other plants. Adults look obscure indeed when their orange hindwings are hidden.

The **Snowberry Clearwing** *(Hemaris diffinis)* is typical of a small group of clear-winged, day-flying sphinx moths that are very good mimics of bumble bees — except that they hover in front of blossoms instead of landing on them. **Abbot's Sphinx** *(Sphecodina abbottii)* also may suggest a large bee as it hovers about flowers at dusk. Another species that appears to be a wasp mimic is the **Nessus Sphinx** *(Amphion floridensis)*, which is often active by day. Its larvae feed on grape, ampelopsis, and Virginia creeper.

The sleek, streamlined **Tersa Sphinx** *(Xylophanes tersa)* is common in the south but strays northward. Like some other sphinx moths, its larvae have two color forms, green and brown.

Terloo Sphinx larva and pupa

The genus ***Proserpinus*** includes six small sphinx moths, mostly patterned with dull green and mostly found in the west. The **Terloo Sphinx** *(Proserpinus terlooii)* reaches the U.S. only in southern Arizona, where its larvae feed on *Boerhaavia.* The **Hog Sphinx** *(Darapsa myron)*, also known as Virginia Creeper Sphinx, is widespread and common in eastern and central parts of the continent. Adults often come to lights, especially in late summer. Larvae feed on Virginia creeper, grape, and ampelopsis.

Small-eyed Sphinx
One-eyed Sphinx
actual size
Blinded Sphinx
Ello Sphinx
Obscure Sphinx
Abbot's Sphinx
Snowberry Clearwing
Tersa Sphinx
Terloo Sphinx
Nessus Sphinx
Tersa Sphinx larvae: 2 variations
Hog Sphinx

OWLET MOTHS

(family Noctuidae), also known as noctuids, make up the largest family of moths. In North America more than 2,900 species, or more than one-fourth of all our Lepidoptera, are classified in this one family. It is hard to generalize about this varied group. Owlet moths range from very small to very large. Most are active at night and wear cryptic patterns of brown or gray, but some fly by day and some have brilliant colors. The first three moths below, formerly regarded as noctuids, are now placed in separate but related families, as classification of moths continues to be revised.

NOLID MOTHS (family Nolidae), formerly included in the Noctuidae, are mostly small and gray. The **Sorghum Webworm Moth** *(Nola sorghiella)* is common in the south, its larvae making communal webs on the seedheads of various grasses; it is sometimes a pest on sorghum crops. The **Confused Meganola** *(Meganola minuscula* or *Meganola phylla)* is one of several small nolids with larvae feeding on oak leaves. **PANTHEID MOTHS (family Pantheidae)** include about two dozen species, mostly medium-sized with dark patterns on gray wings. The **Eastern Panthea** *(Panthea furcilla)* is common and widespread in the east, its larvae feeding on various conifers.

The species from here through p. 254 are now included in the **family Noctuidae.** This large family has been divided into many subfamilies, most not discussed separately here. The genus ***Acronicta*** includes about 75 species of dagger moths. The **Greater Oak Dagger Moth** *(Acronicta lobeliae),* widespread and common, is one of the largest; its larvae feed on oak. The **Funerary Dagger Moth** *(Acronicta funeralis)* occurs coast to coast. As with some other daggers, the larva is more distinctive than the adult. **Harris's Three-spot** *(Harrisimemna trisignata)* is common around wet eastern woods. The beautifully patterned adults often come to lights at night, while the odd black-and-white larvae feed on leaves of many shrubs and trees. The **Green Marvel** *(Agriopodes fallax)* is rather uncommon (as befits a superhero) but widespread, its larvae feeding on viburnum.

Members of many genera in the **subfamily Noctuinae** are called darts. The **Ipsilon Dart** *(Agrotis ipsilon)* is abundant around the world, migrating north in our area in spring and south in fall. Its larva (Black Cutworm) tunnels underground by day and emerges at night, feeding on a vast array of low plants, including many crops. Another abundant species that sometimes achieves pest status is **Dingy Cutworm Moth** *(Feltia jaculifera).* The genus ***Euxoa*** includes more than 150 species, many with strong patterns, such as ***Euxoa auripennis*** and ***Euxoa obeliscoides,*** both common in the west. The **Greater Black-letter Dart** *(Xestia dolosa)* is common in southern Canada and the northern U.S., flying in early summer and again in early fall. Larvae feed on a wide variety of plants, including crops such as corn, tobacco, and barley. The 30 species of ***Abagrotis*** mostly have patterned forewings and dark hindwings. ***Abagrotis glenni*** is western, its larvae feeding on juniper. The **Green Arches** *(Anaplectoides prasina)* is common in conifer forests across northern North America and Eurasia. Its larvae feed on the leaves of blueberries, raspberries, and other shrubs. The **Inclined Dart** *(Richia acclivis)* is widespread but uncommon.

Sorghum Webworm Moth
Confused Meganola
actual size
Eastern Panthea
Greater Oak Dagger Moth
Funerary Dagger Moth
Harris's Three-spot
larva
Ipsilon Dart
Green Marvel
Dingy Cutworm Moth
Greater Black-letter Dart
Euxoa auripennis
Euxoa obeliscoides
Abagrotis glenni
larva
Green Arches
Inclined Dart

Moths have had many admirers over the decades, and one result is that even some noctuids that look undistinguished have colorful names: False Wainscot, Confused Woodgrain, and many more. The **Laudable Arches** *(Lacinipolia laudabilis)* is fairly common in the southeast, flying in early summer and again in early fall. The **False Wainscot** *(Leucania pseudargyria)* is common across the north, extending south to the southern U.S., its larvae feeding on various grasses. Another northern species, extending south in the Appalachians and along the Pacific Coast, is the **Stormy Arches** *(Polia nimbosa)*. Its larvae feed on alders and other plants.

The **Confused Woodgrain** *(Morrisonia confusa)* is very common throughout the east and in parts of the northwest, flying in late spring and early summer. Its larvae feed on basswood, pine, blueberry, and many other trees and shrubs. The **Armyworm Moth** *(Mythimna unipuncta* or *Pseudaletia unipuncta)* occurs almost worldwide. Its larvae feed on a wide variety of grasses and low plants, and they sometimes become so abundant as to earn the common name.

larva of Longior Pinion

The genus ***Lithophane*** includes more than 40 species of long-winged brown or gray moths, commonly referred to as pinions. The **Longior Pinion** *(Lithophane longior)* is widespread in the west, its larvae feeding on juniper. The **Straight-toothed Sallow** *(Eupsilia vinulenta)* doesn't have actual teeth, of course; the odd name refers to microscopic "teeth" on the wing scales (not a field mark!). The **Roadside Sallow** *(Metaxaglaea viatica)* is a fall-flying species, on the wing as late as January in some areas. Sallows of the genus ***Feralia*** are cold-hardy, flying in early spring, even in midwinter in southern areas. **Comstock's Sallow** *(Feralia comstocki)* is widespread across Canada and the northern and central states, its larvae feeding on various conifers.

Species in the genus ***Apamea*** mostly have cryptic patterns. Larvae of most species feed on grasses and other low plants. The genus ***Papaipema*** includes more than 40 species in North America, most with strong patterns on the forewings. Their larvae tunnel in the stems and roots of various plants. The **Ironweed Borer Moth** *(Papaipema cerussata)* is common in the east, its larvae feeding in ironweed.

FORESTER MOTHS (subfamily Agaristinae) make up a small group, but some species are striking and noticeable. The **Beautiful Wood-Nymph** *(Eudryas grata)* does have a beautiful pattern, but when resting on a leaf it looks like a large, messy bird dropping. The little **Grapevine Epimenis** *(Psychomorpha epimenis)* is often seen flying through eastern woodlands and visiting flowers on warm days in early spring. Its larvae feed on grape leaves, folding up the edges to make shelters for themselves. Another common day-flier is the **Eight-spotted Forester** *(Alypia octomaculata)*, widespread in the east. Its larvae, strikingly patterned like the adults, feed on ampelopsis, Virginia creeper, and other plants in the grape family.

Laudable Arches

False Wainscot

Stormy Arches

Confused Woodgrain

Armyworm Moth

Straight-toothed Sallow

larva

Roadside Sallow

Longior Pinion

Ironweed Borer Moth

Apamea antennata

Comstock's Sallow

Eight-spotted Forester

larva

Beautiful Wood-Nymph

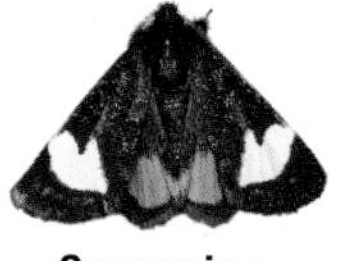

Grapevine Epimenis

Despite its name, the **Corn Earworm** *(Helicoperva zea)* feeds on a wide variety of plants. The species has a global distribution and is a permanent resident in the southern U.S. Adult Corn Earworm Moths invade northward every summer, becoming common north to Canada. They sometimes do considerable damage to crops, but most members of their subfamily **(Heliothinae)** are harmless. These include the beautiful flower moths of the genus ***Schinia,*** which hardly fit the pattern for noctuids: many visit flowers by day, and most have bright colors. We have more than 120 species, with the majority in dry or open habitats, and the larvae usually feed on the flowers, buds, or seedheads of their host plants. The **Clouded Crimson** *(Schinia gaurae)* is common across the southern states, its larvae feeding on gaura. Larvae of the **Primrose Moth** *(Schinia florida)* feed on flowers in the evening-primrose family. ***Schinia sueta*** is common in the far west, using lupines as larval food-plants. The **Arcigera Flower Moth** *(Schinia arcigera)* is widespread and common, its larvae feeding on many plants in the aster family.

Most members of the **subfamily Acontiinae** are small and have strong patterns, but those patterns may help to camouflage them. The **Exposed Bird-dropping Moth** *(Acontia aprica)* does sit out in plain sight on leaves and does look like a bird dropping, perhaps enough that birds fail to recognize it as an insect. The **Black-bordered Lemon** *(Thioptera nigrofimbria)* is abundant and widespread, its larvae feeding on morning glories. Moths in the genus ***Spragueia,*** such as ***Spragueia funeralis*** of the southwest, are mostly very small and dark with bands of red or yellow.

The **Eyed Baileya** *(Baileya ophthalmica),* formerly regarded as a noctuid, is now included among the nolids (p. 246), reflecting the ongoing changes in moth classification. Species of ***Eutelia*** are beautiful as adults and have strange tastes as larvae, feeding on poison sumac and poison ivy.

Loopers form a group **(subfamily Plusiinae)** in which larvae are lacking two pairs of prolegs in the middle of the body, so they move with a "looping" motion suggesting that of the inchworms, or Geometer caterpillars (p. 256). Larvae of most feed on a wide variety of low plants. The **Bilobed Looper Moth** *(Autographa biloba),* **Common Looper Moth** *(Autographa precationis),* and **Celery Looper Moth** *(Anagrapha falcifera)* are all widespread and common. The **Salt-and-pepper Looper Moth** *(Syngrapha rectangula)* is found mainly across the north and is named for its color pattern, not for what it eats! Larvae feed on needles of conifers.

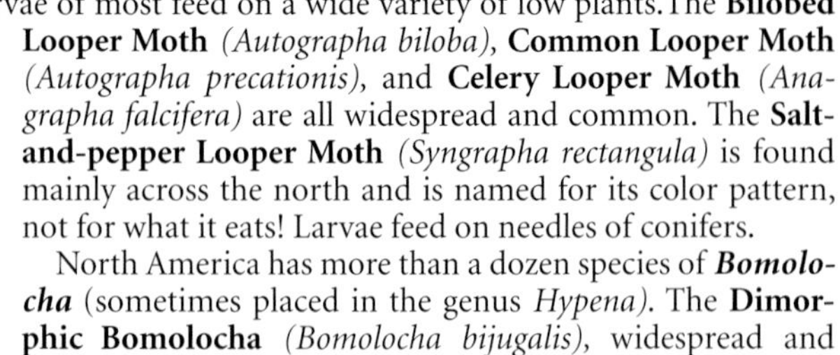

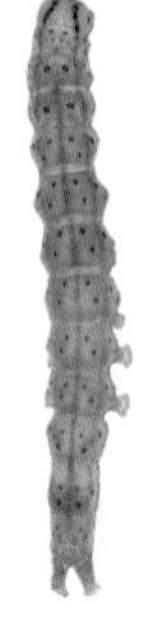

larva of *Bomolocha*

North America has more than a dozen species of ***Bomolocha*** (sometimes placed in the genus *Hypena).* The **Dimorphic Bomolocha** *(Bomolocha bijugalis),* widespread and common, uses dogwood as its larval foodplant. We have about 15 species in the genus ***Zanclognatha.*** Larvae of many are known to feed on dead leaves, but those of **Early Zanclognatha** *(Zanclognatha cruralis)* also feed on living leaves and on foliage of conifers.

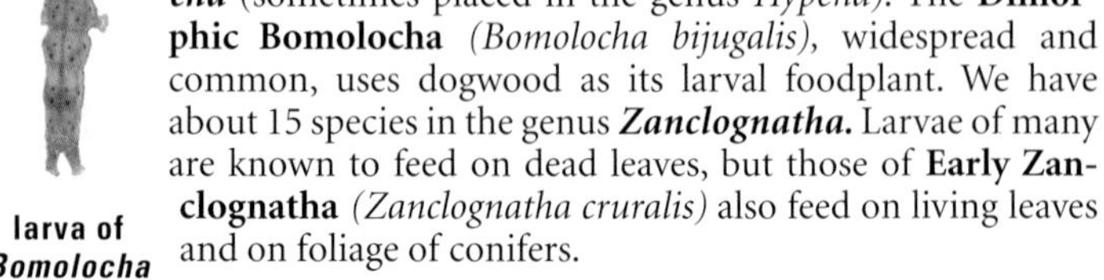

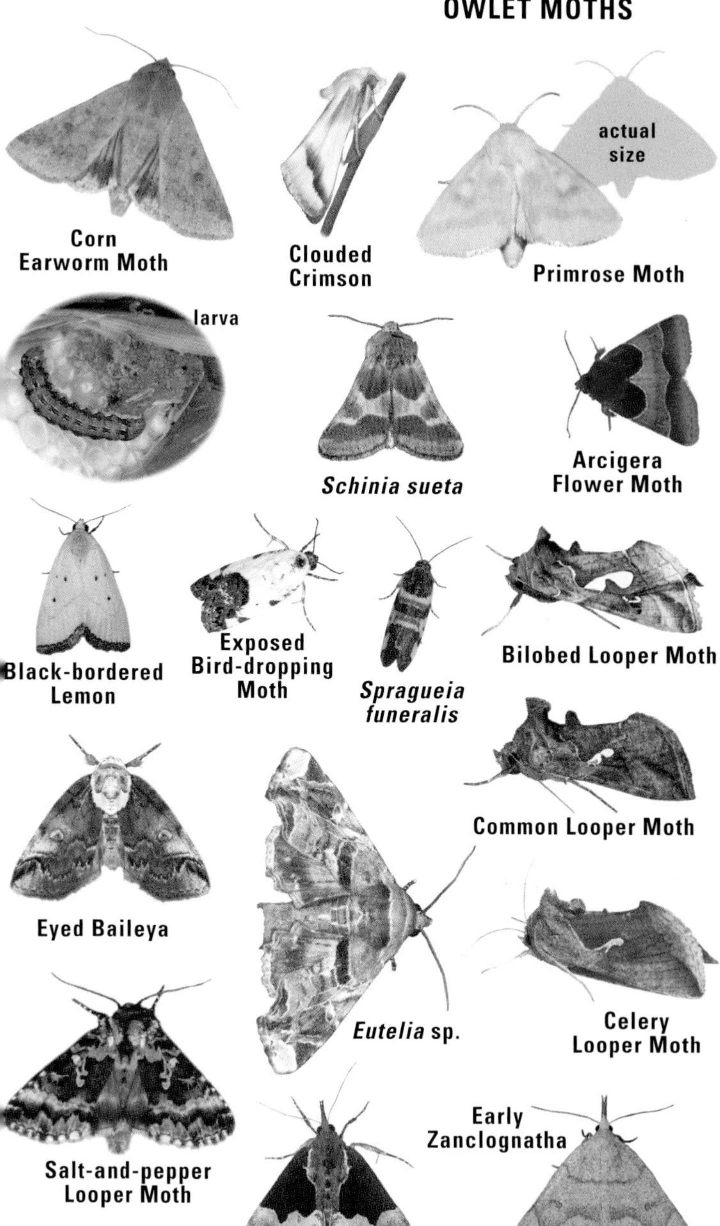

Corn Earworm Moth

Clouded Crimson

Primrose Moth

Schinia sueta

Arcigera Flower Moth

Black-bordered Lemon

Exposed Bird-dropping Moth

Spragueia funeralis

Bilobed Looper Moth

Common Looper Moth

Eyed Baileya

Eutelia sp.

Celery Looper Moth

Salt-and-pepper Looper Moth

Dimorphic Bomolocha

Early Zanclognatha

Members of the genus **Catocala,** the true underwings, have long been favorites with enthusiasts, as reflected in the fanciful or colorful English names applied to many. North America has at least 110 species, all with cryptic forewing patterns that give them superb camouflage as they rest by day on the bark of tree trunks. If they are disturbed into taking flight, the contrasting patterns of their hindwings will flash out, a striking effect that may startle predators and allow the moths to get away. Underwings are sometimes drawn to light but are best lured by "sugaring." Mixing brown sugar, fermented fruit, and stale beer, then painting the concoction on a tree trunk, is well worth a try. Most underwing larvae are rather plain and often have fringes along their sides that break up their outline and make them more difficult to see when they rest on tree bark.

The **Ilia Underwing** *(Catocala ilia)* is one of the most abundant species of *Catocala* in the east and is also widespread in the west, especially in the Pacific states. In most areas it flies from mid- to late summer. Its larvae feed on the foliage of oaks. The **Sweetheart Underwing** *(Catocala amatrix)* is variable in numbers, numerous in some places and scarce in others. Adults are sometimes found hiding inside sheds, cellars, and other dark places in the daytime. The larvae feed on poplar and willow.

larva of Aholibah Underwing

The **White Underwing** *(Catocala relicta)* lives in forests from coast to coast. Despite the name, the color of the forewings varies from mostly whitish to mostly dark gray. Its larvae feed on willow and poplar. Another species that feeds on willow and poplar is the **Darling Underwing** *(Catocala cara).* Adults have green and purple shading on the forewings, and one color form has a yellow spot near the forewing tip. This is another species sometimes found hiding by day in sheltered spots, including inside caves. Common and widespread in the western states is the **Aholibah Underwing** *(Catocala aholibah).* The larvae feed on the foliage of oaks, and adults fly in late summer. The **Girlfriend Underwing** *(Catocala amica)* is a small species with a notably variable forewing pattern. It is common and widespread, its larvae feeding on oaks.

A number of species of *Catocala* have the hindwings contrastingly deep black, usually with a whitish or gray outer fringe. Many of these dark-winged species use hickory or walnut trees as their larval foodplants. Typical of this group is the **Obscure Underwing** *(Catocala obscura),* which is common, especially in southern areas.

The **Large Yellow Underwing** *(Noctua pronuba)* is easily mistaken for a *Catocala* but was introduced to this continent from Europe. First noted in Nova Scotia in the late 1970s, it spread through eastern Canada and the northeastern U.S. during the following 10 or 15 years. Since then its spread has accelerated, with records north to the Arctic Ocean, west to Oregon, and south to Louisiana and Arizona; the species might turn up anywhere in North America. Adults are quite variable in their forewing patterns.

larva

Ilia Underwing

actual size

at rest

Sweetheart Underwing

wings open

Darling Underwing

larva

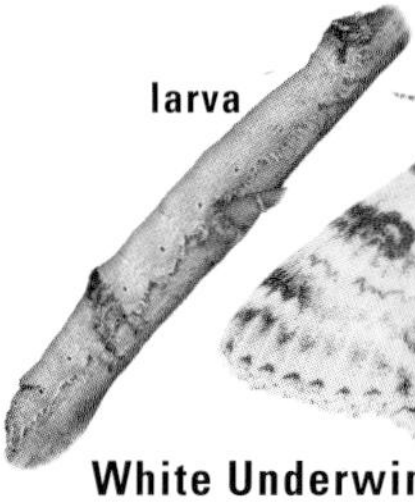

White Underwing

Girlfriend Underwing

Aholibah Underwing

Obscure Underwing

Large Yellow Underwing

larva

OWLET MOTHS: LESSER UNDERWINGS

Most moths on this page are currently classified in the same subfamily **(Catocalinae)** as the typical underwings on the previous page, but only a few of these have contrastingly colorful hindwing patterns—a trait that also turns up in many other groups of moths.

Euscirrhopterus gloveri suggests an underwing moth of some sort but is classified with the forester moth subfamily (p. 248). Common in the southwest, it is usually active at night and often comes to lights.

Lesser underwing moths of the genus ***Melipotis*** are abundant and fairly diverse in the southwestern U.S., but only a few species are widespread farther north. One of the most widely distributed, the **Indomitable Melipotis** *(Melipotis indomita),* is found from coast to coast, often visiting flowers in the daytime. Mesquite is one of its larval foodplants. Several other genera of lesser underwings are somewhat similar, including ***Drasteria, Forsebia,*** and ***Bulia. Drasteria tejonica*** (formerly called *Synedoida tejonica)* is fairly common in the southwest.

The **Decorated Owlet** *(Pangrapta decoralis)* is widespread and numerous, especially in northern woodlands. Its larvae feed on foliage of blueberries. Our 10 species in the genus ***Metalectra,*** generally known as fungus moths, are mostly medium-sized to small as adults, with variegated wing patterns. Larvae, at least in some species, feed on dry fungus growing on trees. The **Black Fungus Moth** *(Metalectra tantillus)* occurs north to Long Island but is most common in the southeast. The **Common Fungus Moth** *(Metalectra discalis)* is widespread in the eastern and central states.

Moths of the genus ***Zale*** (about 35 species) usually sit with the wings spread flat out to the sides and often have continuous color patterns extending across both the forewings and the hindwings. The **Lunate Zale** *(Zale lunata)* is one of the most abundant and widespread and apparently is migratory over the northern part of its range. Its larvae feed on a wide variety of trees and shrubs. The **Horrid Zale** *(Zale horrida)* is not as bad as its name would suggest: *horrida* means "bristly" and refers to the bristles on top of the thorax. Larvae of this species feed on viburnum.

The **Herald Moth** *(Scoliopteryx libatrix)* has a wide distribution in the northern hemisphere, including much of North America. Its larvae feed on willows. Herald Moths overwinter as adults, hiding away in sheltered spots, and large numbers have been found in caves. The **Forage Looper** *(Caenurgina erechtea)* and some similar related species are abundant in open grassy areas and may be active day or night. Their larvae feed on a wide variety of low plants, including grasses and clovers. The elegant little **Short-lined Chocolate** *(Argyrostrotis anilis)* is widespread but uncommon in the east. Its larvae have been reported on marsh-pinks *(Sabatia).*

The huge **Black Witch** *(Ascalapha odorata)* is mostly tropical, but strays wander far to the north every year, sometimes reaching Canada. Amazingly, there are records of large numbers of Black Witches traveling within the eyes of hurricanes. "Fallouts" of these moths have arrived with hurricanes coming ashore from the Gulf of Mexico, where they must have been flying continuously for several days.

actual size

Indomitable Melipotis

Euscirrhopterus gloveri

Black Fungus Moth

Drasteria tejonica

Decorated Owlet

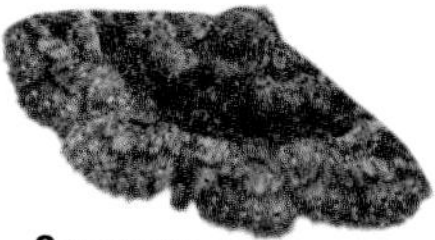

Common Fungus Moth

Lunate Zale

Horrid Zale

Herald Moth

Forage Looper

Black Witch (not to scale)

actual size of Black Witch

Short-lined Chocolate

GEOMETER MOTHS

(family Geometridae) make up one of our largest moth families, with more than 1,400 species in the U.S. and Canada. Adults of many species rest with the wings spread out straight to the sides. Females of some are wingless. Larvae are famed as "inchworms"—their middle pairs of prolegs are lacking or rudimentary so they move forward by looping or "inching" rather than crawling. Many perform incredible disappearing acts when danger threatens. They freeze in a stiff posture angled from their perch, instantly becoming a twig instead of a caterpillar.

Itame guenearia is common in woodlands of the west, its larvae feeding on coffeeberry. Obscure in appearance, but widespread, is the **Common Gray** *(Anavitrinella pampinaria).* Its larvae feed on a wide variety of plants, from trees to low herbs. Many species in the genus ***Semiothisa*** use conifers as their larval foodplants. The **Bicolored Angle** *(Semiothisa bicolorata)* is common, especially in the southeast, its larvae feeding on pines.

larva of Peppered Moth "inching" and Curve-toothed Geometer in twig-imitating posture

The **Common Lytrosis** *(Lytrosis unitaria)* wears a classic geometer pattern, good camouflage as it rests on certain kinds of bark. The **Peppered Moth** *(Biston betularia)* has pale and dark forms. In Britain, the dark form became more common in areas where soot from factories had blackened the tree trunks on which the moths rested. The **False Crocus Geometer** *(Xanthotype urticaria)* is one of several similar species. The larvae, good mimics of twigs, feed on various low plants. The larva of the **Saw-wing** *(Euchlaena serrata)* is even more convincing as a twig mimic. Moths in the genus ***Pero*** are often bicolored, like **Western Pero** *(Pero occidentalis),* which uses conifers for larval foodplants.

Male **Spring Cankerworm Moths** *(Paleacrita vernata)* fly on nights in very early spring; females are wingless. The **Bluish Spring Moth** *(Lomographa semiclarata),* common in the northeast in early spring, is a day-flier that suggests a small butterfly. The **Hollow-spotted Plagodis** *(Plagodis alcoolaria)* is common in forests, its larvae feeding on foliage of many trees. Most green geometers are in a different subfamily (next page), but there are exceptions, including ***Chloraspilates bicoloraria*** of the southwest.

The **Northern Pine Looper Moth** *(Caripeta piniata)* is common in Canada and the northern U.S. Despite its name, the larva of the **Oak Beauty** *(Phaeoura quernaria,* formerly *Nacophora quernaria)* feeds on many trees besides oaks. Also with more varied tastes than its name would suggest is the **Tulip-tree Beauty** *(Epimecis hortaria).* Found from coast to coast in Canada and much of the U.S., the **Pale Beauty** *(Campaea perlata)* flies in summer in wooded areas, its larvae feeding on a wide variety of trees and shrubs. The larva of the **Curve-toothed Geometer** *(Eutrapela clemataria)* feeds on a wide variety of trees and shrubs, both leafy and coniferous, and its twiglike disguise apparently functions well on all of these plants.

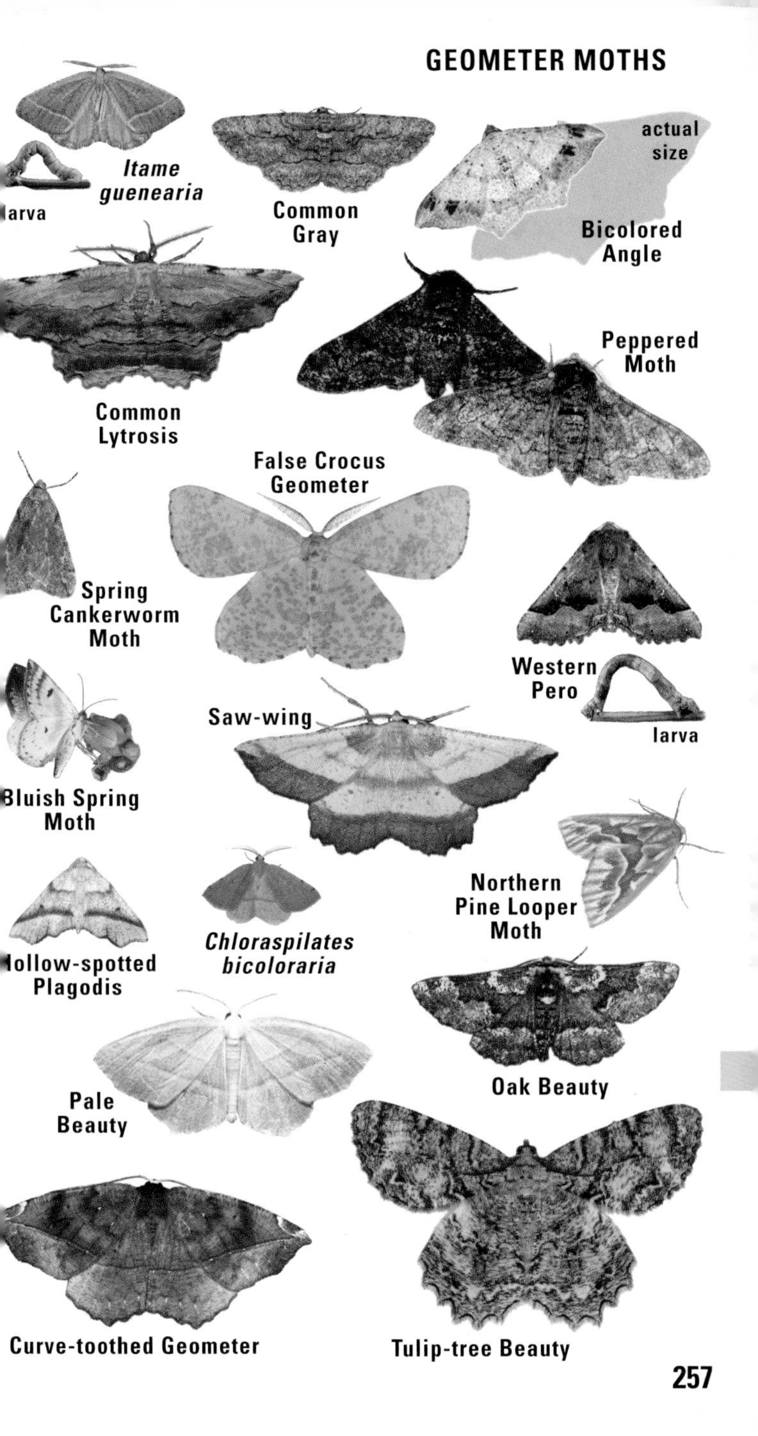
Itame guenearia
larva
Common Gray
actual size
Bicolored Angle
Common Lytrosis
Peppered Moth
False Crocus Geometer
Spring Cankerworm Moth
Western Pero
larva
Saw-wing
Bluish Spring Moth
Northern Pine Looper Moth
Hollow-spotted Plagodis
Chloraspilates bicoloraria
Oak Beauty
Pale Beauty
Curve-toothed Geometer
Tulip-tree Beauty

Flying by day in early spring in forested regions of Canada and the northern U.S. is the **Infant Moth** *(Archiearis infans)*. Its larvae feed on foliage of birches. Males of the **Fall Cankerworm Moth** *(Alsophila pometaria)* may fly from fall to spring in warm climates, mainly late fall in the north, but females are nearly wingless and don't fly at all. Their larvae sometimes are pests on apple and some other fruit trees. The **Large Lace-border** *(Scopula limboundata)* is among about two dozen species of *Scopula* in North America. Its larvae feed on various low plants, shrubs, and trees.

The **Dot-lined Wave** *(Idaea tacturata)* is common in the southeast, flying all year in warmer areas. As in other *Idaea,* the larva feeds on a variety of low plants. The distinctively patterned ***Pigia multilineata*** is locally common in the southwest. Small but colorful, the **Chickweed Geometer** *(Haematopis grataria)* is often seen fluttering across fields by day, or visiting lights at night. It is widespread and common, even abundant in the northern states. Its larvae feed on chickweeds and many other low plants.

One of the most distinctive groups within this family is the **subfamily Geometrinae,** which includes the green geometers. The **Red-bordered Emerald** *(Nemoria lixaria)* is common in the southeast, using red oak as its larval foodplant. Widespread across the southern states is **Southern Emerald** *(Synchlora frondaria)*. Its larvae feed on blackberry and other low plants.

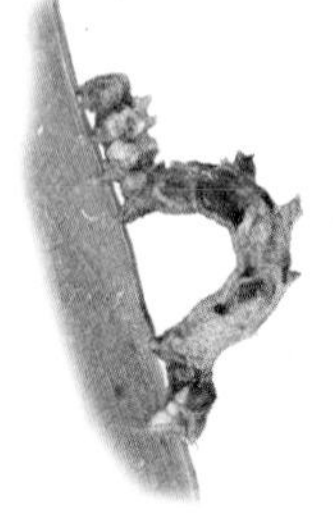

larva of Red-bordered Emerald

Members of the genus ***Dysstroma*** (as well as several related genera) are often known as Carpets. ***Dysstroma formosa*** is common in the west, its larvae feeding on gooseberry. The **Lesser Grapevine Looper** *(Eulithis diversilineata)*, like some related species, often sits in an odd posture with its abdomen curled up over its head. Its larvae feed on leaves of grape and Virginia creeper. Species of ***Stamnodes*** often perch in a butterflylike posture. ***Stamnodes marmorata*** is common and widespread in dry forests of the west.

The genus ***Eupithecia*** includes more than 120 species in North America, commonly called pugs. Most are gray with long narrow forewings spread out flat to the side when at rest, and most are very difficult to identify. Larvae of some *Eupithecia* in Hawaii are predatory, ambushing other insects to feed on them, but those on the North American mainland feed on plants so far as is known. The **Red Twin-spot** *(Xanthorhoe ferrugata)* is widespread and common, especially in the north, its larvae feeding on many low plants. The **Tissue Moth** *(Triphosa haesitata)* is widespread and occasionally common, its larvae feeding on buckthorns and other trees. ***Synaxis cervinaria*** is widespread in the west. Its orange twiglike larvae feed on many trees and shrubs. The genus ***Hydriomena*** includes more than 50 species, many of them difficult to identify, even though they often have colorful markings of green. **Ferguson's Scallop Shell** *(Rheumaptera prunivorata,* formerly *Hydria prunivorata)* has gregarious larvae that feed together on black cherry.

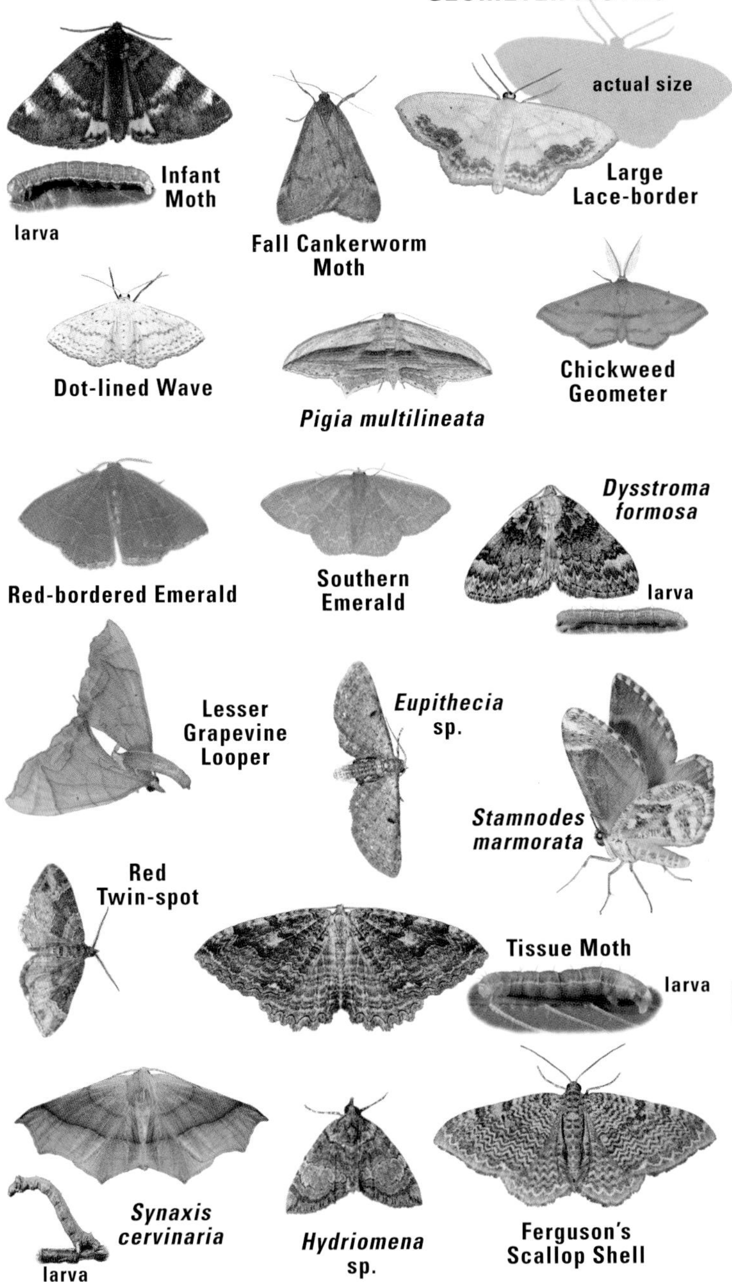
actual size
Infant Moth
larva
Fall Cankerworm Moth
Large Lace-border
Dot-lined Wave
Pigia multilineata
Chickweed Geometer
Red-bordered Emerald
Southern Emerald
Dysstroma formosa
larva
Lesser Grapevine Looper
Eupithecia sp.
Stamnodes marmorata
Red Twin-spot
Tissue Moth
larva
Synaxis cervinaria
larva
Hydriomena sp.
Ferguson's Scallop Shell

TIGER MOTHS

(family Arctiidae) include more than 250 species in North America. Adults of many species are brightly colored or are strongly marked with stripes or other patterns. Their larvae often are densely hairy or spiny, and many are generalists that feed on a wide variety of plants.

Tiger moths in the genera ***Grammia*** and ***Apantesis*** have patterns that look distinctive, but it is challenging to tell the many species apart. The **Ornate Tiger Moth** *(Grammia ornata)* is typical, common in the far west, flying in early summer. The **Virgin Tiger Moth** *(Grammia virgo)* is widespread and common across the north, less numerous southward. Its larvae feed on many low plants. The **Nais Tiger Moth** *(Apantesis nais)* has a general distribution east of the Rockies. Its hindwings are usually yellow, sometimes red. Its larvae feed on grasses, clover, and other low plants.

The five southwestern species in the genus ***Arachnis*** are all beautifully patterned. ***Arachnis picta*** is the most widespread, from California east to Colorado and Texas. The **Great Tiger Moth** *(Arctia caja)* is a hardy northern species, fairly common in the boreal forests of northern Canada, extending southward into the northeastern U.S. and parts of the western states. Its densely hairy larvae feed on a wide variety of plants in fall and spring, with a long break for hibernating through the winter.

The little day-flying **Bella Moth** *(Utetheisa bella)* is fairly common in open fields, drawing attention as it flutters low over the grass, flashing the brilliant pink of its hindwings. The **Clymene Haploa** *(Haploa clymene),* like other members of the genus ***Haploa,*** is conspicuous for its bold chocolate-and-cream forewing pattern. Active day and night, it visits flowers and comes to lights. Its larvae, blackish with side stripes of yellow-orange, feed on a wide variety of plants, including Eupatorium and willow. In its adult stage, the **Isabella Tiger Moth** *(Pyrrharctia isabella)* draws little attention, but its larva is one of our most familiar caterpillars: the "Woolly Bear" or "Black-ended Bear" goes wandering in fall and is often seen crossing roads. Contrary to popular belief, the width of its orange band is no predictor of the severity of the coming winter. The **Salt Marsh Moth** *(Estigmene acrea)* is poorly named: it lives in any open habitat. Its larvae, varying in color from creamy to blackish, feed on a very wide variety of low plants, shrubs, and trees.

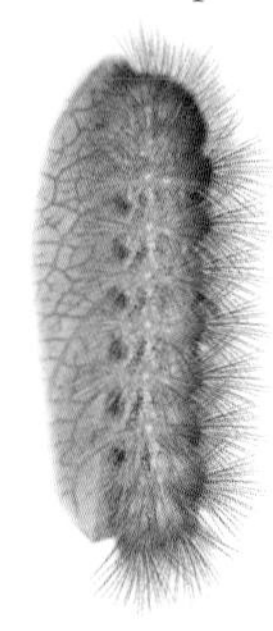

larva of Salt Marsh Moth

Lichen moths **(subfamily Lithosiinae)** are mostly small and brightly colored, and the larvae are among the few moths that feed on lichens. The **Painted Lichen Moth** *(Hypoprepia fucosa)* is widespread and common.

Several western tiger moths are classified in the **subfamily Pericopinae,** including ***Gnophaela latipennis,*** which flies by day in woods of the Pacific states. The only eastern representative of this group, the **Faithful Beauty** *(Composia fidelissima),* reaches our area only in southern Florida. There it flies conspicuously around patches of hardwoods by day. The garish pink larvae feed on plants in the spurge family.

Ornate Tiger Moth

larva

actual size

Virgin Tiger Moth

Nais Tiger Moth

Arachnis picta

Great Tiger Moth

larva

larva ("Woolly Bear")

larva

Bella Moth

larva

Clymene Haploa

Isabella Tiger Moth

Salt Marsh Moth

Painted Lichen Moth

Gnophaela latipennis

larva

Faithful Beauty

Tiger moths, introduced on the preceding page, include several species in the **subfamily Syntominae** that appear to mimic wasps. One of the most common across southern Canada and the northern U.S. is the **Virginia Ctenucha** *(Ctenucha virginica)*, which is active by day but also comes to lights at night. Its larvae, densely tufted with white and yellow, feed on grasses and other low plants. The **Veined Ctenucha** *(Ctenucha venosa)* is common in the central and southwestern U.S. Convincingly similar to a paper wasp is the **Texas Wasp Moth** *(Horama panthalon)*, a common flower visitor in south Texas, sometimes straying northward. The **Polka-dot Wasp Moth** *(Syntomeida epilais)* is common in Florida, where its larvae feed on oleander. They are sometimes considered pests, but most of us would be glad to sacrifice some oleander leaves for this snazzy moth. The **Yellow-banded Wasp Moth** *(Syntomeida ipomoeae)* is another Florida specialty, its larvae feeding on morning glories and other plants.

CLEARWING MOTHS (Sesiidae) include more than 110 species in North America. Most of them lack colored scales on at least part of the wing surface, and many are striking mimics of wasps, flying by day. Larvae bore in the stems or roots of plants. The **Squash Borer Moth** *(Melittia cucurbitae)* is often seen visiting flowers; its larvae bore in the stems of plants in the squash family. ***Melittia grandis*** is a larger species of squash borer found in the southwest. Larvae of the **Peachtree Borer** *(Synanthedon exitiosa)* tunnel into the trunks of peach and other fruit trees. **Riley's Clearwing** *(Synanthedon rileyana)* is common in the southeast, flying in summer and early fall; its larvae bore in horse-nettles. Larvae of the aptly named **Hornet Clearwing** *(Paranthrene simulans)* are borers in various oaks.

Gypsy Moth larva

TUSSOCK MOTHS (Lymantriidae) are represented by only about 32 species in North America, but these include the infamous **Gypsy Moth** *(Lymantria dispar)*. Introduced from Europe to Boston in the late 1860s, it has spread west through the Great Lakes region and south to the Carolinas. Its larvae are known to feed on more than 500 species of plants, and it has defoliated large tracts of forest, temporarily. Probably more damage has been done by the huge amounts of poisons sprayed on forests in ill-conceived attempts to control the moth. Larvae of the **White-marked Tussock Moth** *(Orgyia leucostigma)* are found on a remarkable variety of plants, and they are far more distinctive than the adults, although the adult female is rather odd herself: a wingless creature that lays her eggs all over the cocoon from which she has emerged. The **Southern Tussock Moth** *(Dasychira meridionalis)* is common in the Gulf states, its larvae feeding on oaks.

HOOKTIP MOTHS (Drepanidae) make up a small but distinctive group (now also including the thyatirines, p. 264). The **Arched Hooktip** *(Drepana arcuata)* is widespread across Canada and parts of the northern U.S., its larvae feeding on alder and birch. The **Rose Hooktip** *(Oreta rosea)*, using viburnum as a larval foodplant, is common over much of the east.

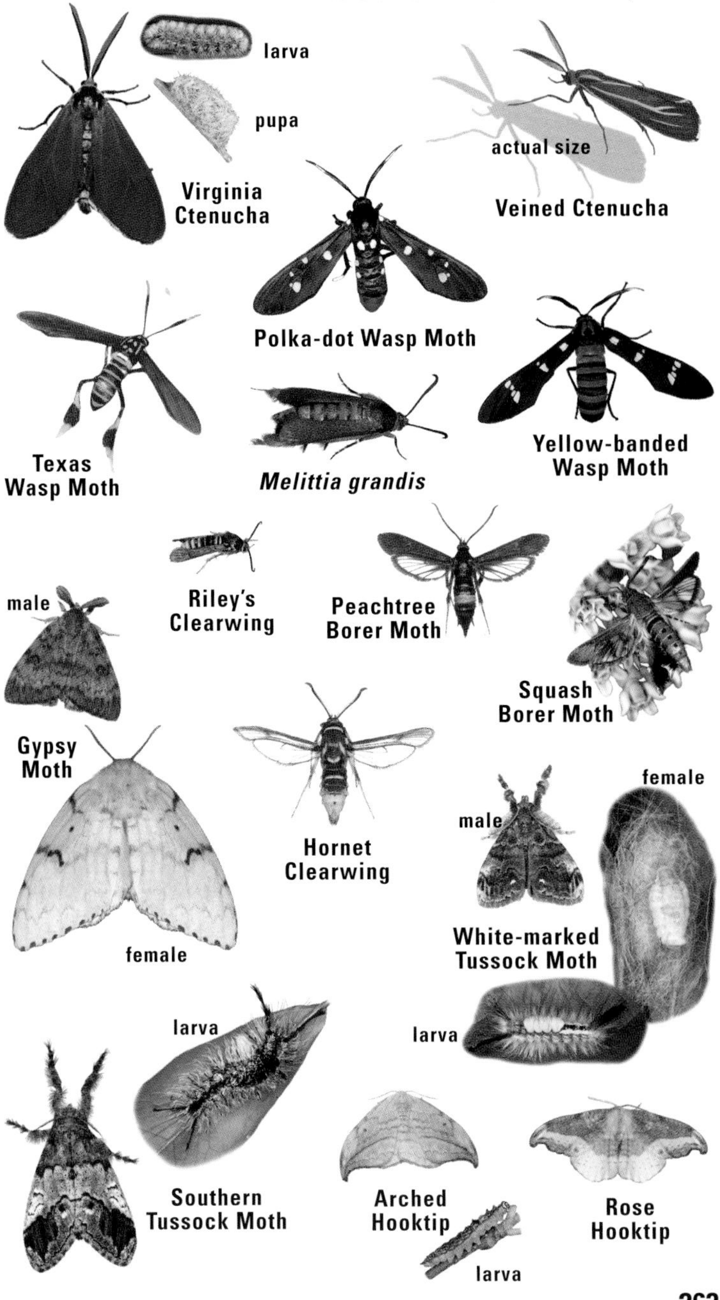

larva

pupa

Virginia Ctenucha

actual size

Veined Ctenucha

Polka-dot Wasp Moth

Texas Wasp Moth

Melittia grandis

Yellow-banded Wasp Moth

Riley's Clearwing

Peachtree Borer Moth

Squash Borer Moth

male

Gypsy Moth

female

Hornet Clearwing

male

female

White-marked Tussock Moth

larva

larva

Southern Tussock Moth

Arched Hooktip

larva

Rose Hooktip

The families on this page are somewhat similar but not all closely related.

PROMINENT MOTHS (family Notodontidae) are mostly stout-bodied, hairy, and medium-sized. They are readily attracted to lights but then usually sit lethargically in one spot. Larvae of many species are strikingly patterned or shaped, and their odd appearance may make them less recognizable to predators. The **White-dotted Prominent** *(Nadata gibbosa)* is common from coast to coast. Its larvae feed on the leaves of oak and other trees. Named for its toothed forewing pattern, the **Double-toothed Prominent** *(Nerice bidentata)* is widespread and common, its larvae feeding on elms.

larva of *Datana* sp.

The several species in the genus ***Datana*** are mostly similar, with dust brown wings crossed by curved lines. Their larvae often have stripes or other strong patterns, and they typically feed in groups. If disturbed they often adopt bizarre poses with both ends of the body raised.

The **White Furcula** *(Furcula borealis)*, common over much of the east, is attractive as an adult, but the larva is more impressive. When disturbed, it can extend its pair of "tails" and thrash them about. It feeds on foliage of wild cherry. The **Black-rimmed Prominent** *(Pheosia rimosa)* is widespread and common in the north, extending well southward in the Appalachians and Rockies. Its larvae feed on poplars and willows.

LAPPET MOTHS and TENT CATERPILLARS (Lasiocampidae) are stout, furry, and sleepy as adults, often sitting quietly under lights where they have been attracted. The **Small Tolype** *(Tolype notialis)*, one of several similar species, is common in the southeast. When adult female tolypes lay eggs, they cover them with black scales from their own bodies. The **Eastern Tent Caterpillar Moth** *(Malacosoma americanum)*, widespread and common, is most often noticed for the "tents" spun by groups of larvae, often in apple, hawthorn, or cherry trees. The larvae crawl out to feed, returning to the web for shelter. The **Forest Tent Caterpillar Moth** *(Malacosoma disstria)* is also common and its larvae also live in large groups, but they do not spin a group web — they are really tentless caterpillars.

The species formerly put in their own family as thyatirid moths are now considered a subfamily **(Thyatirinae)** of the Hooktip moths (previous page). One of the most distinctive is the **Lettered Habrosyne** *(Habrosyne scripta)*, uncommon but widespread across Canada and the northern and central U.S. Its larvae feed on the foliage of blackberry.

The **family Bombycidae** includes the famous Silkworm Moth of Asia (not illustrated), raised by the Chinese for silk production for at least the last 4,700 years. Now classified in this family are the five species formerly placed in the family Apatelodidae. The **Western Apatelodes** *(Apatelodes pudefacta)* and its eastern replacement, **Spotted Apatelodes** *(Apatelodes torrefacta)*, have distinctive wing shapes and a translucent patch near the tip of the forewing. Their furry larvae feed on various trees and shrubs.

actual size

White-dotted Prominent

Double-toothed Prominent

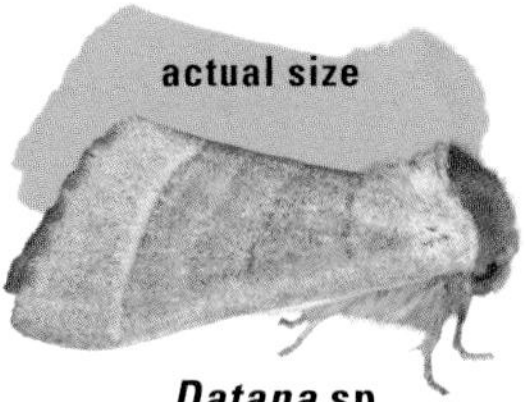

Datana sp.

Black-rimmed Prominent

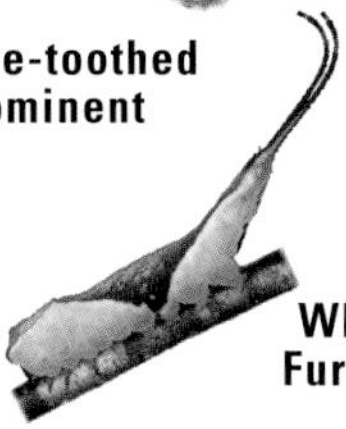

White Furcula

larva

larvae in "tent"

Tolype distincta larva

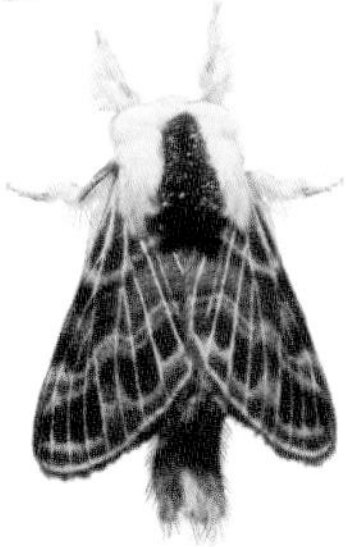

Small Tolype

Eastern Tent Caterpillar Moth

larva

Lettered Habrosyne

Forest Tent Caterpillar Moth

Spotted Apatelodes larva

Western Apatelodes

The order Lepidoptera is often divided into macrolepidoptera (all those on the preceding pages) and microlepidoptera (the many families of moths on pp. 266–271). The "micros" are diverse and important but have received far less attention than the "macros," mostly because the majority of species are quite small and difficult to identify. There are exceptions: a few micros are large, and many are colorful when viewed up close.

Larvae of many micromoths are small enough to feed between the layers of leaves, including those of the **SHIELDBEARER MOTHS (family Heliozelidae).** Adults in this family are tiny, with wingspans usually less than a quarter inch. Adult **FAIRY MOTHS (Adelidae)** are recognized by their very long antennae. Active by day, they may be seen on flowers.

In the **family Prodoxidae,** an amazing relationship exists between **Yucca Moths** *(Tegeticula yuccasella)* and yucca plants. The adult female moth actually gathers a ball of pollen and forcefully pollinates the yucca before laying her eggs, thus guaranteeing that her larvae will have seeds upon which to feed. This family also includes "bogus yucca moths" (genus ***Prodoxus***), which feed on yuccas without pollinating them. The **family Incurvariidae** includes various small dark moths. Larvae of the **Maple Leafcutter Moth** *(Paraclemensia acerifoliella)* start as miners between leaf layers, but later they cut out circular bits of leaf to make a protective case for themselves.

Our 125 species in the **family Tineidae** are called **CLOTHES MOTHS,** but only a few feed on fabrics or woolens. Genuinely destructive at times are the **Webbing Clothes Moth** *(Tineola bisselliella)* and **Casemaking Clothes Moth** *(Tinea pellionella).* **BAGWORM MOTHS (Psychidae)** spend most of their lives inside a bag made of their own silk and bits of debris. After pupating, males fly to locate females, but females remain wingless and lay their eggs inside the bag in which they developed. The **Evergreen Bagworm Moth** *(Thyridopteryx ephemeraeformis)* is common on eastern red cedar. Larvae in the **family Bucculatricidae** are miners between the layers of leaves on trees, usually making narrow twisting mines. Adults are tiny moths with narrow wings. **LEAF BLOTCH MINERS (Gracillariidae)** also mine between leaf layers. Adults are tiny but often colorful.

ERMINE MOTHS (Yponomeutidae) can be brightly marked but small. An exception is the **Ailanthus Webworm Moth** *(Atteva punctella),* larger than most, widespread, common, with a distinctive pattern. Larvae make communal webs on foliage of ailanthus trees. The 50-plus species of ***Argyresthia*** are very small and typically rest with the head down and wings and body angled upward. The **family Plutellidae** includes at least 20 species in North America, the best-known being the **Diamondback Moth** *(Plutella xylostella),* introduced here from Europe before 1850 and now found over most of this continent.

Among the largest micros are the **GHOST MOTHS (Hepialidae),** such as the **Silver-spotted Ghost Moth** *(Sthenopis argenteomaculatus).* These fast-flying moths, active only at dusk, seldom come to lights, so they may go unnoticed, but males sometimes gather in mating swarms. Larvae bore in the roots of trees for one to two years.

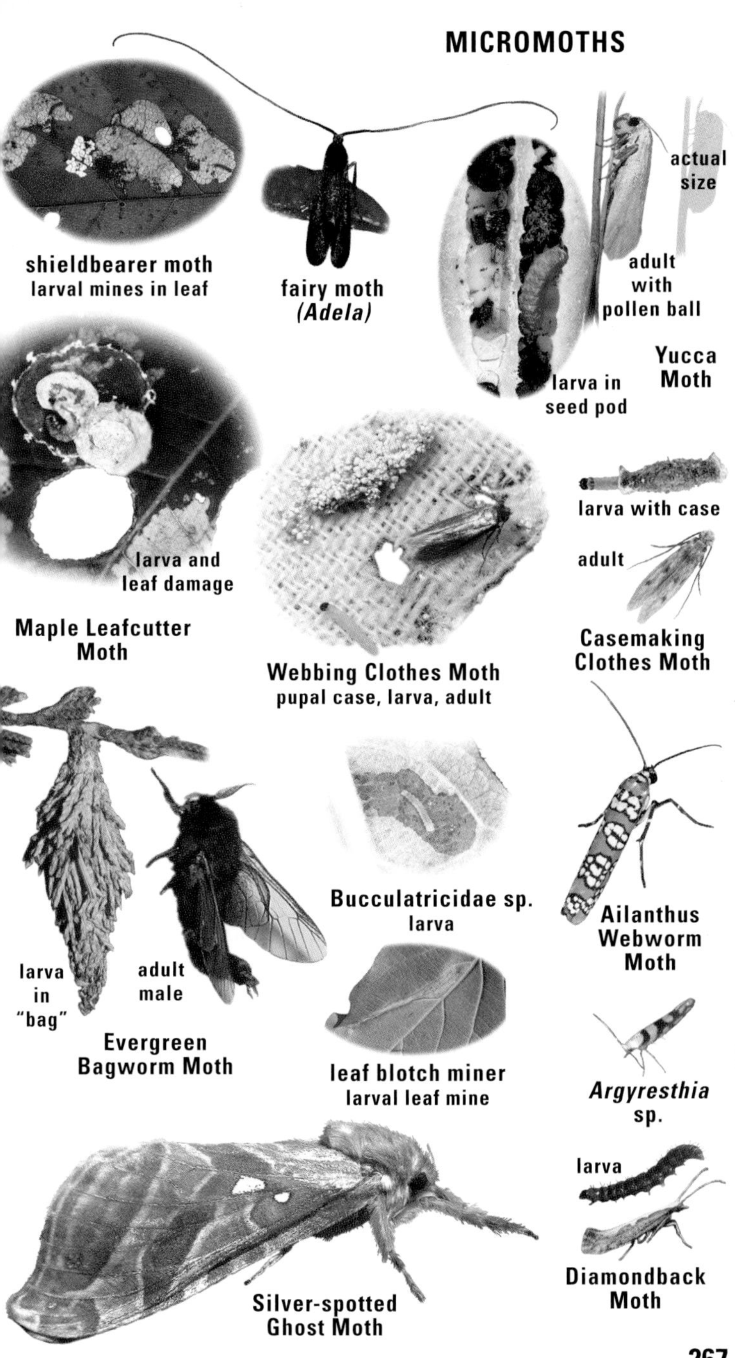

shieldbearer moth
larval mines in leaf

fairy moth
(Adela)

Yucca Moth

Maple Leafcutter Moth

Webbing Clothes Moth
pupal case, larva, adult

Casemaking Clothes Moth

Bucculatricidae sp.
larva

Ailanthus Webworm Moth

Evergreen Bagworm Moth

leaf blotch miner
larval leaf mine

Argyresthia sp.

Diamondback Moth

Silver-spotted Ghost Moth

OECOPHORID MOTHS (family Oecophoridae) vary from tiny to medium-sized. Experts disagree as to exactly which species to include in this family. Adults in the genus ***Antaeotricha*** are among the many moths, from several different families, that resemble bird droppings when resting on leaves. The genus ***Ethmia*** includes more than 40 species, most quite small but with attractive patterns.

CASEBEARER MOTHS (Coleophoridae) are most notable for their larvae, which make portable "cases" for themselves out of bits of foodplant and droppings. Adults are very small and plain. Well over 150 species are known for North America, with many undoubtedly still undescribed. The **Pistol Casebearer** *(Coleophora malivorella)* feeds on buds and leaves of apple and other trees, and in older larvae the case is somewhat pistol-shaped; the adult (as in other species of *Coleophora)* is far less distinctive.

GELECHIID MOTHS (Gelechiidae) include well over 600 species, ranging from small to tiny. Larvae in this family have wildly varied feeding habits: some fold or roll leaves, some feed inside fruits, some make galls. The **Redbud Leaf-folder Moth** *(Fascista cercerisella)* is widespread in the east; its convict-striped larvae spin silk to tie redbud leaves together. The **Goldenrod Gall Moth** *(Gnorimoschema gallaesolidaginis)* is most noticeable for the galls that its larvae form in the stems of goldenrod.

SLUG CATERPILLAR MOTHS (Limacodidae) look most interesting as larvae, although some adults are marked with green. The caterpillars have some of their prolegs replaced by suckers, and they glide along the undersurface of leaves, feeding on many trees and shrubs. The **Hag Moth** *(Phobetron pithecium)* larva, also known as Monkey Slug, is a grotesque creature with 12 hairy "arms" along its sides. Despite its appearance, it is harmless (although some persons may have an allergic reaction to its hairs). Adult females have pale tufts of scales on their legs, while males (not shown) have translucent wings and often fly by day. **Skiff Moth** *(Prolimacodes badia)* larvae are more attractive and much more variable, usually patterned with green and chestnut. Adults are common visitors to lights in the east. The **Saddleback Caterpillar** *(Acharia stimulea)* is among our most easily recognized larvae and well worth recognizing, too, as it bears clusters of highly potent stinging spines. Adults are generally noticed far less often.

Hag Moth larva

COSSID MOTHS (Cossidae) show that "micros" are not always small. The **Carpenterworm Moth** *(Prionoxystus robiniae)* might even be mistaken for a sphinx moth. Its larva bores in the wood of trees. Another species in this family is the "worm" found in bottles of mescal. **SMOKY MOTHS (Zygaenidae)** include about 20 small species; one of the best-known is the **Grapeleaf Skeletonizer** *(Harrisina americana)*, named for the feeding habits of the larva. **PLUME MOTHS (Pterophoridae)** are distinctive as a family, with their "airplane" perching posture, but the more than 140 species are essentially impossible to separate in the field.

Antaeotricha schlaegeri

Antaeotricha leucillana

Ethmia longimaculella

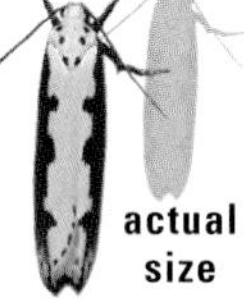

actual size

Ethmia bipunctella

Pistol Casebearer
adult, larval case

Coleophora tiliaefoliella

Coleophora spissicornis

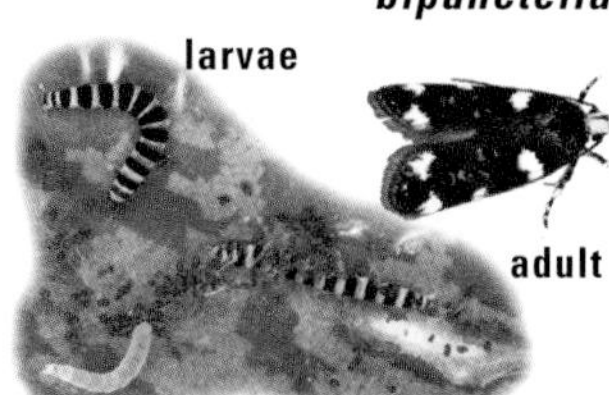

larvae

adult

Redbud Leaf-folder Moth

female

Hag Moth

Skiff Moth
2 larvae and adult

Goldenrod Gall Moth

cutaway view of larva in gall

adult

larva

Saddleback Caterpillar

Carpenterworm Moth

Grapeleaf Skeletonizer

plume moths

TORTRICID MOTHS (family Tortricidae) make up a major family of micromoths, with over 1,100 species in North America. Adults vary from small to tiny, but many have a distinctive shape at rest, with rounded "shoulders" and forewings flared or squared at the tip. Larvae are often out of sight: many spin bits of webbing on foliage to hide themselves, or roll up leaves or tie leaves together, while others may bore inside stems, fruits, or buds. The **Spruce Budworm** *(Choristoneura fumiferana),* common in northern spruce forests, sometimes becomes abundant; populations of songbirds may peak during these budworm outbreaks. The **Fruit-tree Leafroller** *(Archips argyrospila)* feeds on many plants but may become a pest on pear or apple trees. ***Archips purpurana*** also feeds on apple foliage as well as many other plants. The **Oak Leaftier** *(Croesia semipurpurana)* is common, its larvae feeding on oak, tying the edges of the leaves with silk. Three common leafrollers with distinct patterns as adults are ***Sparganothis distincta, Sparganothis pulcherrimana,*** and ***Argyrotaenia quercifoliana.*** The **Codling Moth** *(Cydia pomonella)* is drab as an adult, but larvae are infamous for tunneling into apple, pear, and other fruits. "Worms" in apples are often this species.

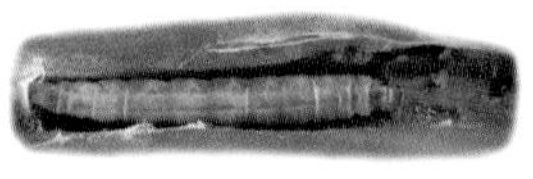

larva of a tortricid
(Sparganothis distincta)

PYRALID MOTHS (Pyralidae) include many small species with varied lifestyles, but the ones that draw the most attention are those few that have an impact on human food supplies. The **Mediterranean Flour Moth** *(Anagasta kuehniella)* is now found worldwide, its larvae feeding on stored grains, flour, and dried plant material, sometimes becoming a serious pest. The same is true for the **Indian-meal Moth** *(Plodia interpunctella)* and the **Meal Moth** *(Pyralis farinalis).* Adapting to the artificially abundant food sources provided by humans, these moths thrive indoors at all seasons and also visit lights outdoors during the warmer months. Pyralid moths of the **subfamily Chrysauginae** are odd little creatures with tufts of scales on their legs; they often stand high in an alert-looking posture.

CRAMBID MOTHS (Crambidae) often have been included among the pyralids and are similarly small and diverse. Species of ***Crambus*** are pale, with narrow wings that they roll up at rest, giving them a sticklike appearance. Their larvae feed mostly on grasses, but larvae of ***Crambus laqueatellus*** feed on mosses. Another grass feeder is the **Sod Webworm** *(Pediasia trisecta);* its larvae spin webs on turf, sometimes killing patches of lawn. The **European Corn Borer** *(Ostrinia nubilalis),* introduced in the early 1900s, has become a major agricultural pest. Its larvae tunnel in corn stalks and also feed on beans, potatoes, and other plants. The **Grape Leaf Folder** *(Desmia funeralis)* is one of several small black-and-white moths. Larvae fold over the edges of grape leaves and secure them with strands of silk, feeding out of sight inside the fold. Common in the east is the **Basswood Leafroller Moth** *(Pantographa limata).* Its larvae roll leaves of oaks and elms as well as basswood. The beautiful ***Conchylodes ovulalis*** is common in the southeast, its larvae feeding on plants in the aster family.

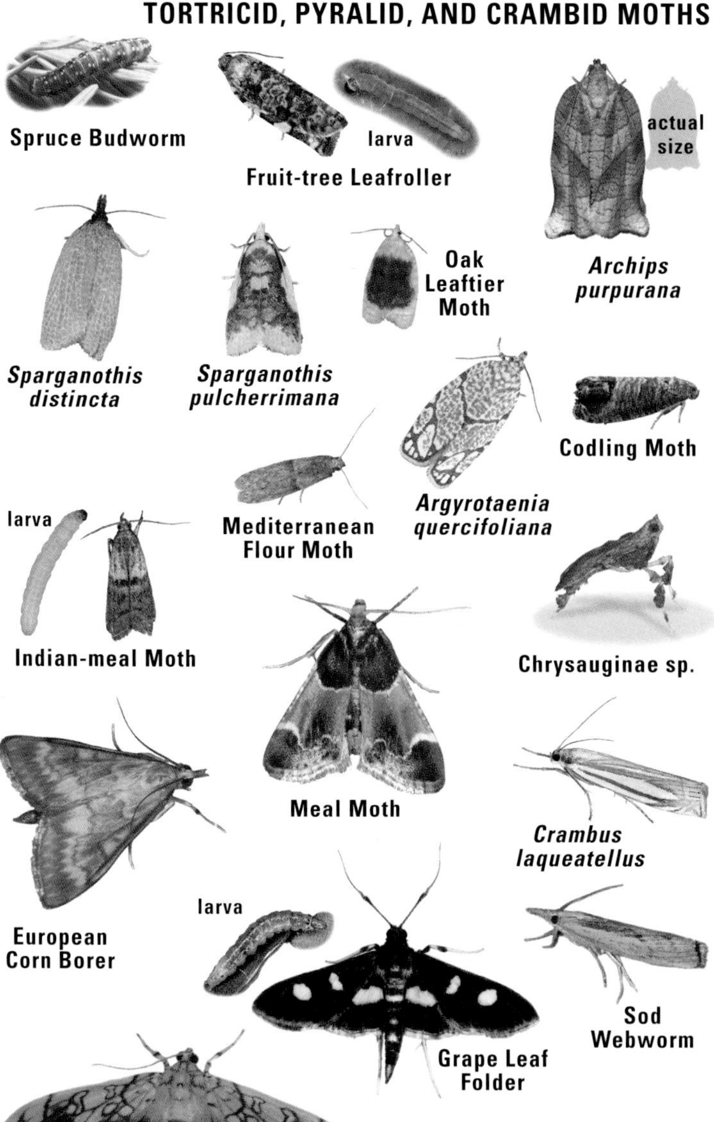

Spruce Budworm

Fruit-tree Leafroller

Archips purpurana

Oak Leaftier Moth

Sparganothis distincta

Sparganothis pulcherrimana

Codling Moth

Argyrotaenia quercifoliana

Mediterranean Flour Moth

Indian-meal Moth

Chrysauginae sp.

Meal Moth

Crambus laqueatellus

European Corn Borer

Sod Webworm

Grape Leaf Folder

Basswood Leafroller Moth

Conchylodes ovulalis

FLIES

(order Diptera) may be our most underappreciated insects. Most human experience is with blood-sucking, sweat-sipping, garbage-infesting, and picnic-harassing species. Meanwhile, we ignore the majority that are directly or indirectly beneficial to us. And, for sheer weirdness of lifestyle, several families of flies are at the top of the list. Not all flies go by the name "fly." Mosquitoes, gnats, greenheads, no-see-ums, midges, bots, and keds are all dipterans. Because of the aerial prowess of insects in general, a great many nonflies bear "fly" as part of the name, such as butterfly, firefly, stonefly, and mayfly. Notice that the names are spelled as all one word. True flies are described by two words, such as mydas fly, robber fly, and soldier fly. Flies are an enormously diverse lot, with approximately 17,000 species recorded on this continent alone. They have exploited nearly every habitat and food source imaginable, including our own flesh and blood. Flies can also be incredibly abundant, with many species forming mating swarms, some visible from long distances and containing thousands of individuals (see midges, p. 280).

Fly, bee, or wasp? Many flies are easily confused with other kinds of insects. A large number are convincing mimics of bees or wasps, both in physical appearance and behavior. However, all flies have only one pair of wings (a few parasitic flies are wingless), whereas bees and wasps have two pairs. In flies, the second pair of wings has, through evolution, become reduced to knoblike organs called halteres, which act like gyroscopes to stabilize the insect as it maneuvers through the air. Flies usually have large prominent eyes and short antennae. Bees and wasps usually have less conspicuous eyes and long, obvious antennae. Flies have sponging, or piercing-sucking, mouthparts, though some species have no mouths at all. Bees and wasps have chewing mouthparts, though these are often complemented by a "tongue" consisting of several different mouthparts that are especially elongated. While some bees and wasps can hover, most hovering insects you encounter will be flies, especially syrphid flies (pp. 298–303) and bee flies (pp. 292–295).

Metamorphosis is complete, flies passing from egg to larva (maggot) to pupa to adult. Pupae are often encapsulated in the last larval skin. Females of many species hatch their eggs internally and "larviposit" their tiny maggot offspring in the appropriate breeding medium. This reproductive strategy reaches its zenith in the louse flies (Hippoboscidae, p. 306), in which the female retains a single larva inside her body until it is ready to pupate. The infamous tsetse flies of Africa share this life cycle. In a few other flies, such as the gall midges of the genus *Miastor,* the larvae themselves are able to reproduce in a phenomenon known as paedogenesis.

Many flies are aquatic in their youth, most notably the mosquitoes and black flies, most midges, shore flies, and many crane flies, deer flies, horse flies, soldier flies, marsh flies, and some syrphid flies and dance flies. Even the pupae of several of these are fully submerged. Some flies do breed in carcasses, dung, and other unsanitary situations. Collectively they are known as "filth flies," but the name seems unnecessarily derogatory. In high densities, these flies can indeed pose a serious threat of spreading disease-causing organisms such as Salmonella. The irony is that most flies are fastidious groomers. Watch them sometime. See how they rub their front legs together, draw their hind legs over their wings, and wipe their eyes. These behaviors give them a somewhat devious persona, which may also heighten our disdain for them.

In their role as decomposers, some species figure prominently in forensic science (see blow flies, p. 304). Many flies are important pollinators of flowers. If you like chocolate, you can thank a species of Ceratopogonidae (p. 280) for pollinating the tropical cacao plant. The genus *Drosophila* (family Drosophilidae, p. 274) is important in genetic research. In 2004, the U.S. Food and Drug Administration approved the maggots of certain calliphorid flies (p. 304) as a "medical device" for situations in which these insects are the most effective way of cleaning and healing wounds after surgery. Not only do the larvae feed exclusively on decaying tissues, but they secrete substances that retard further infection and promote healing.

Unfortunately, some flies visit untold misery on humans as well, in the form of disease microbes and viral agents that they ferry from host to host. In fact, flies may be the world's deadliest organisms for precisely this reason. Malaria, yellow fever, dengue, and encephalitis are but a few of the diseases transmitted by mosquitoes. Black flies and sand flies transmit devastating diseases in the tropics. A few species can infest human flesh in their larval stage, usually exploiting open sores into which the female fly deposits her eggs. This condition is called myiasis, and while not uncommon in the tropics, is rare in North America.

Still other flies attack cereal crops, orchards, pets, and livestock, wreaking economic havoc and personal heartbreak in the process. The Screwworm *(Cochliomyia hominivorax)* is a prime example. Adult female flies lay eggs in festering wounds of livestock, where the maggots mature at the sometimes lethal expense of their hosts. Amazingly, this bane of southern U.S. ranchers is now history, owing to an ingenious plot to eradicate the fly. First discussed in theory in 1937, the release of sterile, captive-bred flies yielded promising results in field trials in 1954. By 1958, millions of flies were being reared in captivity at a facility in Sebring, Florida. The pupae were bombarded with radiation from cobalt 60, rendering them sterile. Dropping boxes of the impotent flies from airplanes led wild females to lay infertile eggs. In 1962, an even larger fly factory was built in Texas, and the resulting production effectively rendered the Screwworm extinct throughout the southern U.S. The program continues in a joint U.S.-Mexico effort to control the fly south of the border.

Flies are certainly not without their natural enemies, from spiders, to wasps, to other flies, even fungi. Flies that you find dead and seemingly adhered to a twig, leaf, or other surface are probably the victims of *Entomophthora muscae.* The fungus actually affects the behavior of the fly, causing it to crawl to an exposed, elevated perch, ensuring it will die in a location where the spores can rain down and infect new victims.

The **House Fly** *(Musca domestica)*, a member of the **family Muscidae** (p. 306), is known to transmit the microorganisms responsible for dysentery, anthrax, typhoid fever, and other diseases. Attention to sanitation is the best way to prevent infestations. In addition, try keeping them at bay by attaching clear plastic bags, half full of water, outside doors and windows. The constant motion of the water interferes with the insect's vision.

POMACE FLIES (family Drosophilidae) are commonly called fruit flies since they are usually seen hovering around fruit on your countertop. True fruit flies are in the family Tephritidae (p. 314). Despite their pesky nature, they are very important in genetics. Thomas Hunt Morgan discovered that pomace flies have large chromosomes in their salivary glands. Individual genes are more easily observed, and mapped, than in any other laboratory-friendly species. The short life cycle and quick turnover of generations also aid genetic research enormously, as visible differences in various traits are almost immediately observable. Mapping of their genome has revealed that we humans share over 20 percent of our DNA with ***Drosophila melanogaster,*** the common species in the lab and in the wild.

DARK-WINGED FUNGUS GNATS (Sciaridae) are commonly encountered indoors, larvae developing in soil of overwatered houseplants. These are the gnats that always seem to drown in the soap dish. Adults live only a few days anyway. There are 170 species in North America, most of which feed as larvae on fungi. These larvae sometimes migrate en masse over lawns and sidewalks, creating quite a spectacle. **SCUTTLE FLIES (Phoridae)** are tiny, resembling pomace flies. Common species run in a jerky fashion, fly a short distance, and run again. They lead all manner of lives in the larval stage, some species being scavengers, others herbivores, others predators or parasites. They are highly diverse, and only a fraction have been described by scientists. At last count there were over 370 species recorded for North America. **MOTH FLIES (Psychodidae)** include the small hairy insects found perched on the bathroom wall or in the kitchen basin. Outdoors, their favorite haunts are moist, shady habitats. Larvae of some species feed on the gunk in the drain trap, hence the presence of adults in the sink. Some species can reach pestiferous densities at sewage treatment facilities. Also included in this family are "sand flies," biting insects known for transmission of leishmaniasis in the tropics.

MARCH FLIES and LOVE BUGS (Bibionidae) are often abundant springtime insects, noted for their large mating swarms. The sexes are dimorphic, males having bulbous eyes that take up most of the head. They are generally small, stout-bodied insects. Following mating, females burrow into the soil and lay eggs. As larvae, march flies are sometimes injurious to the roots of turf and crops. There are nearly 80 species north of Mexico. The genus ***Bibio*** includes more than 50 species in North America. ***Plecia nearctica,*** the notorious "Love Bug," is a road hazard in the southeast, especially in Florida, where swarms overwhelm the windshields of innocent motorists. Male and female often fly in tandem. They have two generations, one in spring, one in fall. One other *Plecia* species occurs in Newfoundland.

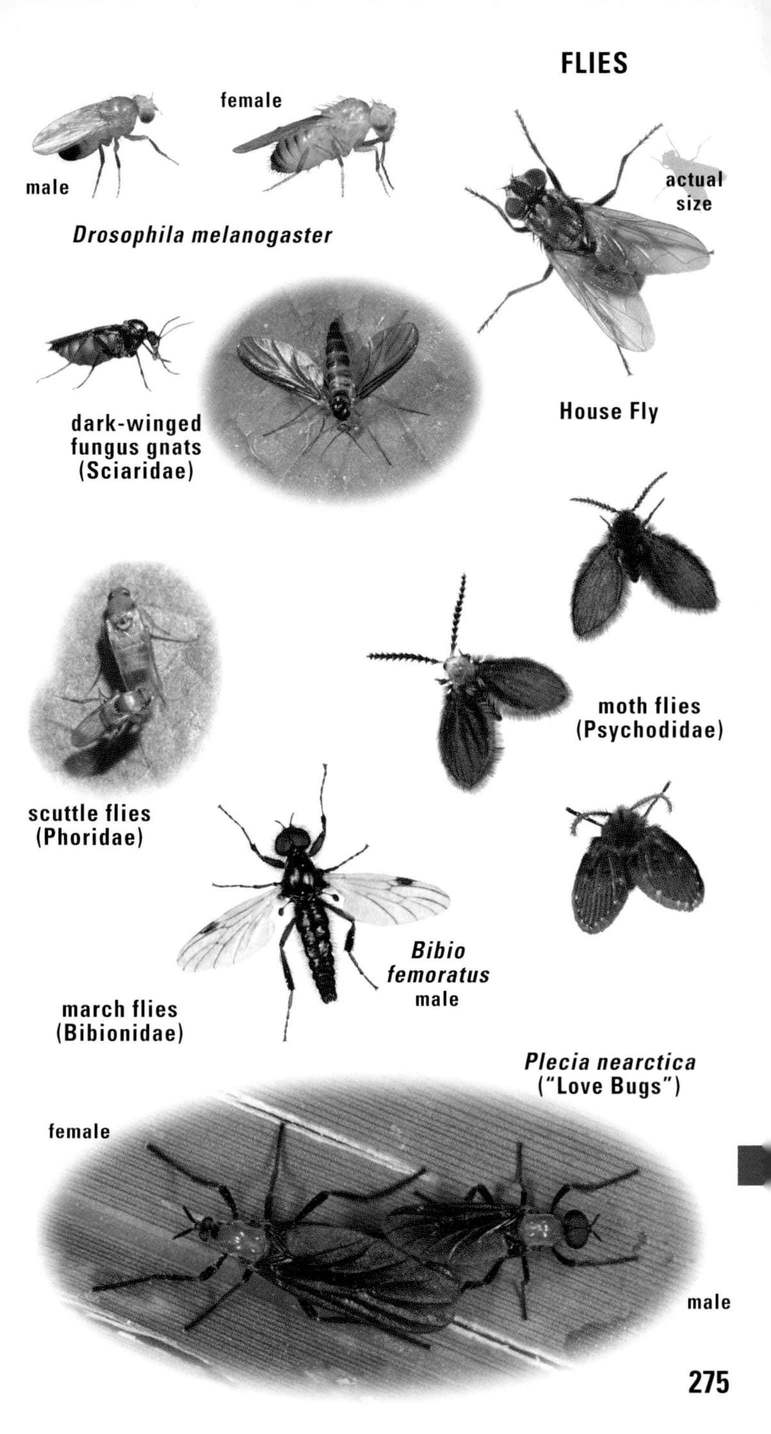

Drosophila melanogaster

House Fly

dark-winged fungus gnats (Sciaridae)

moth flies (Psychodidae)

scuttle flies (Phoridae)

Bibio femoratus male

march flies (Bibionidae)

Plecia nearctica ("Love Bugs")

CRANE FLIES

(family Tipulidae) are often mistaken for giant mosquitoes. "Daddy long-legs" and "gully nippers" are other colloquial monikers for these gangly, spooky, but harmless insects. Often abundant and extremely diverse, they are impossible for anyone but an expert to identify beyond the family level. Much of the taxonomy relies on subtle differences in wing venation to separate the 64 North American genera. Many scientists have now elevated some subfamilies to family level, yielding the Limoniidae, Cylindrotomidae, and Pediciidae in addition to Tipulidae. Individual crane flies often have fewer than six legs, shedding them easily under stress. Other insects frequently confused with crane flies include hangingflies (Bittacidae, p. 38). While they frequently suspend themselves from vegetation in the manner of hangingflies, crane flies have only one pair of wings, not two pairs, and they lack the long spurs at the tip of each tibia (a leg segment) that hangingflies have. In Canada and the northern U.S., small wingless crane flies are active on snow in the dead of winter. The largest land invertebrate in the Arctic is one such species. Adults of many winged species are attracted to lights at night. Terrestrial crane fly larvae are known as leatherjackets, but many species thrive in wet situations or are totally aquatic. Larvae of most species feed on fungi, decaying organic matter, or the roots of sod grasses and cereal crops. Some crane fly species have predatory larvae. In turn, most types of larvae are a major source of food for various long-billed sandpipers and related shorebirds, especially in Arctic tundra but also in other wetland situations. Larvae that manage to escape predation usually overwinter (though there may be more than one generation per year in some species), pupating the following spring, mostly on or in the soil. The life cycle varies from six weeks up to four years in some Arctic species. There are roughly 1,600 species of Tipulidae north of Mexico.

The genus ***Nephrotoma*** includes 40 species in North America. In Europe and Australia these are known as tiger crane flies for the often contrasting colors of the adults. Larvae of several species occur in dry soil in lawns, pastures, and rangeland habitats. Members of the genus ***Ctenophora*** are the wood-boring tipulids, represented by four species in North America. Larvae feed on fungal mycelia in decaying hardwood logs. Adults are more robust than most crane flies, with proportionately shorter legs. Females possess a spikelike ovipositor for laying eggs in dead wood. ***Holorusia rubiginosa*** is a giant western species with a wingspan that can reach 70 mm (nearly 3 inches). Its size, and the white "racing stripes" on the thorax, help identify it. Larvae live in moist soils, moss, leaf mold, and sediments at the edges of streams.

Tipula is an enormous genus with over 480 species in North America. Larvae may be aquatic or terrestrial, depending on the species. Aquatic larvae burrow in the bottom of streams or still waters, scavenging decaying organic matter. The subgenus ***Platytipula*** includes 11 species, collectively widespread. Subgenus ***Lunatipula*** includes 160 species north of Mexico.

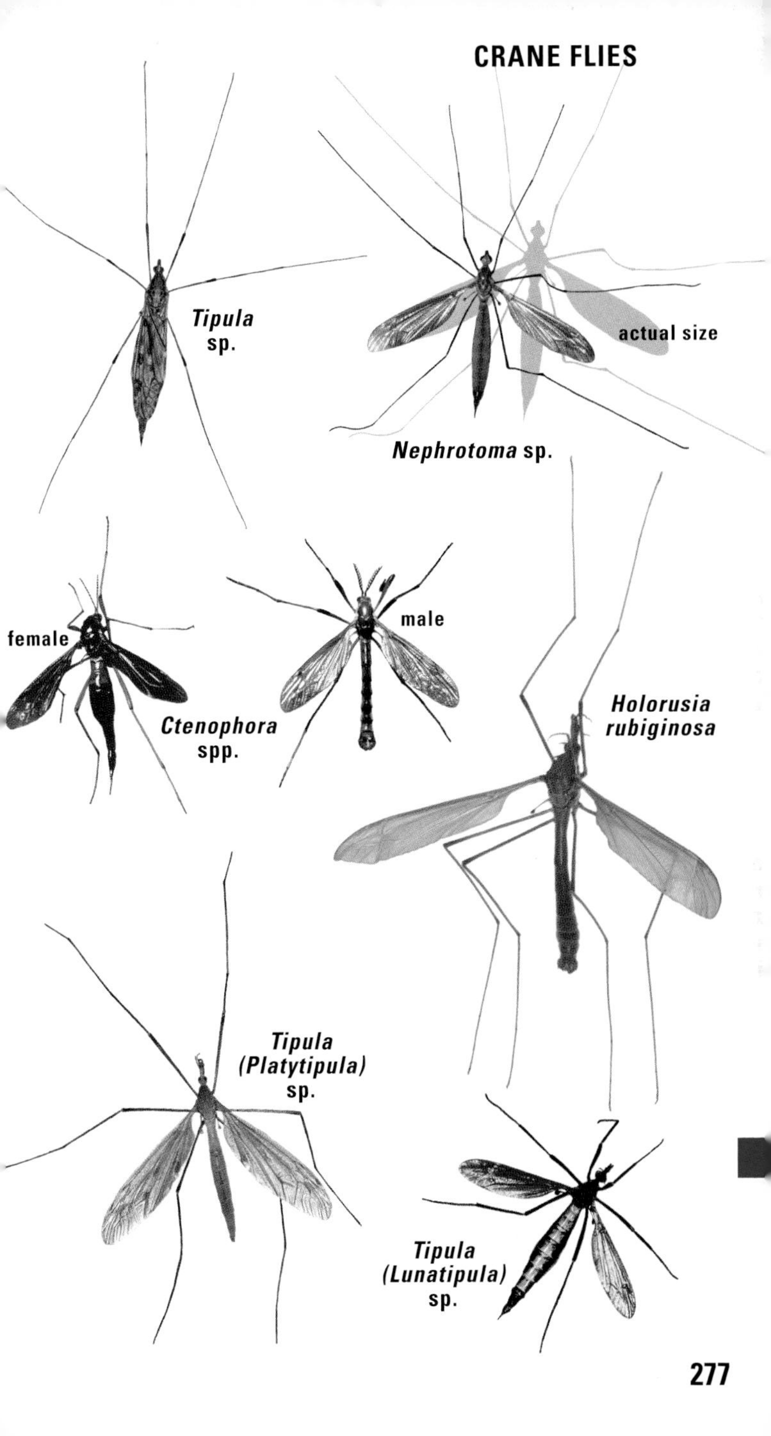
Tipula
sp.
actual size
Nephrotoma sp.
male
female
Ctenophora
spp.
Holorusia
rubiginosa
Tipula
(Platytipula)
sp.
Tipula
(Lunatipula)
sp.

The first three species below are among the typical crane flies **(family Tipulidae)**, introduced on the preceding page. Two other families of similar flies are illustrated here also.

The genus ***Limonia*** includes about 125 species in North America. These are small to medium-sized insects. Some species have a beaklike face. They can sometimes be found sipping nectar from flowers. Larvae of different species are found in a wide range of aquatic and terrestrial habitats, ranging from flowing fresh water to the margins of ponds and lakes, even marine intertidal zones on the Atlantic and Pacific coasts. They usually burrow in bottom sediments, sprawl on the surface, or cling to submerged rocks and logs where they graze on macroscopic algae. Terrestrial species are found in rotting wood, mosses and liverworts, and fungi. One subgenus includes leaf miners.

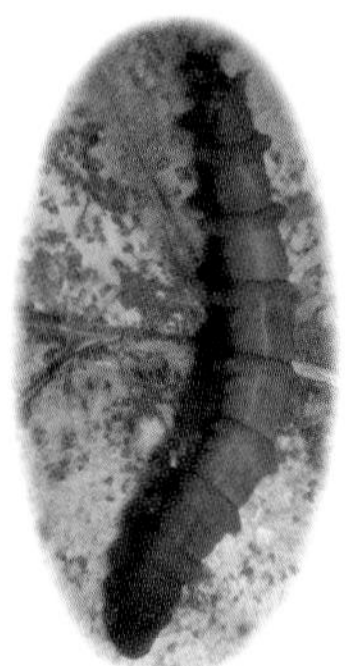

crane fly larva
(Tipula)

Hexatoma albitarsis is one of 35 species in its genus and ranges over most of the eastern U.S. Collectively, the genus is widespread. Larvae are aquatic, usually living in the shallows of sandy or gravelly streams. A few species occupy standing bodies of water. All are predatory, feeding on oligochaetes and the larvae of other flies they find in bottom sediments and detritus. They actively burrow, and they sprawl on the bottom. The subgenus ***Eriocera*** (not illustrated) includes small adult crane flies seen bouncing up and down on the underside of leaves in moist wooded habitats.

The genus ***Gnophomyia*** includes three species north of Mexico, collectively widespread, especially east of the Rockies. Larvae feed on fungal mycelia in rotting hardwood logs and stumps.

PHANTOM CRANE FLIES (Ptychopteridae) include the bizarre black-and-white ***Bittacomorpha*** species. The flared "ankles" allow these flies to ride light breezes, dancing in and out of the shadows. Who says bell-bottoms are out of fashion? The genus ***Ptychoptera*** includes 10 species in North America, collectively widespread. The larvae of both genera live in saturated mud at the edges of streams, ponds, and swamps, where they feed on decaying organic matter. Larvae can be over 40 mm long, owing to a long respiratory tube extending from the posterior.

WINTER CRANE FLIES (Trichoceridae) are generally smaller than true crane flies, distinguished by the presence of simple eyes (ocelli) at the crown of the head (true crane flies lack ocelli). Winter crane flies are commonly seen on warm, sunny days in late fall or early spring in eastern North America. In the Pacific northwest, adults are sometimes seen in the dead of winter. Look for swarms of males at the entrance of caves or mineshafts, in dark cellars, and hollow trees. Larvae are scavengers on a variety of plant matter and are especially common in compost, manure, and fungi. There are 29 species, 27 in the genus ***Trichocera*** alone. Two other genera are found north of Mexico.

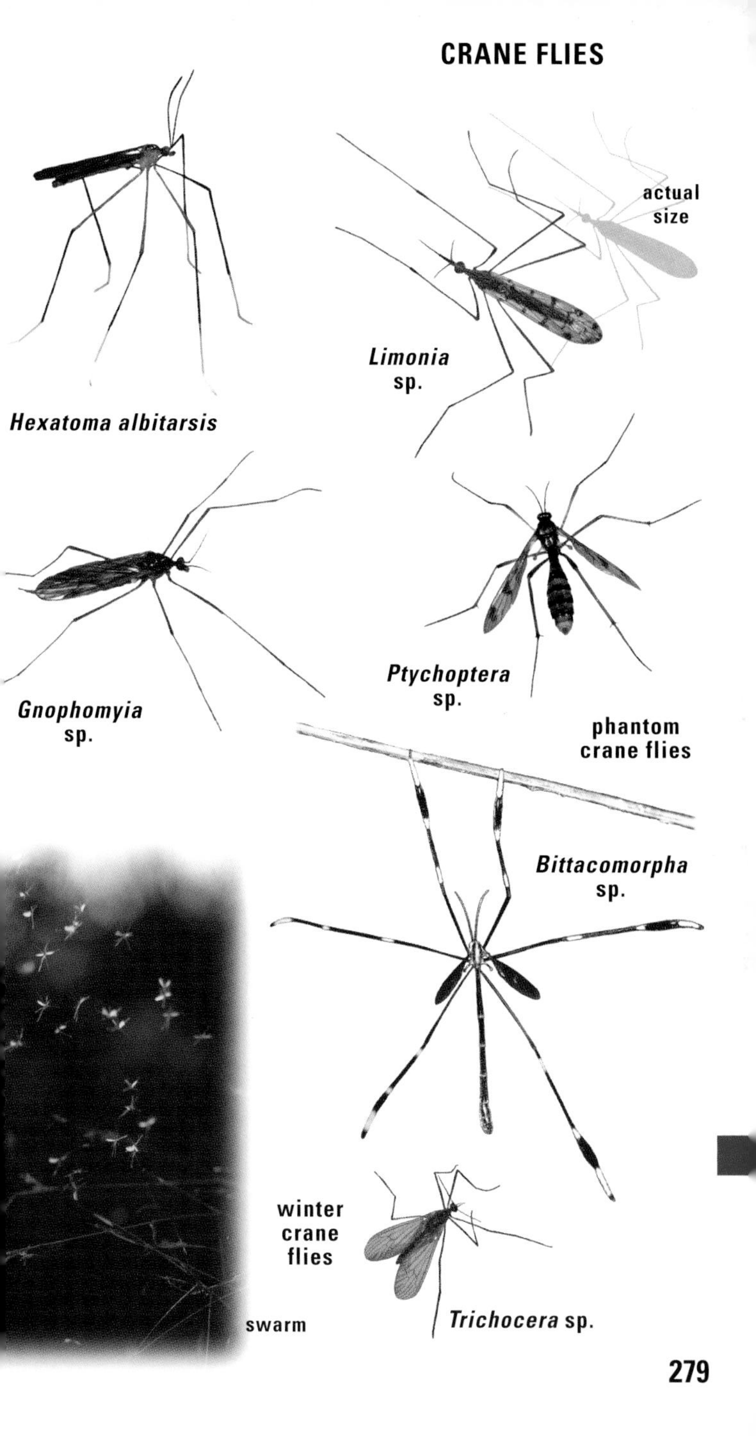
actual
size
Limonia
sp.
Hexatoma albitarsis
Gnophomyia
sp.
Ptychoptera
sp.
phantom
crane flies
Bittacomorpha
sp.
winter
crane
flies
swarm
Trichocera sp.

MIDGES (family Chironomidae) can be among the most abundant of insects and are frequently mistaken for mosquitoes. Their resting posture, with front legs held up and forward, identifies them (mosquitoes often lift the *hind* legs at rest). Many midges form large swarms, often at dusk, usually males only, over lakes and above tall objects. The feathery antennae of males are sensitive to the high-pitched wingbeats of females. At night both sexes often flock to lights. Most larval midges live in aquatic habitats, even at the bottom of deep lakes (including down to 1,000 feet in Lake Superior). They typically feed on organic debris or microscopic plants and animals. Many species construct a tubular retreat of fine particles glued together with mucus secretions. Some are colored bright red by hemoglobin, earning the name "bloodworms." Larvae are a major food source for fish. Larval development may last up to three years, while adults are on the wing for at most a few weeks. The genus ***Polypedilum*** includes over 40 species in North America. ***Chironomus*** is widespread in North America, with more than 25 species. ***Orthocladius*** is represented in North America by at least 29 species. ***Paratanytarsus*** is a widespread genus in Canada, with five species there. In the tribe **Pentaneurini,** larvae are predatory.

BLACK FLIES (Simuliidae) are also known as buffalo gnats for their humpbacked appearance. They are the bane of outdoorsmen and ranchers, as they bite humans and livestock relentlessly. Eight of nine Arctic species can get along without blood meals, females being able to lay viable eggs without the added protein. Larvae are aquatic, anchoring themselves to rocks in fast-flowing rivers and streams, using featherlike mouthparts to strain microorganisms from swift current. There are roughly 165 species north of Mexico, most in northern latitudes. Certain tropical species are the vectors of "river blindness." ***Prosimulium*** includes about 42 species in North America, while ***Simulium*** has about 66 species.

BITING MIDGES or "NO-SEE-UMS" (Ceratopogonidae), also known as punkies, have a big bite for their size. Thankfully, many species in this family suck the blood of other insects, such as dragonflies, mantids, crane flies, even mosquitoes. The banks of rivers, shores of lakes, and the seashore are areas frequented by these insects. Simply moving a few yards away from the place they are annoying you can provide instant relief. Larvae are aquatic. This is a large family with 580 species in North America.

DIXID MIDGES (Dixidae) are most often encountered in the aquatic larval stage. Larvae of many species assume an inverted U-shaped posture, the midsection of their body resting out of the water, with both ends submerged. Young larvae filter microorganisms from the water with brushlike mouthparts. Males swarm at dusk along the banks of rivers or ponds.

PHANTOM MIDGES (Chaoboridae) are named for their transparent larvae. The hook-shaped appearance of the face is created by the prehensile antennae, used to grab prey. The silvery ovals visible through the cuticle are the swim bladders, organs that help regulate the buoyancy of the insect. Instead of surfacing for air, larvae obtain oxygen from the water via anal gills. They live only in still waters. Adults resemble chironomid midges.

MIDGES AND BLACK FLIES

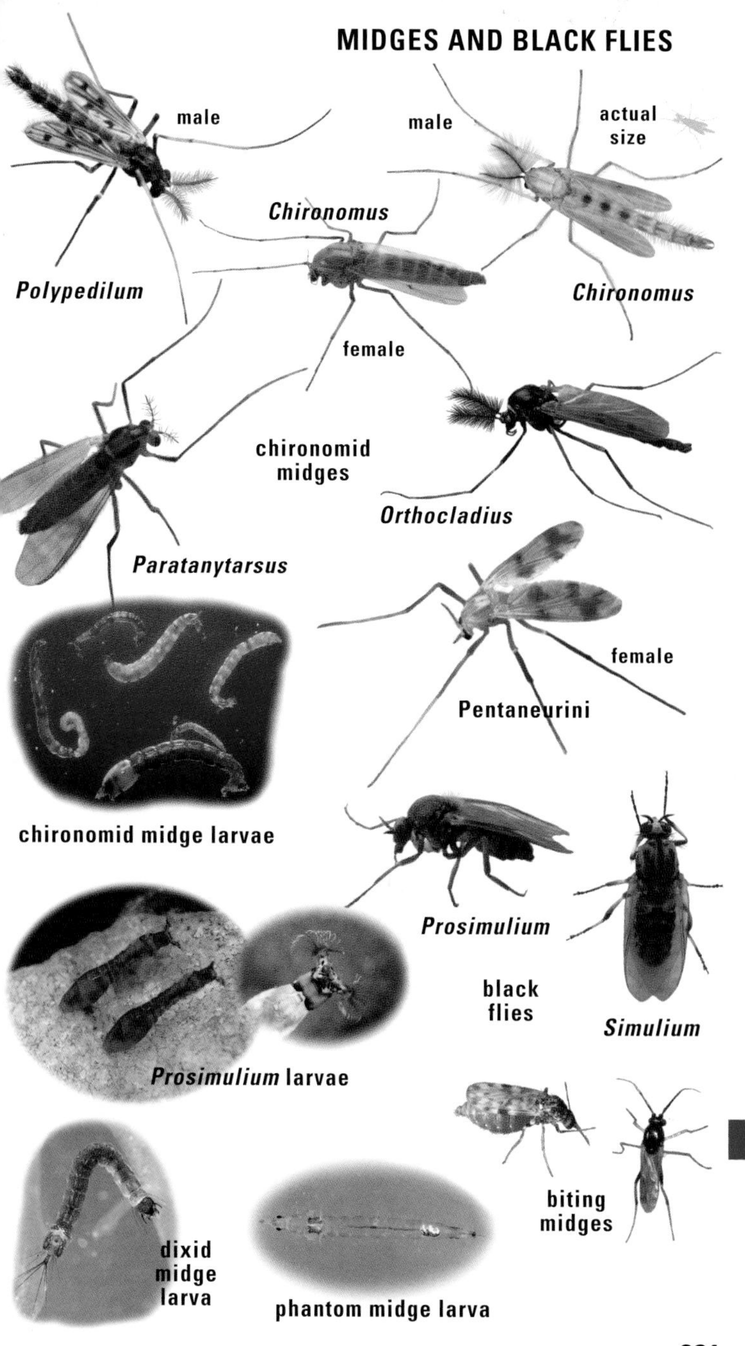

MOSQUITOES

(family Culicidae) may be the world's most dangerous insects given the lethal diseases they can transmit. Only females feed on blood (and not all species do so), usually requiring the protein to produce eggs. Humans are generally the host of last resort. Most mosquitoes feed on other animals, especially birds, and a few specialize on amphibians. Eggs are laid singly, or in floating "rafts" on the water's surface. All known species are aquatic as larvae ("wrigglers") and pupae ("tumblers"). An absurdly small amount of water can breed mosquitoes, so dispose of any water-collecting objects, clear gutters, and drain any other shallow puddles. Adults are actually quite attractive insects, many species being covered in iridescent scales. Males have amazing plumose antennae, which aid in detecting the species-specific whine of a female's wingbeats.

Anopheles is a genus with 14 species in North America. Adult females rest in a headstanding position. Larvae are readily identified by their horizontal posture at the water's surface. The **Common Malaria Mosquito** *(Anopheles quadrimaculatus)* of the eastern U.S. is occasionally found inside houses. ***Anopheles punctipennis*** ranges across the entire U.S. Larvae thrive in cool water. Adults are among the first mosquitoes of spring and are on the wing throughout the summer in the northern states.

Toxorhynchites rutilus ranges along the Gulf and Atlantic coasts. Adults do not bite but use their *bent proboscis* to sip nectar. Larvae are predatory on the larvae of other mosquitoes. ***Ochlerotatus*** was recently split from the genus *Aedes* and now contains the majority of species (59) once considered part of *Aedes*. The **Eastern Tree-hole Mosquito** *(Ochlerotatus triseriatus)*, found east of the Rockies, is a pesky biter in wooded areas.

Our eight species of ***Aedes*** are attractive mosquitoes, often patterned in black and white. The **Inland Floodwater Mosquito** *(Aedes vexans)* occurs over the entire continent. It readily feeds on humans and domestic animals, and it may be active at night or it may bite by day in shady areas. Winters are spent in the egg stage. The **Asian Tiger Mosquito** *(Aedes albopictus)* was introduced to Houston in 1985 in piles of discarded tires imported from Japan. It also showed up in Los Angeles in 2001 in water-laden containers of "lucky bamboo" shoots from Taiwan. This species is a known vector of dengue fever in the tropics. Its current established range is most of the eastern U.S. south of the Great Lakes and New Jersey.

The four species of the genus ***Wyeomyia*** (not illustrated) breed only inside pitcher plants, overwintering as larvae frozen solid in the icy reservoir of the plant. ***Psorophora*** is a genus with 13 representatives in eastern and central North America. These can be real giants. Larvae prey on other insects, including other mosquito larvae, in temporary pools.

The genus ***Culex*** includes 27 species in North America. The **Northern House Mosquito** *(Culex pipiens)* is the main vector of West Nile viral encephalitis in the northeast U.S. Birds and equines are the primary hosts of the virus, but effective vaccinations are available to treat horses. The adult females hibernate, often in human structures.

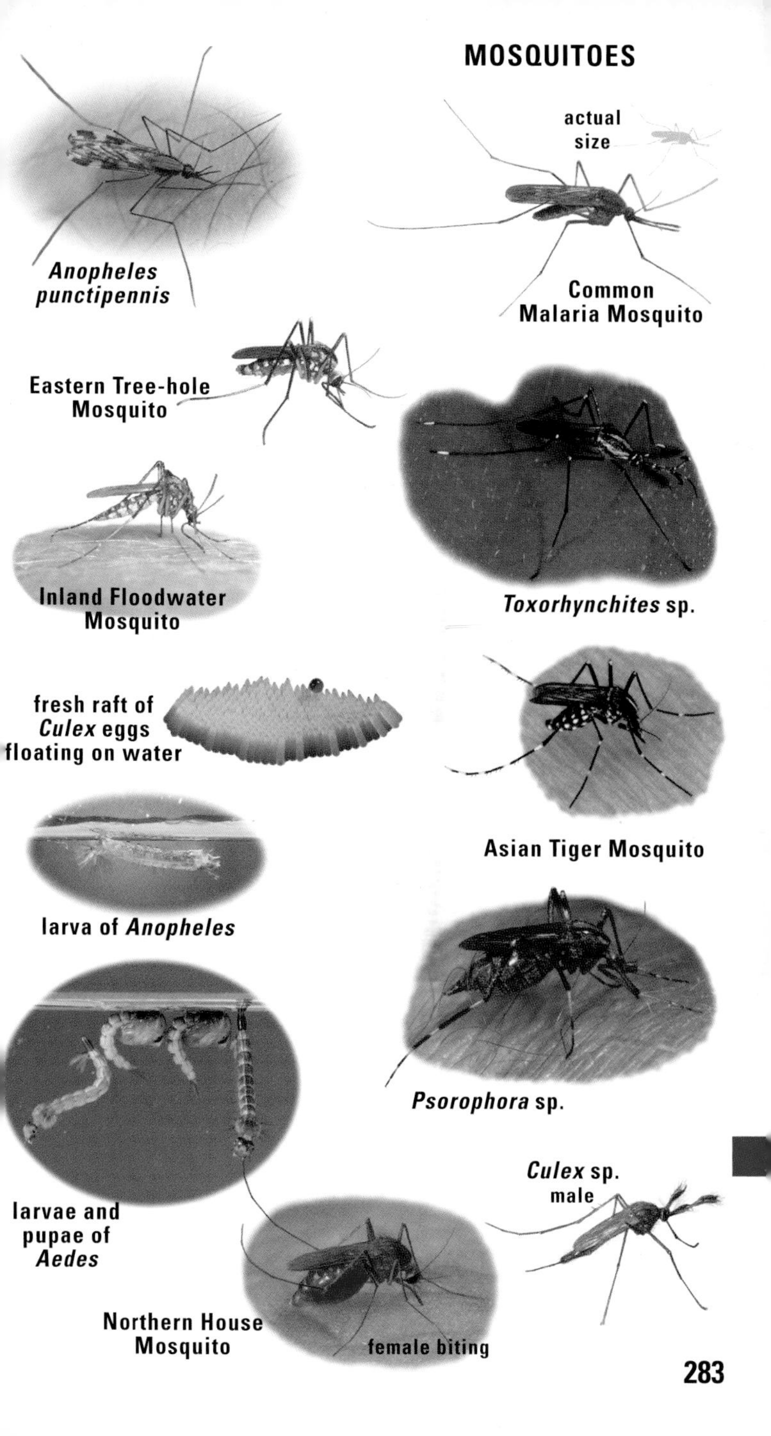

actual size
Anopheles punctipennis
Common Malaria Mosquito
Eastern Tree-hole Mosquito
Inland Floodwater Mosquito
Toxorhynchites sp.
fresh raft of Culex eggs floating on water
Asian Tiger Mosquito
larva of Anopheles
Psorophora sp.
Culex sp. male
larvae and pupae of Aedes
Northern House Mosquito
female biting

DEER FLIES AND HORSE FLIES

(family Tabanidae) are also known as greenheads or bulldogs. We have about 350 species in about 30 genera in North America, and many have dazzling, rainbow-striped or spotted eyes. Allow one to mesmerize you and you will get a painful bite. Females of most species are blood feeders, using knifelike mouthparts to slice and dice, then slurping up the flow with a spongelike proboscis. Wounds continue to ooze after the flies depart, sometimes leading to significant blood loss in afflicted livestock. Tabanids are vectors of tularemia and anaplasmosis in North America, but the risk is slight. Only the females bite; males sip flower nectar and hover in open areas or over prominent objects to intercept females. Both sexes have large eyes, but the eyes meet on top of the head on males and are separated on females. Typically the eggs are laid in dense masses on leaves above the water. Larvae live mostly in water or in wet mud and are predatory on other insects and worms.

The typical **deer flies** *(Chrysops)* include about 100 species in North America. They average smaller than the horse flies, and they usually have blotched wings and often spotted eyes, helping to identify them. Widespread over much of the continent, they are most common in wet wooded areas of the east and northeast. Deer flies often circle your scalp, so wearing a hat helps deter them.

Goniops chrysocoma can be fairly common in moist forests of the eastern U.S., north to southern Ontario. Females lay eggs beneath a leaf and guard them, buzzing loudly at potential threats. Larvae drop to the ground and seek prey among leaf litter and other debris.

Members of the genus ***Hybomitra*** are greenheads, with 60 species on our continent, most diverse and abundant in Canada and the northern U.S. These are midsized (10–20 mm), usually with striped eyes. Still, they may be confused with many *Tabanus* species. ***Hybomitra lasiophthalma*** is abundant across Canada, ranging south to Utah in the west and over most of the eastern U.S. except the extreme southeast.

horse fly larva

The genus ***Tabanus*** includes the true horse flies, with 101 North American species. Their eyes may be either solidly colored or striped. Females wait on vegetation and pursue large, dark moving objects, including motor vehicles. The **Black Horse Fly** *(Tabanus atratus)*, which can be over an inch long, occurs from southern Ontario and Quebec south to Texas and Florida. This may be the "blue tail fly" of folk song fame. ***Tabanus americanus*** is another species often referred to as greenhead, owing to its solidly bright green eyes. It is widespread and common over much of North America, except for parts of the far west. ***Tabanus trimaculatus*** occurs over much of the eastern two-thirds of the continent. The **Striped Horse Fly** *(Tabanus lineola)*, part of a complex of very similar species, is widespread and common in the eastern U.S.

DEER FLIES AND HORSE FLIES

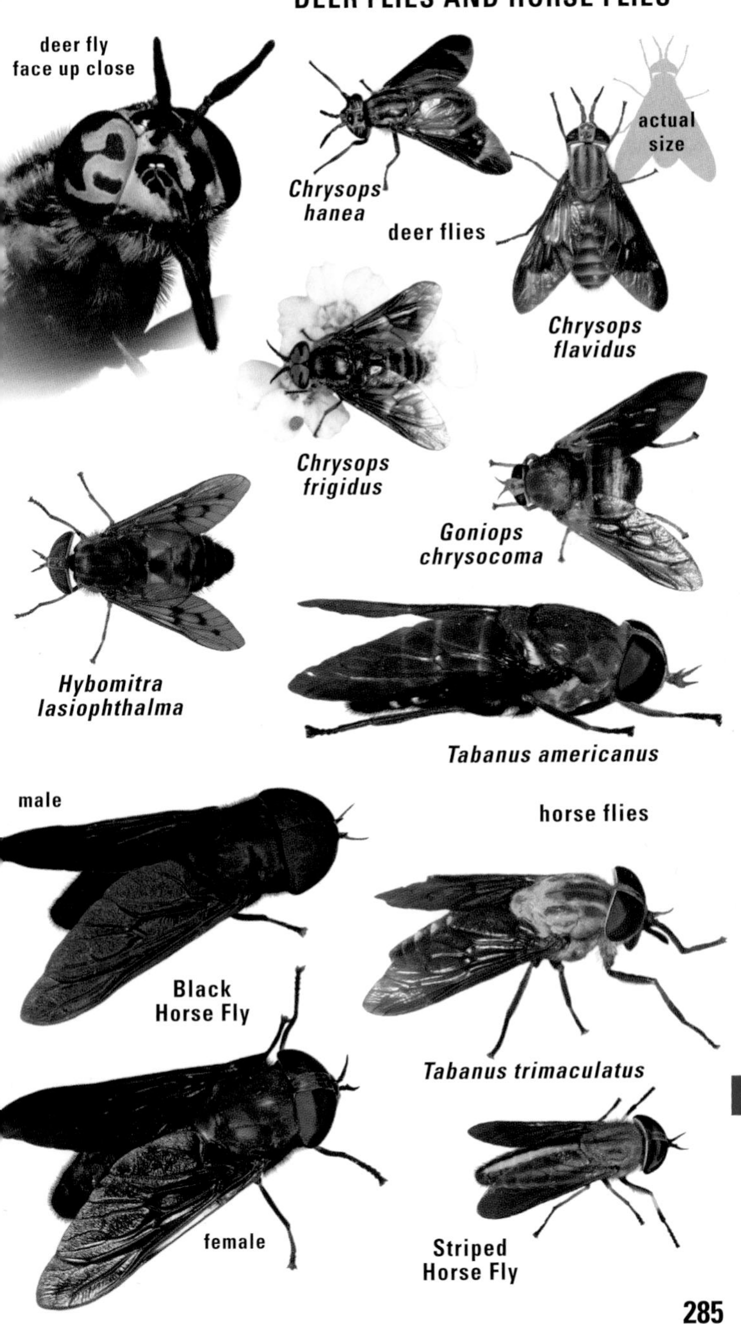

SOLDIER FLIES, SNIPE FLIES, AND OTHERS

SOLDIER FLIES (family Stratiomyidae) are striking wasp and bee mimics. Many resemble flower flies (pp. 298–303) but are usually less common. The Y-shaped or T-shaped position of the antennae of many species is distinctive. The genus ***Hermetia*** includes 12 species found in the south and southwest, but the **Black Soldier Fly** *(Hermetia illucens)* is more widespread in the east. The genus ***Sargus*** includes six wide-ranging species in North America. ***Ptecticus*** includes four widespread species in North America. Look for them around decaying piles of lawn clippings, compost heaps, and other decaying vegetation. ***Stratiomys*** is a widespread genus in North America with at least 26 species. The genus ***Odontomyia*** has at least 31 species in North America, many of them recognizable by their bold black and green pattern. ***Nemotelus*** are small flies, 12 species of which occur north of Mexico.

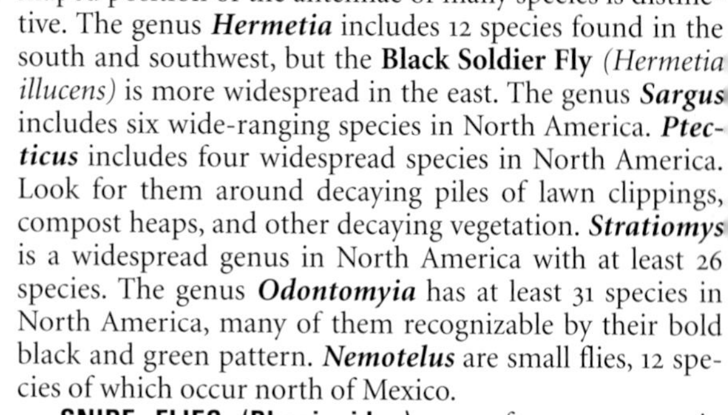

aquatic larva of soldier fly

SNIPE FLIES (Rhagionidae) are often common in wooded areas. Both larvae and adults are predatory; larvae mostly live in decaying wood, while adults are usually seen on foliage. Adults of a few western species may bite humans. The genus ***Chrysopilus*** includes 32 species in North America, collectively widespread. ***Chrysopilus ornatus*** is a common spring woodland species in eastern North America, west to Kansas and Manitoba. The **Golden-backed Snipe Fly** *(Chrysopilus thoracicus)* is also common in eastern forests. ***Rhagio*** is represented by 25 species north of Mexico; ***Rhagio mystaceus*** is typical.

STILETTO FLIES (Therevidae) are usually grayish, with a pointed abdomen, and may resemble robber flies (next page). There are about 140 species in North America, but they are generally uncommon.

SMALL-HEADED FLIES (Acroceridae), small and humpbacked, are generally rare, but some species visit flowers and others may be found lurking near the haunts of spiders. Their larvae develop as internal parasites of spiders. Most of the more than 50 species live in the southwest.

XYLOMYIDS (Xylomyidae) mimic wasps such as sawflies (p. 320) and ichneumons (p. 326). Look for them on foliage along forest edges. Larvae live under bark on logs and are predatory or eat decaying organic matter. There are about 11 species in North America. ***Macroceromys americana*** (formerly *Xylomya americana)* is one common species.

FUNGUS GNATS (Mycetophilidae) are small flies found mostly in damp forests. They are recognized by the enlarged coxae (basal segment of each leg) and the long spines at the tip of each tibia ("shin" segment). Look for adults by day around hollow logs, eroded cavities under tree roots, and similar situations. Larvae of most feed in fungi, including mushrooms and woody bracket fungi. A few species are bioluminescent, like those on the ceiling of Waitomo Cave in New Zealand. The North American ***Orfelia (Platyura) fultoni*** (not shown) glows from beneath overhanging banks of shady brooks. Both species belong to the subfamily Keroplatinae, all of which secrete slimy salivary threads to snare prey.

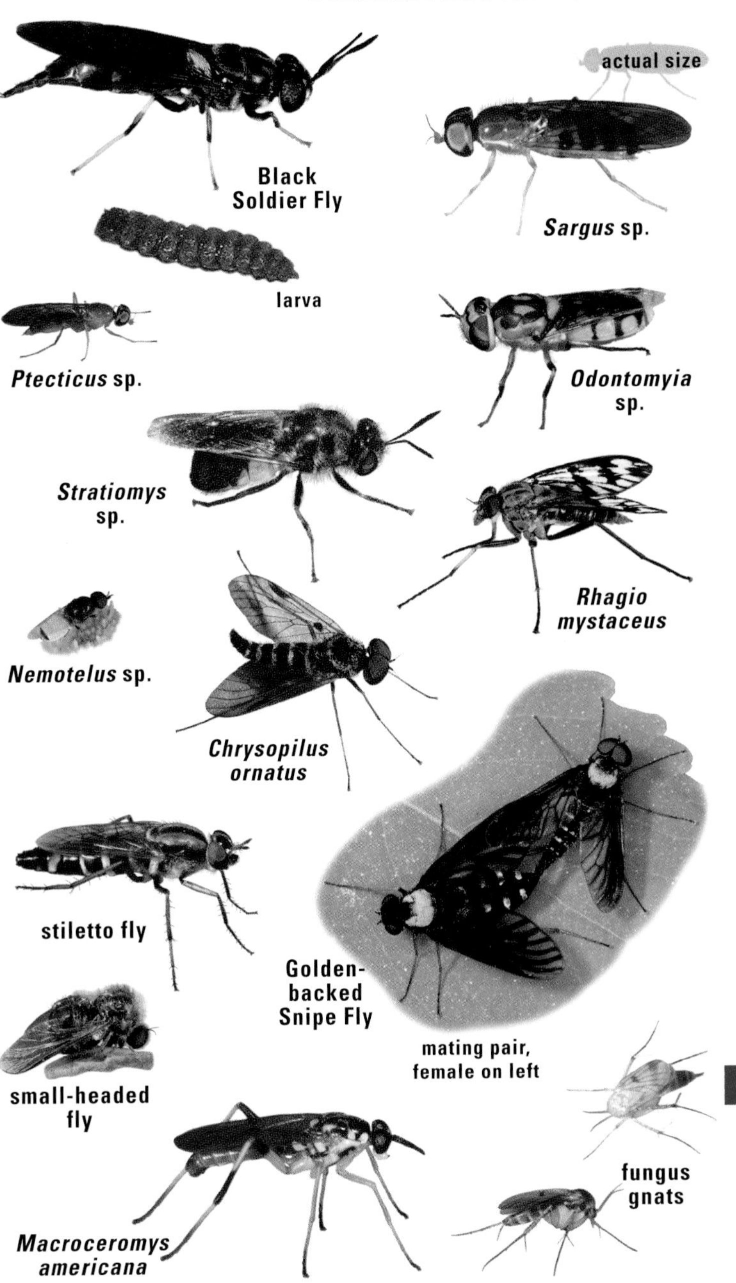
actual size
Black Soldier Fly
larva
Sargus sp.
Ptecticus sp.
Odontomyia sp.
Stratiomys sp.
Rhagio mystaceus
Nemotelus sp.
Chrysopilus ornatus
stiletto fly
Golden-backed Snipe Fly
mating pair, female on left
small-headed fly
fungus gnats
Macroceromys americana

ROBBER FLIES

(family Asilidae) are to other insects what falcons are to other birds: swift predators on the wing. They vary greatly in size (3–50 mm long) and shape (long and slender to compact and robust). They are most diverse in the western and central U.S. All can be recognized by the usually "bearded" face and the concave top of the head between their eyes. Many act like flycatchers, perching on logs, foliage, twigs, or the ground, cocking their heads at insects passing overhead, and dashing out to apprehend a victim. In the few cases in which their life histories are known, robber fly larvae are external parasites of beetle grubs or larvae of other insects. There are close to 1,000 species in North America.

The genus ***Laphria*** includes remarkable bumble bee mimics. There are at least 63 species on the continent. Look for them in forests where the canopy opens or there is dappled sunlight. The eastern ***Laphria grossa*** often alights on stumps of oak and elm. It is known to feed on Japanese beetles. The smaller ***Laphria thoracica*** is quite common in the northeast U.S. It sometimes stakes out apiaries and feeds on the honey bees. The color pattern varies greatly among individuals. ***Laphria saffrana*** probably mimics queens of the Southern Yellowjacket (p. 356). Look for it in pine forests and coastal plains of the southeast. ***Laphria astur*** ranges in the northwest. ***Laphria index*** is probably a complex of several species, transcontinental in Canada and the northern U.S., marked with metallic copper or orange on the thorax and abdomen. ***Laphria canis*** probably represents another species complex occurring mostly in the east.

Efferia is an enormous genus with 100 species in North America, many of them very common, especially in the west. Watch for them in dry, open habitats such as fields, thickets, and forest edges, where they land on posts and other exposed perches. Males have conspicuous bulbous claspers on the tip of the abdomen and are marked with dazzling silver bands. Females use their swordlike ovipositor to deposit eggs in dead flower heads, cracks in the soil, and other crevices.

Promachus is a widespread genus in North America, with 21 species here, including ***Promachus hinei*** of the central states. You might hear this species before you see it, as it departs its perch with a loud, buzzing flight, quickly alighting again nearby, usually on a vertical branch or twig.

Members of the genus ***Mallophorina*** are recognized by their bright green eyes. The 15 species in North America prey almost exclusively on bees and wasps. ***Mallophora*** includes 14 species in North America, bee mimics that prey mostly on honey bees, bumble bees, and wasps.

The genus ***Proctacanthus*** includes 18 North American species. They are large flies, capable of taking prey the size of grasshoppers. They are seen mostly in grasslands and on shrubs along the margins of woods and swamps, but a few species hang out on riverbanks.

Asilus is a large genus, with 74 species in North America. ***Asilus sericeus*** ranges over much of the eastern U.S. This is one of the first robber flies of spring in the southern part of its range, but adults are on the wing through August. Look for them resting on twigs, weeds, or open ground.

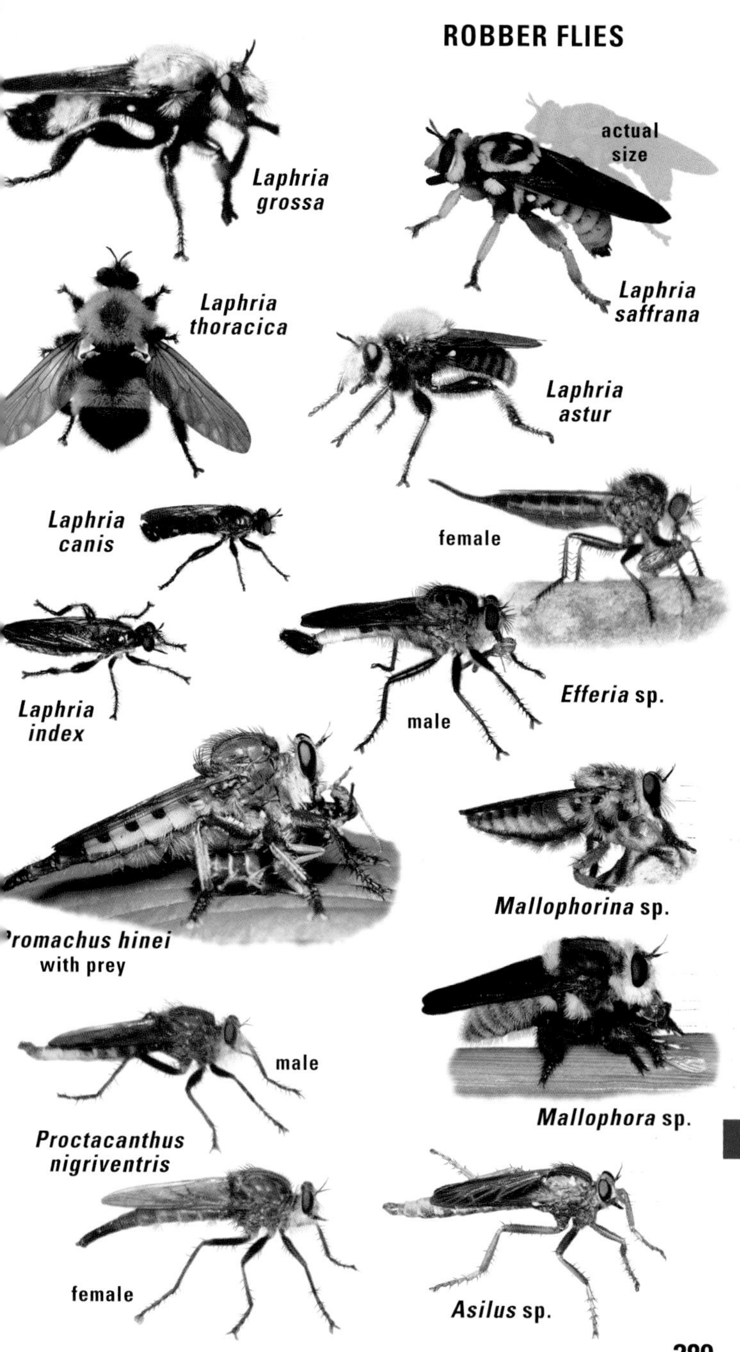
actual
size
Laphria grossa
Laphria saffrana
Laphria thoracica
Laphria astur
Laphria canis
female
Laphria index
Efferia sp.
male
Mallophorina sp.
Promachus hinei with prey
Mallophora sp.
male
Proctacanthus nigriventris
female
Asilus sp.

The genus ***Diogmites*** includes 26 North American species, collectively widespread. Sometimes called hanging thieves, they habitually suspend themselves by their front or middle legs, manipulating prey with the remaining legs. Their prey includes wasps; some individuals have even been observed feeding on smaller species of *Diogmites.* Look for them in old fields, marshy areas, and other open habitats, where they perch close to the ground on low vegetation.

The genus ***Lasiopogon*** is represented by at least 118 species in North America, found over most of the continent as far north as Alaska but especially diverse in the west. Half that number have yet to be formally described by entomologists. These little insects fly early (often March to May) and, at least in the northeast, are the first robber flies to be seen in spring. Look for them in riparian habitats, especially near the banks of streams lined with deciduous trees. Adults usually perch on bare ground or on rocks and logs. Larvae develop in the soil, and the life cycle in at least one species takes up to two years.

Stichopogon includes 10 small species, collectively found across much of North America. They frequent the banks of streams and rivers, resting on rocks, gravel, or open sand, often in association with tiger beetles. ***Laphystia*** includes 30 North American species, collectively widespread in distribution. They are often found on sandy beaches adjacent to streams, lakes, and the sea, as well as on alkali flats in the desert. The three species of ***Holcocephala*** are relatively tiny and are sometimes called gnat-ogres. Look for them sitting on the very tips of weeds and grass blades along forest edges. Adults fly mostly in June and July.

The genus ***Dioctria*** includes eight species in North America, all western with the exception of one Florida species. These are relatively small robber flies that are weak fliers, hunting from perches on leaves and low undergrowth in forest openings or along woodland margins. ***Cyrtopogon*** numbers 70 species in North America, mostly western. They seem to favor coniferous forests in mountainous regions. Watch for them on logs in sunlit openings. Many species have elaborate courtship rituals, males of some species sporting ornamented legs that they display to the female. All his pomp and circumstance still rarely wins her favor.

FLOWER-LOVING FLIES (family Apioceridae) are found in arid areas of western North America. Essentially nothing is known of the biology of this family, though they have been observed nectaring at flowers, and females are known to lay their eggs singly, in sand. The infamous **Delhi Sands Flower-loving Fly** *(Rhaphiomidas terminatus abdominalis)* is a highly endangered species known only from a few isolated dune systems in southern California. Its federal listing may or may not spell economic doom for certain small towns, depending on whom one listens to. Its genus includes eight other species. The genus ***Apiocera*** includes 20 species, collectively ranging from British Columbia to Texas. ***Apiocera chrysolasia*** occurs only in southern California and northern Mexico.

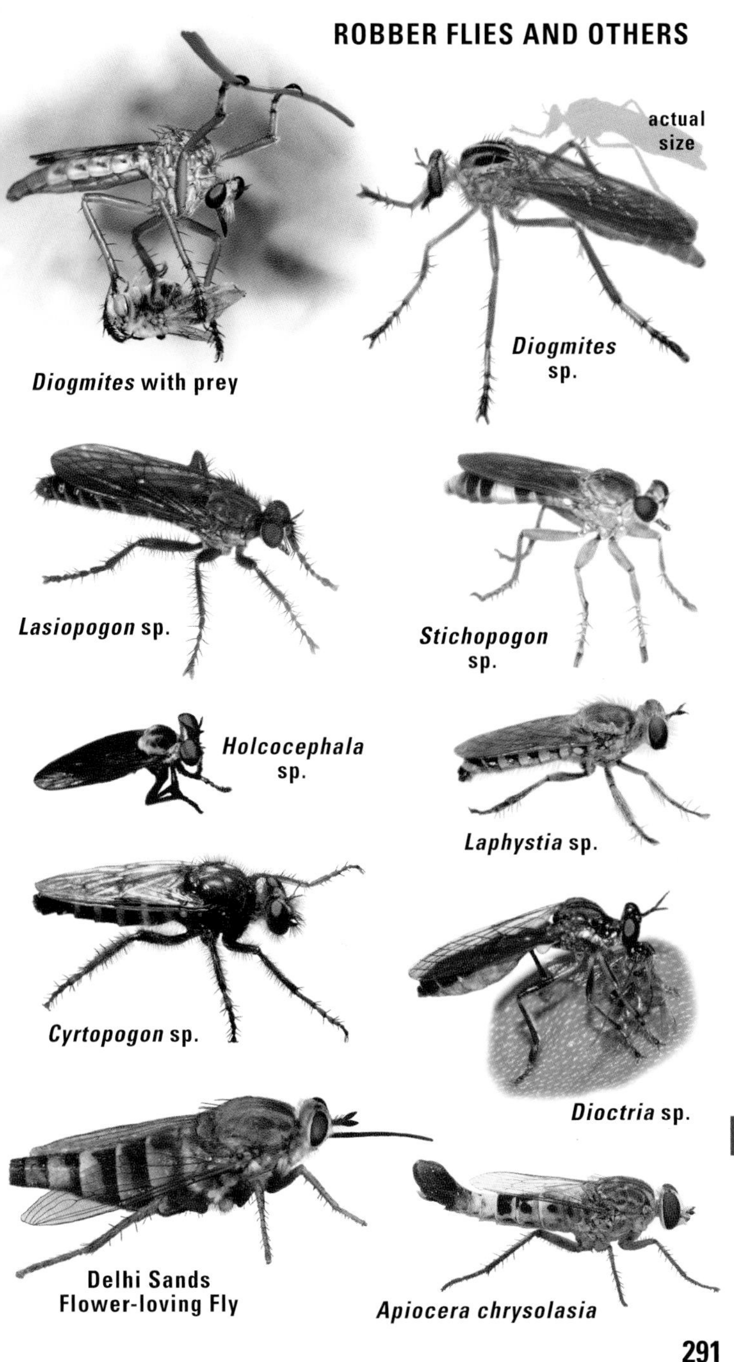

Diogmites with prey

Diogmites sp.

Lasiopogon sp.

Stichopogon sp.

Holcocephala sp.

Laphystia sp.

Cyrtopogon sp.

Dioctria sp.

Delhi Sands Flower-loving Fly

Apiocera chrysolasia

BEE FLIES

(family Bombyliidae) might be mistaken for large, fuzzy mosquitoes because of the long proboscis on some species, but the beak is for sipping nectar, not blood. Most species are covered in long hairs and/or delicate, often silvery scales that wear off rapidly as the insect ages. Many species have ornate dark patterns on the wings. Their hovering flight helps reinforce their resemblance to bees, hence the family's common name. Larvae are parasitic on the immature stages of other insects.

The genus ***Anthrax*** includes 37 species, collectively found across the continent. Most species are parasites of solitary wasps that build mud nests or nest in preexisting cavities. Adults can often be observed hovering around potential nest sites of their hosts, such as under the eaves of structures and the trunks of dead trees. Females are known as bombers for their practice of lobbing eggs down the tunnel or cavity of a host nest while hovering in front of it. Both sexes may hover around people also, even alighting on them, but they are harmless. ***Anthrax albofasciatus*** and ***Anthrax georgicus*** are two typical species of the southeastern U.S. ***Anthrax irroratus*** is commonly seen in the southwest. ***Anthrax analis*** (not illustrated) is a parasite of larval tiger beetles (p. 130).

The **Tiger Bee Fly** *(Xenox tigrinus,* formerly *Anthrax tigrinus)* is a parasite of large carpenter bees *(Xylocopa,* p. 344). Most common in eastern North America, it occurs as far southwest as Arizona.

The genus ***Exoprosopa*** is represented by 43 species, collectively occurring throughout North America. Adults are large, various species ranging from 6–22 mm for the entire genus. Larvae are apparently parasitic on other parasites of bee, wasp, or beetle larvae living in the soil. Most species are found in arid regions, where the adults are seen nectaring on a variety of flowers, especially those in the aster family.

The genus ***Poecilanthrax*** includes 33 North American species. Adults range from 8–14 mm in length. Larvae are parasites of a variety of soil-dwelling cutworms and armyworms in the owlet moth family (Noctuidae). Adults can be found commonly on flowers, especially on rabbitbrush in the arid west. ***Poecilanthrax californicus*** and ***Poecilanthrax poecilogaster*** are two species that range widely in the western U.S. ***Poecilanthrax nigripennis*** occurs in the southeastern U.S.

The 25 species of ***Hemipenthes*** collectively range across North America. Look for them at composite flowers (flowers in the aster family). Larvae are parasitic on tachinid fly larvae, ichneumon wasp larvae, and other parasitic species that have caterpillar hosts. ***Hemipenthes sinuosa*** is a common species. ***Hemipenthes jaennickeana*** is often seen in the southwest.

The genus ***Villa*** includes 37 species in North America. Look for them nectaring on sumac and various composite flowers. Most species are parasitic on moth caterpillars. ***Villa lateralis*** is one typical example of this genus. The 27 North American species of ***Chrysanthrax*** are most diverse in the southwestern U.S. Look for adults on low-growing composite flowers. The larvae are parasitic on "white grubs," the larvae of *Phyllophaga* scarab beetles.

Anthrax albofasciatus

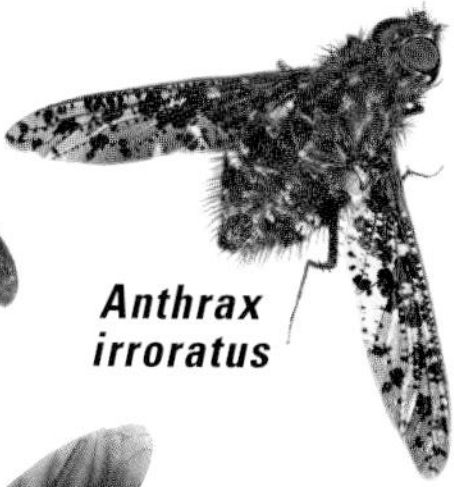

Anthrax irroratus

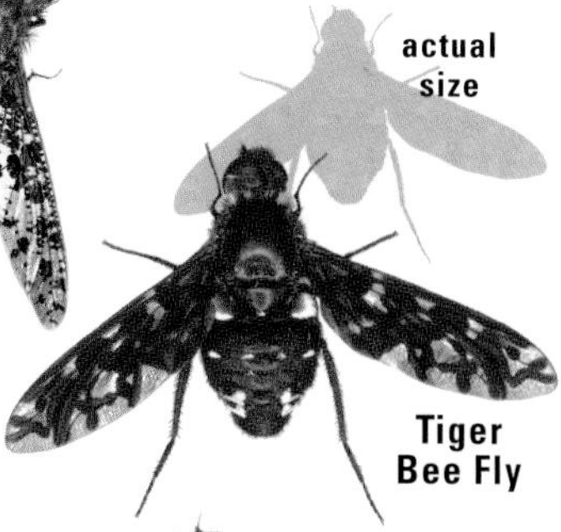

actual size

Tiger Bee Fly

Anthrax georgicus

Exoprosopa spp. (2 examples)

Poecilanthrax nigripennis

Poecilanthrax californicus

Poecilanthrax poecilogaster

Hemipenthes jaennickeana

Hemipenthes sinuosa

Villa lateralis

Chrysanthrax sp.

Chrysanthrax cypris

BEE FLIES AND MYDAS FLIES

Fuzzy flies that often hover in front of flowers, the bee flies **(family Bombylidae)** are introduced on the previous page.

The genus ***Bombylius*** includes 35 species in North America. They measure up to 15 mm in length. Not all species have been studied yet, but those species for which the life cycle is known are parasitic on various solitary ground-nesting bees. Adults are attracted to a variety of flowers. ***Bombylius major,*** one of the most conspicuous and distinctive bee flies, is abundant and widespread. Look for it nectaring at lilac and plum blossoms. ***Bombylius pygmaeus*** is another widespread species.

Systoechus is a genus with five species in North America. Adults average 6–9 mm long. Members of this genus are predators of grasshopper egg pods in the larval stage. ***Systoechus vulgaris*** is one widespread species. ***Systoechus solitus*** occurs in the southeast. Adults of all species are fast fliers, expert at hovering.

In the southwestern U.S., members of the genus ***Anastoechus*** can be found in large numbers nectaring on flowers in late summer and fall. As larvae, these flies are predators on the eggs of grasshoppers.

Lepidophora is a genus with three North American species, collectively widespread. Adults have a distinctive "broken-backed" appearance and a bouncing flight. Look for them nectaring on the blooms of coneflowers *(Rudbeckia)*. Larvae live in the nests of wasps. The genus ***Toxophora*** is represented by seven species in North America. They are small, ranging in length from 6–12 mm. Larvae are parasitic on wasps. Look for the adults nectaring on a variety of composite flowers. ***Toxophora vasta*** is one typical species.

The genus ***Phthiria*** includes 31 species, most diverse in the western U.S. but collectively widespread. They are small, ranging from 2–8 mm long. The larvae are predatory on grasshopper egg pods. Adults nectar on many kinds of composites but have been collected in large numbers on sunflowers in the southwest.

Members of the genus ***Systropus,*** less beelike than most bee flies, are convincing mimics of thread-waisted wasps. As larvae they are parasitic on caterpillars of moths in the family Limacodidae.

MYDAS FLIES (family Mydidae) include our largest flies (some tropical species are up to 60 mm, or more than 2 inches long). Despite their often imposing size, these flies are harmless to humans. They may be mistaken for wasps, especially spider wasps (p. 352). Larvae live mostly in decaying wood and apparently are predatory on beetle grubs. There are about 50 species in North America, most diverse in the southwest. The genus ***Mydas*** includes 17 species, widespread in the eastern U.S. and southern Ontario, also the Pacific states. ***Mydas tibialis*** is a common eastern species, while ***Mydas clavatus*** is common and generally distributed.

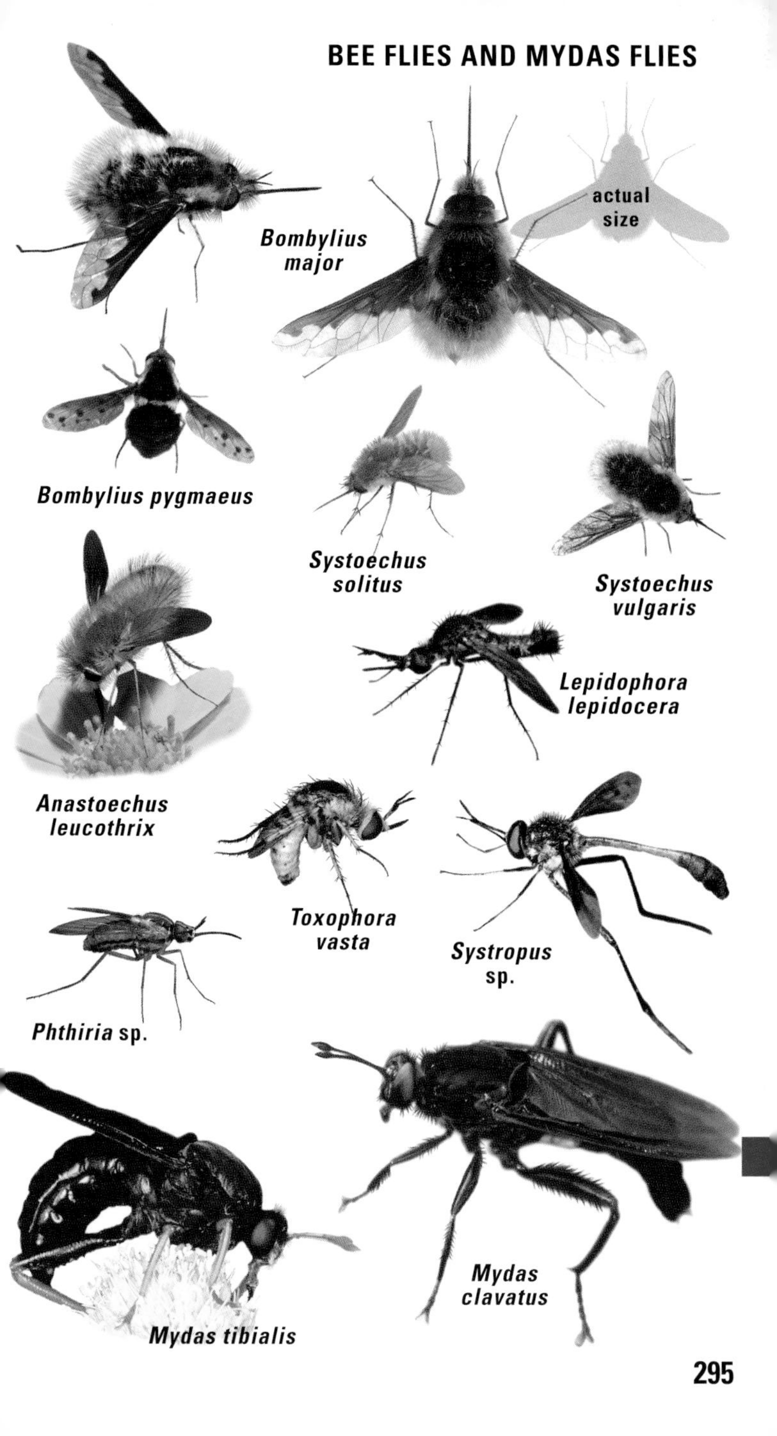

Bombylius major

Bombylius pygmaeus

Systoechus solitus

Systoechus vulgaris

Anastoechus leucothrix

Lepidophora lepidocera

Toxophora vasta

Systropus sp.

Phthiria sp.

Mydas clavatus

Mydas tibialis

DANCE FLIES (family Empididae) are *highly* diverse, with over 760 species on our continent. Many resemble tiny robber flies. Nearly all are predatory, mostly on other flies. Dance flies often gather in the hovering, "dancing" swarms you see backlit in forests and meadows, though many other families of flies exhibit this behavior as well. Much more unusual is the practice of "gift-giving" by males of some species, which bring a dead insect to the female for her to dine on while they mate. Males of some species of *Hilara* and *Empis* from the Pacific northwest wrap their gifts in a silken balloon. In a few devious species, males simply offer an empty balloon. Cheapskates! Look for adult dance flies on foliage, or on flowers, since many feed on nectar. Larvae of only a few of our species are known; all are predators. Some feed on the larvae or pupae of black flies. ***Rhamphomyia*** includes about 150 species in North America, many in alpine habitats. Adults of some prey mostly on mosquitoes. In the **Long-tailed Dance Fly** *(Rhamphomyia longicauda),* females are completely dependent on male food gifts for nourishment. Since males bring prey only to females they perceive to have well-developed eggs, females inflate sacs on their abdomens to make themselves appear egg-laden, whether or not they are. In the genus ***Empis,*** males of most species gather in loose swarms, enticing females with prey or other "gifts." March flies (p. 274) are a favored prey item.

LONG-LEGGED FLIES (Dolichopodidae) include the small metallic green, blue, copper, or bronze flies seen darting across leaves. Other members of this large family (1,275 species in North America) are drab and less animated. The family name translates loosely to "decorated foot," referring to males of some genera that sport flaglike patches of hairs on the front tarsi, used in courtship displays. These flies are predators, especially on mites, aphids, and smaller flies. Immature stages are poorly known, though some larvae have been taken from rotting plant matter. All are probably predatory, except some known stem borers. The genus ***Condylostylus*** includes 40 species in North America, many with dark markings on the wingtips. ***Dolichopus*** includes more than 300 North American species. Look for adults running across the surface of water, where they prey on mosquito larvae. ***Chrysosoma*** includes three species from eastern North America. ***Pelastoneurus vagans,*** one of 29 North American species in its genus, occurs over most of the continent. ***Hydrophorus chrysolagus*** is found from the northeastern U.S. to Alaska. Adults of this genus skate across the surface of lakes and other large bodies of water.

LAUXANIID FLIES (Lauxaniidae) are chiefly associated with deciduous woods, especially maple and cherry trees. Larvae graze on the surface of decaying leaves, consuming either fungal spores, microbes, or both. Look for adults on the foliage of understory plants, especially along the banks of streams, and in marshy habitats. Some species can be found commonly on flowers or at lights at night. There are more than 150 North American species. The genus ***Minettia*** is widespread in North America. ***Homoneura*** includes 49 species north of Mexico. Look for adult flies in swamps.

actual size

inflated female in flight

Long-tailed Dance Flies in courtship

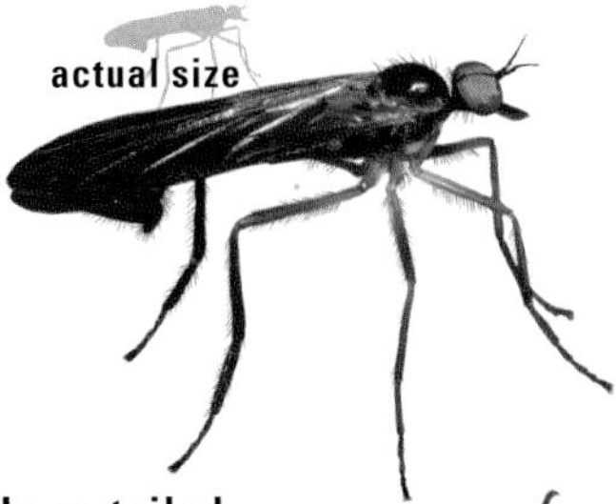

male flying with prey for female

Long-tailed Dance Fly

Empis sp.

Dolichopodidae sp.

Dolichopus latipes

Condylostylus sp.

Hydrophorus chrysolagus

Pelastoneurus vagans

Condylostylus sp.

Minettia obscura

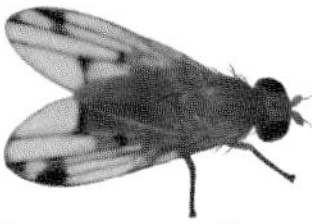

Homoneura fraterna

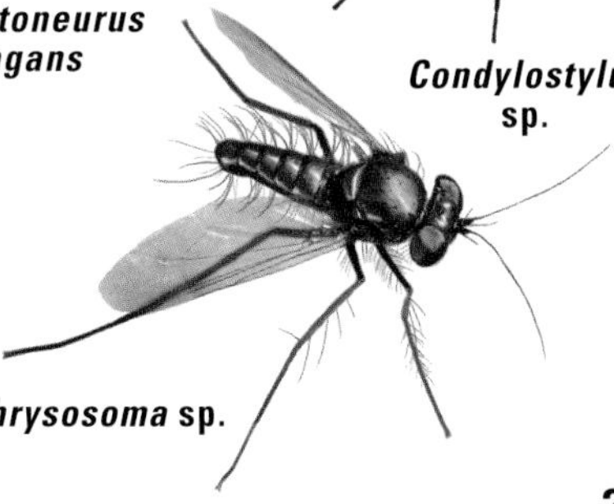

Chrysosoma sp.

FLOWER FLIES: YELLOWJACKET MIMICS

Flower flies **(family Syrphidae)** are dead ringers for bees or wasps. In Europe they are known as hover flies because some species habitually hover in midair. Despite their appearance, not only are they harmless, they are valuable pollinators of flowers. The larvae of many species prey on aphids. There are more than 870 species in North America. For easier reference, they are categorized here as "yellowjacket mimics," "bee mimics," and "wasp mimics." To see how convincing their mimicry can be, check them against their stinging counterparts, illustrated on various pages following p. 318. Yellowjackets and their kin fold their wings longitudinally at rest, and many of the mimicking flies illustrated here have the front margin of each wing darkly pigmented to imitate this fold.

The genus ***Chrysotoxum*** includes seven North American species, all similar in appearance and 9–17 mm in length. Larvae are suspected predators that have been found in ant nests, hollow trees, compost heaps, and under rocks. ***Chrysotoxum ypsilon*** is commonly seen in the southwest, while various other species are widespread across the north or in regions east of the Rockies. The genus ***Sericomyia*** includes 11 species on this continent. Larvae live as scavengers in soggy habitats. Some species in this group are large enough to be effective mimics of yellowjacket queens.

Sphecomyia is a genus represented in North America by eight species. Adults have longer antennae than most syrphids, and they behave so much like yellowjackets that they are difficult to identify in the field. ***Sphecomyia vittata*** occurs in the east. It flies in the characteristic low, slow, side-to-side manner of a yellowjacket queen investigating potential nesting sites. The wingbeat frequency of this and other syrphid flies even makes them *sound* like wasps. ***Milesia virginiensis*** is one of three North American species in its genus. Look for adults buzzing in sunny spots in open woodlands or along forest edges in eastern and central regions of the continent. This is the "news bee" of folklore, named for its occasional habit of hovering in front of a person and "giving them the news." This lovely fly also made it onto a series of insect-themed postage stamps issued in 1999.

The genus ***Spilomyia*** includes 11 species north of Mexico. They extend their front legs to imitate the longer antennae of yellowjackets. ***Spilomyia sayi,*** formerly known as *Spilomyia quadrifasciata,* is fairly common in southeastern Canada and the northeastern U.S., flying from midsummer through early fall. In addition to visiting flowers, males fly to hilltops to await females. ***Spilomyia longicornis*** is a common species in a variety of eastern habitats. Look for adults on flowers of goldenrod, aster, and wild carrot. ***Spilomyia fusca*** is a striking mimic of the Bald-faced Hornet (p. 356) in the eastern U.S. and southeastern Canada.

The genus ***Temnostoma*** is represented by 10 species in North America. Larvae are found in damp, rotten logs. ***Temnostoma balyras*** occurs from eastern and central Canada south to Mississippi. Adults mimic mason wasps (p. 358). The fly even extends and waves its front legs to mimic the longer antennae of a wasp. A highly successful species, ***Temnostoma vespiforme*** is widespread in Europe and Asia as well as in North America.

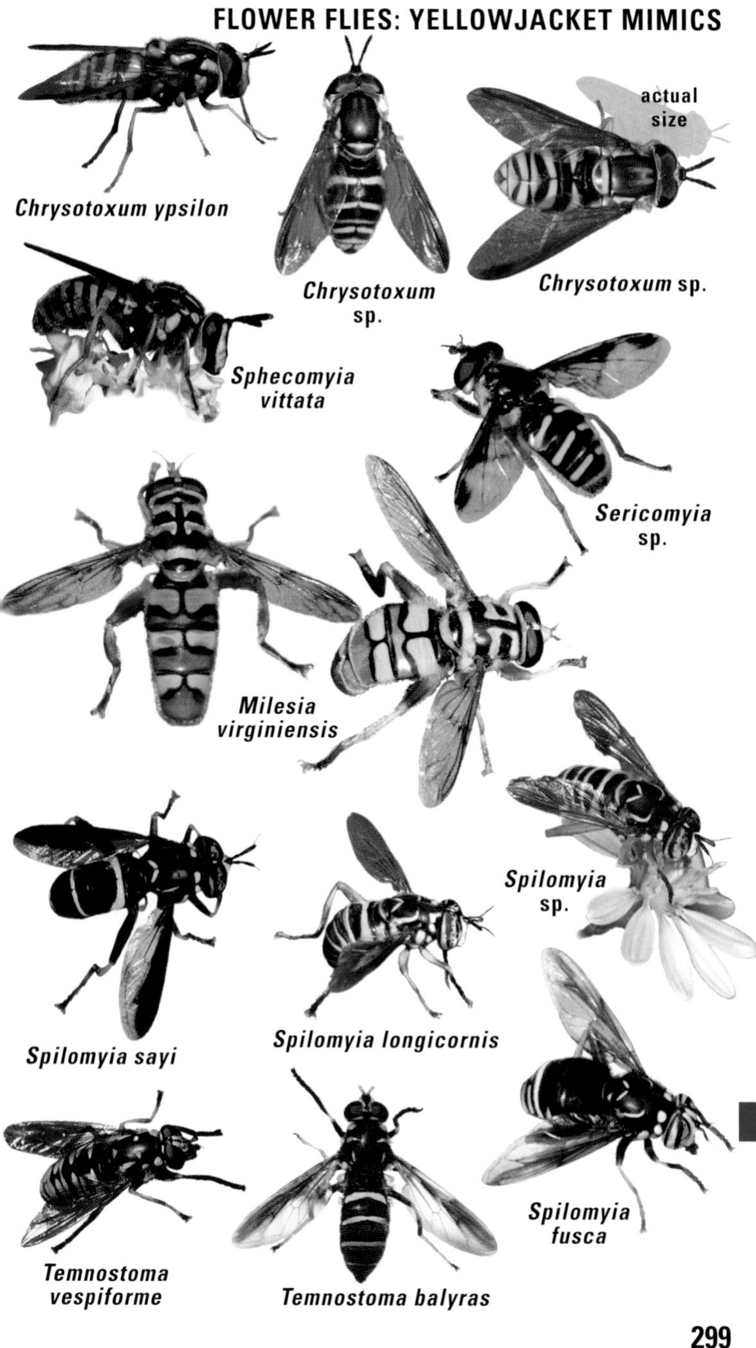

Chrysotoxum ypsilon

Chrysotoxum sp.

Chrysotoxum sp.

Sphecomyia vittata

Sericomyia sp.

Milesia virginiensis

Spilomyia sp.

Spilomyia sayi

Spilomyia longicornis

Temnostoma vespiforme

Temnostoma balyras

Spilomyia fusca

Microdon is a genus with 29 North American species, occurring over much of the continent. Larvae are bizarre, sluglike creatures that dwell in ant nests, feeding initially as predators of ant pupae and later as scavengers. Their appearance is odd enough that they were originally described by scientists as molluscs, then as scale insects.

Members of the genus ***Mallota*** are mimics of large bees. Note heavy "thighs" and bowed "shins" on hind legs of males. Eleven species occur in North America. Larvae have been found in rotten material from tree holes. ***Mallota bautias*** is common in open woodlands east of the Rockies. Look for adults on elderberry flowers.

The genus ***Copestylum*** is represented by at least 39 species in North America, including 22 species that were formerly placed in the genus *Volucella*. Adults range from 7–19 mm. The larvae of most species are scavengers in the nests of bees, wasps, or ants; ***Copestylum mexicana*** (not illustrated) is common across the southern U.S., where it breeds in decaying cacti.

The **Narcissus Bulb Fly** *(Merodon equestris)*, not illustrated, is a European import and is one of the few pests in what is mostly a harmless or beneficial family of flies. The larva feeds inside the bulb of daffodils. Adults are bumble bee mimics. The range of this species now extends across most of the U.S. and southern Canada.

Members of the genus ***Eristalis*** are abundant flies, with 18 North American species (including those formerly placed in the genus *Eoseristalis*). Larvae are known as rat-tailed maggots, for an extensible tube used to connect each aquatic larva to the water's surface. They scavenge in polluted water or wet carcasses. The **Drone Fly** *(Eristalis tenax)* is a convincing and widespread honey bee imposter. ***Eristalis arbustorum*** is a European immigrant now widespread in the eastern U.S. and eastern Canada. ***Eristalis dimidiatus*** ranges east of the Rockies, south to Kansas and North Carolina. ***Eristalis flavipes*** occurs from Alaska to California and North Carolina. ***Eristalis transversa*** is widespread in the eastern three-quarters of the continent. ***Eristalinus aeneus*** occurs in Europe, Africa, and much of North America.

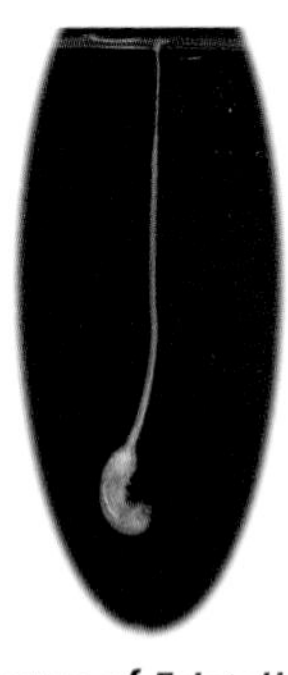

larva of *Eristalis* (rat-tailed maggot)

The genus ***Tropidia*** includes eight species, collectively found across most of this continent. ***Tropidia quadrata*** is a common eastern species.

The genus ***Criorhina*** includes 14 species of bumble bee mimics in Canada and the northern U.S. Most are western and range from 12–21 mm in length, depending on the species.

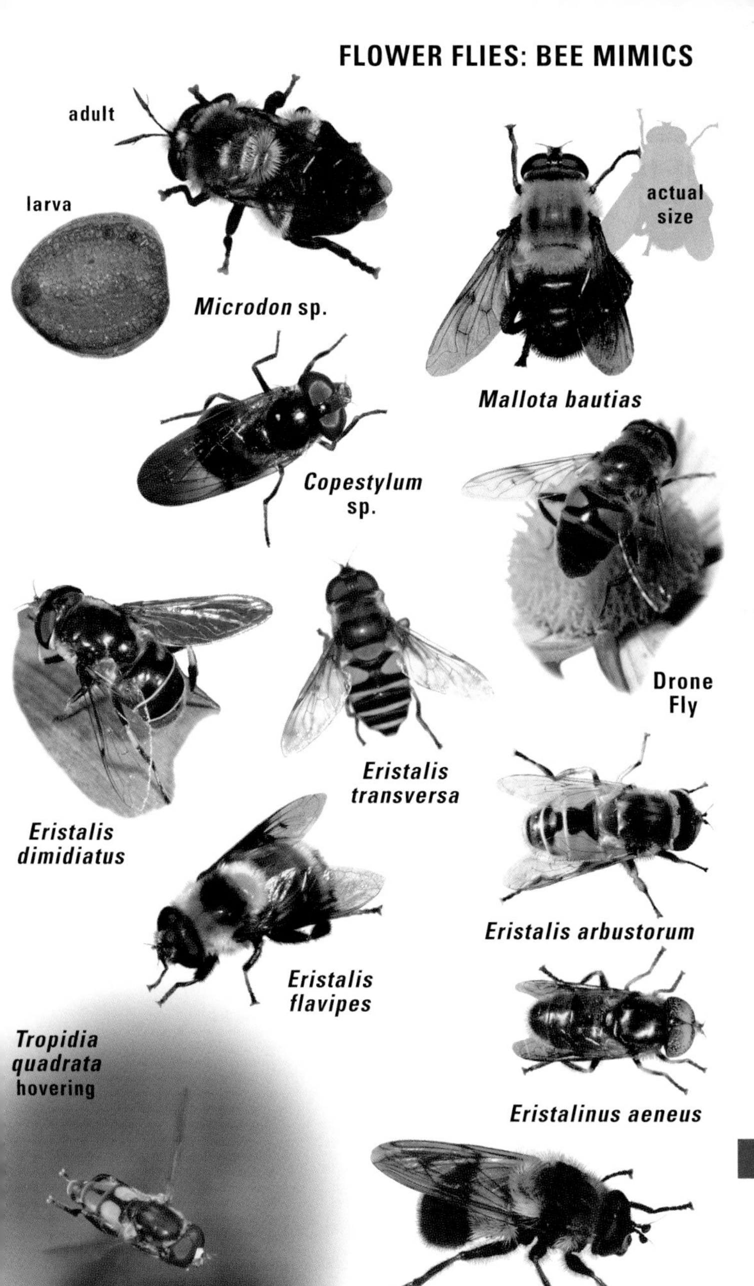
adult
larva
Microdon sp.
actual size
Mallota bautias
Copestylum sp.
Drone Fly
Eristalis dimidiatus
Eristalis transversa
Eristalis arbustorum
Eristalis flavipes
Tropidia quadrata hovering
Eristalinus aeneus
Criorhina sp.

The genus ***Syrphus*** includes 11 species in North America. Larvae are welcome predators of aphids and related insects. Adults feed on nectar, pollen, and the honeydew secreted by aphids. ***Syrphus ribesii*** occurs over most of this continent, plus Europe, Asia, and Central America.

Platycheirus is a large genus with 70 species in North America, including five species formerly placed in *Melanostoma*. Adults are small, varying from 5–10 mm depending on the species. Larvae are both predatory on other small insects and scavengers in rotting vegetation. ***Toxomerus*** includes 17 small species (adults 5–9 mm in length). The biology of their larvae is largely unknown. ***Toxomerus geminatus*** is widespread east of the Rockies. ***Toxomerus polita*** is widespread in the east and southwest. The larvae are recorded as feeding on corn pollen.

The genus ***Ocyptamus*** is represented by 13 species in North America, including most that were formerly in the genus *Baccha*. Adults are small with a distinctively slender, delicate appearance. Larvae prey on aphids and scale insects. ***Ocyptamus fascipennis,*** recognized by the dark patch on each wing, ranges over the eastern half of the U.S. and parts of southeastern Canada. ***Scaeva pyrastri*** occurs in Europe and Africa as well as North America. Here it is widespread in western and central regions. Larvae feed on aphids. ***Xanthogramma flavipes*** occurs in the east, from Quebec to Nebraska and North Carolina.

The genus ***Chalcosyrphus*** has 27 species occurring north of Mexico. Some of these are striking mimics of certain wasps. For example, ***Chalcosyrphus violascens*** (not illustrated) bears a remarkable resemblance to the Blue Mud Dauber (p. 334), even flicking its wings while crawling over logs or the forest floor. It ranges widely from Quebec through the eastern U.S. Others in this genus appear to mimic *Astata* wasps (p. 338). ***Xylota*** is represented by 29 North American species. ***Xylota quadrimaculata*** is widespread in southern Canada and the eastern U.S. ***Xylota angustiventris*** occurs over the eastern third of the U.S. and adjacent southern Canada. Adults may be mimics of *Trypoxylon* wasps (p. 338).

The genus ***Allograpta*** includes five species in North America. Larvae prey on aphids. ***Allograpta obliqua*** occurs throughout most of the U.S. and into Quebec. ***Sphaerophoria*** is represented in North America by 13 species. Larvae are among our important allies in aphid control. ***Sphaerophoria contigua*** ranges throughout most of the U.S. and southern Canada. The genus ***Helophilus*** includes 10 species north of Mexico, some ranging as far north as Alaska. They can often be recognized by the pattern of lengthwise stripes on the thorax. ***Syritta pipiens*** ranges over most of the U.S. and southern Canada and also occurs in Eurasia and Africa. Note the heavy "thighs" on the hind legs of adults. Larvae are scavengers on wet, well-decayed organic matter. ***Somula decora*** is one of two North American species in its genus. Adults may mimic scoliid wasps (p. 350), down to the fringe of hairs on each abdominal segment. This species ranges across the eastern U.S. and into New Brunswick. Larvae are known to occur in wet debris in tree holes.

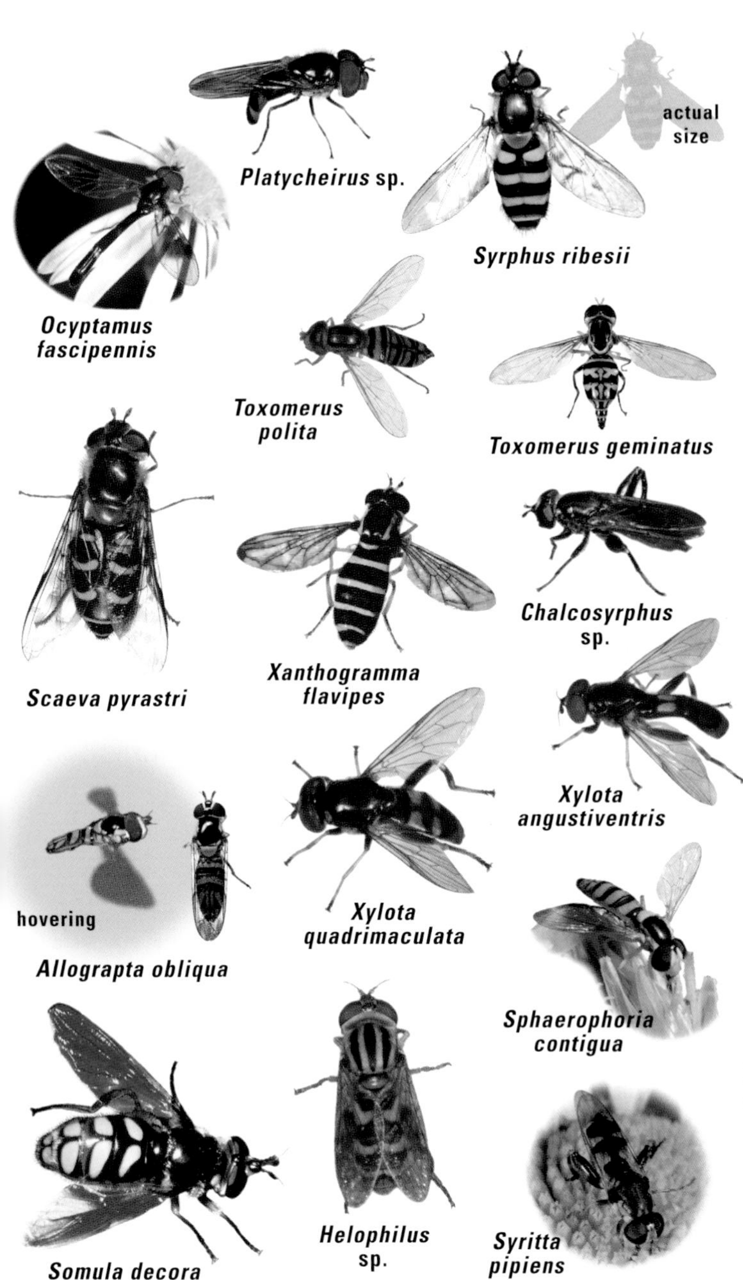

Platycheirus sp.

Syrphus ribesii

Ocyptamus fascipennis

Toxomerus polita

Toxomerus geminatus

Chalcosyrphus sp.

Scaeva pyrastri

Xanthogramma flavipes

Xylota angustiventris

Xylota quadrimaculata

Allograpta obliqua

Sphaerophoria contigua

Somula decora

Helophilus sp.

Syritta pipiens

BLOW FLIES (family Calliphoridae) include familiar bluebottles and greenbottles, named for their metallic colors. Some are instrumental in helping crime scene investigators solve homicides. Carrion-breeding blow flies arrive at a carcass in a predictable sequence. By determining which species are present, and at what stage of development, detectives work backward to estimate time of death. Larvae of other species are blood-sucking parasites of nestling birds. Many adults are competent pollinators of flowers. Often seen sunning on walls and sidewalks, these are the most common "house flies" in the southwest, far outnumbering Muscidae.

The **Screwworm** *(Cochliomyia hominivorax)*, not illustrated, is a devastating livestock pest named for the shape of its larvae. Females lay eggs in wounds of cattle and other animals; the larvae feed on the flesh and invite infection. Once found across the southern U.S., it is now scarce. Beginning in 1958, this insect was reared in captivity and the pupae bombarded with radiation, rendering the males sterile. Release of massive numbers of sterile flies drove populations to minimal levels. The **Secondary Screwworm** *(Cochliomyia macellaria)* ranges north to Oregon and Quebec. Larvae of the **Black Blow Fly** *(Phormia regina)* are used by surgeons on occasion to clean deep bone infections untreatable by other means. Maggots feed only on dead tissue, secreting antibacterial chemicals to retard further infection. They are even approved by the Food and Drug Administration.

Greenbottle flies *(Lucilia)* are abundant coast to coast. All are wholly metallic green, bronze, or blue. ***Lucilia sericata*** is a cosmopolitan species. ***Calliphora*** includes bluebottles of 10 species in North America. ***Calliphora vicina*** lives on every continent except Antarctica. The dull, metallic-blue abdomen helps separate this genus from the similar ***Cynomya. Cynomya cadaverina*** is a widespread bluebottle found from Alaska to Georgia. The metallic blue-green abdomen is distinctive. The **Cluster Fly** *(Pollenia rudis)* occurs throughout the U.S. and southern Canada. Adults often hibernate in large numbers indoors. Larvae are parasites of earthworms.

BLACK SCAVENGER FLIES (Sepsidae) are small black flies found near dung or carrion that "row" their wings while walking. Larvae develop in manure, decaying vegetation, or carrion in advanced stages of decomposition. Some occur in rotting seaweed. There are about 29 species in North America. ***Sepsis punctum*** is one of six species in its genus. **HELEOMYZID FLIES (Heliomyzidae)** are small, usually bristly flies. North America has more than 100 species. The few known larvae occur in mammal burrows, bird nests, or bat caves, or in carrion, dung, or rotting fungi. Adults occur in damp, shaded woodlands. Members of the genus ***Hylemya*** are often recognized by the black band across the thorax. The genus ***Suillia*** includes eight species in North America, collectively widespread. **DUNG FLIES (Scathophagidae)** are predatory as adults, while larvae of many feed on mammal droppings. In other species, larvae are leaf miners or plant feeders. North America has nearly 150 species, most diverse in the north. The most conspicuous species are fuzzy golden yellow flies in the genus ***Scathophaga,*** found around fresh dung and also on spring flowers.

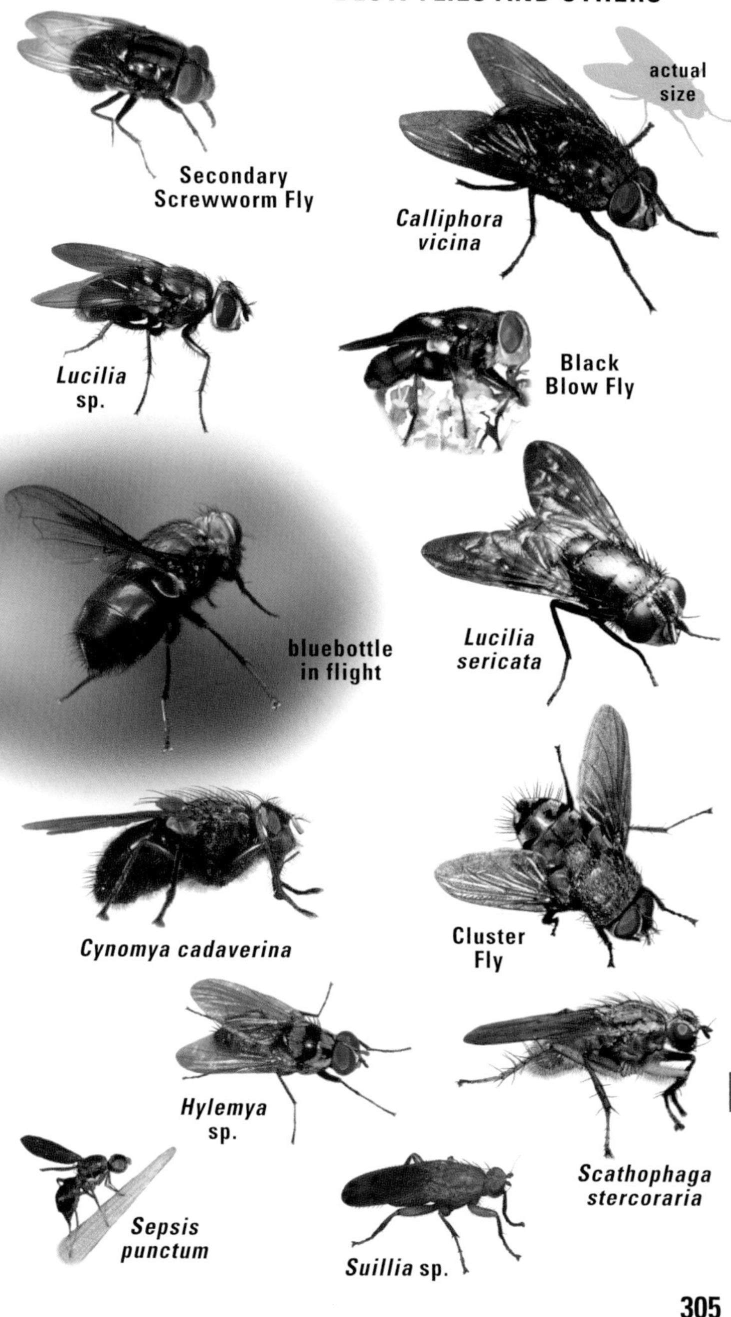

Secondary Screwworm Fly

Calliphora vicina

Lucilia sp.

Black Blow Fly

bluebottle in flight

Lucilia sericata

Cynomya cadaverina

Cluster Fly

Hylemya sp.

Scathophaga stercoraria

Sepsis punctum

Suillia sp.

HOUSE FLIES AND ALLIES (family Muscidae) are usually lumped in the "filth flies" category, and it is true that some species are mechanical vectors of serious diseases. Larvae of most species feed in manure or decaying organic matter. Some feed on living plant tissue, while others are predators of other insect larvae and small invertebrates. Adults feed on a variety of liquid food and regurgitate a "vomit drop" to liquefy solid morsels. A few of these flies are predatory on other insects, and another two are blood feeders. There are more than 600 species in North America. The **House Fly** *(Musca domestica)* is the best-known species in this family (see p. 274). The **Stable Fly** *(Stomoxys calcitrans)* uses its proboscis to tap blood vessels of humans and livestock alike. It breeds most readily in piles of decaying straw. ***Mesembrina solitaria*** is associated with bison dung and is consequently rare today. Once it was common in boreal latitudes and high elevation grasslands where the "buffalo" roamed. ***Graphomya maculata*** is one of nine species in its genus in North America, widely distributed.

FLESH FLIES (family Sarcophagidae) defy generalization. Few are easily recognized as adults, and habits of the larvae are all over the map. Some breed in carrion, others in open wounds on vertebrates, others in dung, and still others are parasitic on other insects, earthworms, millipedes, spiders, snails, or turtle eggs. Larvae of one genus feed on insects trapped inside pitcher plants. There are roughly 250 species north of Mexico. Members of the genus ***Sarcophaga*** are recognized by their scarlet eyes and "tail." Our 79 North American species vary a great deal in habits, but our common species are mostly associated with carrion. Females arrive later at a carcass than blow flies but lay live offspring (larvae) to make up the time lag.

BOT FLIES (family Oestridae) resemble bees but are not frequently encountered as adults. Larvae are internal parasites of mammals, some being harmful to livestock. Males of many species congregate at high points, a behavior known as "hilltopping." From these overlooks they intercept females. Adults of many species do not feed and, in fact, lack mouths. There are 41 species in North America. The genus ***Gasterophilus*** includes four species of horse bots introduced from Europe, now widespread in North America. Depending on the species, the larvae develop in the stomach, duodenum, or rectum of the host. Mature larvae are eliminated by the host and pupate in the soil. ***Cuterebra*** is a genus with 26 native species of rabbit and rodent bots. Larvae are parasites of rabbits, squirrels, and other small mammals, creating a tumorlike bulge beneath the skin. The genus ***Cephenemyia*** includes six species that infest the sinus cavities of deer.

LOUSE FLIES (family Hippoboscidae) are parasitic insects that resemble ticks more than flies. Aside from one livestock pest, these creatures are rarely seen, except by hunters who find them while dressing their kills. Females rear one offspring at a time, the larva feeding in utero from special "milk" glands. The mature larva is "born alive" and immediately pupates in the soil (or on the host in some cases). Most are host specific on bird species, with a few occurring on mammals. The **Squab Fly** *(Pseudolynchia canariensis)* is a parasite of pigeons and other birds.

HOUSE FLIES AND OTHERS

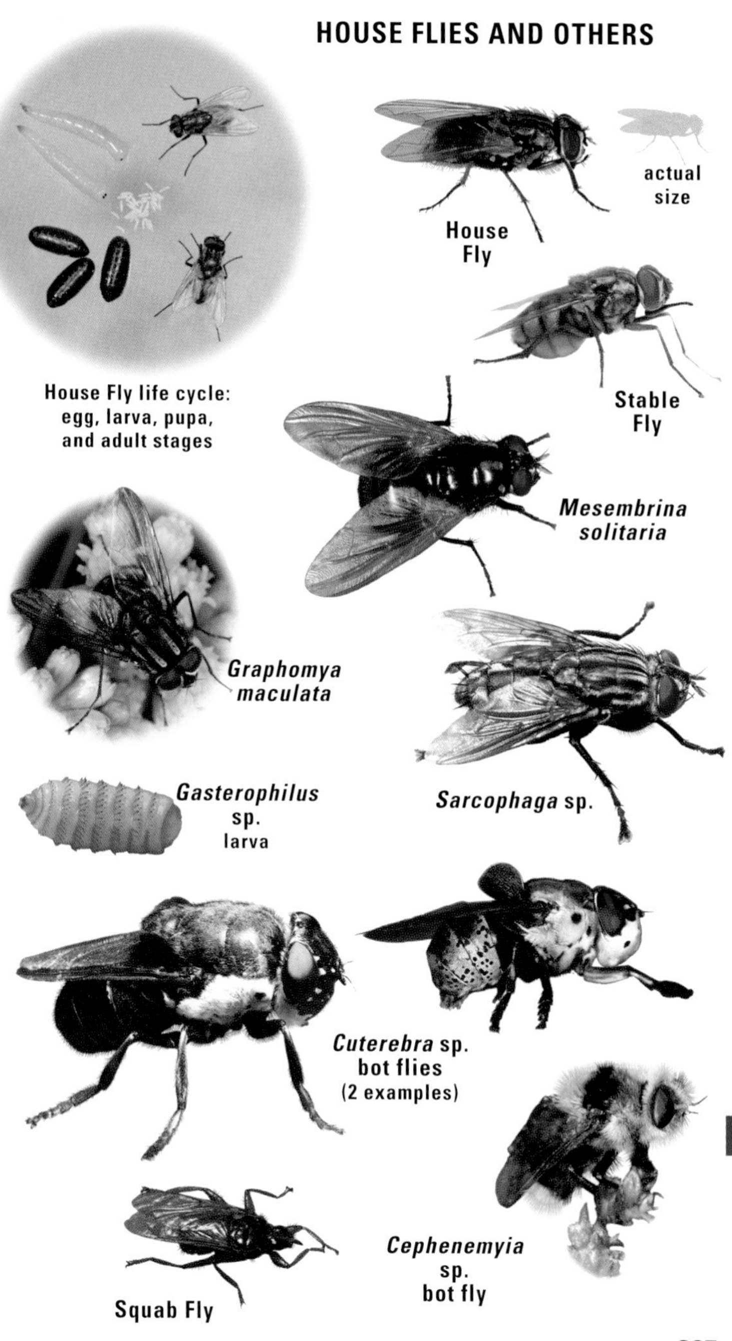

House Fly life cycle: egg, larva, pupa, and adult stages

House Fly

actual size

Stable Fly

Mesembrina solitaria

Graphomya maculata

Sarcophaga sp.

Gasterophilus sp. larva

Cuterebra sp. bot flies (2 examples)

Cephenemyia sp. bot fly

Squab Fly

TACHINID FLIES

(family Tachinidae) are highly diverse and almost exclusively internal parasitoids of other insects, especially caterpillars. In contrast to true parasites, parasitoids usually kill the host organism. Most conspicuous tachinids are robust flies with spiny abdomens. Look for them on flowers. Sixteen species have been imported to the U.S. as biological control agents of pest insects. That tactic occasionally backfires. *Compsilura concinnata* (not shown), introduced to control the Gypsy Moth, is such a generalist parasitoid that it is equally happy with the larvae of virtually any moth, butterfly, or sawfly. It has been especially harmful to native giant silkmoths. Many female tachinids lay eggs directly on the host insect, or forcibly inject them, but some "broadcast" hundreds of tiny eggs on foliage. The ova hatch only if they are consumed (with foliage) by a caterpillar. In other species, the eggs hatch into bizarre, flattened larvae called planidia that actively stalk a host or patiently wait to latch onto a passing one. Females of a few tachinid species deposit live maggots that wait or search for hosts. All tachinid larvae feed inside the host, studiously avoiding vital organs until the last stage of their own development. There are over 1,300 species known from North America.

Our 14 species of ***Belvosia*** are not uncommon on late summer or fall wildflowers. Their hosts are mostly caterpillars and pupae of sphinx moths and giant silkmoths, although other moths and skippers have also been recorded. ***Tachinomyia*** includes 10 North American species, parasites of various moths as well as at least one sawfly. The genus ***Gonia*** is represented by 29 species north of Mexico. The adult flies emerge early in spring and can be found sipping sap flowing from tree wounds. Females lay large numbers of tiny eggs on foliage, some of which are then consumed by the intended hosts, the caterpillars of owlet moths (Noctuidae). There are also records of parasitism of May beetles and blister beetles. ***Leschenaultia*** is represented by 10 species in North America. Known hosts are the caterpillars and pupae of various moths. The caterpillars ingest the tiny tachinid eggs in the course of feeding.

The genus ***Gymnosoma*** includes seven North American species, all parasites of stink bugs and shield bugs. Females lay an egg on the exterior of the host. The larva that hatches enters the bug through the membrane between abdominal segments. There it feeds on the fat bodies of the host, which ironically renders it sterile but usually spares its life. Adult flies are often seen on the flowers of wild carrot. Our six species of ***Trichopoda*** resemble bees, often with a fringe of hairs on the hind legs that resembles a "pollen basket." Look for adults on flowers. Larvae are parasites of leaf-footed bugs and squash bugs and also stink bugs. Females adhere an egg to the top of a host insect where it cannot be wiped off by self-grooming. Mature larvae usually exit the host (which may still be alive, albeit barely) and pupate in the soil. The genus ***Cylindromyia*** includes 18 species of slender-bodied flies that resemble sphecid wasps at first glance. Hosts vary with the species and include adult stink bugs, larvae and pupae of giant silkmoths and owlet moths, and short-horned grasshoppers.

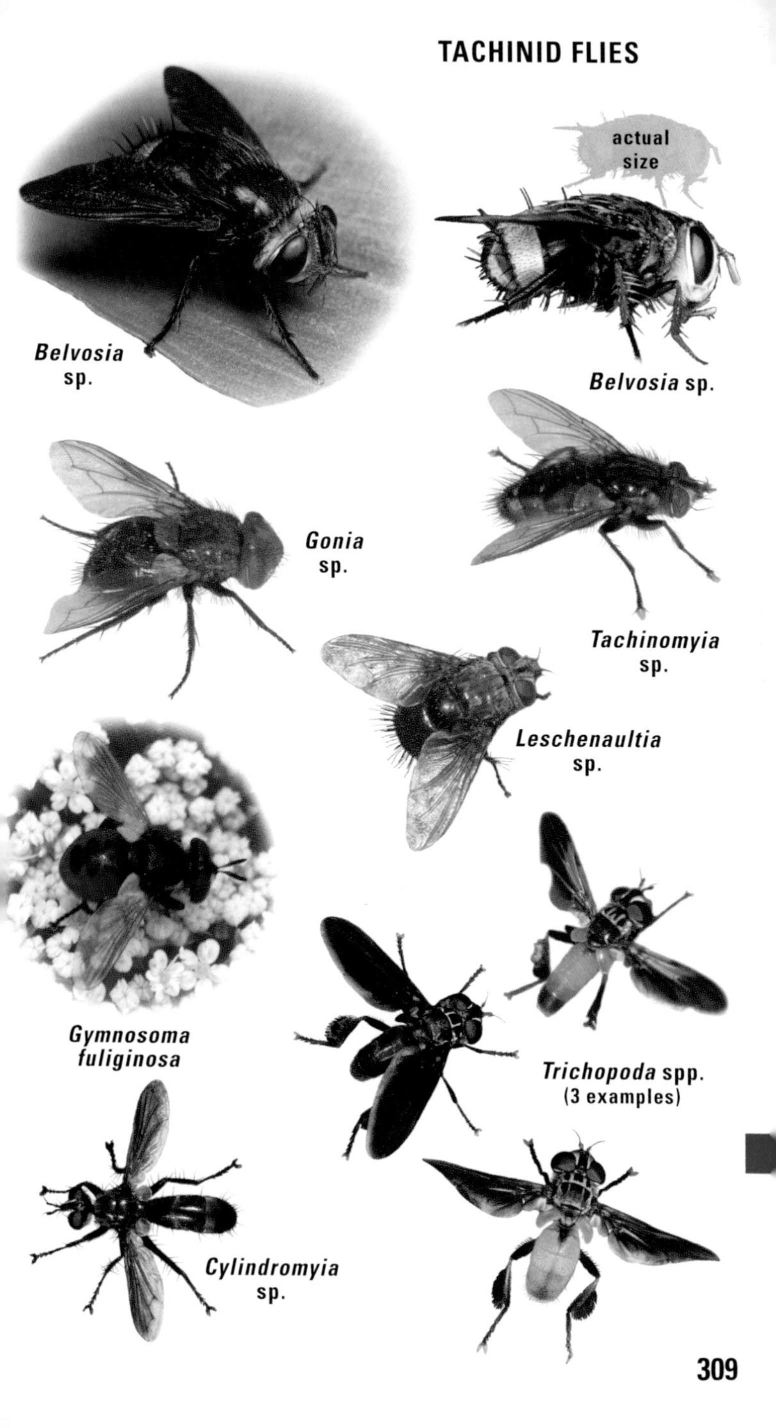
actual
size
Belvosia sp.
Belvosia sp.
Gonia sp.
Tachinomyia sp.
Leschenaultia sp.
Gymnosoma fuliginosa
Trichopoda spp. (3 examples)
Cylindromyia sp.

Hystricia abrupta (formerly in the genus *Bombyliopsis*) is a large, very spiny fly, not uncommon on flowers in forest openings and wet glades in summer and early fall. It ranges across southern Canada and most of the U.S. except the midwest. Its hosts are the caterpillars and pupae of various tiger moths (Arctiidae, p. 260).

The genus ***Tachina*** includes 39 species in North America, collectively found over most of the continent. Hosts vary with the different species and include the caterpillars of a variety of moth species, walkingsticks, larval leaf beetles, sawflies, and bumble bees.

Members of the genus ***Archytas*** are often abundant on fall flowers such as asters. The gray thorax and contrasting black abdomen help identify them instantly. There are 13 species in North America. Known hosts include the caterpillars and/or pupae of tiger moths, owlet moths (including cutworms and armyworms), prominent moths, tent caterpillars, bagworm moths, and flannel moths. The fly larvae hatch from eggs scattered in the habitat of the hosts and then patiently await the approach of a victim. These "planidia" are tiny, flattened creatures that latch onto a caterpillar with a gluelike oral secretion that adheres them to the body. Within about 15 minutes, they penetrate the caterpillar's cuticle (or "skin").

Juriniopsis adusta is one of four North American species in its genus, all occurring in the southern U.S. Its hosts are caterpillars of certain tiger moths and skippers. ***Adejeania vexatrix*** ranges from British Columbia and Montana south to California and New Mexico. This large reddish fly is easily recognized by the long proboscis it uses to sip nectar from flowers.

Epalpus signifer is one of three North American species in the genus, collectively ranging across most of the continent. The thin, irregular yellow line across the otherwise black abdomen is characteristic of the species. The only known host is a certain owlet moth (one of the pinion moths, genus *Lithophane*), of which the caterpillar stage is parasitized.

Cordyligaster septentrionalis is the only species of its genus found in North America, occurring over much of the eastern U.S., at least as far west as Kansas. Its larval hosts are unknown.

The genus ***Zelia*** includes 12 species north of Mexico, collectively widespread. They are easily recognized by the long legs and pointed abdomen. The larval hosts are apparently the grubs of beetles. Occasionally the adult flies are attracted to lights at night. By day, look for them on the ground, or on low plants, in open habitats.

Voria ruralis is a parasite of the various owlet moths (Noctuidae) and possibly other moths. Several fly larvae may develop inside one caterpillar, as evidenced by the illustration here. The maggots pupate inside the empty husk of the host. Long regarded as just one species, insects included under this name actually represent a complex of species with a wide North American distribution.

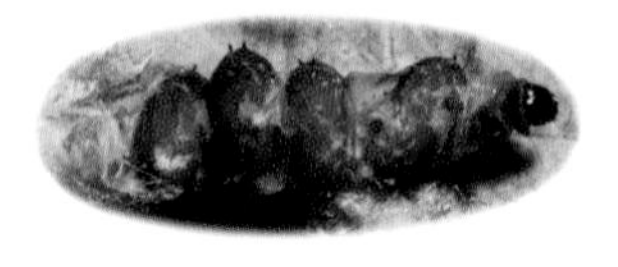

skin of dead caterpillar with *Voria* pupae inside

Hystricia abrupta

Tachina sp.

Juriniopsis adusta

Epalpus signifer

Archytas apicifer

Cordyligaster septentrionalis

Adejeania vexatrix

Voria ruralis

Zelia sp.

VARIOUS SMALL FAMILIES OF FLIES

THICK-HEADED FLIES (family Conopidae) are no more stubborn or dense than any other insect; the name comes from the disproportionately large head. They are mostly parasites of adult solitary bees, and sometimes wasps. The female fly assaults the host in midair, often forcing it to the ground and ramming an egg between the victim's abdominal segments before releasing it. There are about 66 species in North America. These can be common, though not abundant, flies. The genus ***Physocephala*** includes seven North American species. They can be recognized by the way the hind femur ("thigh") is somewhat swollen at its base. ***Physoconops*** is represented by 12 species in North America. They closely resemble *Physocephala,* but the hind femur is slender throughout its length. Host species for *Physoconops* include leafcutter bees. ***Myopa*** includes 14 species in North America, most being reddish brown.

RUST FLIES (Psilidae) are small insects, relatively nondescript. For species in which the life cycle is known, the larvae feed in the living stems or roots of plants. A few are pests, such as the **Carrot Rust Fly** *(Psila rosae),* not illustrated. About 30 species are found north of Mexico.

STILT-LEGGED FLIES (Micropezidae) mimic wasps, walking with front legs extended like antennae. A few are convincing ant mimics. Look for them in wooded areas, usually crawling slowly about the base of trees, or on low foliage. There are 27 species in North America, many more in the tropics. ***Rainieria antennaepes*** is the only North American species in its genus. ***Taeniaptera trivittata*** is an ant mimic. Dark bands on its wings serve to emphasize the distinct body divisions of the ants it mimics.

CACTUS FLIES (Neriidae) are immediately recognized by the elongated head and streamlined body. They are restricted to the southwestern U.S., where they occur around decaying cacti.

LEAF MINER FLIES (Agromyzidae), in the larval stage, bore in various parts of living plants. Most make species-specific mines between the layers of leaves, but others damage seeds, stems, roots, or young twigs and trunks of trees. Most are very host specific, so knowing the plant in which a larva is found can narrow the list of suspects. There are hundreds of species, many still awaiting formal description by scientists.

GRASS FLIES (Chloropidae) are often abundant in fields and in sedges along pond edges. They are tiny, but some are strikingly patterned. A few species are a nuisance, flitting about the eyes or ears of people and livestock, and thus known as "eye gnats." In the larval stage some species are also pests, feeding in stems of crop plants. Others eat decaying vegetation, while others breed in fungi or are predators on smaller insects. A few form plant galls. There are about 270 species known from North America.

FLUTTER FLIES (Pallopteridae) include nine species in North America. They are considered rare, but look for them in moist woodlands. Adults are 3–5 mm and usually seen on flowers or low branches of shrubs and trees. Larvae of some species have been associated with the galleries of bark beetles (p. 216), while others have been reared from the flower heads and stems of composite and umbelliferous plants.

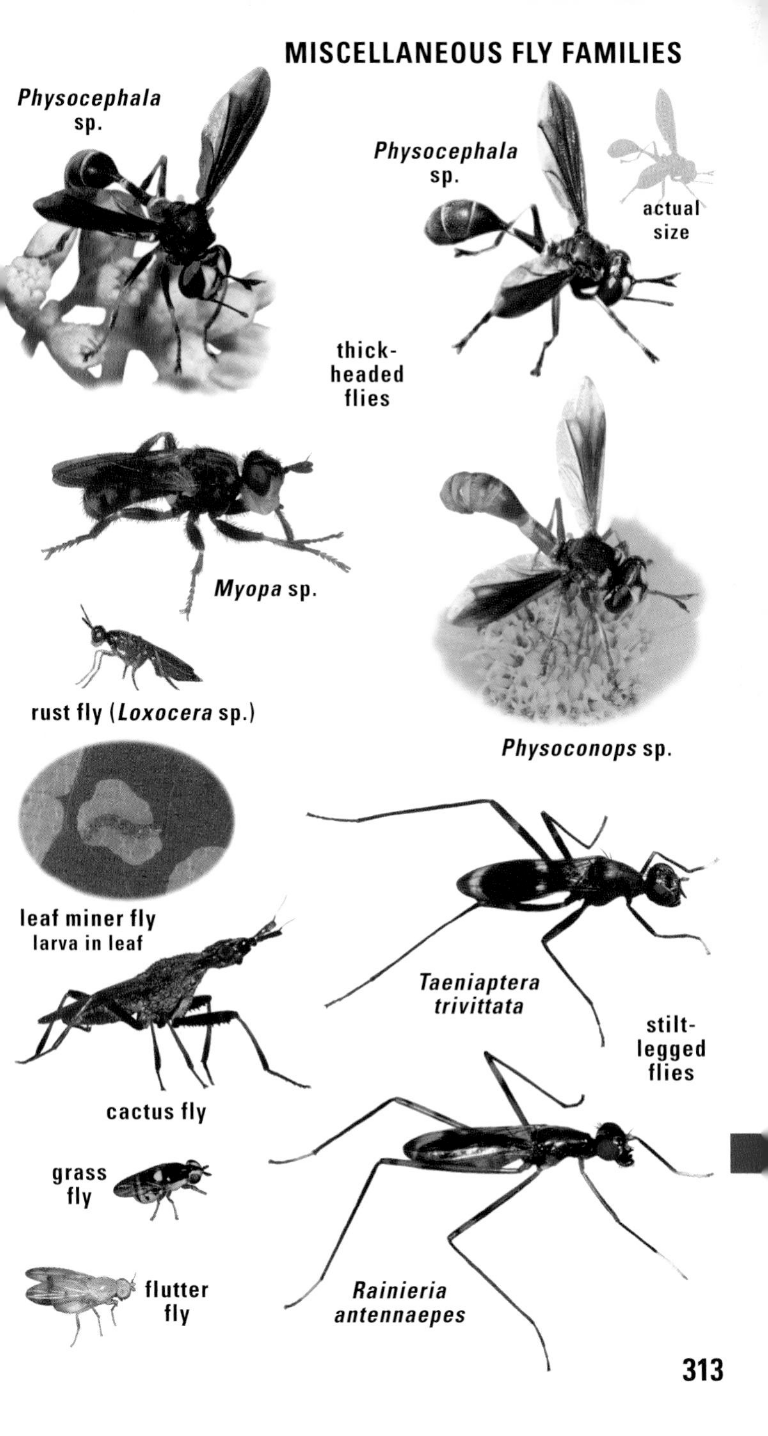
Physocephala sp.
Physocephala sp.
actual size
thick-headed flies
Myopa sp.
rust fly (*Loxocera* sp.)
Physoconops sp.
leaf miner fly larva in leaf
Taeniaptera trivittata
stilt-legged flies
cactus fly
grass fly
flutter fly
Rainieria antennaepes

FRUIT FLIES (family Tephritidae) are sometimes called peacock flies because of the wing patterns, used in courtship dances. This family includes 300 species north of Mexico, some introduced from elsewhere, with more foreign species the agriculture industry is trying to keep at bay. A few are allies, introduced to control noxious weeds. The genus ***Eutreta*** includes 17 North American species, all with light dots on dark wings, some known to form galls on the host plant. ***Urophora cardui,*** native to Europe, was imported to North America to control Canada thistle (actually native to Eurasia). Five native ***Urophora*** species occur in the west, and two others have been introduced to control invasive weeds. The larval host of ***Zonosemata vittigera*** is a native nightshade. Larvae of the **Sunflower Maggot Fly** *(Strauzia longipennis)* bore in stems of the host and feed on the pith.

The **Mediterranean Fruit Fly** *(Ceratitis capitata)* is the scourge of citrus growers everywhere, especially in California. Aerial spraying goes into effect at the mere suggestion that a female "Med Fly" may be on the loose. ***Rhagoletis*** includes 21 species here, several of which are pests in orchards. Adults in this genus mimic jumping spiders, the dark bands on the wings suggesting the first two pairs of legs on such arachnids. Real jumping spiders have been observed to be fooled into terminating their stalking behavior of these flies. The **Apple Maggot Fly** *(Rhagoletis pomonella)* is a notorious pest of cultivated apples but feeds on hawthorn and other plants "in the wild." The **Walnut Husk Fly** *(Rhagoletis completa)* bores in husks of walnuts, sparing the nutmeat but making the husks harder to remove. ***Neotephritis finalis*** is one of the most commonly seen fruit flies in North America. Larvae develop in flower heads of plants in the aster family.

LONCHAEID FLIES (Lonchaeidae) are small, mostly blue-black, with clear wings, and not easily identified in the field. Larvae appear to follow in the wake of other insect larvae that bore in plants, though a few may initiate feeding on plants themselves. Some can be found under the bark of dead or dying trees, often in the galleries of bark beetles.

PLATYSOMATID FLIES (Platysomatidae) might be mistaken for picture-winged flies or fruit flies, but they often have metallic bodies. North America has about 40 species, most in the genus ***Rivellia,*** and their biology and life histories are essentially unknown.

PYRGOTID FLIES (Pyrgotidae) resemble large fruit flies but are nocturnal parasites of adult May beetles. Female flies accost flying May beetles in midair, forcing an egg between the beetle's abdominal segments. The larva that hatches from the egg kills the beetle in about two weeks. We have six species, all in the east; ***Pyrgota undata*** is one common example.

PICTURE-WINGED FLIES (Otitidae or **Ulidiidae)** are named for the often banded or spotted wings. There are about 133 species known from North America. The larvae of most are scavengers on decaying organic matter, but a few are plant feeders. ***Idana marginata*** is sometimes seen at sap flows in the northeastern U.S. and southeastern Canada. The genus ***Tritoxa*** includes five species, ranging from southern Canada to Mexico. ***Delphinia picta*** is widespread from the Great Plains eastward.

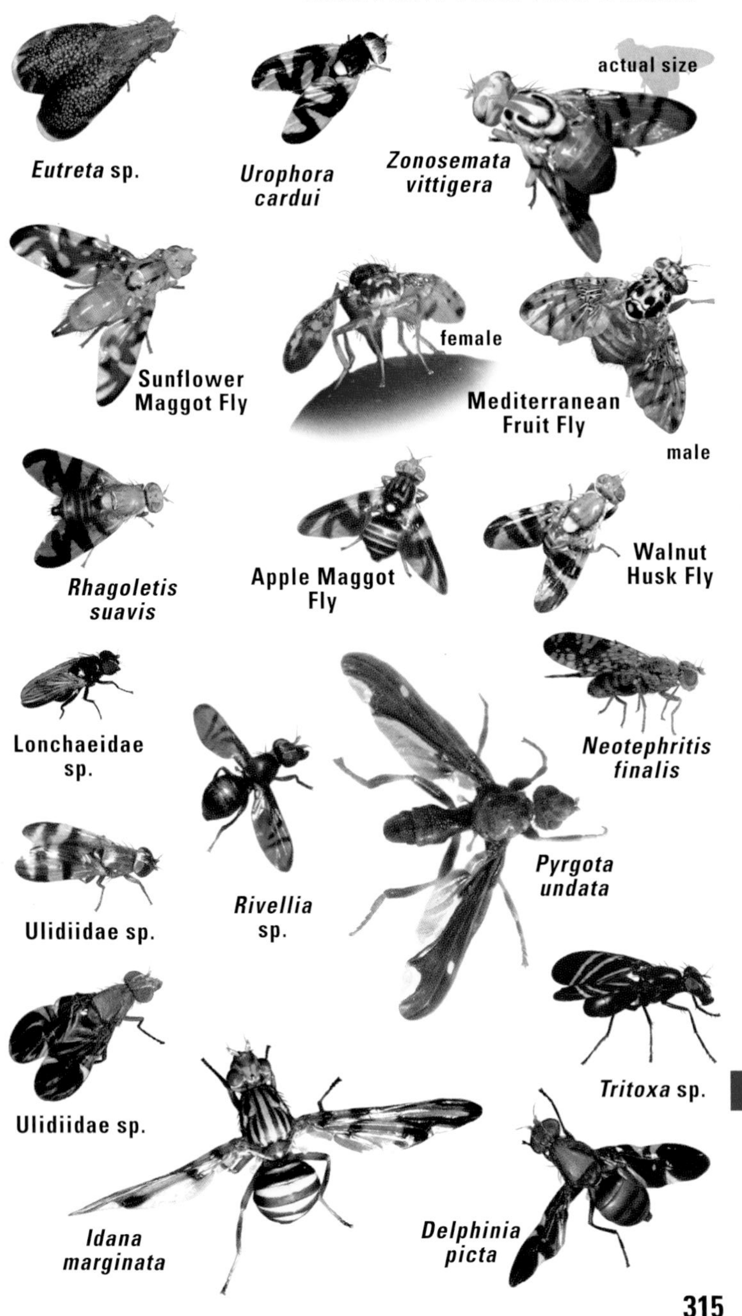
actual size
Eutreta sp.
Urophora cardui
Zonosemata vittigera
Sunflower Maggot Fly
female
Mediterranean Fruit Fly
male
Rhagoletis suavis
Apple Maggot Fly
Walnut Husk Fly
Lonchaeidae sp.
Neotephritis finalis
Pyrgota undata
Ulidiidae sp.
Rivellia sp.
Ulidiidae sp.
Tritoxa sp.
Idana marginata
Delphinia picta

MARSH FLIES AND SHORE FLIES

MARSH FLIES (family Sciomyzidae) are modest-sized flies recognized mostly by a concave face and elongate antennae. They usually frequent the margins of wetlands. Larvae of all known species are predators or parasites of molluscs. Most afflict freshwater snails, but a few attack terrestrial snails, snail eggs, or slugs. Some tropical species are highly beneficial in controlling the snail hosts of the parasites causing leishmaniasis.

Tetanocera plebeja is one of 30 species in its genus found here. Adult females scatter their eggs singly on vegetation. Larvae are parasites/predators of a species of slug, *Deroceras laeve*, initially entering the host under its mantle. After killing the first host, a larva moves on to kill three or more slugs in succession. Larvae of other species follow similar patterns but may enter the host via its mouth or by boring into an eyestalk. Eeew, gross!

The genus ***Sepedon*** includes 16 species in North America. Look for them in a variety of swampy habitats, even roadside ditches. Females lay eggs in clusters on vegetation. The larvae prey on pulmonate freshwater snails. ***Antichaeta*** includes nine species in the western and northern U.S. and southern Canada. Female flies deposit their eggs on the eggs of snails. Adults can be found in a variety of habitats but especially around marshes. ***Euthycera arcuata*** ranges east of the Rockies in the U.S. and southern Canada. Larvae apparently prey on terrestrial snails, overwintering in shells of partially eaten victims. The genus ***Limnia*** includes 17 widespread species north of Mexico; their biology is poorly known.

SHORE FLIES (family Ephydridae) are very diverse. There are at least 425 species recorded from North America. Many species are incredibly abundant around alkaline lakes and are an important source of food for waterfowl. The genus ***Ephydra*** includes 16 species. The **Alkali Fly** *(Ephydra hians),* widespread in the west, swarms on the shores of Mono Lake, California. The **Brine Fly** *(Ephydra cinera)* is the common species on the shores of the Great Salt Lake, Utah. Females lay their eggs underwater, climbing down any emergent surface and trapping an air bubble under their wings. To surface, they merely let go, and the bubble literally pops them to the surface. Larvae feed on organic matter filtered from the lake bottom ooze. ***Helaeomyia petrolei*** (not illustrated) is the "petroleum fly," an insect that lives, as a larva, in perhaps the most inhospitable habitat in North America: tar pits in southern California.

The genus ***Hydrellia*** includes 57 species in North America, at least two of which are introduced. Larvae of most species are leaf miners in aquatic and semiaquatic plants. ***Hydrellia pakistanae*** was introduced to Florida in 1987 to control an invasive aquatic weed. The genus ***Notiphila*** includes about 50 North American species. Males defend territories from other males, using a spinelike process on the tip of the abdomen to try to overturn a foe.

SEAWEED FLIES (family Coelopidae), not illustrated, can be abundant around rotting seaweed along the strand line on the beaches of the Pacific and Atlantic coasts. There are two genera, with five species, on our continent. **Kelp flies** (genus *Coelopa*) are the most likely to be noticed.

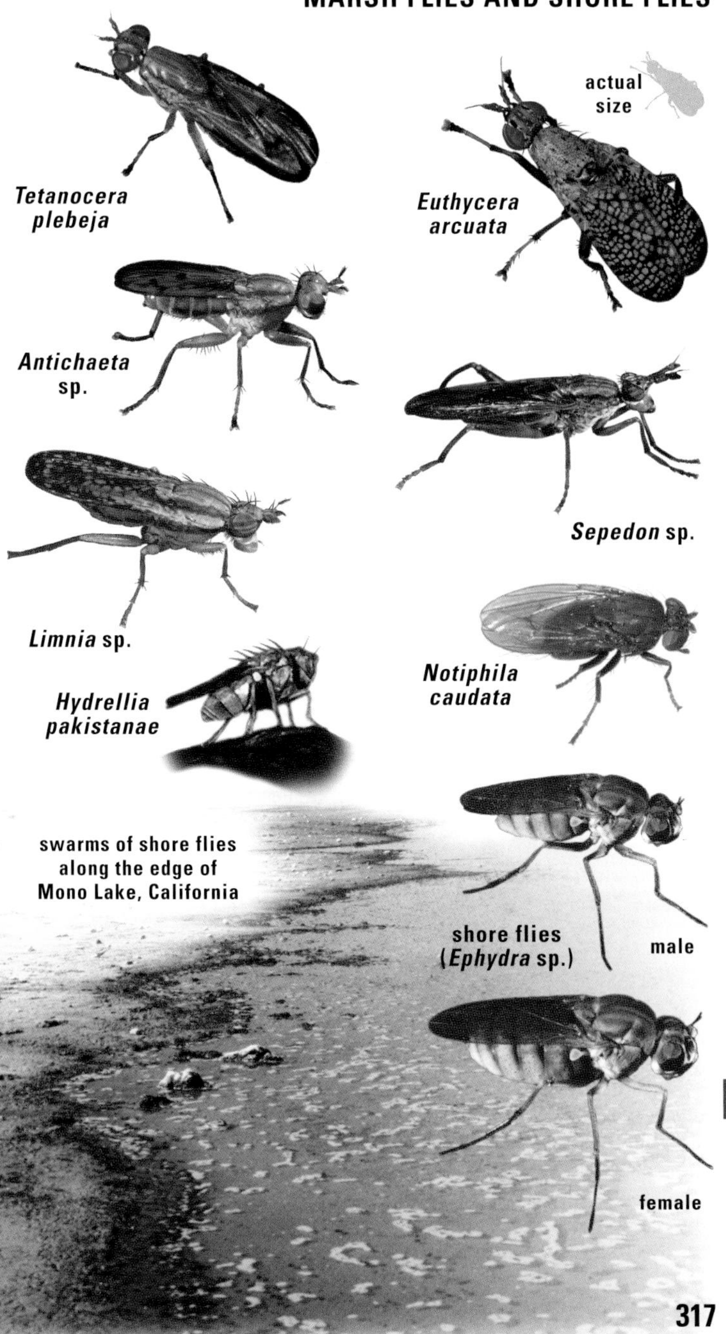
actual
size
Tetanocera plebeja
Euthycera arcuata
Antichaeta sp.
Sepedon sp.
Limnia sp.
Notiphila caudata
Hydrellia pakistanae
swarms of shore flies along the edge of Mono Lake, California
shore flies (*Ephydra* sp.)
male
female

WASPS, BEES, ANTS

(order Hymenoptera) are often referred to as social insects, even though the vast majority lead solitary lives. They are sometimes called stinging insects as well, but many species do not sting, and of those that do, only females are capable of inflicting pain. The long, intimidating "tails" of some wasps are not stingers but egg-laying organs called ovipositors. As a whole, members of this group may be identified by their two pairs of membranous wings, the hind pair smaller (the worker castes of ants are wingless, as are females of some wasps), and often a narrow constriction of the body between the thorax and abdomen. There are approximately 17,000 species of Hymenoptera in North America. They include our smallest insects (parasitic "fairy flies" that could theoretically fly through the eye of a needle) as well as some of the largest (tarantula hawks, p. 352). Why are such seemingly disparate organisms as sawflies, wasps, ants, and bees all grouped in the same order? Mostly it has to do with their evolutionary relationships. Sawflies and their kin are generally regarded as the most primitive examples, with complex wing venation and vegetarian larvae, while bees are considered "advanced," with simpler wing venation but often complex social behavior.

Members of this order are indispensable to human enterprise. Were it not for bees and certain wasps, many agricultural crops would go unfertilized. Many parasitic wasps are employed to control pest insects. Ants also take their toll on pests, and their subterranean activities help turn and aerate the soil. In the yard and garden, all of these insects, including yellowjackets, help to pollinate flowers and to keep down populations of pest insects.

All species in the Hymenoptera go through a complete metamorphosis. With the exception of bees, sawflies, woodwasps, gall wasps, and pollen wasps, the larvae are carnivorous. Adults of most species fuel their energetic activities on flower nectar and/or honeydew exuded by aphids and related insects. The immature stages of Hymenoptera are almost always concealed deep in nests and rarely seen. In parasitic species, eggs and larvae are hidden inside the host organism, the pupae erupting from the host at the end of the life cycle. Nests are often subterranean burrows or existing cavities in the ground or dead wood. Some build freestanding structures of mud, paper, resin, or other organic materials. Still other species excavate in rotten wood or stimulate the growth of galls on plants. Many bees secrete wax or other substances used to construct or line cells.

These insects certainly exhibit some of the most complex behaviors of all invertebrates. Despite the ability of some species to sting, they are generally loath to do so and can be easily observed as they single-mindedly go about the business of foraging, hunting, courtship, or nest-building. Watching ants swarm, sand wasps dig burrows, mud daubers build nests, or cicada killers subdue and transport their prey are rewarding experiences for the observer.

The degree of parental and sibling care demonstrated by social and subsocial species is unsurpassed by any other type of invertebrates. Division of labor among the worker caste of social insects can be based on age of the individual or its body type, especially in ants when "major" workers function as soldiers or heavy laborers. Communication in social hymenopterans is largely chemical, taking the form of mutual feeding (trophallaxis) and secretion of an array of subtle odors (called pheromones). Individual pheromones can suppress ovarian development in workers, signal alarm, orient workers along foraging trails, and serve other functions.

Bee or wasp? This question is often asked with the implication that bees are "good" and wasps are "bad." While bees pollinate flowers, so do many wasps. We get honey from honey bees, but pest control services are provided by most wasps. Generally speaking, most bees are hairy (each body hair is branched) and wasps less so (body hairs are not branched). Bees have modifications for collecting pollen, while wasps do not (neither do parasitic bees, however). But as molecular research on insects continues, the division between bees and wasps becomes ever more blurred. It is quite conceivable that some bees and some wasps will eventually be lumped together at the level of family.

As if it were not difficult enough to tell the difference between bees and wasps, many other kinds of insects mimic them. Chief among these are various flies, moths, and beetles. Be sure to check other places in this guide for look-alikes. Only a few examples are shown below:

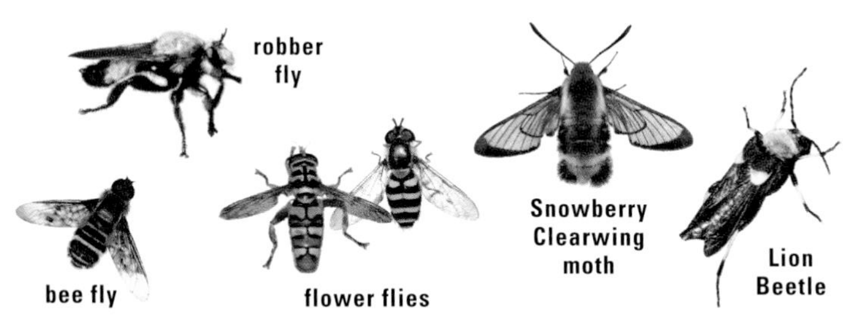

Attracting bees and wasps? Yes! The services they provide make it worth your while to plant your garden with native flowers that attract bees and other native pollinators. Furthermore, some bees and wasps may suffer from a decline in available real estate. Those solitary species that nest in preexisting cavities can use housing help. Bundles of paper straws or cardboard tubes, prunings from sumac and bamboo, or blocks of wood with holes of varying diameters and depths drilled into them, make attractive nest sites. Make sure to provide a roof to keep the openings dry in rainy weather. Place the bundle or block on a south-facing surface at least three feet off the ground. You and your family can then sit back and watch the hardworking female insects take up residence.

SAWFLIES

are represented by several families of stingless wasps. Females use a sawlike organ to deposit their eggs in foliage or twigs, hence the common name. Larvae are often conspicuous and are easily mistaken for caterpillars of moths and butterflies.

COMMON SAWFLIES (family Tenthredinidae) are represented by about 730 species in North America, most diverse in the northern half of the continent. Larvae of most species resemble caterpillars of butterflies or moths, but they have more than five pairs of prolegs (the stubs running down the length of the larva's body). When alarmed, some thrash their back ends in the air. Larvae of a few species are sluglike or are leaf miners, gall makers, or stem borers. Adults appear to mimic stinging wasps. ***Tenthredo*** is the largest genus in the family, with at least 118 species in North America; four examples are illustrated here. Adults may be important pollinators, and they feed on other insects as well. ***Nematus*** is a genus with 60 North American species. The **Willow Sawfly** *(Nematus ventralis)* is widespread east of the Rockies. The imported **Currant Sawfly** or **Gooseberry Sawfly** *(Nematus ribesii)* is native to Europe but now found in the northern U.S. and southern Canada. The hosts are species of *Ribes* such as gooseberry or currant.

Willow Sawfly larvae

Larvae of the **Dusky Birch Sawfly** *(Croesus latitarsus)* line up along the edges of birch leaves, curling their bodies into an S shape at rest or when disturbed. ***Dolerus*** is a widespread genus with 36 North American species. Larvae feed on horsetails, grasses, sedges, and rushes, sometimes damaging field crops. They pupate in the soil. Adults are common in early spring near their hosts and on flowers of hawthorn, cherry, and willow. ***Dolerus asper*** is widespread across northern Eurasia and North America. ***Caliroa*** includes 14 North American species with sluglike larvae that feed on the underside of leaves. The **Pear Sawfly** or **Pear Slug** *(Caliroa cerasi)* is probably native to Europe, now with a nearly global distribution thanks to commerce. Hosts include chokeberry, hawthorn, and mountain ash.

Larvae of ***Aneugmenus flavipes*** feed on ferns. Ash trees are host plants for the **Black-headed Ash Sawfly** *(Tethida barda)* of the east. Adults play dead when threatened. The genus ***Eriocampa*** includes two North American species. ***Eriocampa ovata*** ranges in the northern U.S. and adjacent Canada but was probably introduced from Europe. Red alder is the larval host plant. Larvae of the **Butternut Woollyworm** *(Eriocampa juglandis)* are covered in waxy filaments resembling shredded coconut. ***Taxonus*** is a genus with nine North American species. The few known host plant records are in the rose family. ***Taxonus terminalis*** occurs from eastern Canada to Texas. ***Macrophya*** includes 46 species in North America. ***Macrophya formosa*** and ***Macrophya flavicoxae*** are two common eastern species. ***Lagium atroviolaceum*** ranges from Quebec to Kansas. Elderberry and viburnum are the larval hosts. Adults are black, or black and red, and mimic spider wasps, flicking their wings and bobbing their antennae.

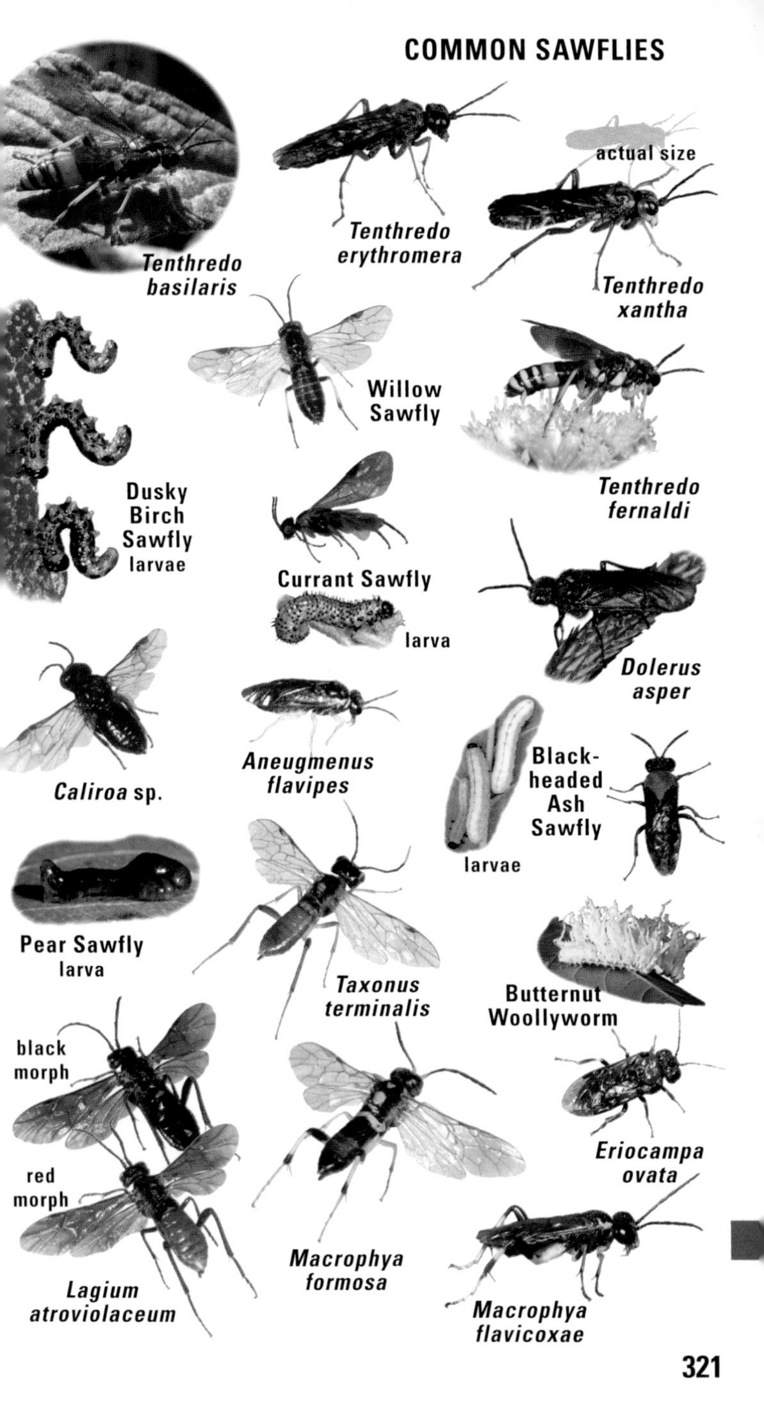
Tenthredo basilaris
Tenthredo erythromera
actual size
Tenthredo xantha
Dusky Birch Sawfly larvae
Willow Sawfly
Tenthredo fernaldi
Currant Sawfly
larva
Dolerus asper
Caliroa sp.
Aneugmenus flavipes
Black-headed Ash Sawfly
larvae
Pear Sawfly larva
Taxonus terminalis
Butternut Woollyworm
black morph
red morph
Eriocampa ovata
Lagium atroviolaceum
Macrophya formosa
Macrophya flavicoxae

WEBSPINNING SAWFLIES (family Pamphiliidae) live as larvae in webs or rolled leaves fastened with silk. Adults of most species are 8–15 mm long and found around their host plants in spring. There are 72 species in five genera in North America. ***Acantholyda*** includes 34 species, widespread wherever their conifer hosts occur. ***Acantholyda brunnicans*** ranges west of the Rockies. Pines are the larval host. ***Onycholyda*** (formerly in *Pamphilus*) are eight species that feed on deciduous trees and shrubs. ***Onycholyda luteicornis*** is widespread in eastern and central regions, its larvae feeding on *Rubus* (raspberry and related plants).

ARGID SAWFLIES (Argidae) are recognized by the *long terminal segment of the antennae,* U-shaped in males of some species. Adults measure 8–15 mm. Most larvae feed on leaves of the host plant. The family is widespread but many of our 59 species (in eight genera) occur in the southwest. ***Arge*** is a genus with 10 North American species, ranging well north into Canada and Alaska. The eastern ***Arge humeralis*** feeds on poison ivy. The **Birch Sawfly** *(Arge pectoralis)* of the eastern U.S. and southern Canada also feeds on alder, hazel, hawthorn, and willow. ***Sphacophilus*** includes 22 species, most in the southwest, but ***Sphacophilus cellularis*** is eastern. Sweet potato and morning glory are the host plants.

sawfly cocoon

CIMBICID SAWFLIES (Cimbicidae) may measure a robust 18–25 mm, or up to an inch long. The *antennae are clubbed.* The caterpillar-like larvae feed on foliage of the host plant and spin tough, fibrous cocoons in which they pupate. There are 12 species in three genera in North America. The four ***Zaraea*** species are associated with plants in the honeysuckle family. ***Zaraea lonicerae***, introduced from Europe, now occurs from New York to Maryland and Ohio. ***Trichiosoma*** includes four species. Adults sometimes girdle twigs with their large, strong jaws. ***Trichiosoma triangulum*** is widespread in Canada and the northern U.S. Look for adults and larvae on alder, ash, poplar, willow, birch, and *Prunus.* The **Elm Sawfly** *(Cimbex americana)* is one of four North American species in its genus. Adults vary in color and form and may girdle twigs. The species occurs from Alaska to Newfoundland, south to North Carolina and Oregon. Host trees include maple, alder, birch, basswood, poplar, and willow.

CONIFER SAWFLIES (Diprionidae) include 48 species in six genera, widespread wherever their hosts occur. *Antennae are serrated* (females) or *comblike* (males). Adults are small, short-lived, and seldom seen. Larvae feed in groups on needles when young, dispersing as they age. Sporadic outbreaks can stunt growth and sometimes kill trees. These sawflies are occasional pests in nurseries and tree farms and on ornamental conifers. The genus ***Neodiprion*** includes 38 species, all native to North America. ***Neodiprion lecontei*** occurs in the eastern half of the continent, its larvae feeding on pines.

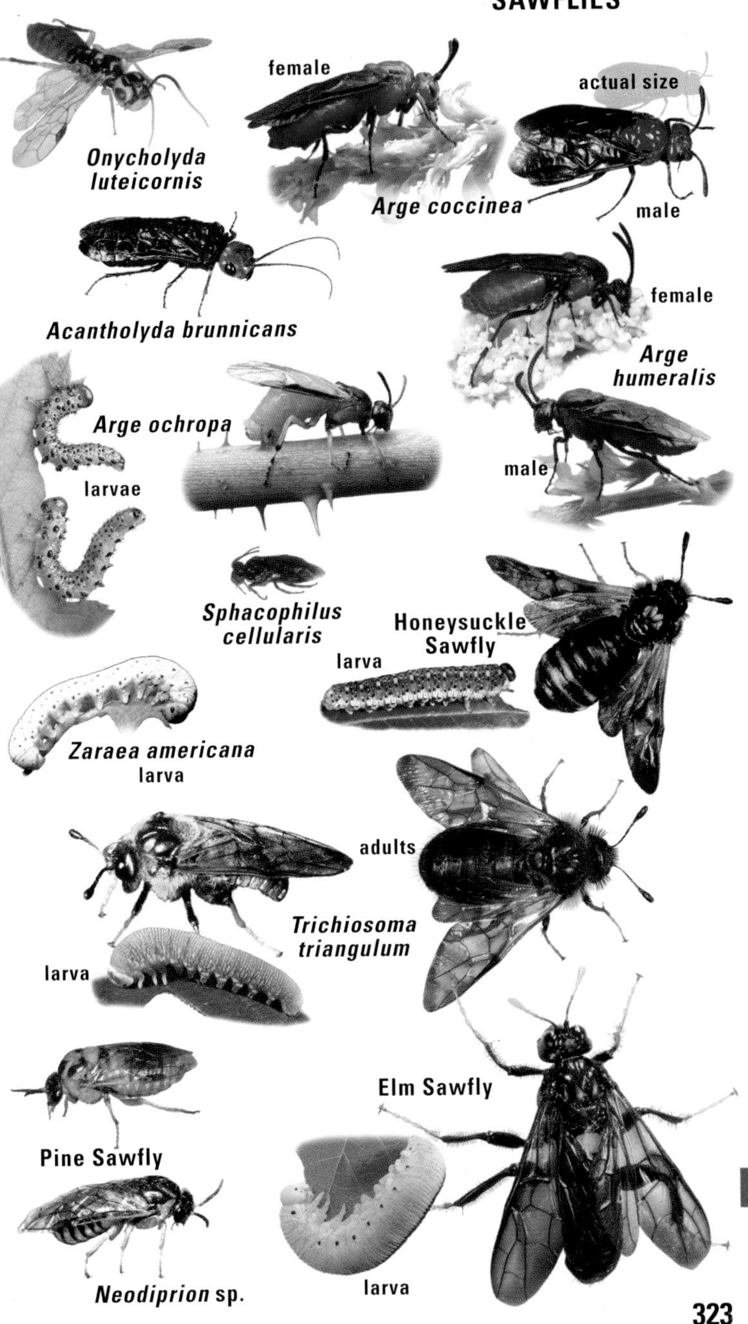
female
actual size
Onycholyda
luteicornis
Arge coccinea
male
Acantholyda brunnicans
female
Arge
humeralis
Arge ochropa
larvae
male
Sphacophilus
cellularis
Honeysuckle
Sawfly
larva
Zaraea americana
larva
adults
Trichiosoma
triangulum
larva
Elm Sawfly
Pine Sawfly
Neodiprion sp.
larva

HORNTAILS (family Siricidae) are *large cylindrical wasps* found in forested areas. Their appearance is intimidating, but they are actually harmless. Lumberjacks have a profane name for these insects, reflecting the female's habit of poking her drill-like egg-laying organ (ovipositor) into stumps, logs, and snags. Larvae are wood borers. Even deep inside a bole, they are not safe from ichneumon wasps of the genus *Megarhyssa* (next page), which parasitize these larvae through equal feats of egg-laying deep within wood. Adult horntails sometimes emerge indoors from structural timber. There are about 19 species of horntails north of Mexico, in five genera. The **Pigeon Tremex** *(Tremex columba)* is a very distinctive horntail found east of the Rockies and southwest to southern California. The female lays eggs in dead or dying maple, elm, oak, beech, and other hardwoods. She also deposits a wood-rotting fungus that allows the larvae to bore more easily. The life cycle takes one to two years. The genus ***Urocerus*** includes five North American species. Males of some species practice "hilltopping," congregating atop promontories to wait for females. ***Urocerus gigas*** is widespread in the northern hemisphere. Its North American subspecies, ***flavicornis,*** occurs across much of Canada and the western U.S., its larvae feeding in pine and other conifers. An adult female lives for two to four weeks and can lay up to 350 eggs, each laid singly in a hole that she pierces in the wood. The total life cycle takes about three years.

WOODWASPS (Xiphydriidae) are represented in North America by ***Xiphydria,*** with six species. They resemble small horntails and have similar habits. Adults may emerge from firewood indoors. ***Xiphydria maculata*** is common in the northeast U.S. and southern Canada, with scattered records farther west and south. Hosts are mostly dead or dying maples. Females deposit eggs in tandem with a wood-rotting fungus. The life cycle takes one year.

ORUSSID WASPS (Orussidae) are rather small cylindrical wasps, with round heads and the *antennae low on the face.* Larvae are probably parasites of wood-boring beetles or horntails. Look for adults on logs or snags, pacing in short, jerky bursts. There are three genera, with six species, in North America, but only ***Orussus*** (four species) is likely to be seen. ***Orussus sayii*** occurs in the eastern U.S. and in Ontario. ***Orussus occidentalis*** is a similar species that occurs in the west.

larva of stem sawfly in tunnel (*Janus* sp.)

STEM SAWFLIES (Cephidae) are miners in stems, shoots, canes, or twigs as larvae, and some species are numerous enough to cause damage to grain crops. Adults are thin cylindrical insects seen near host plants or on yellow flowers. There are six genera in our fauna, with 12 species. The **Wheat Stem Sawfly** *(Cephus cinctus)* is transcontinental in southern Canada and the northern and central U.S. Spring barley, rye, timothy, and native grasses are also hosts in addition to wheat.

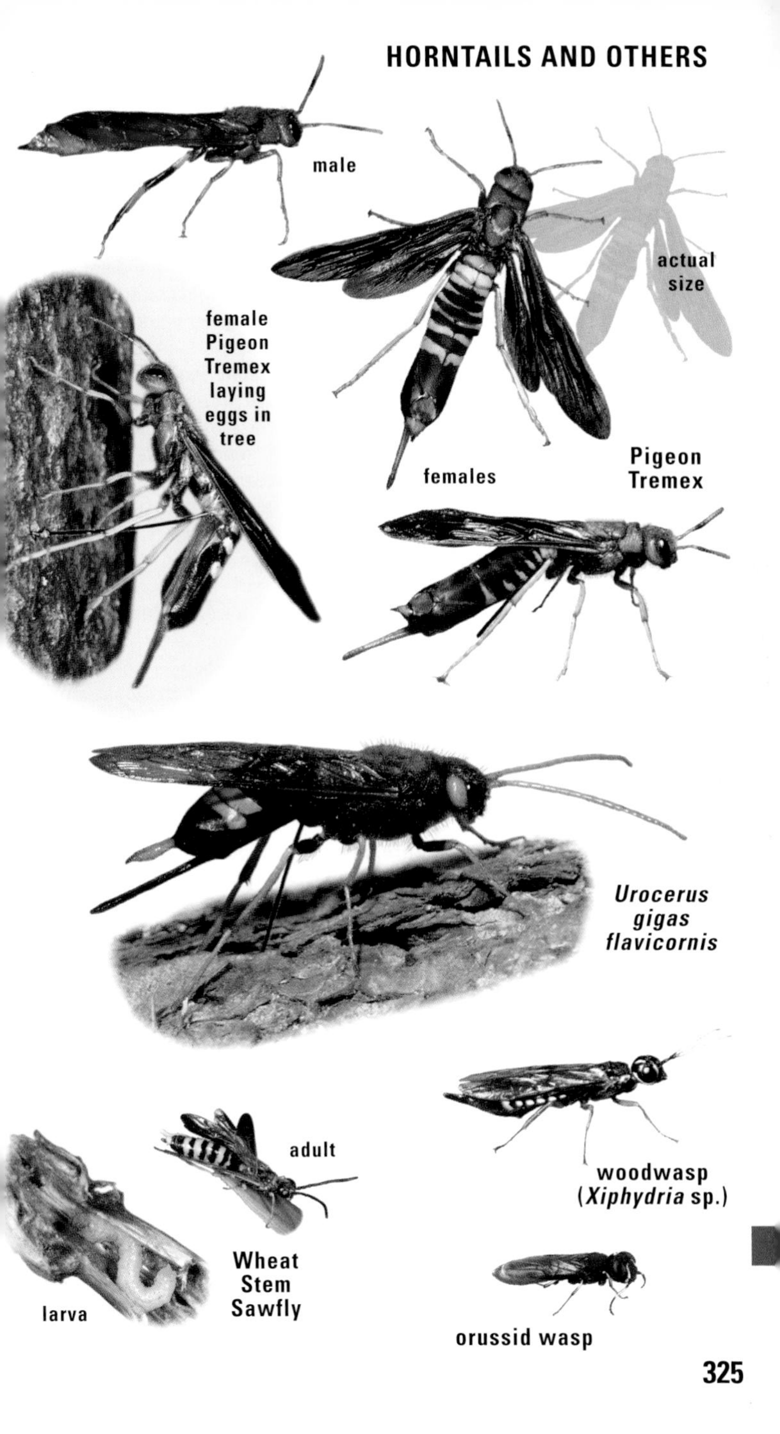
male
actual
size
female
Pigeon
Tremex
laying
eggs in
tree
females
Pigeon
Tremex
Urocerus
gigas
flavicornis
adult
woodwasp
(*Xiphydria* sp.)
Wheat
Stem
Sawfly
larva
orussid wasp

ICHNEUMON WASPS

Think warblers are difficult? They are a cakewalk compared to identifying ichneumon wasps **(family Ichneumonidae).** This is by far the largest family in the order Hymenoptera. Well over 3,300 species are known north of Mexico, and there are probably many more that remain undescribed to science. Most images submitted for this book could not be identified even by experts. There are, however, some characters that can help place a given specimen in this family. Ichneumons have a very great number of antennal segments. They often have a broad white or yellow band on each antenna. Female ichneumons may sport a long, tail-like appendage. This is not a stinger but an ovipositor, an organ used to lay eggs. Some species have short ovipositors that do function as stingers. Similar families include Aulacidae, Gasteruptiidae, and Pelecinidae (next page), and Braconidae (p. 330). The majority of ichneumons are parasitic as larvae in the caterpillars of moths and butterflies or in sawfly larvae.

The **subfamily Pimplinae** includes species that are parasites of the pupae of moths and butterflies, or on caterpillars that are about to pupate. However, ***Dolichomitus irritator*** of eastern North America parasitizes the larvae of wood-boring beetles. ***Acrotaphus wiltii*** is widespread across the eastern U.S. Its only known host is a single species of orb-weaving spider. The female wasp stings the spider into a brief paralysis, allowing her to place an egg on the host. The larva that hatches feeds as an external parasite. Members of the genus ***Pimpla*** are widespread across the U.S.

The **subfamily Anomaloninae** includes species found in relatively dry habitats. Collectively they are parasites of immature Lepidoptera and beetles. The female wasp inserts her egg in the larva of the host, but emergence of a new adult wasp is always from the pupa of the host. ***Theronia atalantae*** is transcontinental in Canada and parts of the northern U.S.

Members of the **subfamily Xoridinae** are external parasites of immature wood-boring beetles and horntail wasps. Larvae of these insects are the usual hosts, but sometimes pupae or even preemergent adults are attacked. Included here is the genus ***Xorides.*** Look for the females as they prospect for hosts on the surface of logs. Members of the **subfamily Acaenitinae** are typically internal parasites of immature wood-boring insects. ***Arotes amoenus*** is one of four species in its genus in the eastern U.S. and southern Canada; all are parasites of beetles.

The **subfamily Rhyssinae** features our largest ichneumons. The genus ***Megarhyssa*** includes four North American giants parasitic on the larvae of wood-boring horntail wasps (preceding page). Females somehow "divine" the location of a host, then use their long, whiplike ovipositor to penetrate solid wood (either drilling or winding through cracks and crevices, no one is certain which), laying an egg on the horntail grub. During this delicate process, the wasp is vulnerable to predation, and one often finds long "hairs" embedded in logs or trees that are in fact the ovipositors of the doomed adults. Meanwhile, the ichneumon larva lives on as an external parasite of its host.

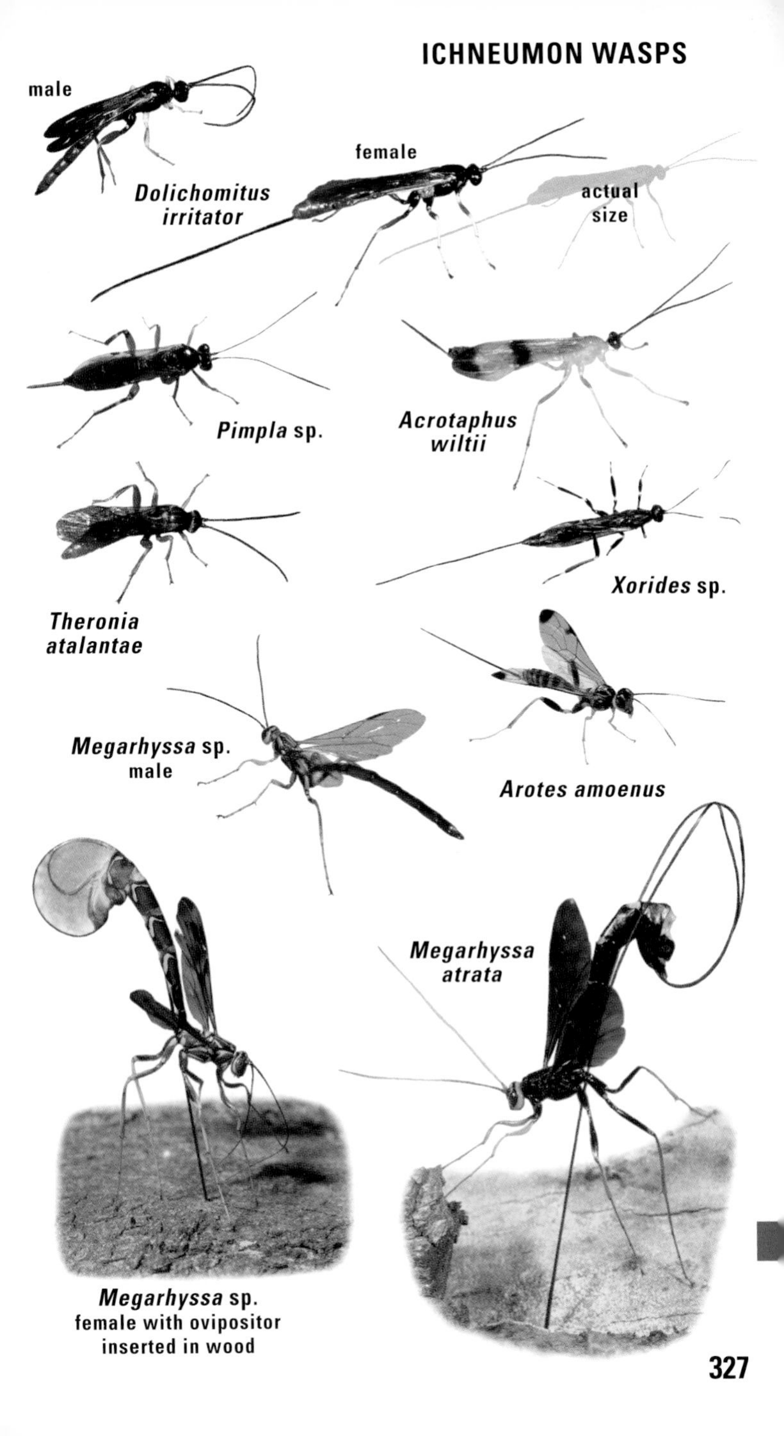

Megarhyssa sp.
female with ovipositor inserted in wood

ICHNEUMONS AND OTHER WASPS

The ichneumons, a huge family, are introduced on the previous page. Also treated below are members of four small families of wasps.

The **Ichneumoninae** make up the second largest subfamily of ichneumons, and one of the most diverse. All are internal parasites of caterpillars but emerge as adult wasps from the pupa of the host. Look for females seeking their quarry on foliage or among leaf-litter. Males and females of the same species are often colored differently. Females typically have broad pale bands on dark antennae. In males, these bands are subdued or even absent. Females have venom glands and some can sting weakly. Females of some species hibernate as adults under loose bark on logs and in other protected situations. The genus ***Eutanyacra*** includes more than 15 species in North America. ***Gnamptopelta obsidianator*** is our only species in its genus. The genus ***Trogus,*** with six species, is found widely across this continent.

The **subfamily Ophioninae** includes large, gangly, pale brownish or yellowish species that often come to lights at night. Genera from other subfamilies can be nearly identical, however, in appearance and nocturnal habits. Ophionines are typically internal parasites of large caterpillars. New adults emerge from the pupa or cocoon of the host. Curiously, one species of ***Ophion*** is parasitic on white grubs of *Phyllophaga* May beetles.

The **subfamily Phygadeuontinae** (or Cryptinae) is the largest subfamily in the Ichneumonidae. Most of its members are external parasites of the pupae of other insects with complete metamorphosis. Some parasitize the egg sacs of spiders and pseudoscorpions. Still others can develop as parasites of parasites (secondary parasites or hyperparasites). Some species in the genus ***Gelis*** (not illustrated) are wingless and resemble ants.

ENSIGN WASPS (family Evaniidae) are small, with tiny flaglike abdomens and wings folded flat across the back while they crawl over surfaces. They are parasites of the egg capsules of cockroaches. There are four genera in eastern North America, with 11 species. Look for them on tree trunks.

AULACIDS (Aulacidae) resemble ichneumons, but the abdomen stems high off the thorax. They are parasites of wood-boring beetles or wasps. There are two genera with 29 species in North America, collectively widespread. ***Pristaulacus flavicrurus*** occurs in the northeast, south to South Carolina. ***Pristaulacus fasciatus,*** which appears to mimic a spider wasp, ranges widely in the east.

Adult **GASTERUPTIIDS (Gasteruptiidae)** often frequent flowers of the parsley family. The "neck" connecting the head to the thorax, and the swollen hind "ankles," are distinctive. As larvae they are parasites of solitary bees or wasps that nest in wood, consuming the larval hosts and/or their provisions. ***Gasteruption*** is our only genus, with 15 species.

PELECINIDS (Pelecinidae) are represented in North America by ***Pelecinus polyturator*** of the eastern half of the U.S. Females are recognized by the extremely long abdomen and swollen hind "ankles." Males are much smaller and are rare, or at least rarely noticed. In the larval stage, these insects live as internal parasites of ground-dwelling scarab beetle grubs or of beetle larvae found in rotting wood.

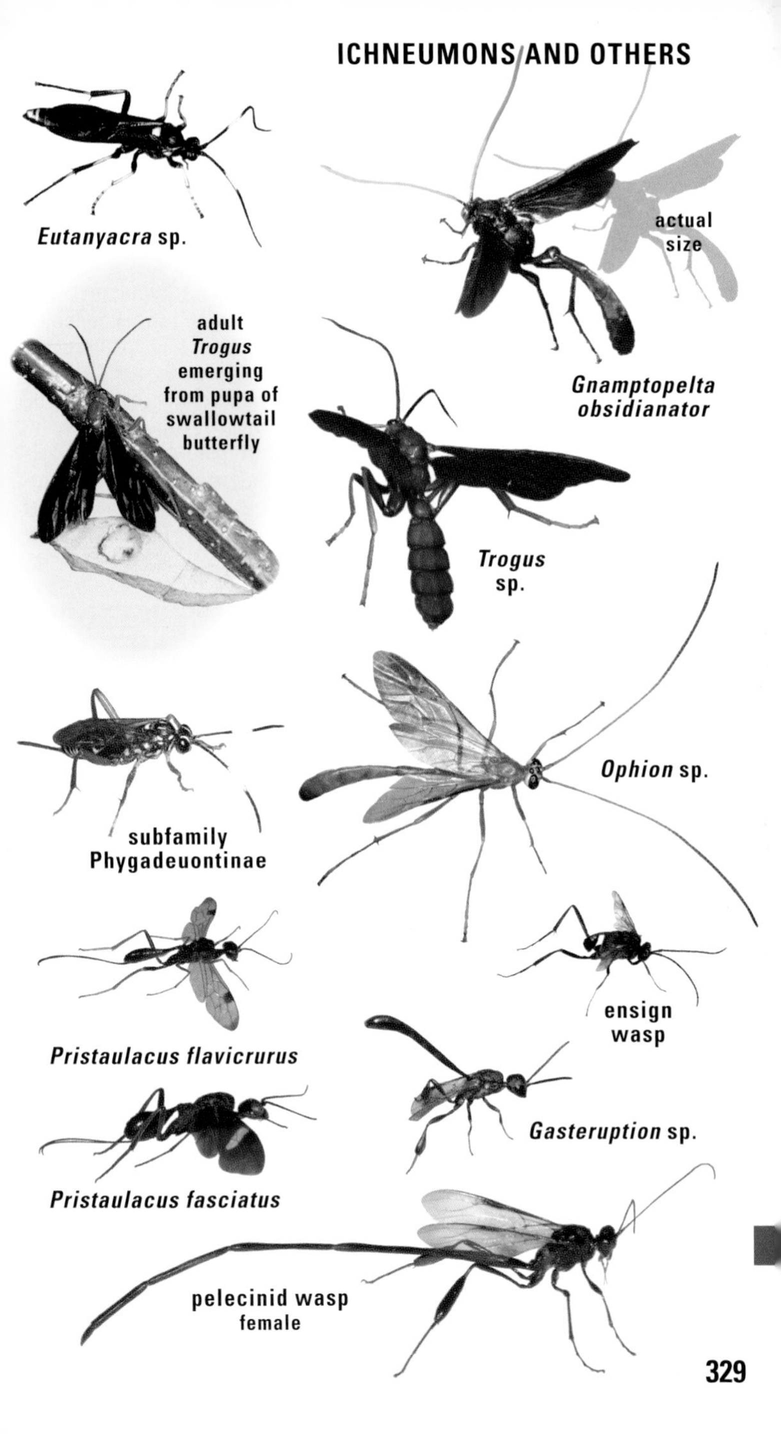
Eutanyacra sp.
actual size
Gnamptopelta obsidianator
adult *Trogus* emerging from pupa of swallowtail butterfly
Trogus sp.
Ophion sp.
subfamily Phygadeuontinae
ensign wasp
Pristaulacus flavicrurus
Gasteruption sp.
Pristaulacus fasciatus
pelecinid wasp female

BRACONID WASPS (family Braconidae) are parasites of a variety of other insects. We have more than 1,900 species; most are quite small, but larger species may be confused with ichneumon wasps. With a close study, braconids are recognized by the fusion of the first and second dorsal abdominal segments. Other distinguishing characters, including wing venation, are equally invisible in the field. All known species, in the larval stage, are parasitic on or in other insects. Many are parasitic in caterpillars. A few species are capable of polyembryony, whereby a single egg can yield hundreds of offspring. Caterpillars that seem to be sprouting "eggs" or "balloons" from their backs are actually sporting the cocoons of braconid pupae. The host is usually doomed to die before completing its own metamorphosis. The host-specific tendencies of braconids have made them great allies in the ongoing war against agricultural pests. Some are raised in captivity for mass release in cultivated fields.

The genus ***Meteorus*** includes about 40 North American species. Most are parasites of caterpillars. ***Leiophron uniformis*** is one of 19 North American species in its genus. Its only known host is a species of plant bug. The genus ***Atanycolus*** includes 36 species. They are external parasites of woodboring beetles. ***Spathius elegans*** is widespread in the eastern U.S. Members of the genus ***Dinocampus*** are common parasites of adult lady beetles (pp. 154–159). Mature larvae exit the comatose beetle and spin a cocoon beneath it. ***Dinocampus coccinellae*** is the common species. The genus ***Cotesia*** includes many species, some introduced from elsewhere. All are internal parasites of larval butterflies and moths. They are most evident in the pupal stage, as mature larvae erupt from the host caterpillar, spinning white silken cocoons on the outside of the host's body. Members of the genus ***Agathis*** are solitary, internal parasites of caterpillars. The female wasp lays an egg in the egg of its host, but the larva that hatches allows its host caterpillar to continue development before eventually killing it. The braconid larva may spin its own cocoon inside the cocoon spun by its host.

Most **CHALCIDIDS (family Chalcididae)** are parasitic in pupae of moths and butterflies, or in maggots or pupae of flies. Adults usually have swollen "thighs" on the hind legs. More than 100 species are known for North America. Our two species of ***Acanthochalcis*** are giants of the family at about half an inch long; most chalcidids are much smaller. **TORYMIDS (Torymidae)** include over 600 species in North America. Most are bright metallic green, blue, or purple. ***Torymus*** species are mostly parasitic in galls formed by other wasps and gall midges.

SEED CHALCIDS (Eurytomidae) are named for a minority of species that develop as larvae in seeds or grass stems. Of our more than 240 species, at least an equal number are parasitic on other insects. The 63 species in the genus ***Tetramesa*** feed as larvae in grass stems. Related genera include "joint worms" that cause swellings in the stems of cereal crops. **EULOPHIDS (Eulophidae)** include more than 500 species in North America. Hosts include leaf-mining and leaf-rolling insects, wood-boring beetles, and Neuroptera. The genus ***Tetrastichus*** includes more than 100 species here.

actual size

Meteorus sp.

Leiophron uniformis

Atanycolus sp.

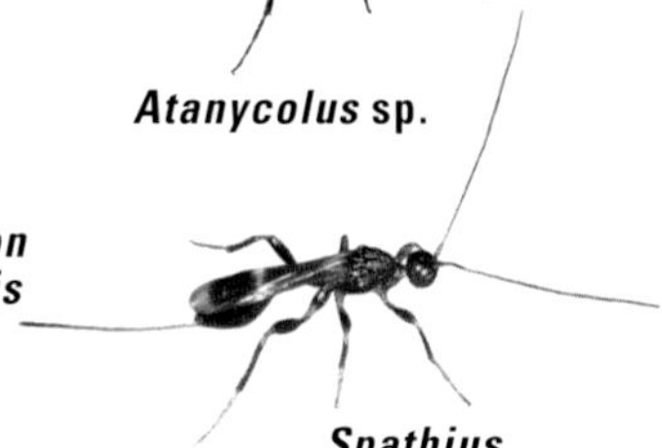

Spathius elegans

Dinocampus coccinellae parasitizing a lady beetle

Agathis sp.

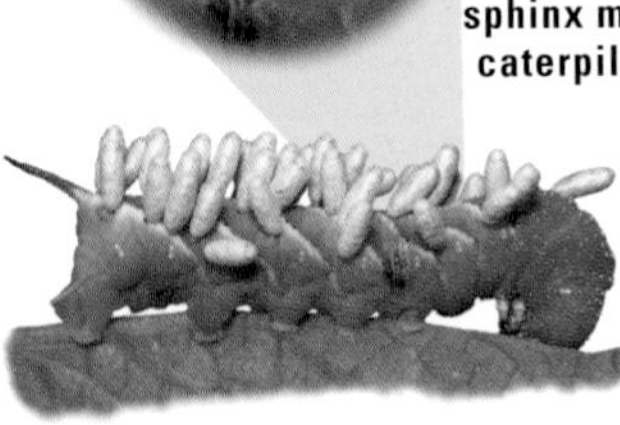

Cotesia pupae on sphinx moth caterpillar

Acanthochalcis sp.

Chalcididae sp. (2 examples)

a eulophid (*Tetrastichus*) laying egg in gall

Tetramesa sp.

Torymus sp.

CUCKOO WASPS (family Chrysididae) are also called gold wasps. Small, metallic, and heavily armored, they are mostly parasites in the nests of other wasps or bees. The female slips into the nest of the host, laying an egg inside. After the egg hatches, the cuckoo wasp's larva eats the rightful occupant and/or in many cases the food stored in the nest. One group parasitizes the larvae of sawflies, and a few drab cuckoo wasps parasitize walkingstick eggs. Lacking a sting, adults frequently curl into a ball for defense. Many of our 245 species visit aphid colonies or flowers, or seek hosts around old barns and dead trees. Some sweat bees (p. 340) look similar.

Many of our cuckoo wasps are classified in ***Chrysis,*** a complex genus of at least 77 North American species. Members of this group have four to six tiny "teeth" on the hind edge of the abdomen. The genus ***Holopyga*** includes eight species in North America. ***Holopyga ventralis*** is common throughout the U.S. and Canada, wherever its hosts, *Bicyrtes* sand wasps, can be found (p. 336). Look for *Holopyga* visiting flowers.

Members of the genus ***Parnopes*** have long tongues. Our seven species are parasites of sand wasps. ***Parnopes edwardsii*** is common on umbrella-shaped flowers west of the Rockies, from Alaska to Mexico. Known hosts for this species are various sand wasps belonging to the genera *Bembix* and *Steniolia* (p. 336).

PTEROMALIDS (family Pteromalidae) may represent the largest family of the micro-Hymenoptera, but many scientists consider this to be a highly artificial assemblage. As such, it is difficult to characterize. Many of the groups currently treated as subfamilies may deserve family status. Most species for which the biology is known are parasites of the larvae of other insects, especially various flies, beetles, lepidopterans, scale insects, and other bees and wasps. Some are hyperparasites: that is, they are parasitic on flies or wasps that are themselves parasites of caterpillars or other insects.

Formerly considered a subfamily of the Pteromalidae, but now treated separately, are the **PERILAMPIDS (Perilampidae).** Members of the genus ***Perilampus*** (24 species, widely distributed) are brilliantly metallic and might be mistaken for cuckoo wasps. Perilampids develop as parasites or hyperparasites. Their eggs hatch into planidial larvae, small and flat creatures that can wait for long periods without feeding until they have the chance to attach themselves to a passing host insect.

LEUCOSPIDS (Leucospidae) may be mistaken for mason wasps at first glance. The wings are folded lengthwise at rest, the "thighs" of the hind legs are swollen, and the female's ovipositor curls over the back. North America has about half a dozen species, widespread but usually uncommon. Adults are generally seen on flowers. Their larvae develop as parasites of the larvae of solitary bees and wasps.

cuckoo wasps
(Chrysididae)
3 examples

Parnopes edwardsii

Holopyga ventralis

Leucospis sp.

Perilampus sp.

SOLITARY WASPS

(family Sphecidae) include mud daubers, sand wasps, and related hunting wasps. Females are typically specialized predators of other insects, storing the paralyzed bodies of their victims in burrows, existing cavities, or previously built mud cells. Perhaps our most undervalued insects, many are gentle and colorful, with interesting behaviors to observe.

Our two species of ***Chlorion*** prey on field crickets. ***Chlorion aerarium*** occurs over the entire U.S. in blue, green, and violet forms. They dig nest burrows in sandy areas, often near railroad beds. Members of the genus ***Podium*** prey on wood roaches. ***Podium luctuosum*** ranges over the east, usually in forests where it scours the ground and logs for prey.

The genus ***Sphex*** includes 12 North American species of regal wasps. Females dig burrows in soil, each tunnel terminating in several cells. Katydids are the principal prey. At rest, *Sphex* hold their wings flat over the abdomen. The **Great Golden Digger** *(Sphex ichnumoneus)* averages almost an inch long. Look for it on flowers. ***Sphex jamaicensis,*** restricted to Florida, nests in aggregations of 20–40 individuals. The **Great Black Wasp** *(Sphex pensylvanicus)* occurs throughout the U.S. except the northwest. The **Blue Mud Dauber** *(Chalybion californicum)* is abundant throughout the U.S. and into British Columbia. It renovates abandoned nests of *Sceliphron* wasps (below), carrying water to the nest to remold it, then stocking the cells with spiders, sometimes including Black Widows. Mud nests with a bumpy look are the work of this wasp. Females of the **Black-and-yellow Mud Dauber** *(Sceliphron caementarium)* gather mud balls they form into cells, frequently plastered beneath bridges and barn roofs. They pack large numbers of various spiders into each chamber.

Some species of **grass carrier wasps** in the genus ***Isodontia*** close their nests with plugs of dry grass. Tree crickets are their preferred prey. At rest, *Isodontia* hold their wings splayed at an angle. ***Isodontia apicalis*** is common in parts of the east, while ***Isodontia elegans*** occurs mostly in the west. The 10 species of ***Palmodes*** nest in the ground, females digging their own burrows and storing katydids for their offspring. Females of ***Palmodes dimidiatus*** vary in color from an all-black to all-red abdomen. The genus ***Prionyx*** includes six species that use grasshoppers as food for their young. In ***Prionyx parkeri,*** males sleep in "bachelor parties." Our 19 species of ***Podalonia*** are valued for dispatching cutworms. One caterpillar feeds a single wasp larva in a simple burrow.

Eremnophila aureonotata is the only species in its genus north of Mexico. The female stocks her burrow with a single caterpillar. ***Ammophila*** is a genus of at least 61 species of thread-waisted wasps, most diverse in desert regions. The simple, single-celled burrows are provisioned with caterpillars. Some species hold a pebble or other object in their jaws and use it to pack down the soil to close a completed burrow. White sweet clover is a favorite nectaring flower. At dusk they settle down to sleep, often in loose aggregations, on dry vegetation. ***Ammophila nigricans*** is the largest eastern species. The **Common Thread-waisted Wasp** *(Ammophila procera)* ranges widely across the U.S.

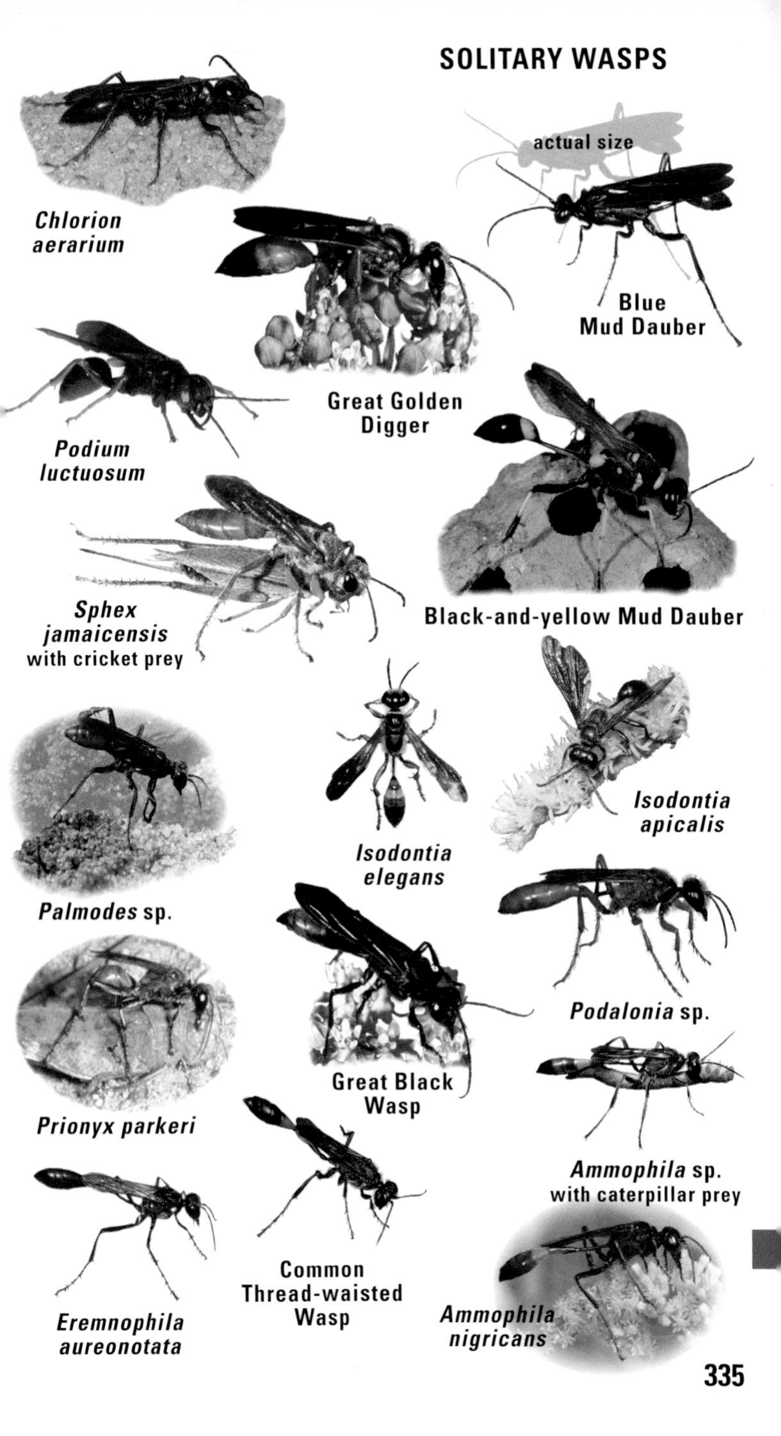

Chlorion aerarium

Great Golden Digger

Blue Mud Dauber

Podium luctuosum

Black-and-yellow Mud Dauber

Sphex jamaicensis with cricket prey

Isodontia elegans

Isodontia apicalis

Palmodes sp.

Podalonia sp.

Prionyx parkeri

Great Black Wasp

Ammophila sp. with caterpillar prey

Eremnophila aureonotata

Common Thread-waisted Wasp

Ammophila nigricans

Beewolves in the genus ***Philanthus*** take other stinging insects as prey for their offspring. A European beewolf captivated famed entomologist Jean Henri Fabre, and ethologist Nikolaas Tinbergen won a share of a 1973 Nobel Prize in part for his landmark experiments with that insect. Our species are no less fascinating. Males of some are territorial, scent-marking perches that they defend from other males. Females of many species are gregarious, excavating burrows close together in patches of sandy soil or dirt banks. Each tunnel usually has multiple cells. Several paralyzed bees or wasps are placed in each cell as food for the larval wasp.

Weevil wasps in the genera ***Cerceris*** and ***Eucerceris*** hunt beetles as food for their young. Adults have blunt abdomens with constrictions between the segments. Females tend to nest in firm soil along the edges of paths, neglected baseball diamonds, and similar situations. ***Cerceris compacta*** ranges in the eastern and southwestern U.S. Its only known prey is a species of leaf beetle. ***Eucerceris tricolor*** occurs in the southwest.

Our 18 species of ***Bembix*** are *the* sand wasps that usually nest in dunes, beaches, even golf course bunkers. Individual females often form dense nesting aggregations. Females prey on flies and practice "progressive provisioning," feeding the developing larvae as needed, much as adult birds do for their chicks. ***Stictiella*** (10 species) and ***Glenostictia*** (19 species) are very similar to *Bembix* and also nest in sandy areas. *Stictiella* species prey on moths or butterflies, while *Glenostictia* provision their nest burrows with flies. ***Steniolia*** are ornately marked western insects similar to other sand wasps. Males have long "tongues" and visit thistles regularly. Females nest in powdery soil, feeding the growing larva on paralyzed flies.

Hoplisoides includes 17 North American species. They paralyze treehoppers, leafhoppers, or fulgorids for their offspring. ***Bicyrtes*** hunt bugs, particularly nymphs of stink bugs and leaf-footed bugs. The adult wasps are less alert and nervous than other sand wasps. The 16 species of ***Gorytes*** paralyze leafhoppers, spittlebugs, treehoppers, or fulgorids as food for their larvae. Females dig nest burrows with one to four cells. ***Alysson*** includes nine North American species that prey mostly on leafhoppers.

The **Horse Guard** *(Stictia carolina)*, widespread in the south and east, is named for the behavior of females, which hover around equines to catch horse flies. These large wasps may treat humans in a similar manner (as fly lures) and so seem more intimidating than they are. Females nest in sand, sometimes in dense colonies occasionally numbering in the thousands. Paralyzed flies are fed progressively to the single larva in each burrow. The tunnel is closed each time, the wasp using subtle landmarks to find it again. Meanwhile, we cannot remember where we parked the car.

Cicada killers *(Sphecius)* are four species of spectacular wasps. "Screeching" cicadas, deviating from their normal monotonous buzz, are often being attacked by these giant wasps. Nests are excavated in soil, the tunnels usually branching and terminating in cells, with an average of nearly 16 cells per nest. Look for burrows of cicada killers in your garden. The entrance mound might be mistaken for the diggings of a small rodent.

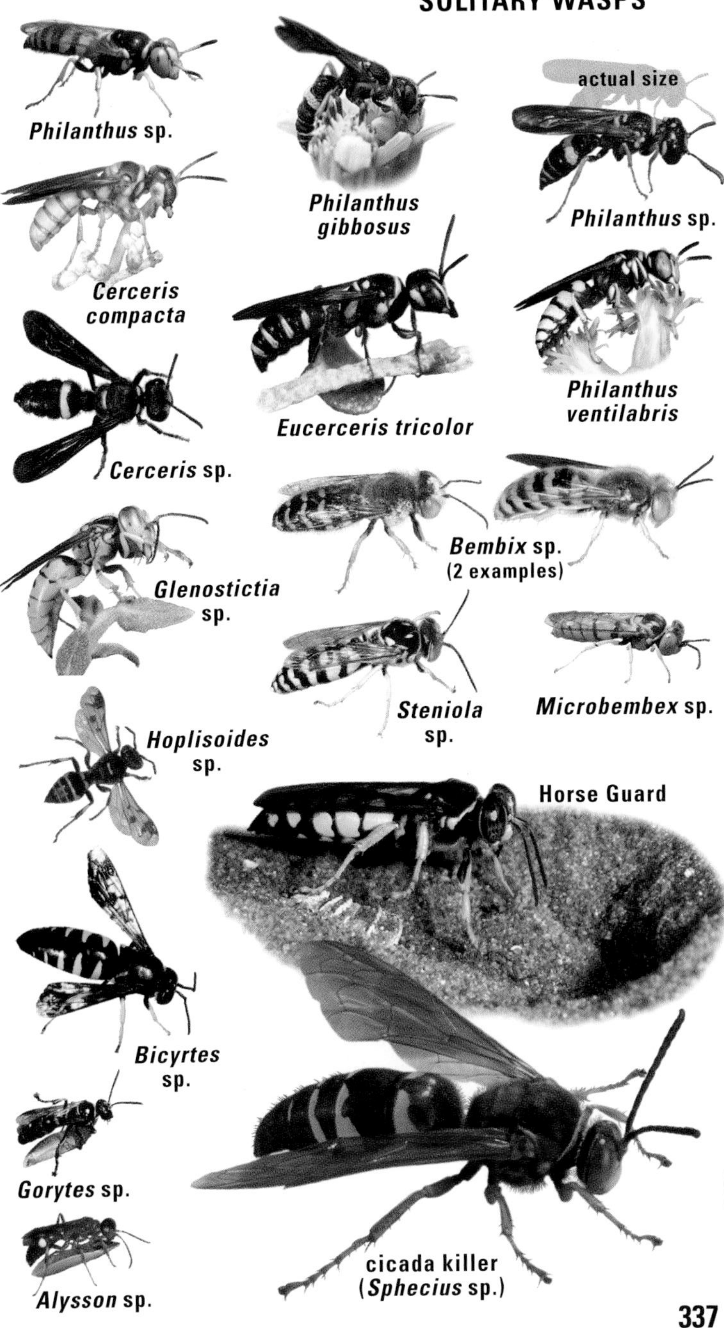

Philanthus sp.

Philanthus gibbosus

Philanthus sp.

Cerceris compacta

Eucerceris tricolor

Philanthus ventilabris

Cerceris sp.

Bembix sp. (2 examples)

Glenostictia sp.

Steniola sp.

Microbembex sp.

Hoplisoides sp.

Horse Guard

Bicyrtes sp.

Gorytes sp.

Alysson sp.

cicada killer (*Sphecius* sp.)

Mimesa is a genus of 120 species of leafhopper predators. Adult females nest in the ground, excavating a burrow with a handful of cells branching off. Our 12 species in the genus ***Pemphredon*** are aphid hunters frequently seen around aphid colonies on trees and weeds. Females nest in pith of cut twigs, in abandoned beetle borings in solid wood, and in rotten wood. They partition the cavity into a series of cells, each stocked with aphids for a single offspring. The genus ***Astata*** includes 14 species of stink bug predators in North America. The sexes are very different, males having broader hindwings and eyes that meet at the top of the head. Males perch atop weeds and other elevated objects in open areas, periodically darting out and returning. The slower-flying females collect prey in a complex, multicellular burrow, with several victims and one egg per cell.

The genus ***Larropsis*** includes 25 species north of Mexico, most in the western U.S. Females provide for their offspring by paralyzing camel crickets and burying them in burrows that may be the crickets' own. At least 42 species of ***Tachysphex*** occur in North America. The sharply pointed abdomen is distinctive, and many species are bicolored black and red.

Members of the genus ***Trypoxylon*** are long-bodied wasps that prey on spiders. Most of our 27 species are small and nest in preexisting cavities in wood, but the **Pipe Organ Mud Dauber** *(Trypoxylon politum)* is large and builds mud nests. The male has a hook on the underside of his abdomen and guards the nest against parasites while the female is away seeking prey or gathering mud. Our 29 species of the genus ***Crabro*** are small, slender wasps that stock paralyzed flies for their offspring in burrows each female digs in the soil. Males have shieldlike plates on their "wrists," used to cover the eyes of the female during mating. This does not blind his prospective mate, but filters light in a unique pattern that may help her judge the male's fitness. Members of the genus ***Ectemnius*** are small to medium-sized wasps that usually nest in the soft pith of twigs, partitioned into several cells. Paralyzed flies are cached for the larvae to consume.

TIPHIID WASPS (family Tiphiidae) are parasites of beetle larvae in their own larval stage. Some species have even been imported to help combat the Japanese Beetle and other pest scarabs. Females of some species are wingless, and the genders of some differ markedly in body form. Antennae of females are often short and coiled. The genus ***Tiphia,*** with 95 North American species, is typical of the family. Species vary in their seasonal timing, but all look essentially identical and can be identified only by experts. The genus ***Paratiphia*** is most abundant and diverse in arid habitats of the west, with 27 species north of Mexico. ***Paratiphia robusta*** is typical, ranging from British Columbia to Texas. ***Myzinum*** are the largest and most colorful members of the family to be found here. This mostly tropical genus is represented by 17 species in North America. The sexes differ radically in appearance. The slender males sport a "pseudostinger" at the tip of the abdomen. The robust females have the real thing. Species of ***Methocha*** (not illustrated) are parasites of predatory tiger beetle larvae, which takes some daring on the part of the wingless and antlike females.

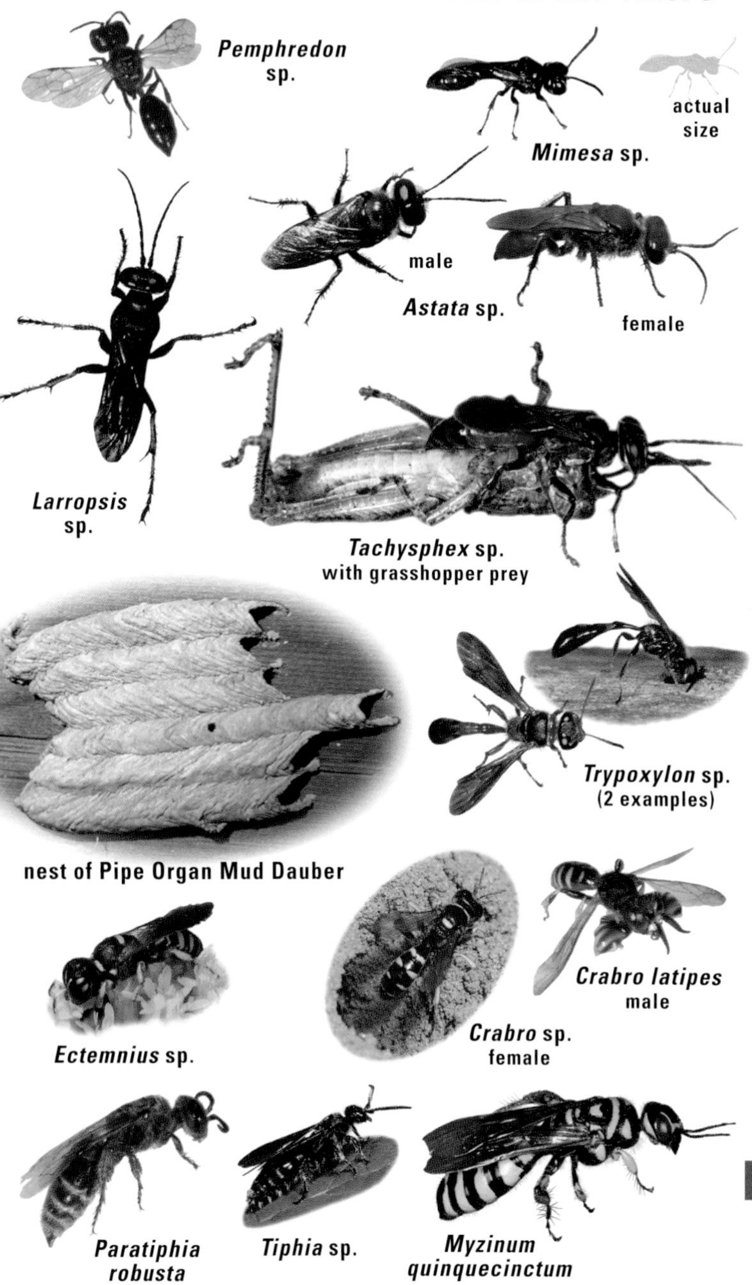

Pemphredon sp.
Mimesa sp.
actual size
male
Astata sp.
female
Larropsis sp.
Tachysphex sp. with grasshopper prey
Trypoxylon sp. (2 examples)
nest of Pipe Organ Mud Dauber
Crabro latipes male
Ectemnius sp.
Crabro sp. female
Paratiphia robusta
Tiphia sp.
Myzinum quinquecinctum

PLASTERER BEES (family Colletidae) are named for the habit of some females of lining their nests with a waterproof substance. Members of the genus ***Colletes*** (not illustrated) are "polyester bees," some of which are among the first bees of spring. The 96 species collectively occur over most of North America, often nesting in dense aggregations. Each female digs her own burrow in sandy soil, lining the cells with a natural plastic secreted from abdominal glands. This polymer bag holds a pool of nectar and pollen, keeping it fresh and fungus-free. An egg is suspended from the ceiling, and the larva that hatches feeds on the sweet store of food. The genus ***Hylaeus*** includes yellow-faced or "masked" bees, named for their pale faces, especially in males. These small insects lack body hair and pollen baskets and are frequently mistaken for wasps. Females usually nest in hollow twigs or stems. There are about 50 species north of Mexico.

MINING BEES (Andrenidae) include about 1,200 species of solitary bees in North America. Members of the genus ***Andrena*** are mostly medium-sized, abundant springtime bees. Some even emerge before all snow has vanished. Many species of *Andrena* are best identified by association with the flowers they pollinate. Nests are burrows in the soil, the entrance often hidden beneath a fallen leaf or other litter. Like any good homebuilder, the female waterproofs the walls of her brood cells, using a secretion from a gland in her abdomen. The genus ***Perdita*** includes at least 500 species of small, nearly hairless bees. Especially abundant in arid habitats, many depend on only a few closely related species of flowers for pollen and nectar. Many species nest communally, or at least more than one female uses a single burrow entrance. Two southwestern species gather pollen from evening primroses from dusk until just after dark.

SWEAT BEES (Halictidae) are generally small to medium-sized. The family exhibits behaviors ranging from solitary to semisocial. Nests are burrows in the soil or sometimes rotting wood. Over 500 species range north of Mexico. Species of ***Halictus*** often resemble *Andrena,* typically with a banded abdomen. All known species nest in soil. Most exhibit some degree of sociality, the usual pattern being for daughters to remain with their mother, expanding the nest and helping to rear additional generations of her offspring throughout the nesting season. Males are normally produced as part of the last generation, in late summer or fall.

Augochlora, Augochlorella, and ***Augochloropsis*** are all similar, brilliant metallic green or blue. *Augochlorella* can also be copper or bronze. They pollinate many flowers but can also be found around aphid colonies, feeding on honeydew secreted by those insects. Our 14 species of ***Agapostemon*** are at least partly metallic green, often with a banded abdomen. ***Lasioglossum*** are the "sweat bees" most often alighting on your arm. They are small, usually glossy, weakly metallic, with patterns of short hairs on the abdomen. About 280 species live north of Mexico. Our 74 species of ***Sphecodes*** are generally recognized by the all-red abdomen of females and the usually red-and-black abdomen of males. They are small to medium in size, and all are parasites of other bees, especially other halictids.

Hylaeus sp.

Andrena sp.
(3 examples)

Perdita sp.

Halictus sp.
(3 examples)

Lasioglossum sp.

Augochlora sp.

Augochlorella sp.

Agapostemon sp.

Augochloropsis sp.

Sphecodes sp.

LEAFCUTTER, MASON, AND RESIN BEES

(family Megachilidae) may be recognized by the "pollen brush" on the underside of the female's abdomen, not "baskets" on the hind legs as in most bees. More than 600 species are known in the U.S. and Canada.

Resin bees in the genera ***Anthidiellum*** (three species) and ***Dianthidium*** (20 species) are medium-sized chunky insects. Their nests are lumps of sand grains or pebbles glued together with plant resins and attached to stems or under rocks or ledges. Each nest is a single cell from which one adult bee will eventually emerge. ***Anthidiellum erhorni*** is southwestern, while ***Dianthidium subparvum*** is more widespread in the west.

Anthidium (27 species) and ***Trachusa*** (14 species, not illustrated) are larger versions of the two preceding genera, found mostly in arid habitats of the west. Male *Anthidium* have large teeth on the posterior segments of the abdomen and use them to battle other males for territories. Females excavate burrows in the soil. *Anthidium* females fashion cells lined with plant fibers, earning them the name "cotton bees" or "carders." ***Anthidium illustre*** occurs from Oregon to the Baja peninsula, and east to Utah.

Our 45 species of ***Coelioxys*** are recognized by the pointed abdomen in females. Males have obvious teeth on the tip of the conical abdomen. Both sexes sleep singly, gripping the tip of a grass stem with their jaws. In North America, these insects are cleptoparasitic in nests of *Megachile,* the larva consuming the pollen and nectar stores intended for the offspring of its host. ***Coelioxys octodentata*** ranges from southern Canada to Mexico.

Leafcutter bees in the genus ***Megachile*** include about 140 species north of Mexico. Females shear neat ovals and perfect circles from leaves, using the pieces to construct barrel-like cells. The cells are stacked end-to-end inside preexisting tunnels in wood or burrows in the ground. The **Alfalfa Leafcutter Bee** (*Megachile rotundata,* not shown) is a Eurasian species introduced here in the 1940s to aid in pollination of that crop. The females nest in "bee boards" stacked in portable sheds provided by farmers. The **Giant Resin Bee** *(Megachile sculpturalis)* is a very large species native to Japan and east Asia. First recorded here in 1994 in North Carolina, it is now widespread in the eastern U.S. It is easily confused with large carpenter bees but is more slender. It nests in abandoned tunnels bored by carpenter bees and also in dead bamboo stems and similar cavities.

Members of the genus ***Lithurge*** closely resemble *Megachile.* The five species occur throughout the U.S. except the Pacific northwest. Species of ***Hoplitis*** nest in preexisting cavities, hollow pith of sumac stems, etc. This genus is most common in the west, with about 25 species.

Mason bees in the genus ***Osmia*** include some introduced "orchard bees" employed to pollinate apples, pears, and other fruits. Most are dark metallic green or blue and quite hairy. They nest in preexisting tunnels in wood, partitioning the tube into cells and sealing the entrance with mud when finished. Bundles of paper straws or cardboard tubes, or blocks of wood with drilled holes of varying diameters and depths, make attractive nest sites for the females. Provide a roof for shelter from rain and place the bundle on a south-facing surface at least three feet off the ground.

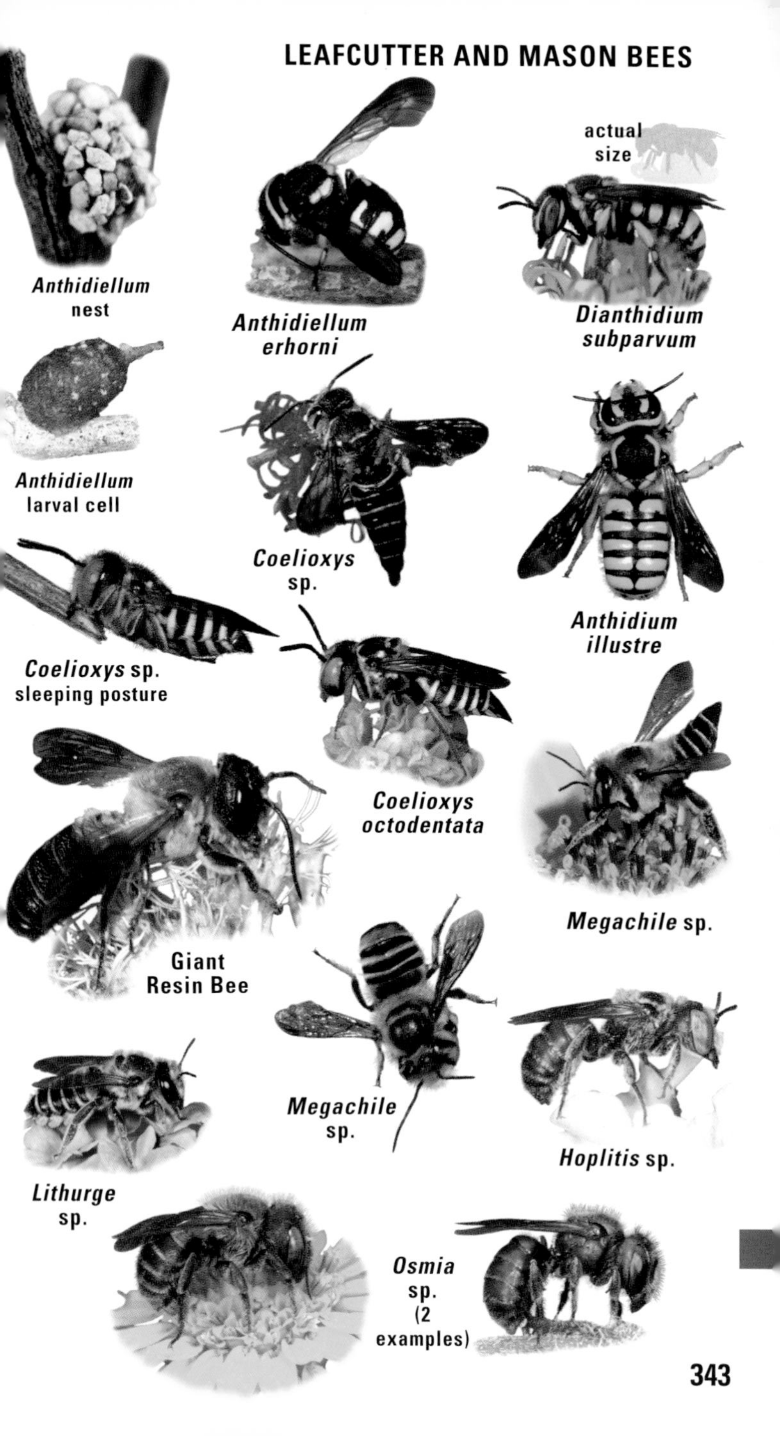

Anthidiellum nest

Anthidiellum erhorni

Dianthidium subparvum

Anthidiellum larval cell

Coelioxys sp.

Anthidium illustre

Coelioxys sp. sleeping posture

Coelioxys octodentata

Giant Resin Bee

Megachile sp.

Megachile sp.

Hoplitis sp.

Lithurge sp.

Osmia sp. (2 examples)

BUMBLE BEES AND CARPENTER BEES

Bumble bees **(genus Bombus)** are our only native social bees. The dense blanket of hair helps to insulate these essentially warm-blooded bees, allowing them to fly at cooler temperatures than most pollinators. Bumble bees sometimes "rob" nectar by chewing a hole at the bottom of a flower, bypassing the anthers and pistil, but in most cases they are highly effective at pollinating. They "buzz pollinate" some blossoms, like those of tomato, nightshade, and eggplant, generating a frequency of vibration that causes pollen grains to be released onto the bee. Bumble bees are sometimes placed in special nest boxes inside greenhouses to pollinate hothouse tomatoes. In nature, nests are often built in abandoned rodent burrows, but almost any soft, cozy situation will do. The wax structure contains brood cells where the larvae are reared and "honeypots" for the storage of nectar. The only bumble bees to survive through winter are young queens that mated the previous fall. In spring, each queen starts a new nest and raises a first generation of female worker bees. Males are not produced until late in the year. An insect that looks like a bumble bee may not be: many other insects mimic them, including robber flies, syrphid flies, certain sphinx moths, even various beetles.

The **Yellow Bumble Bee** *(Bombus fervidus)* ranges over the continent except Texas and the southeast and can be common in cities and farmland. The **American Bumble Bee** *(Bombus pensylvanicus)* is widespread, but declining, over most of the U.S. Look for it in open fields. The **Yellow-faced Bumble Bee** *(Bombus vosnesenskii)* ranges from Alaska to New Mexico. ***Bombus lapponicus*** is widespread in the Old World and also in our far west. The **Two-spotted Bumble Bee** *(Bombus bimaculatus)* lives east of the Rockies and north into Canada. The life cycle of a colony is short, and males can be found by midsummer in some localities. The **Red-belted Bumble Bee** *(Bombus rufocinctus)* ranges over most of the continent except the southeast. The **Tricolored Bumble Bee** *(Bombus ternarius)* inhabits Canada and the northeastern U.S., while the **Common Eastern Bumble Bee** *(Bombus impatiens)*, widespread in the east, has been introduced into California. Cuckoo bumble bees (genus ***Psithyrus,*** often considered a subgenus of *Bombus)* are social parasites of true bumble bees, displacing a resident queen in a nest and recruiting the workers to raise her offspring. They resemble true bumble bees but lack pollen-carrying structures on the hind legs. (Not illustrated.)

Large carpenter bees (genus ***Xylocopa***) only rarely weaken beams in homes and other wooden structures. At least a few species are semisocial, with overlapping generations of females sometimes using the same nest. Males of the common **Eastern Carpenter Bee** *(Xylocopa virginica)* often hover close to the nests of females, chasing away rival males and other perceived threats. Their yellow faces let you know they don't pack a sting. The **California Carpenter Bee** *(Xylocopa californica)* has several subspecies, all of them metallic green or blue-black. In the desert southwest, this bee nests in old agave or sotol stalks. In the **Valley Carpenter Bee** *(Xylocopa varipunctata)* of the southwest, females are black and males are golden.

Yellow Bumble Bee

American Bumble Bee

Two-spotted Bumble Bee

Yellow-faced Bumble Bee

Bombus lapponicus

Common Eastern Bumble Bee

Tricolored Bumble Bee

Red-belted Bumble Bee

California Carpenter Bee

Valley Carpenter Bee

Eastern Carpenter Bee

HONEY BEES

All true honey bees are native to the Old World, including the species found throughout North America—currently one of our most intensively studied, most economically important, and most familiar insects.

The **Honey Bee** *(Apis mellifera)* was first introduced to North America by colonists in Jamestown in 1622. Pollination of introduced crops, and production of beeswax and honey, made them indispensable allies of European settlers. Today, apiculture (beekeeping) is a multimillion-dollar business in the U.S., where migrant beekeepers truck their colonies to various locations for the pollination of almonds, apples, alfalfa, and other crops. The industry currently suffers from devastating outbreaks of bee tracheal mites and varroa mites, the infiltration of the more aggressive Africanized strain of Honey Bee, and the overuse of pesticides in agricultural systems and mosquito abatement.

These are highly social insects, each colony consisting of one queen and several hundred female workers. Most Honey Bee colonies in North America are in man-made hives, although feral colonies may be found nesting in tree cavities. The cells in the nest are arranged in vertical combs made of beeswax manufactured by the bees. Young workers are "nurse bees" that stay inside the nest, tending to the larval bees and cleaning the wax cells. They become "foragers" as they age, leaving the hive to seek nectar and pollen. Males ("drones") and new queens are produced once a year. Mating takes place in midair. The queen may mate with several males, increasing the genetic diversity of her eventual offspring.

Foraging Honey Bees may do some random searching for nectar sources, but once they locate productive patches of flowers, they communicate the locations of those patches to their nestmates. This communication, elucidated by the studies of Karl von Frisch in the 1960s, involves a "dance" put on by the returning worker bee that indicates the direction and the approximate distance to the flowers. Mutual feeding, called "trophallaxis," is yet another form of communication.

Worker Honey Bees (all females, and sisters of each other within a given hive) commit suicide in the act of stinging, as the barbed stinger holds fast to the victim. The departing bee has vital organs torn out of her body as she flees the scene. The venom sac continues pumping poison into the wound long after the bee has left.

The infamous "killer bee" responsible for so much recent paranoia is just another strain of the familiar Honey Bee. In the 1950s, beekeepers in Brazil brought in an African strain to try to crossbreed it with the mild-mannered European strain, but the African bees escaped and established themselves in the wild. They have since spread through South and Central America and into the southern U.S. These aggressive bees respond to any disturbance of the nest by attacking instantly, and they may chase humans or animals for several hundred yards. Since these "Africanized" bees look the same as other Honey Bees, it's best to steer clear of all bee colonies, especially those nesting in wild situations.

actual size

drone (male)

worker (female) bee

beekeeper at work, with protective gear

mass of Honey Bees on hanging combs

Varroa mites (pests in beehives)

queen Honey Bee surrounded by workers

Not as well known as the honey bees, bumble bees, and carpenter bees on the preceding pages, these are classified in the same family, **Apidae.**

Digger bees in the genus ***Anthophora*** (about 70 species north of Mexico) are mostly medium-sized bees that nest in the ground, often in large aggregations. Many species fashion a mud chimney around the entrance to their burrows. They are fast-flying insects expert at hovering, alighting on blossoms only briefly. ***Anthophora abrupta*** is a common eastern species that burrows in vertical clay banks. Males have a "mustache" of special hairs they use to absorb fragrant substances (parsnip sap in one instance) to mix with mandibular gland secretions used in marking territories.

Members of the genus ***Centris*** are large hairy bees of the west, especially the southwestern deserts, with 22 species recorded in the U.S. Look for them pollinating blossoms of palo verde or desert willow.

Classified in the genus ***Diadasia*** are small to medium-sized, very hairy bees found in the western half of North America. ***Diadasia rinconis*** occurs from Texas to California. It is one of the "cactus bees" that pollinates the blossoms of prickly pear cacti but may visit other flowers as well. ***Diadasia enavata*** ranges from Kansas to California. Look for this species almost exclusively on sunflowers. About 25 species occur north of Mexico, several named for the plant they favor pollinating. ***Melitoma taurea*** is a medium-sized bee, gray with a striped abdomen. It ranges from Illinois to Florida and pollinates only the flowers of wild sweet potato *(Ipomoea).* Look for nests in soil clinging to roots of blown-down trees. The nest entrance can be a long aboveground tunnel with a narrow slot along its length.

The widespread genus ***Melissodes*** includes over 100 species of medium-sized bees, usually very common. Males typically have long antennae and yellow faces. One common eastern species, ***Melissodes bimaculata,*** is black with two white spots on the abdomen. ***Melissodes trinodis*** occurs from southeastern Canada to Kansas and Georgia. It visits only composite flowers, especially sunflowers.

Our three species of ***Xeromelecta*** are cleptoparasites of digger bees: the larva eats the pollen and nectar provisions intended for the host's offspring. ***Epeolus*** (52 species north of Mexico) and ***Triepeolus*** (about 102 species) are cleptoparasites of other bees. The larvae feed on pollen and nectar stored for the host's offspring. These bees are common across the continent. The two genera look nearly identical. Like many bees, they are fond of mint blossoms. Bees in the genus ***Nomada*** resemble small wasps, usually yellow or reddish and nearly hairless. Look for them in spring along forest edges. Females lay their eggs in the burrows of bees in the families Andrenidae and Halictidae. The larvae of these cleptoparasites rob the pollen and nectar stores of their hosts, starving the rightful larval occupants if not killing them first. Nearly 290 species live north of Mexico.

The **small carpenter bees** (genus ***Ceratina***) are so much smaller than the well-known large carpenter that they do not even appear to be related. They are cylindrical, dark, weakly metallic, and might be mistaken for sweat bees. Look for them on a variety of flowers, especially thistles.

DIGGER BEES AND OTHERS

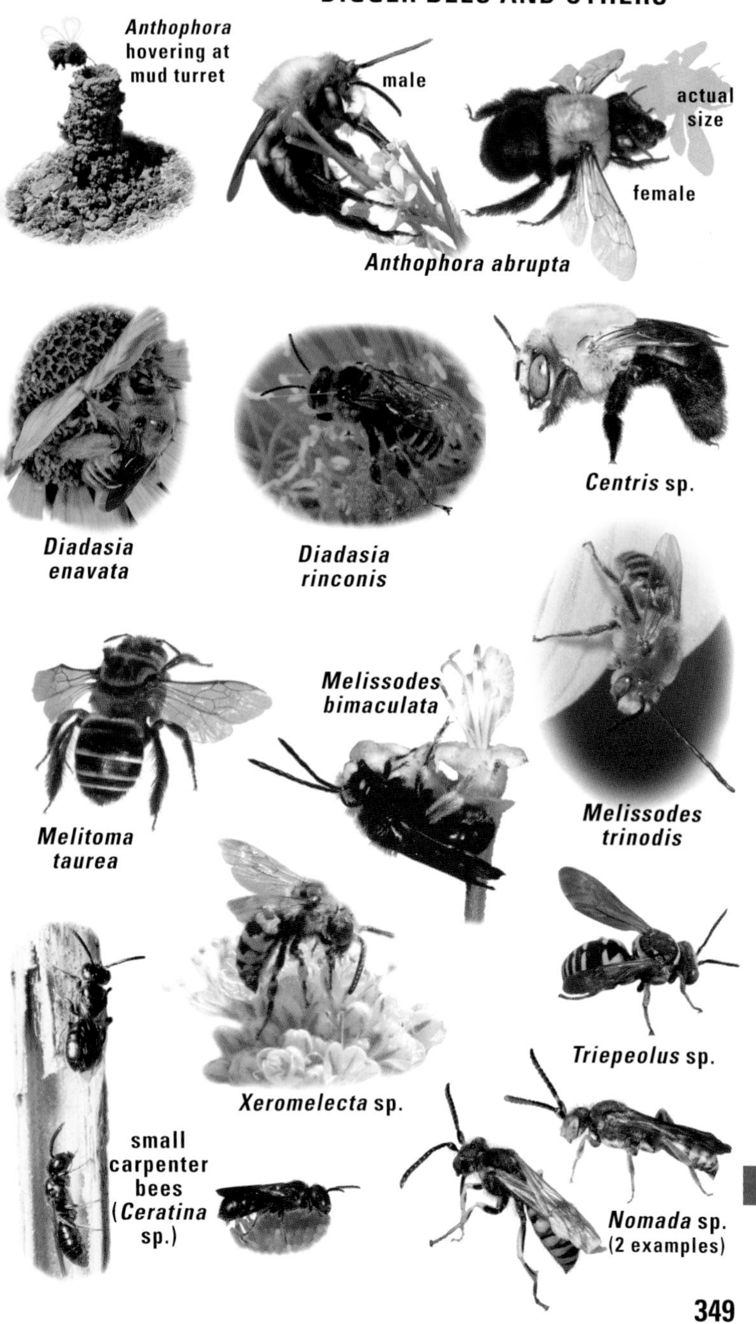

VELVET ANTS AND SCOLIID WASPS

VELVET ANTS (family Mutillidae) are actually solitary wasps, the females of which are wingless and often bright red, orange, or yellow. Take heed of the warning colors: females pack perhaps the most painful sting of any North American insect. Males of most species are winged and often colored differently than females. Both genders can "squeak" by rubbing their abdominal segments together. Individuals within a species vary drastically in size depending on the size of the larval host. Most velvet ants are nocturnal or crepuscular (active at dusk and/or dawn) and seldom seen. Diurnal (day-active) species can be highly conspicuous. All are parasites of other insects, usually other wasps or solitary bees. Hosts are unknown for the majority of velvet ants. They are most diverse in arid habitats such as deserts, dunes, and prairies. Many other insects appear to mimic velvet ants, including some checkered beetles (p. 206) and ground beetles (p. 136).

The genus ***Dasymutilla,*** with about 140 species in North America, includes the most commonly seen velvet ants, many with long, brightly colored hairs. The **Eastern Velvet Ant** *(Dasymutilla occidentalis)* is the so-called "cow killer." Folklore has it that the female's sting is painful enough to kill livestock. This species is transcontinental in the U.S., more common in the southeast where one host, the Horse Guard (p. 336), occurs. Bumble bees are also documented as hosts. ***Dasymutilla magnifica*** is one of several species in the southwestern U.S. with a black head and thorax and a red, orange, or yellow abdomen. The **Thistledown Velvet Ant** *(Dasymutilla gloriosa)* appears to mimic windblown seeds of creosote bushes in the southwest. ***Dasymutilla nigripes*** is another common eastern species ranging as far west as Alberta and Arizona. ***Dasymutilla quadriguttata*** is widespread and common, especially in the midwest.

Timulla vagans is one of 30 North American species in its genus. Females are easily mistaken for true ants as they run across barren ground. Members of the genus ***Pseudomethocha*** are small but common, with 43 species north of Mexico. The velvet ants in the genus ***Sphaeropthalma*** are generally active only at night, and most are rather dull in color.

SCOLIID WASPS (family Scoliidae) are large, usually hairy wasps that are parasitic on the larvae of scarab beetles. The sexes often look radically different. Females dig to uncover a scarab grub, sting it if necessary to subdue it, lay an egg on it, and leave the scene. Males of some species gather in small groups to sleep, curling their bodies around a stem or grassblade in a peculiar and distinctive posture. There are five genera and 22 species in North America. ***Scolia*** is a genus of dark-winged wasps with eight species here. ***Scolia bicincta*** ranges over the eastern half of the U.S. Look for adults in late summer on flowers of thoroughworts. ***Scolia dubia*** ranges south of Massachusetts and west to Arizona and California. This species is also a late-summer visitor to flowers. ***Scolia ardens*** ranges from Texas to California. The genus ***Campsomeris*** includes nine U.S. species, the sexes of each being very different from each other in appearance.

actual size
Dasymutilla nigripes
Thistledown Velvet Ant
Dasymutilla magnifica
Dasymutilla sp.
Timulla vagans
male
Dasymutilla quadriguttata
Eastern Velvet Ant
Pseudomethocha oculata
male
Dasymutilla sp.
male
Sphaeropthalma sp.
female
Scolia ardens
Scolia bicincta
Scolia dubia
male scoliid wasps sleeping
female
Campsomeris sp.
male

SPIDER WASPS

(family Pompilidae) are solitary, high-strung hunters. Paralyzing a victim with its sting, the female stashes the helpless spider in a burrow, cavity, or mud cell (depending on species), as food for a single offspring. Many visit aphid colonies for honeydew secreted by those insects. Others drink flower nectar. There are roughly 290 species in North America.

The 14 species of ***Pepsis*** are the famous "tarantula hawks" of the southern U.S. Witnessing an epic battle between a female wasp and her gargantuan arachnid adversary is an experience not soon forgotten. Look for both sexes visiting flowers. ***Pepsis thisbe*** occurs from Nebraska and Texas to California. Fresh specimens have scarlet wings owing to bright scales that wear off as the insect ages, leaving the wings much duller.

The genus ***Entypus*** includes seven species. ***Entypus unifasciatus*** hunts large wolf spiders in fields over the eastern half of the U.S. ***Entypus fulvicornis,*** not shown, has all-black wings and a similar range. Both species visit flowers for nectar and aphid colonies for honeydew. Females dig nesting burrows, usually with five to ten cells, prior to hunting for spider prey. The wasp drags her victim backward, over the ground, to the nest.

The genus ***Priocnemis*** includes 12 species in North America. These are among the first spider wasps seen in spring. ***Priocnemis minorata*** is common in the eastern U.S. and southern Canada. The females excavate burrows, with up to seven cells, in soil before searching for prey. The genus ***Dipogon*** includes at least 19 species north of Mexico. Look for them on tree trunks, where they may be mistaken for ants as they search bark crevices for their prey. Their wings are often spotted or banded, as in the widespread ***Dipogon sayi.*** They nest in preexisting cavities, such as old beetle borings, and apparently prey exclusively on crab spiders.

Members of the genus ***Auplopus*** build mud cells to store prey. They may amputate some or all of the legs of victims prior to transport back to the nest. Jumping spiders and clubionid spiders are the usual prey. ***Auplopus mellipes*** is common in woodlands in the eastern half of the U.S. The little barrel-shaped cells are usually arranged end-to-end, often inside the abandoned nests of Pipe Organ Mud Daubers (p. 338). ***Auplopus architectus*** is transcontinental in the U.S., except the northern Rockies. These wasps rarely visit flowers but are sometimes seen around aphid colonies.

Our 12 species of ***Episyron*** prey on orb weavers and nest in loose sand. Females usually fly their victim back to the nest, often jamming the paralyzed spider in the crotch of a plant upon arrival, then inspecting the burrow unburdened.

The two species of ***Tachypompilus*** are big red wasps that tackle large wolf spiders and fisher spiders. ***Tachypompilus ferrugineus*** nests in crevices in rock walls and other situations. Watch for them in open habitats and along forest edges, where they are sometimes seen on flowers.

Anoplius is a genus of 43 species, mostly very similar. Most prey on wolf spiders, excavating cells in the walls of preexisting earthen cavities. ***Anoplius atrox*** is one of only two large species with red markings. It nests in soft soil or sand in areas east of the Rockies.

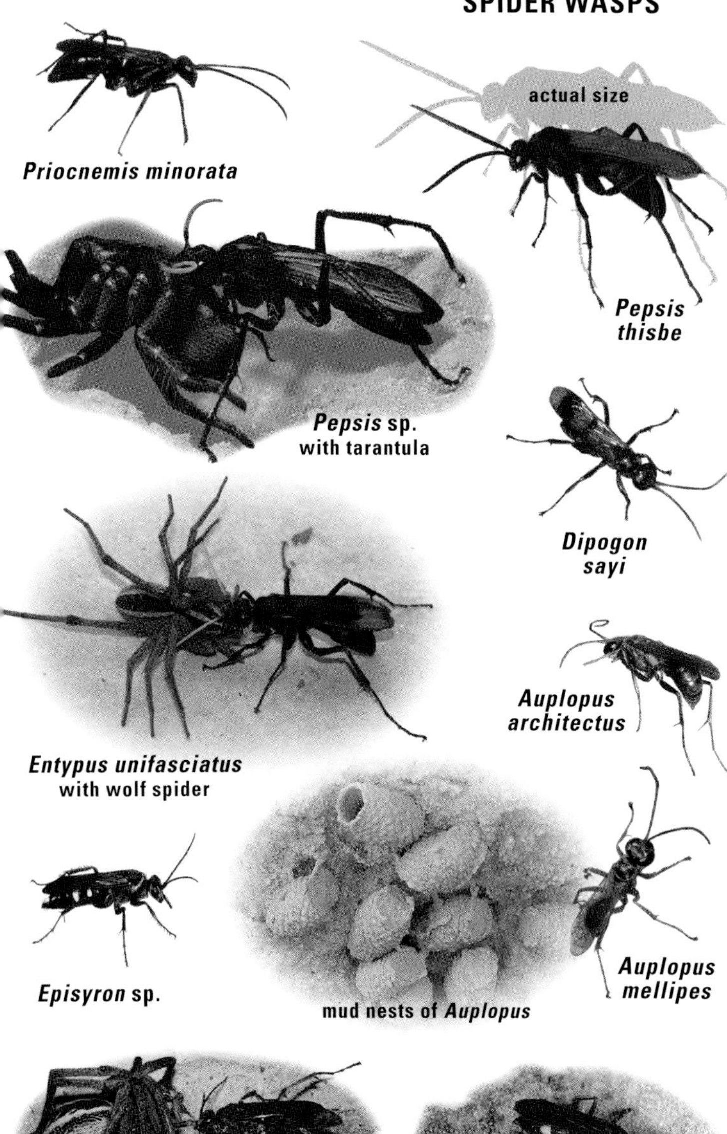

Priocnemis minorata

Pepsis thisbe

Pepsis sp. with tarantula

Dipogon sayi

Entypus unifasciatus with wolf spider

Auplopus architectus

Episyron sp.

mud nests of *Auplopus*

Auplopus mellipes

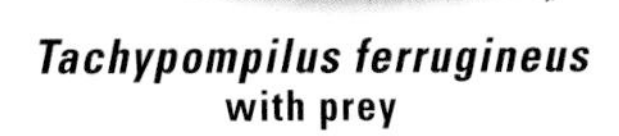

Tachypompilus ferrugineus with prey

Anoplius atrox at burrow

VESPID WASPS

(family Vespidae) include both solitary and social species. They are arranged here by subfamily. All but the pollen wasps share one easily seen character: the wings are folded lengthwise at rest. These are valuable insects that dispatch large numbers of pests, especially caterpillars.

PAPER WASPS (subfamily Polistinae) are also called umbrella wasps for the shape of their uncovered combs, frequently suspended beneath the eaves of houses. Generally placid, they are efficient predators of caterpillars. A new colony is established in spring, often by more than one female foundress. Further interactions yield one "queen," as subordinates suppress their ovarian potential. Nests mature at under 200 individuals. Twenty species occur in North America. ***Polistes dominulus*** is native to Europe and China. Discovered in Massachusetts in 1980, it has spread almost nationwide. It is displacing some native species, nesting in birdhouses and other enclosed spaces. ***Polistes fuscatus*** is native, abundant, and widely distributed. ***Polistes metricus*** is a widespread eastern species nesting in shrubbery. ***Polistes exclamans*** is mainly southern in distribution, reaching as far north as Nebraska and New Jersey. ***Polistes annularis*** occurs in the eastern half of the U.S., as far north as Connecticut and South Dakota. ***Polistes flavus*** is widespread in the southwestern quadrant of the continent, while ***Polistes comanchus*** has a more restricted southwestern range.

Mischocyttarus is a mostly tropical genus with two species here. A *long segment connecting thorax and abdomen* distinguishes them from the *Polistes* wasps. ***Mischocyttarus flavitarsis*** ranges over the western half of the U.S. and north into British Columbia.

TWISTED-WING PARASITES (Strepsiptera or **Stylopidae)** are not related to other insects in this section of the guide, but they are often observed on the bodies of adult *Polistes* wasps. Female stylopids are sedentary, capsule-like creatures seen protruding from beneath the warped abdominal segments of the wasp (stylopids cause similar deformities in other types of wasps, also various other insects). Male stylopids, rarely seen, have amazing fanlike hindwings and stumplike forewings. The classification of stylopids is problematic. They were formerly treated as a family (Stylopidae) of beetles. They are now usually placed in their own order (Strepsiptera), but their relationship to other orders of insects is uncertain.

male stylopid

female stylopids in *Polistes* wasp (detail on left)

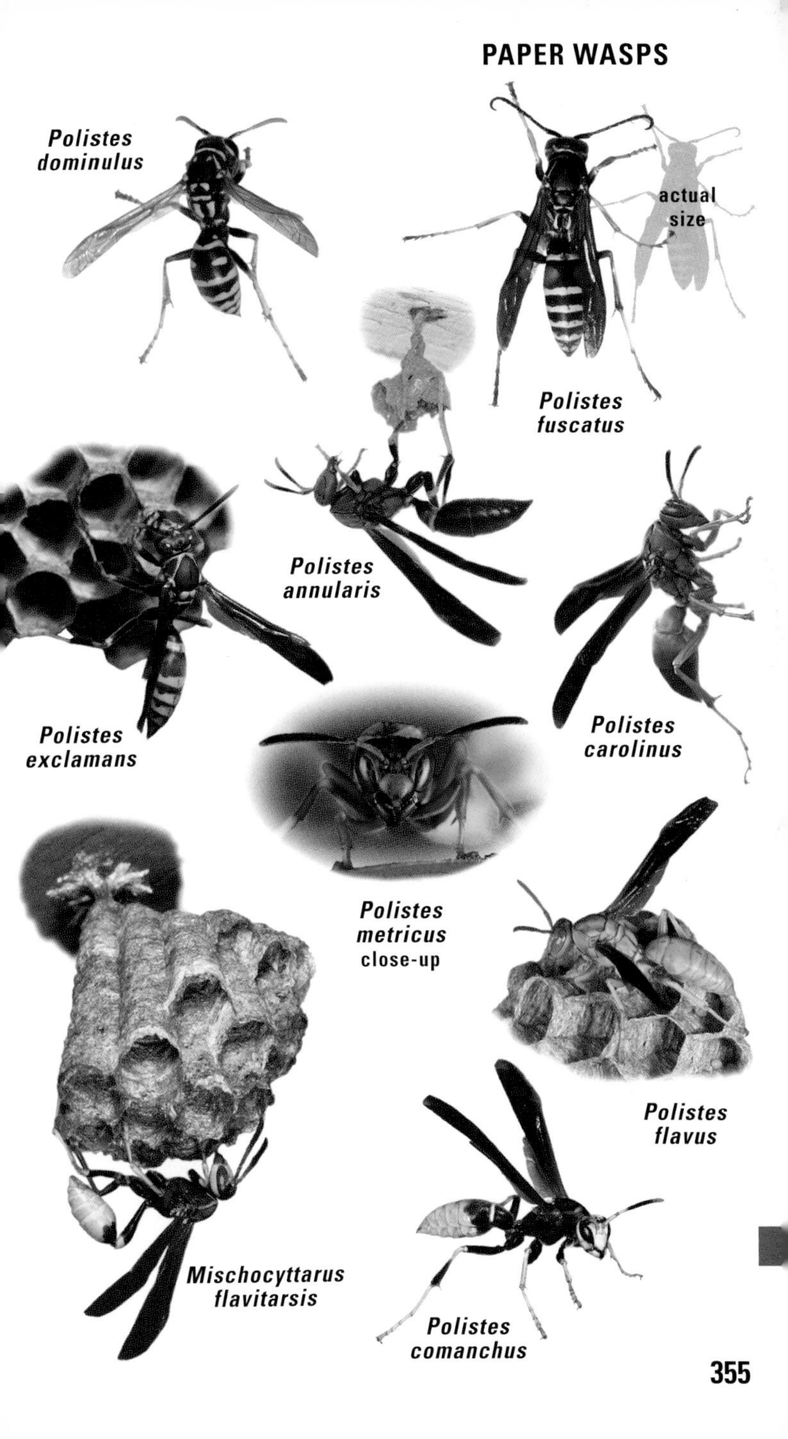
Polistes dominulus
actual size
Polistes fuscatus
Polistes annularis
Polistes exclamans
Polistes carolinus
Polistes metricus close-up
Polistes flavus
Mischocyttarus flavitarsis
Polistes comanchus

YELLOWJACKETS AND HORNETS

(subfamily Vespinae) build paper nests of chewed wood and plant fibers. Yellowjackets are most diverse in the north. Colonies typically consist of one reproductive female (queen), many female workers, and, in the fall, males and new queens. Nests are seasonal and are not reused. Only queens survive the winter. Adults visit flowers for nectar, or aphid colonies for the honeydew secreted by those insects, especially in the fall.

Members of the genus ***Vespula*** are yellowjackets that nest underground, usually in abandoned rodent burrows, earning the misnomer "ground bees." Most of our 12 species are helpful predators, but a few have a tendency to crash cookouts uninvited. When dining outdoors, use care in chugging canned beverages, lest you drink in an angry wasp. Nests, founded by a single queen, can contain up to 5,000 wasps at their peak. Skunks, bears, and other mammals dig out nests to eat the juicy larvae. The **German Yellowjacket** *(Vespula germanica),* not illustrated, native to Europe, has been introduced in North America, where it continues to expand its range. Its scavenging behavior and tendency to nest between walls in buildings make it a bona fide pest. The **Eastern Yellowjacket** *(Vespula maculifrons)* is native throughout the eastern U.S. and southeastern Canada. The **Western Yellowjacket** *(Vespula pensylvanica)* occurs in the western third of the U.S. and adjacent Canada, where it is the principal pest species. The **Southern Yellowjacket** *(Vespula squamosa)* ranges south of the Great Lakes in the east. Queens of this species are known to invade young nests of *V. maculifrons,* killing the queen and coercing workers into babysitting its own offspring. In contrast to the aforementioned urban species, others dwell in remote areas in small short-lived colonies. The **Blackjacket** *(Vespula consobrina),* transcontinental in Canada and south to Appalachia, the Rockies, and the Pacific states, is strictly predatory.

Dolichovespula wasps build nests aboveground. Five species occur here. The **Bald-faced Hornet** *(Dolichovespula maculata)* is a beneficial species that kills great numbers of flies and sometimes other yellowjackets. Nests of this species are large familiar structures but are easily overlooked during the seasons they are occupied by 200–300 workers. This species ranges widely but is absent from much of the midwest, southwest, and Great Basin. ***Dolichovespula arenaria*** occurs throughout most of Canada, Alaska, the western and northeastern U.S., and Appalachia.

True hornets of the genus ***Vespa*** are represented here by the **European Hornet** *(Vespa crabro).* This species was introduced to New York in the mid-1800s. Well established in the northeast, they may be expanding their range to the northwest and south. At up to an inch long (18–25 mm) they are much larger than yellowjackets. Average nests hold 200–400 workers and are built in hollow trees and other cavities. Workers are active day and night and may be attracted to lights at night. They prey on a variety of insects, sometimes raiding beehives to rob honey.

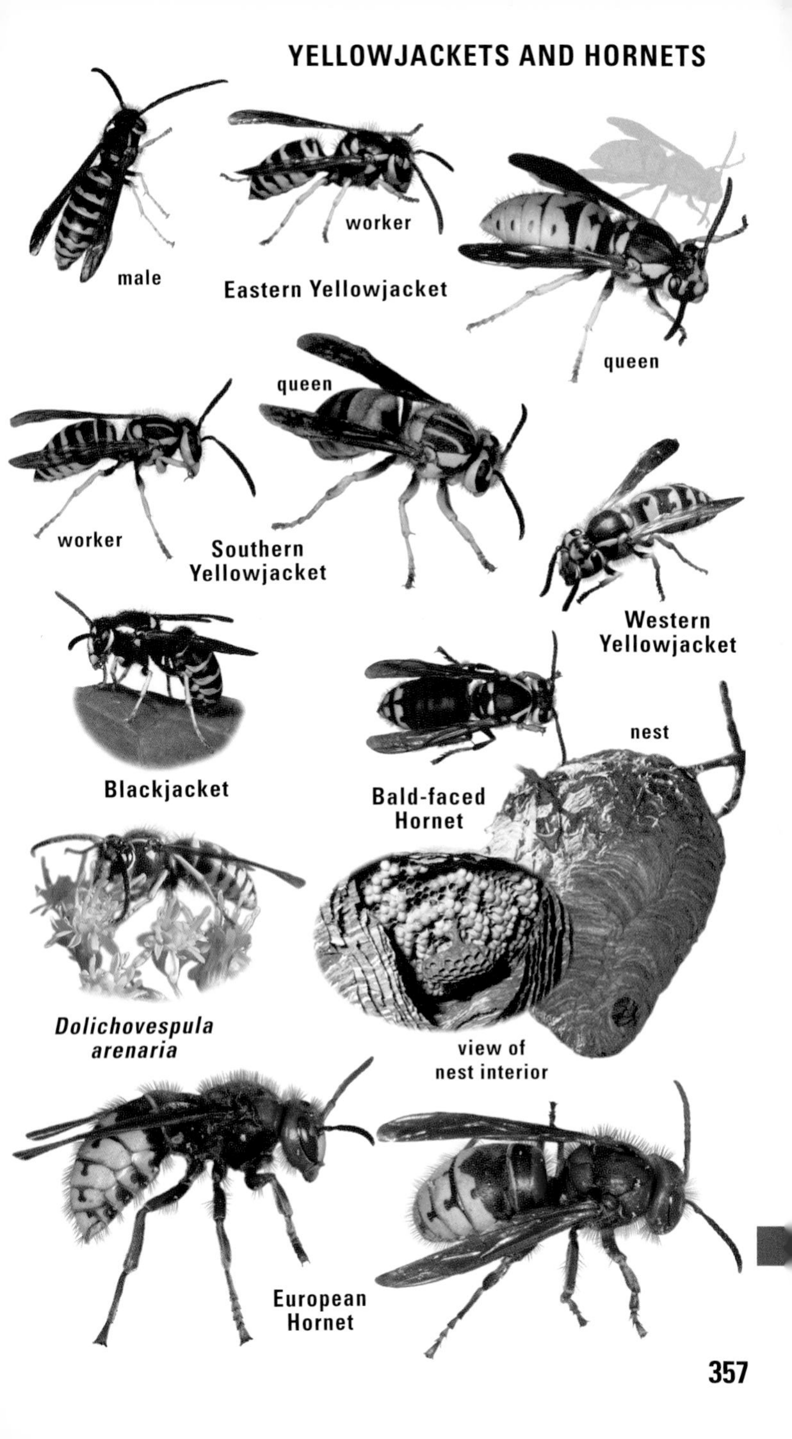
male
worker
Eastern Yellowjacket
queen
queen
worker
Southern Yellowjacket
Western Yellowjacket
nest
Blackjacket
Bald-faced Hornet
Dolichovespula arenaria
view of nest interior
European Hornet

POTTER WASPS AND MASON WASPS

(subfamily Eumeninae) are nearly all solitary, not social. Females hunt and paralyze caterpillars, stockpiling them for their offspring. A few construct mud nests, but the majority use preexisting cavities, like beetle borings in dead wood. The tunnel is partitioned into cells, one egg and several caterpillars in each. Look for the adult wasps on flowers and around aphid colonies. More than 260 species occur north of Mexico.

The genus ***Euodynerus*** includes 30 species, many with a chunky appearance. Most nest in borings in wood, but a few burrow in soil, make mud nests, or utilize abandoned nests of mud daubers. ***Euodynerus leucomelas*** occurs in the northern half of the U.S. and adjacent Canada, nesting in hollow sumac twigs. ***Euodynerus hidalgo*** ranges continentwide. It nests in holes in wood or in abandoned mud dauber nests. ***Euodynerus foraminatus,*** found across the continent, nests in a variety of situations including hollow sumac twigs and old mud dauber nests. ***Monobia quadridens*** ranges in the eastern U.S. southwest to New Mexico. Females nest mostly in abandoned borings of carpenter bees. They are sometimes mistaken for the Bald-faced Hornet (preceding page). Most members of this genus are tropical. ***Pseudodynerus quadrisectus*** resembles *Monobia* but averages slightly smaller, with more white markings. It ranges over the eastern U.S. south of New Jersey and Illinois, females nesting in cavities in wood. There are two generations per year, as is the case for most mason wasps.

The genus ***Parancistrocerus*** includes 26 species north of Mexico. Each species hosts a corresponding species of mite. The wasps have built-in "parking slots" for the mites, on the second abdominal segment, under the overhanging lip of the first segment. Female wasps nest in a variety of cavities, including empty plant galls and abandoned mud dauber nests. ***Parancistrocerus bicornis*** ranges from New York south to Florida and west to southern Arizona. The genus ***Ancistrocerus*** includes 17 species, collectively ranging across the continent, including Alaska. Most nest in preexisting cavities or in the pith of twigs. Others use abandoned galls or old burrows of ground-nesting bees or wasps. A few build nests of mud. ***Ancistrocerus tuberculocephalus*** occurs from British Columbia to California, east to Wyoming and New Mexico. It uses old mud dauber nests.

Members of the genus ***Eumenes*** are the potter wasps, with eight species in the U.S. and southern Canada. Females fashion a graceful, marble-sized urn of mud for each offspring. ***Eumenes fraternus*** is the most common species in the eastern U.S. and Ontario. ***Eumenes bollii*** occurs in the western U.S., north and east to Alberta and Minnesota. ***Zethus spinipes,*** widespread in the east, is the most common of our five species in its genus. It strongly resembles a potter wasp.

POLLEN WASPS (subfamily Masarinae) are mostly an Old World group, with one genus, ***Pseudomasaris,*** found in the western U.S. and adjacent southern Canada. Our 14 species *resemble yellowjackets with clubbed antennae.* They can be confused with cimbicid sawflies (p. 322). Females visit only violet-colored penstemon or phacelia flowers, storing pollen and nectar in cells they fashion from mud. A single larva develops in each cell.

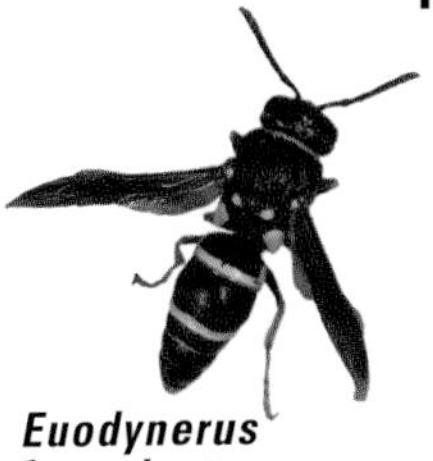

Euodynerus foraminatus

Euodynerus leucomelas

actual size

Euodynerus hidalgo

Pseudodynerus quadrisectus

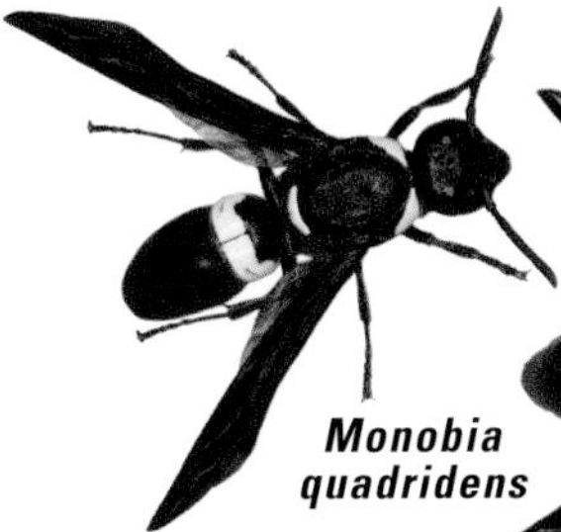

Monobia quadridens

female

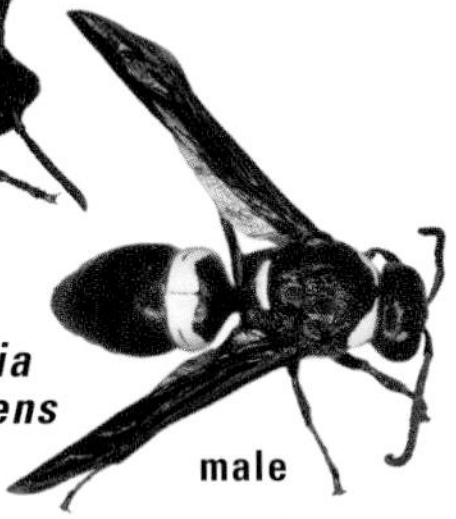

male

Parancistrocerus bicornis

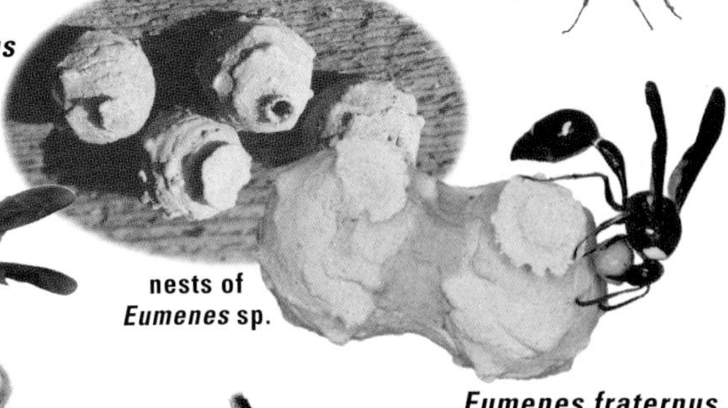

Ancistrocerus tuberculocephalus

nests of *Eumenes* sp.

Eumenes fraternus nest-building

Eumenes bollii

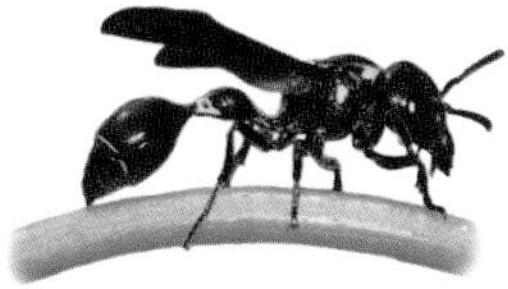

Zethus spinipes

nest

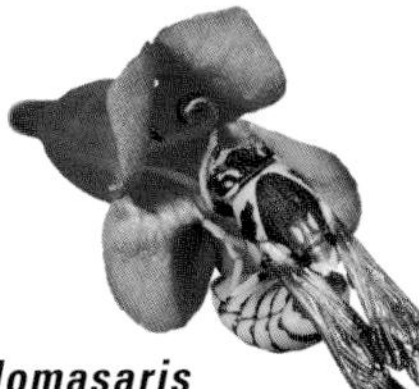

Pseudomasaris sp.

ANTS

(family Formicidae) may be the pinnacle of insect evolution. Nearly all species are colonial, with at least one queen and multiple workers. New reproductives (winged "alates") usually swarm once a year. They are often confused with flying termites (p. 32) but are identified by elbowed antennae, two unequal pairs of wings (hind pair smaller), and distinctly segmented bodies. Other arthropods resemble ants, through mimicry or shared evolutionary heritage. These include velvet ants, various true bugs, even some spiders. Many ant species can sting. Several species are social parasites of other ants. Colonies often host other insects, including tiny crickets, rove beetles, even roaches. Ants may also bring aphids or certain caterpillars into the nest to feed on sweet secretions from those insects. Many plants depend on ants for seed dispersal.

The **Giant Hunting Ant** *(Pachycondyla villosa)* is tropical, reaching the U.S. only in southern Texas. Workers are active by day, running about in search of insects on which they feed. Colonies are usually in soil or in logs or stumps. ***Amblyopone pallipes*** is one of three U.S. species in its genus. It occurs throughout most of the U.S. These predators specialize on soil centipedes and nest in soil or rotten wood. Mature colonies are tiny, averaging 12 adults. Three species of **trapjaw ants** *(Odontomachus)* occur in the southern U.S. Workers walk around with the mandibles held 180 degrees apart. Their jaws literally have a "hair trigger," and any contact between a victim and the sensitive setae causes the mandibles to close swiftly and violently. Colonies are small, usually in soil or rotting stumps or logs.

Legionary ants *(Neivamyrmex)* include 25 species in the U.S., most in the southwest. Workers are blind. Most species conduct nocturnal raids, often on nests of other ants, carting off larvae, pupae, and eggs as food for their own offspring. The nomadic cycle of a typical species lasts about three weeks, the colony resting by day in a bivouac in a natural cavity or the nest of another ant colony they have destroyed. Everybody comes along, larvae and pupae toted by some workers while others hunt. ***Labidus coecus*** is one of the most widespread army ants, occurring from the southern U.S. to Argentina. It forages mostly underground or under leaf-litter, often attacking colonies of other ants. ***Forelius pruinosus*** is abundant in the southwest, ranging to Florida and New York. The tiny workers form narrow foraging columns in the heat of the day, easily observed on urban sidewalks.

The **Argentine Ant** *(Linepithema humile)*, native to South America, is now found globally. It likely voyaged to New Orleans aboard Brazilian coffee ships before 1891. Intolerant of neighboring colonies in its native haunts, it can form "supercolonies" of interconnected nests in adopted homelands. One such extended from Italy to Spain, over 6,000 kilometers. This species is driving out native ants in occupied territories. The **Odorous House Ant** *(Tapinoma sessile)* defends its colony in part by emitting butyric acid, smelling of rancid butter or rotting coconut. This native occupies a variety of habitats, usually nesting in soil beneath stones and other objects. Colonies may contain several queens and nearly 10,000 workers. The species ranges across southern Canada and most of the U.S.

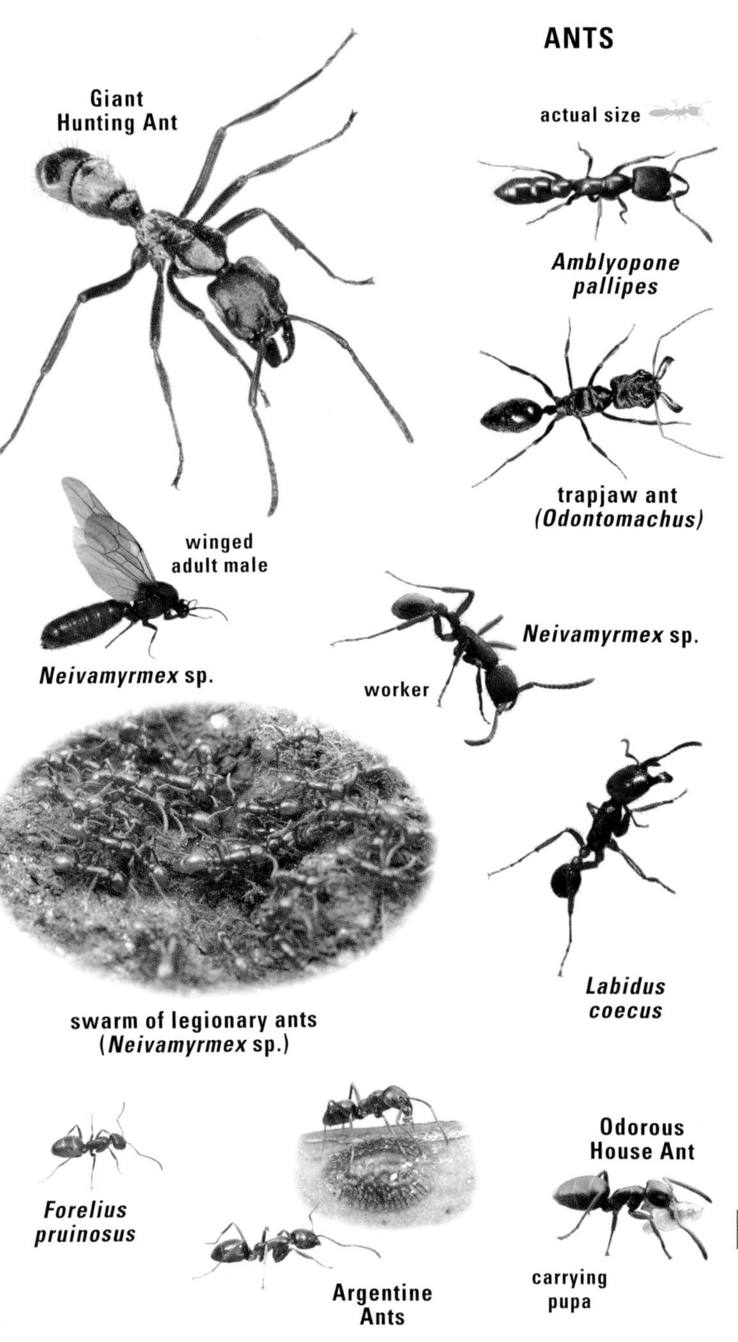
Giant
Hunting Ant
actual size
Amblyopone pallipes
trapjaw ant
(*Odontomachus*)
winged
adult male
Neivamyrmex sp.
Neivamyrmex sp.
worker
Labidus coecus
swarm of legionary ants
(*Neivamyrmex* sp.)
Forelius pruinosus
Argentine
Ants
Odorous
House Ant
carrying
pupa

of the **subfamily Myrmicinae** have two nodes ("humps") on the petiole linking the abdomen to the thorax. Many species can sting.

Harvester ants ("Pogos"), genus ***Pogonomyrmex,*** are desert and prairie ants. Our 28 species harvest seeds, occasionally damaging crops. The **Western Harvester Ant** *(Pogonomyrmex occidentalis)* ranges from the Great Plains westward. The area around the nest mound typically is stripped of vegetation. Mature colonies average 3,000 workers that aggressively defend their nests with stings and bites. In ***Pogonomyrmex rugosus*** of the southwest, mating flights occur in summer following afternoon thunderstorms. Colonies average nearly 8,000 workers. Our 66 species of **big-headed ants** *(Pheidole)* are named for the soldier caste. Most species are dimorphic, with "major" and "minor" workers. These ants typically harvest seeds.

Leafcutter ants of the genus ***Atta*** make up a distinctive tropical group with two species reaching our southwest. Worker "parasol ants" cut large pieces of foliage from plants, holding them triumphantly aloft in their jaws as they parade back to the nest. The leaves are chewed into a medium that breeds a unique fungus in special underground gardens. This fungus feeds the ants. Mature colonies may cover half an acre, with multiple entrances surrounded by low mounds. ***Acromyrmex versicolor*** ranges in the southwest. As with *Atta,* workers cut pieces of leaves that they masticate into "mulch" to breed a fungus found only in *Acromyrmex* colonies.

Acrobat ants (genus ***Crematogaster***) include 31 species north of Mexico. When threatened, workers arch the abdomen, ooze formic acid from the tip, and smear it on the attacker. The heart-shaped abdomen is distinctive. Most species are omnivores. ***Temnothorax longispinosus*** prefers to nest in chunks of rotten wood on the forest floor but will even utilize old acorns hollowed out by weevils. Colonies average fewer than 50 workers. The genus ***Monomorium*** has a global distribution. Many urban U.S. species are nonnative, but the **Little Black Ant** *(Monomorium minimum)* is native. Nests are in cracks in masonry, in woodwork, or in soil or rotted wood. The **Pavement Ant** *(Tetramorium caespitum)* is abundant throughout temperate North America. Slow-moving, these ants forage only about 40 square meters around the nest. Neighboring colonies may engage in ritual combat, resulting in conspicuous pileups on sidewalks.

Fire ants in the genus ***Solenopsis*** include several native species, but the **Red Imported Fire Ant,** or RIFA *(Solenopsis invicta),* came to the U.S. from Brazil. RIFA currently occupies the Gulf Coast to North Carolina, Tennessee, and Oklahoma, with isolated populations in southern California. RIFA is reviled for its burning stings and economic impacts. Nests include soil mounds up to 18 inches in diameter, dense enough to damage combines. Workers are polymorphic, ranging from tiny "minors" to large "majors," with several intermediate forms. Mature colonies average 220,000 adults. Ground-nesting birds and rodents may be routed from large areas by these ants. Colonies in recreational areas, schoolyards, and on private property compromise quality of life and impact public health through loss of productivity and even death from severe reactions to stings.

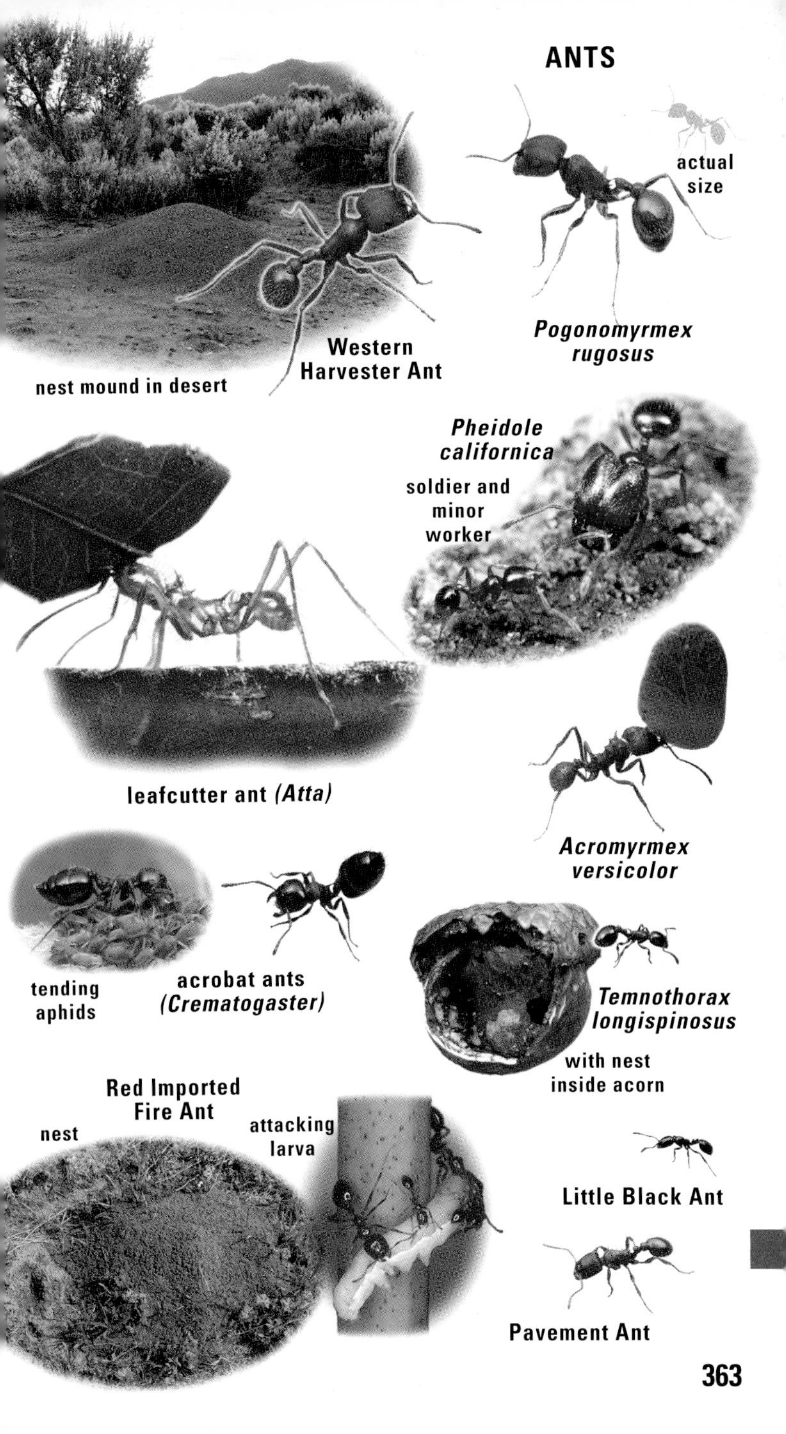
actual
size
Western
Harvester Ant
nest mound in desert
Pogonomyrmex rugosus
Pheidole californica
soldier and
minor
worker
leafcutter ant *(Atta)*
Acromyrmex versicolor
tending
aphids
acrobat ants
(Crematogaster)
Temnothorax longispinosus
with nest
inside acorn
Red Imported
Fire Ant
nest
attacking
larva
Little Black Ant
Pavement Ant

of the **subfamily Formicinae** are generally compact, stingless ants with one node ("hump") on the petiole joining the abdomen to the thorax.

Carpenter ants ***(Camponotus)*** include some of our largest ants. Not all tunnel in wood. Those that do are not eating wood, simply hollowing out nest chambers. They prey on live insects or scavenge dead ones. Mature colonies usually include satellite nests connected by trails or tunnels to the parent nest. Swarms of males and queens occur in spring. The **Black Carpenter Ant** *(Camponotus pennsylvanicus)* is the common species in the east. Workers are largely nocturnal. Nests are in wood, with satellite nests in timbers, beneath insulation, and in wall voids. Occupants can number 10,000 or more. The **Western Carpenter Ant** *(Camponotus vicinus)* nests in various situations. Workers aggressively defend the colony. Mature nests may exceed 100,000 individuals, including multiple queens.

The 15 species of "citronella ants" or "lemon ants" ***(Acanthomyops)*** are named for the odor given off by aggravated colonies. The ants tend root aphids or mealybugs underground, subsisting mostly on honeydew secreted by those insects, with additional foraging aboveground at night.

The genus ***Formica*** includes 86 species here. Workers swarm over the huge mounds, thrusting abdomens forward to spray any intruder with a dose of formic acid. ***Formica subsericea,*** widespread and abundant in the east, usually nests in soil beneath stones or leaf-litter. The **Allegheny Mound Ant** *(Formica exsectoides)* of the northeast builds mounds in grassy clearings. Large nests may harbor 270,000 workers and 1,400 queens. **Amazon ants** in the genus ***Polyergus*** are obligatory social parasites of *Formica* ants. An Amazon queen forces her way into the host's nest and kills the resident queens. Oddly, she is immediately adopted by the host workers, which then raise her offspring. Worker Amazons cannot even feed themselves: their jaws are designed strictly for piercing their victims in battle.

Our 18 species of ***Lasius*** are among our most abundant ants. In some locations, the **Cornfield Ant** *(Lasius alienus)* is codependent with the corn root aphid, gathering eggs of the aphid and housing them underground all winter. The eggs hatch in spring, the nymphs chauffered to corn roots by the ants. The ants get honeydew from the aphids in return. ***Lasius umbratus,*** of eastern North America, nests under stones in moist soil.

Honeypot ants ***(Myrmecocystus)*** are common in the arid west. Special workers called "repletes" are able to expand their abdomens to an incredible degree. Hanging from the ceiling of nest chambers underground, they serve as living storage casks of nectar for times when other food is scarce. Some species forage during the day; others are nocturnal, often pale with large eyes. Honeydew secreted by various insects is the major food source for some species. Colonies are often located adjacent to *Pogonomyrmex occidentalis* nests, the worker honeypot ants scavenging dead "Pogo" workers as food. The **Winter Ant** *(Prenolepis imparis)* is active at cold, even freezing, temperatures. Most colonies estivate during the hot summer, resuming activity when the weather cools. This species is among the first ants to swarm in early spring.

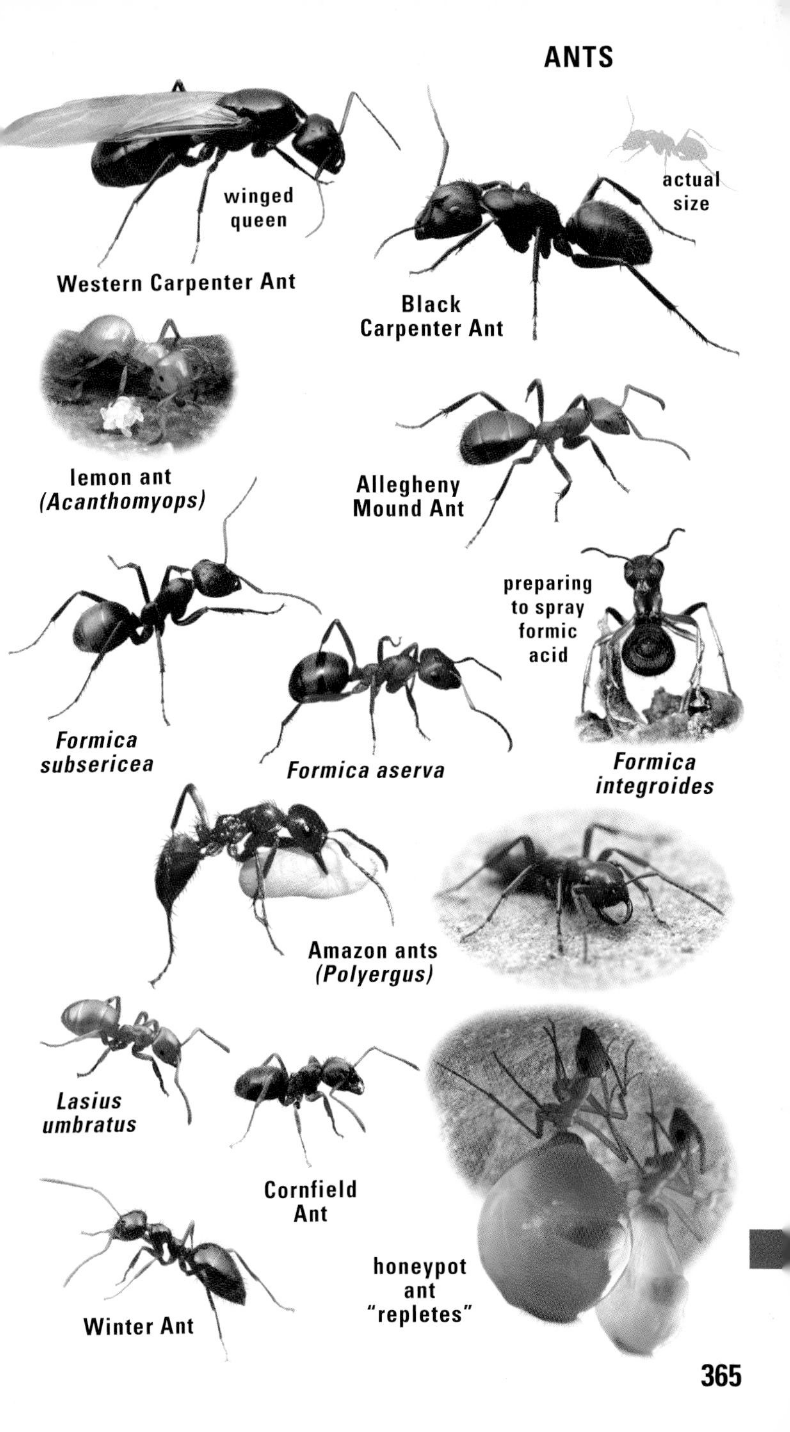

Western Carpenter Ant

Black Carpenter Ant

lemon ant *(Acanthomyops)*

Allegheny Mound Ant

Formica subsericea

Formica aserva

Formica integroides

Amazon ants *(Polyergus)*

Lasius umbratus

Cornfield Ant

Winter Ant

honeypot ant "repletes"

PHOTOGRAPHER CREDITS

Only about 280 of the photos used for images in this book were taken by Kenn Kaufman. The rest, more than 80 percent, were provided by the outstanding photographers listed below. Plates are listed by page number (in bold), followed by a dash and the number of the image. Images on each page are numbered from top to bottom and from left to right.

John C. Abbott: **2**-10; **3**-2; **4**-28, 32; **29**-5, 8; **35**-8, 9, 10; **42**-1; **43**-2; **45**-7; **46**-1; **49**-5; **50**-1; **52**-1; **53**-8; **61**-1, 9; **75**-14; **79**-13; **105**-4; **107**-4; **117**-3; **121**-3, 12; **125**-15; **141**-11; **147**-5; **149**-10; **151**-3, 8; **153**-3, 4, 6, 8, 10; **169**-11; **199**-16; **209**-7; **211**-1, 8; **220**-1; **221**-8, 9, 10; **223**-2, 7; **225**-3, 8; **279**-6; **284**-1; **289**-12; **291**-1; **309**-7; **331**-11; **354**-1
John H. Acorn: **13**-1; **131**-6, 8, 11; **133**-4, 10; **134**-1; **135**-10; **137**-1; **141**-7, 14; **143**-8; **201**-5
Jeff Adams / Xerces Society: **281**-9
Thomas Ames: **3**-6; **57**-5, 9; **59**-9, 14, 15, 18; **61**-8
Robert L. Anderson / USDA: **215**-8
Sue Anderson: **349**-1
Doug Backlund: **199**-1
Kevin Baldwin: **2**-17; **35**-2, 3
G.R. Ballmer: **4**-2; **39**-5; **71**-6, 14; **95**-1; **97**-7; **101**-7, 10; **103**-15; **119**-6; **209**-11; **287**-12; **291**-9, 10; **295**-8, 9; **325**-9; **331**-8; **333**-4; **337**-4, 6, 7, 8; **341**-4; **343**-2, 3, 4, 8, 12, 14, 15; **349**-15; **359**-9
Steven Barney: **149**-2
Troy Bartlett: **5**-20; **23**-3; **81**-8, 10; **87**-6; **93**-1, 14; **95**-18; **97**-6; **113**-15, 16; **115**-4; **117**-19; **123**-3, 16, 17; **125**-20; **165**-9; **191**-4; **207**-2; **215**-6; **249**-13; **295**-10; **313**-2; **315**-13; **335**-14; **339**-11; **351**-14
Scott Bauer / USDA ARS: **2**-13; **4**-5; **33**-5; **86**-6; **99**-2; **101**-9; **103**-12; **159**-9; **281**-14; **315**-5, 6; **347**-5; **363**-4
Giff Beaton: **5**-14; **45**-3; **47**-4, 6; **71**-1, 4, 7; **75**-5, 8; **111**-3; **213**-7; **289**-1, 6, 11, 13; **291**-2, 4, 6; **293**-1, 6, 7, 11, 14; **295**-4, 7, 12; **299**-6; **301**-6; **309**-8; **319**-1, 3; **327**-7
Bill Beatty / Visuals Unlimited: **91**-2; **125**-16; **155**-6; **189**-9; **199**-5; **201**-2; **299**-10
R.A. Behrstock / Naturewide Images: **43**-4; **45**-1, 2; **47**-8, 9; **49**-4, 8; **53**-10; **225**-9; **227**-7, 8, 9; **279**-1
C. L. Bellamy: **187**-7, 8; **189**-7, 11
Ric Bessin: **101**-13; **163**-7; **267**-11
Nic Bishop: **123**-13; **141**-10; **171**-5; **289**-3; **293**-3; **295**-2; **323**-19; **327**-8; **337**-17
Michael Boone: **353**-5
Rick and Nora Bowers / BowersPhoto.com: **5**-32; **27**-3, 5; **37**-10; **51**-3, 5, 6; **69**-6; **75**-1, 7; **97**-3, 12; **121**-1, 2; **125**-5; **155**-1; **159**-5; **169**-2; **173**-6; **181**-8; **185**-9; **187**-10; **203**-10; **207**-9; **223**-5; **231**-3, 4, 7, 9, 10, 16; **233**-10; **235**-7, 8; **237**-7; **241**-16; **243**-1, 2; **293**-2, 8, 9, 13; **295**-1; **337**-11; **341**-10; **343**-13; **349**-6; **355**-9; **359**-7, 8
Jim P. Brock: **233**-3, 8; **237**-14, 15
Brokaw / Visuals Unlimited: **152**-1
S. L. Brown: **349**-2; **351**-15
Peter Bryant: **4**-8; **5**-26; **23**-9; **109**-5, 10; **125**-10; **157**-9; **159**-8; **183**-13; **215**-2; **221**-2, 4; **225**-5; **226**-1; **275**-1, 2; **300**-1; **333**-1; **335**-6; **337**-10; **339**-5; **343**-6
Gay Bumgarner / Visuals Unlimited: **227**-6
Nathan Burkett: **283**-1, 2, 3, 4, 8
Scott Camazine: **35**-4; **307**-9; **347**-1, 3
Steven J. Cary: **237**-5
David Cavagnaro / Visuals Unlimited: **105**-9
Duncan Champney: **285**-9; **305**-7
William M. Ciesla / ForestryImages.org: **103**-7
Sarah Clark: **67**-2; **359**-4
Clemson University / USDA Coop: **35**-11; **99**-9; **209**-4, 5; **267**-6; **307**-1
Tom Coates: **23**-7; **109**-2; **157**-10; **289**-9; **297**-5; **315**-17; **325**-4; **333**-6; **335**-13; **351**-7
Patrick Coin: **4**-31; **5**-18; **79**-6; **93**-7; **121**-13; **133**-3, 5; **135**-3; **139**-11; **143**-5; **173**-4; **199**-12; **215**-17; **225**-11; **281**-2; **299**-9; **309**-6; **329**-2
James Cokendolpher: **2**-4; **3**-17; **15**-1; **23**-1; **25**-10; **29**-10; **63**-1, 3, 7; **65**-4; **69**-7; **71**-13; **73**-6; **75**-15; **77**-6; **81**-3; **85**-3, 4; **89**-4, 5; **99**-6; **109**-4; **117**-10; **119**-9; **127**-3; **129**-1; **137**-5; **141**-3, 5; **143**-9; **151**-7, 11; **155**-3; **169**-4; **181**-4; **187**-3; **188**-1; **193**-4; **195**-1, 15; **197**-2, 4; **199**-4, 13; **201**-10; **205**-4; **207**-3; **209**-10; **217**-11;

225-4; 283-7, 9; 331-7; 347-6; 353-10; 357-8; 363-10

Ray Coleman / Visuals Unlimited : 64-1

Gerald and Buff Corsi / Visuals Unlimited: 67-3

Jay Cossey: 31-6; 123-14; 137-2; 239-9; 247-9; 249-15; 285-1; 287-5, 11; 299-11, 13; 309-2, 10; 329-3, 4; 337-16; 341-15

Whitney Cranshaw / InsectImages.org: 37-1, 5; 101-2, 3; 103-6; 119-7

Stephen Cresswell: 127-4; 147-13; 167-9; 177-7; 297-4; 313-8

Gyorgy Csoka / InsectImages.org: 179-10; 213-6

Rob Curtis / The Early Birder: 2-1; 3-24; 4-3, 9, 13, 14, 18, 20; 5-19, 22, 25, 28, 31, 33; 23-6, 8; 25-6; 27-4, 6; 33-3; 35-14; 41-7; 59-17, 19; 61-6, 14; 65-2, 6, 8, 9; 67-7; 71-2; 73-9; 75-11; 77-9; 83-5; 86-3, 4; 87-5; 89-6, 10; 91-1; 93-11, 16; 95-4, 7, 8, 9, 10, 11, 12, 15, 16, 17; 97-2, 8, 9, 10, 11, 13, 15, 17; 99-7; 111-2, 5; 113-3, 4, 5, 6, 7, 8, 9, 10, 12, 13; 115-1, 5; 117-1, 7; 119-4, 10, 11; 121-4, 5, 14; 123-2, 9, 12, 15; 125-1, 3, 7, 8, 12, 13, 22; 127-2, 8, 12; 128-5; 135-1, 5, 6, 8; 137-6, 9, 10; 139-9; 143-3, 4, 11; 147-9; 149-9; 155-10; 157-3, 4; 161-3, 15; 163-1, 6; 165-1, 6, 7, 8; 167-5; 169-3, 6, 8, 9; 173-8, 11; 175-8; 177-5; 181-1, 5; 183-5; 185-6; 189-1, 2, 4, 5; 191-5, 8; 193-12; 195-2, 4, 6, 11; 197-5, 10, 11, 12, 13; 199-9, 14, 15, 17; 201-9, 11, 13; 203-2, 3, 14; 205-2, 3; 207-4; 211-2; 213-3, 9, 14, 16, 17; 215-3, 9, 10, 12, 13; 217-2; 219-4; 225-6; 240-1; 245-11; 251-17; 259-20; 261-10; 263-8, 10, 12, 22; 267-2; 271-15; 275-5, 7, 9, 10, 11; 277-1, 2, 6, 7; 279-3, 4; 281-3; 283-5, 12; 285-2, 3, 4, 5, 7, 10, 11; 287-1, 7, 8, 13, 14; 289-5, 7, 8; 291-3, 5; 293-5, 12; 295-11; 297-6, 8, 9, 13; 299-8; 301-1, 3, 8, 11; 303-1, 2, 3, 6, 7, 9, 10; 305-1, 4, 8, 12; 307-4, 5; 309-5, 9; 311-3; 313-4, 9, 10; 315-4, 7, 10, 11, 12, 18; 317-1, 2, 7; 318-1; 319-2, 4; 321-2, 3, 9, 19; 323-3, 10, 14; 327-1, 2, 6; 329-1, 5, 10; 331-1, 2, 6; 333-7; 335-3, 4, 5, 8, 11, 15; 337-5, 9, 14, 15; 339-2, 8, 9, 10, 15; 341-2, 3, 5, 6, 8, 13, 14; 343-5, 11; 345-1, 2, 3, 6, 7, 8, 12; 349-8, 9, 13; 351-8, 16, 17; 353-7; 355-2, 6; 357-1, 3, 10, 11; 359-1, 2, 3; 361-3

Rob Curtis / The Image Finders: 81-6; 117-12; 357-12

Les Daniels: 91-3, 4, 5, 6, 7, 8

T.W. Davies / California Academy of Sciences: 159-1; 203-1; 323-12

Jerald E. Dewey / USDA Forest Service: 271-1

Tony Diterlizzi: 97-4, 5; 201-14

John Dooley / USDA APHIS: 37-3

Sherri Doust: 127-10

Sidney W. Dunkle: 71-5, 16, 17; 73-3, 4, 12; 109-6; 317-8; 343-1

Michael Durham / Visuals Unlimited: 305-5

K. Dean Edwards: 247-13; 259-3, 8; 267-4, 10; 269-7, 18

Randy Emmitt: 235-18; 237-16

Arthur V. Evans: 4-26; 53-1; 63-9, 12, 13; 65-10; 77-2, 11; 79-10; 81-4; 107-5, 6; 117-5, 14; 123-1, 5; 127-6, 13; 131-5; 133-6; 139-4, 5, 8, 13; 141-8, 9, 13, 15; 145-1, 11; 147-2, 4, 7, 8; 149-1, 4, 5, 7; 151-5, 13; 161-10; 167-6, 7; 171-2; 173-2; 179-1; 183-11; 185-5, 8; 187-11; 191-3, 7, 10, 12; 193-1, 2, 6, 7, 8; 195-5, 8, 9; 197-1, 199-10; 200-1; 203-8, 15; 207-1, 6, 10, 13, 16; 209-9; 213-13; 227-2; 319-5

Howard Evans / Colorado State University: 65-7; 335-10

Shane Farrell: 271-12

William Ferguson: 2-8, 9; 3-23; 4-11, 22; 25-1; 27-1; 29-4; 33-1, 6, 7, 8, 9; 39-4; 63-8; 83-2, 3, 4; 86-12; 103-4, 9, 10, 17; 107-3; 109-3, 8; 113-1; 115-2, 8, 9; 119-5; 121-7; 151-2; 153-7; 155-11; 157-2; 161-5, 9, 12; 163-3, 8, 11, 13, 15; 165-3, 5; 171-1, 7, 10; 173-13; 175-7; 181-9; 183-2, 6; 186-1; 207-11, 12; 209-8; 211-5; 215-4; 217-6, 7, 10, 12; 219-5; 223-6, 8; 224-1; 277-4; 286-1; 289-4; 291-7; 305-6; 307-8, 10; 310-1; 311-8; 323-11; 331-9, 10, 13; 337-12; 347-4; 349-12; 351-9, 10, 11; 357-6; 363-6

David H. Funk: 2-15, 18; 3-8, 9, 13; 5-17, 23; 31-9; 39-2, 3; 41-4, 5; 57-1, 2, 3, 4, 6, 7, 11, 12; 59-4, 5, 6, 8, 10, 11, 13, 16; 61-2, 3, 4, 7, 15, 17; 71-10; 77-4, 5; 79-1, 2, 3, 4, 7, 8, 9, 11, 14, 15; 81-7, 9; 83-6, 7, 8, 9, 11; 85-2; 89-3; 97-14; 99-1; 109-7; 111-1; 165-10; 185-4; 191-6, 9; 213-1; 217-1, 3; 221-3, 6; 277-3; 278-1; 279-5; 281-1, 4, 5, 6, 7, 8, 10, 11; 283-6, 10, 11; 287-2, 4, 6, 9, 10; 297-1, 2, 3, 10, 12; 309-

1; 313-1, 3, 7; 315-16; 317-3, 4, 5, 9, 10; 321-16; 327-9, 10; 354-2, 3
William Grenfell / Visuals Unlimited: 181-6
Joyce Gross: 95-5; 97-1; 107-1; 125-9; 155-9; 213-8, 12; 215-15; 219-9; 299-3
Fran Hall: 101-8, 11; 103-11; 322-1
Charles Harrington / USDA Agricultural Research Service: 177-6
Glenn Hayes / KAC: 77-10
Vincent J. Hickey: 357-9
Amy Hill: 59-7
John Himmelman: 239-6, 7, 13; 247-4; 249-6; 251-6; 261-11, 13; 267-15; 269-1, 2, 5
Jarmo Holopainen: 209-2
Edward H. Holsten / USDA Forest Service: 325-5
Steve Hopkin: 31-1, 2, 3, 5, 7, 8, 10, 11, 12, 13, 14, 15
Gerald Hoveling: 299-5; 339-14; 341-1, 9; 357-7
Howe / Visuals Unlimited: 159-13
Dustin Huntington: 45-4; 311-7
Lacy Hyche / ForestryImages.org: 217-4; 321-13
Bob Jensen: 147-15; 353-2
Bill Johnson: 2-2, 11; 23-4; 27-2; 29-3; 31-4; 59-3; 69-10; 73-15; 75-4; 76-1; 86-8, 9, 10; 89-9; 95-6, 13; 99-4; 101-1, 12; 103-5; 105-3, 5, 13; 107-2; 113-2; 115-3; 117-16; 121-11; 127-1; 131-1; 137-3; 139-7, 12; 145-3; 149-3; 151-1; 155-4, 7; 161-8, 11; 165-11, 12; 169-1; 173-3, 10; 175-4; 179-3; 183-4; 187-5; 189-10; 191-1, 2; 193-10; 195-7, 12; 196-1; 203-7, 12; 205-8; 211-3; 213-11; 215-11; 219-7; 295-5; 301-4, 5, 7; 303-5, 15, 16; 307-2; 311-5; 313-6; 320-1; 321-4, 8; 323-4; 325-1; 335-7; 351-1, 5, 6; 357-2
Jim Kalisch: 119-8
Ronald Kelley / ForestryImages.org: 267-5
Ian Kimber: 267-13, 17
Takumasa Kondo: 4-7; 5-13; 86-11; 105-2, 6, 7, 8, 10, 11, 12, 14, 15, 16, 17; 275-3; 363-11
Dennis Kunkel / Visuals Unlimited: 25-8, 9; 35-1, 6, 7
Greg Lasley: 3-5, 20; 45-5, 6, 8; 47-1; 49-1, 6; 51-2; 53-6; 71-12
Alan Chin-Lee: 263-5
Richard Leung: 95-3; 125-17, 18; 201-4; 279-2, 7; 331-12; 337-3; 353-4, 9; 359-11
Charles Lewallen: 67-5; 93-4; 103-3; 121-15; 155-14; 163-14; 167-4; 207-7; 219-3; 337-2; 351-13; 355-3, 5
David Lightfoot: 39-6, 7; 71-8, 11; 73-7, 8, 11; 75-2, 3, 6, 12; 99-8; 123-8; 177-3; 193-11; 305-2; 345-5
Robert Lindholm / Visuals Unlimited: 93-15
Michelle Mahood: 59-12; 61-5
Bruce J. Marlin: 115-6; 181-2; 353-1
Steve Marshall: 3-25; 69-1, 8, 9; 71-3, 15; 135-7; 139-10; 143-6, 12; 151-9; 153-5; 155-8, 13; 157-1, 13; 159-3, 4, 6, 10, 11; 163-10, 12; 169-7; 173-14; 175-1; 177-4; 179-6, 11; 183-1; 187-10; 189-6; 195-14; 201-7; 207-5; 211-4; 219-1, 2, 8; 299-4, 7, 12; 301-9, 12; 303-13; 321-14; 323-13, 17; 325-6
Joe McDonald / Visuals Unlimited: 93-13; 109-1
Robin McLeod: 117-13; 313-5
Gary Meszaros / Visuals Unlimited: 151-4; 179-7
Jeffrey Miller: 5-16, 30; 179-8; 189-8, 12, 13; 199-7; 215-14; 217-8, 9; 239-3, 5; 241-10, 11, 13, 15; 244-1; 245-3; 247-5, 8, 11, 12, 14, 15, 16, 17; 248-1; 249-8, 12; 251-5; 252-1; 253-8; 256-1; 257-1, 4, 10, 11; 259-1, 2, 10, 11, 16, 17, 18, 21; 261-1, 3, 16, 17; 263-11, 13, 14; 265-8, 11; 301-2; 307-7; 315-9; 323-15; 324-1; 331-5
Tom Murray: 5-15; 35-12; 99-3; 115-7, 10; 123-6; 125-23; 127-5; 155-12; 159-14; 171-8; 197-8; 281-15; 287-15; 289-14; 291-8; 295-3; 299-2; 301-10; 303-8, 14; 305-9; 313-11; 315-8; 321-6, 11, 21; 323-5, 16; 327-4; 331-4; 333-5; 335-9, 16; 337-18; 339-4; 343-9; 353-6
Phil Myers: 63-5; 89-8; 147-12; 211-6; 315-14
Hannah Nendick-Mason: 349-11
Brent Newman: 151-6, 12; 337-1; 353-8
Blair Nikula: 43-3, 5, 6, 7; 47-5, 7; 49-3, 7, 9; 51-1; 53-2, 3, 5, 7, 9
Rolf Nussbaumer / KAC: 143-2; 203-11
Jim Occi: 173-7
Cheryle O'Donnell: 2-16; 37-2, 4, 6, 7, 9
M.D. Overton: 133-7; 147-3; 201-1
Robert Parks: 307-11; 335-12; 343-7; 345-4, 11; 349-4, 5; 351-12; 355-7, 8; 359-13
Herbert A. Pase III / ForestryImages.

org: 323-18
Michael Patrikeev: 71-9; 73-5; 77-8; 325-3; 339-6; 353-11;
Jerry Payne / USDA ARS: 213-4
R.K.D. Peterson: 41-3; 83-10; 127-9; 135-11; 139-2; 325-7, 8
Frank Phillips: 221-5; 271-4; 305-3
Mark Plonsky: 305-11; 307-3
Rick Poley / Visuals Unlimited: 105-1
William Preston: 83-12; 111-6
Mike Quinn: 155-2, 5
David Rentz: 85-5
Bryan Reynolds: 25-7; 29-9; 65-11; 67-4; 73-10; 109-9; 121-10; 183-7, 10; 203-13; 289-10; 293-10; 295-6; 297-7, 11; 299-1; 305-10; 307-6; 315-2, 3; 327-3, 5; 339-3; 341-12; 343-10; 355-4
Lance Risley: 267-1
Joshua Rose: 47-2
Edward S. Ross: 133-8, 9; 135-9; 139-1; 141-12; 143-10; 145-2, 9, 10; 147-1, 10; 149-6, 8; 153-1, 2, 9; 157-5; 159-2; 163-16; 165-2; 169-12; 174-1; 175-3; 177-10; 181-7; 185-3; 187-1, 2; 195-10; 197-3, 6; 199-11; 201-3, 12; 205-7
Jane Ruffin: 233-16
Ryan Sawby: 113-14; 345-10; 349-10
Mark Schwarzlaendar / ForestryImages.org: 215-16
Lynn Scott: 245-1, 7; 247-6; 249-2, 3, 4, 11; 251-3, 11, 14; 253-2, 7; 255-6; 257-5, 6, 8, 13, 14, 18; 259-15, 19; 261-4; 263-9; 265-4, 12; 267-14; 269-3, 4, 6; 271-19, 20
David Scriven: 287-3
Judy Semroc: 285-6
John Shaw: 233-2, 6, 14; 235-11; 237-9
Leroy Simon: 125-19; 127-7; 161-6; 173-1, 12
Leroy Simon / Visuals Unlimited: 41-6; 179-9; 233-15; 241-9; 243-14; 261-5, 7; 271-7
Ann and Rob Simpson: 23-5; 29-7; 33-2; 63-11; 65-3; 75-13; 83-1; 101-5; 103-13; 119-2; 341-7, 11
Rob and Ann Simpson / Visuals Unlimited : 101-14; 177-2; 181-3; 225-10
James Solomon / USDA: 177-8
Alton N. Sparks Jr. / ForestryImages.org: 37-8
Bill Stark: 61-10, 11, 16, 18
Sparky Stensaas: 111-4; 169-5
Bob Stewart: 231-2
Louis Tedders / USDA ARS: 125-11
Texas Ag. Extension Service: 217-5
Tierney / Visuals Unlimited: 179-4
John and Gloria Tveten: 3-16, 27; 4-29; 5-12, 34; 7-1; 23-11; 35-13; 41-1, 2; 57-10; 63-2, 4, 6, 10; 69-5; 75-16; 77-7; 79-5; 81-5; 85-1, 6; 86-5, 7; 93-5, 10; 95-14; 97-16; 99-5; 101-6; 103-2, 14; 117-4, 6, 8, 11, 15; 119-12, 13; 121-6; 123-4, 7, 10, 11, 18; 125-2, 4, 6; 131-12, 14; 133-2; 135-2; 137-12; 139-3, 6; 141-1, 2, 4; 143-7, 13; 147-6, 11, 14; 151-10; 161-4, 14; 167-2; 171-4, 6, 9; 175-2, 5, 6, 9; 177-1; 179-2, 5; 181-10; 183-3, 8, 9, 12; 185-1, 2, 7, 10; 187-4, 6; 189-3, 14, 15; 191-11; 193-5, 9; 195-13; 197-9; 199-8; 201-6, 8; 203-5, 6; 213-5, 10, 15; 215-7; 223-1, 3, 4; 225-7; 227-3, 4; 233-4, 7, 11; 235-2; 238-1; 239-1, 4, 8, 10, 11, 12, 14, 15; 241-3, 5, 6, 8, 12; 243-3, 5, 6, 7, 9; 245-4, 5, 8, 10, 13; 247-1, 2, 7; 249-1, 5, 7, 9, 10, 17; 250-1; 251-4, 8, 10; 253-1, 6, 9, 12; 255-3, 10, 12; 256-2; 257-3, 19; 259-5, 13; 260-1; 261-2, 6, 8; 263-1, 3, 7, 15, 16, 17, 18, 19; 264-1; 265-5, 7, 10, 13, 14, 15; 267-3, 7, 8, 12; 269-8, 9, 10, 11, 12, 13, 14, 15, 16, 20; 270-1; 271-2, 3, 5, 6, 8, 9, 10, 11, 17, 21, 22; 285-8; 289-2; 293-4; 315-15; 319-4; 321-12; 325-2; 329-6, 8; 331-3; 339-7; 345-9; 351-3; 357-4, 5; 359-5, 6, 10, 12; 361-1
S. J. Upton: 35-5
USDA ARS archives: 207-14,15; 209-1, 3, 6; 211-7; 317-6
Matt Wallace and Lew Dietz: 93-2, 3, 6, 8, 9, 12
Robyn Waayers: 3-21; 73-1, 14; 141-6; 169-10; 215-1; 351-4; 353-3
Richard Walters / Visuals Unlimited: 131-2
Michael Wigle: 3-7, 10, 11, 12, 14; 55-1, 2, 3, 4, 5, 6, 7, 8, 9, 10, 11, 12, 13; 57-8; 59-1, 2; 61-12, 13, 19; 281-12, 13
Bev Wigney: 103-8; 321-1; 323-2, 6, 7, 8
Alex Wild: 5-24; 39-1; 63-14; 67-6; 275-4; 333-2, 3; 349-14; 361-2, 4, 5, 6, 7, 8, 9, 10, 11; 363-1, 2, 3, 5, 7, 8, 9, 12, 13; 365-1, 2, 3, 4, 5, 6, 7, 8, 9, 10, 11, 12, 13
Eric Williams: 167-3
Kevin Williams: 335-1
Bob Wilson / Visuals Unlimited: 245-9
Doug Wilson / USDA ARS: 267-16
Hartmut Wisch: 173-5
Dale and Marian Zimmerman: 69-3; 75-9; 79-12; 351-2

COMPREHENSIVE INDEX

In many cases the short index on p. 392 will be all you need. Here we list all scientific and English names used in the guide. Scientific names of species and genera are in *italic* type. In seeking English names, look under group names; for example, the Rhinoceros Beetle is listed under "B" with other beetles, not under "R." Some large subgroups have their own headings; for example, the Convergent Lady Beetle is listed under "L" with other lady beetles.

Short Index for KAUFMAN FIELD GUIDE TO INSECTS OF NORTH AMERICA

(For complete listings, see the main index starting on p. 370.)

Short Index for KAUFMAN FIELD GUIDE TO INSECTS OF NORTH AMERICA

(For complete listings, see the main index starting on p. 370.)